ELDRIDGE

TIDE AND PILOT BOOK 2014

Our One Hundred Fortieth Year of Continuous Publication

CONTENTS

Origin of ELDRIDGE 2
About the 2014 Edition, Supplement 4
Letter from the Publishers 5
How to Use the Tables and Charts 11
Story Contest Rules for 2015 220
Story Contest Winner 2014 221

EMERGENCY FIRST AID 243-245
Hypothermia and Cold Water Immersion 242

RULES AND REGULATIONS
Inland Navigation Rules 6
The Rule of the Road - Poem 6
Federal Safety Equipment Requirements 7
Navigation Lights 8
Sound Signals for Fog and Maneuvering 9
Cape Cod Canal 44-45
Chesapeake & Delaware Canal 146-147
IALA Buoyage System 213
Nautical Chart & Buoyage Guide 214
Federal Pollution Regulations 260
Pumpout information - Sources by State 261

TIDE AND CURRENT TABLES
Time of High Water 12-20
Time of Current Change 22-29
Portland High & Low Water 30-35
Boston High & Low Water 38-43
Cape Cod Canal Currents 46-51
Woods Hole Currents 52-57
Pollock Rip Channel Currents 60-65
Newport High & Low Water 78-83
The Race Currents 86-91
Bridgeport High & Low Water 98-103
Kings Point High & Low Water 104-109
Hell Gate Currents 110-115
The Narrows Currents 116-121
The Battery High & Low Water 122-127
Sandy Hook High & Low Water 134-139
Delaware Bay Entrance Currents 140-145
Chesapeake & Delaware Canal Curr 148-153
Baltimore High Water 156-159
Miami Harbor Entrance High Water 160-163

CURRENT CHARTS & DIAGRAMS
Boston Harbor 37
Buzz. Bay, Vineyard & Nantucket Sounds 66-77
Narragansett Bay 84
Long Island & Block Island Sounds 92-97
New York Bay 128-133
Upper Chesapeake Bay 154
Relationship of High Water & Ebb Current 155

NOTES ON TIDES AND CURRENTS
Why Tides and Currents Often Differ 10
Piloting in a Cross Current 21
Smarter Boating in Currents 36
Eldridge Letter & "Graveyard" Chart 58-59
Holding a Fair Current, L.I. to Nantucket 85
Tidal Heights & Depths 199
Tide Cycle Simplified, Rule of Twelfths 201
Coping with Currents 222

LIGHTS, FOG SIGNALS and OFFSHORE BUOYS
Light Characteristics Diagram 166
Nova Scotia to Key West and Bermuda 167-197

FLAGS AND CODES
Yacht Flags & How to Fly Them 216
International Signal Flags and Morse Code 217
International Code of Signals, Diver Flags 218
U.S. Storm Signals 219

ASTRONOMICAL DATA - Standard Time
Sunrise & Sunset - Boston 224
Sunrise & Sunset - New York 225
Sunrise & Sunset - Jacksonville 226
Sunset at Other Locations 227
Local Apparent Noon 229
Sun's True Bearing 230
Sun's Declination 231
Moonrise & Moonset - Boston 232
Moonrise & Moonset - New York 233
Phases of the Moon 234
Daily Moon Phase Calendar 235
Tides, the Moon and the Sun 236
Visibility of the Planets 237

RADIO NAVIGATIONAL AIDS & DATA
What your GPS Can and Cannot Do 205
Atlantic Coast DGPS Stations 207
Racons 208-209
Dial-A-Buoy 210-211
VHF and Marine Communications 238, 241
Marine Emergency and Distress Calls 239
U.S. Coast Guard Contact Information 240
Time Signals 241

WEATHER
NOAA Marine Weather Forecasts 241
Forecasting with Wind & Barometer 246
Keys to Predicting Weather 246
Weather Signs in the Sky 247
Wind Chill 248
Beaufort Scale 249
Hurricanes 250-254
Historic East Coast Hurricanes Chart 251
Saffir-Simpson Hurricane Wind Scale 252
Weather Notes - Maine to Chesapeake 255
Humidity & Dew Point Table 256

MISCELLANEOUS
Anglers United by One Fish 164-165
Fish Sense: Mastery and Mystery 168-170
Distance between Ports 192,194
Heaving the Lead 198
Using GPS to Create a Deviation Table 202
Using GPS to Adjust Your Compass 203
Ship's Bell Code 204
Portland Head Light 215
Distance of Visibility 219
Table for Turning Compass Points to Degrees 223
Celestial Navigation - Not Yet a Lost Art 228
Hurricane Sandy - 2012 250
Beachcombing 257
What Time Is It? 258-259
Converting Angular and Linear Measure 262
Table of Equivalents 263
Cruising Charts 264-265
Ferry Service Information 266
Where to Buy ELDRIDGE 267-269
What Would You Do? 270
Index to Advertisers 271

Publishers: **Robert Eldridge White, Jr.** and **Linda Foster White**
P.O. Box 775, Medfield, MA 02052 Tel. 617-482-8460
ebb2flood@gmail.com, www.eldridgetide.com Fax: 617-482-8304

Printed in U.S.A

ISBN 978-1-883465-20-9

THE ORIGIN OF ELDRIDGE

In 1854 George Eldridge of Chatham, a celebrated cartographer, published "Eldridge's Pilot for Vineyard Sound and Monomoy Shoals." The book had 32 pages, a grey paper cover and no recorded price. Its pages were devoted to "Dangers," embellished with his personal observations, and to Compass Courses and Distances, etc. This volume was the precursor of the Tide and Pilot Book, which followed 21 years later.

In 1870 George Eldridge published another small book, called the "Compass Test," and asked his son, George W. Eldridge, to go to Vineyard Haven and sell it for him, along with the charts he produced.

Son George W. Eldridge was dynamic, restless, and inventive. He was glad to move to the Vineyard, for Vineyard Haven was at that time an important harbor for large vessels. The number of ships passing through Vineyard and Nantucket Sounds was second only to those plying the English Channel. As the ships came into the harbor (frequently as many as 100 schooners would anchor to await a fair current), George W. would go out to them in his catboat to sell his father's charts and the "Compass Test." He was constantly asked by mariners what time the current TURNED to run East or West in the Sound. He then began making observations, and one day, while in the ship chandlery of Charles Holmes, made the first draft of a current table. Shortly after, with the help of his father, he worked out the tables for places other than Vineyard Sound, and in 1875 the first Tide Book was published. It did not take long for mariners to realize the value of this information, and it soon became an indispensable book to all who sailed the Atlantic Coast from New York east. Gradually George W. added more important information, such as his explanation of the unusual currents which caused so many vessels to founder in the "Graveyard."

Captain George W. Eldridge based the tables on his own observations. In later years, knowing that the government's scientific calculations are the most accurate obtainable, the publishers have made use of them; some tables are directly taken from government figures and others, which the government does not give in daily schedules, are computed by the publishers from government predictions. Since the Captain's day there have been many changes and additions in the book to keep abreast of modern navigational aids.

In 1910 Captain George W. Eldridge transferred the management of the book to the next generation of his family, as he was interested in developing his chart business and inventing aids to navigation. At his death in 1914, his son-in-law became Publisher. Wilfrid O. White, an expert in marine navigation and President of Wilfrid O. White & Sons Co., compass manufacturers, served as Publisher until his death at 1955. Wilfrid's son Robert Eldridge White then became publisher, and with great help from his wife Molly he expanded the coverage of the book and significantly increased its readership. On Bob's death in 1990, Molly continued to expand the book's scope and circulation, with valuable assistance from her son Ridge and daughter-in-law Linda. On Molly's passing in 2004 the book moved once again into the hands of the next (fourth) generation. At every generational transfer the new Publisher was well prepared, each having apprenticed for years.

Whether new to ELDRIDGE or a longtime reader, we welcome you aboard! Please continue to offer your suggestions and, where necessary, corrections. Your sharp eyes keep us on course. We hope, as did Captain George W. Eldridge, that this book might ensure for you a "Fair Tide" and the safety of your ship.

Robert Eldridge White, Jr.
Linda Foster White
Publishers

Yours for a fair tide
Geo. W. Eldridge

About the 2014 ELDRIDGE, our 140th edition

☆ **NOTE:** ***The information in this volume has been compiled from U. S. Government sources and others, and carefully checked. The Publishers cannot assume any liability for errors, omissions, or changes.***

Publishers: Robert Eldridge White, Jr. - Linda Foster White
Editorial and Advertising: Tel.: 617-482-8460, Fax: 617-482-8304
Email: ebb2flood@gmail.com, **Web:** www.eldridgetide.com
Mailing Address: P.O. Box 775, Medfield, MA 02052

New or Noteworthy Articles in the 2014 Edition

p. 5 **Letter from the Publishers** – Eldridge at 140
p. 21 **Piloting in a Cross Current** – Three ways to stay on course
pp. 164-165 **Anglers United by One Fish** – Lou Tabory reminisces
p. 168-170 **Fish Sense: Mastery and Mystery** – Zach Harvey
p. 215 **Portland Head Light** – A notable landmark for Down East cruisers
p. 221 **Story Contest Winner** – A Dog and a Dinghy Story
p. 222 **Coping with Currents** – Back by popular demand
p. 228 **Celestial Navigation – Not Yet a Lost Art** – Andy Sumberg
p. 243-245 **Emergency First Aid** – expanded, revised CPR
p. 250 **Hurricane Sandy 2012** – Only a Category 1 storm, but devastating
p. 251 *Improved* **Hurricane Chart for Historic East Coast Hurricanes**
p. 257 **Beachcombing** – Peter Spectre on the pleasures of the hunt
p. 258-259 **What Time Is It?** – Jan Adkins makes the complex simple
p. 270 **What would you do?** – Two Scenarios to Provoke Discussion

FREE SUPPLEMENT ON REQUEST - available after May 15, 2014
Changes and updates through May 1, 2014. To obtain a free Supplement:

1) Download a pdf from: www.eldridgetide.com *or* www.robertwhite.com (click on Eldridge)
2) Mail a stamped, self-addressed envelope, with your request, to:

ELDRIDGE TIDE & PILOT BOOK
P.O. Box 775
Medfield, MA 02052

I enclose a **stamped, self-addressed envelope**. Please send me a free Supplement, updating data through May 1, 2014.

Name ____________________

Address ____________________

City, State, Zip ____________________

Letter From the Publishers

To our readers: ELDRIDGE at 140

This edition marks the 140th year of our publication, and with that milestone comes a reason to look back at where we came from. The world was a very different place in 1875. It was only ten years after the Civil War; steam propulsion was just starting to make headway against square-rigged ships; Alexander Graham Bell made the first voice transmission; Edison and others were experimenting with the light bulb. And George W. Eldridge published his first current tables.

The first readers of ELDRIDGE were mostly captains of coastal schooners, seeking a fair current to assist their slow vessels. This book was small and simple. Courses and bearings were listed in compass points, not degrees, and the only reference port for tides and currents was Vineyard Haven, home of the publisher. Since 1875 this book has grown to the present 272 pages, with 17 reference ports covering the East Coast from Maine to Florida. Today our readers cover the spectrum from pleasure boaters and beachgoers to racing skippers and saltwater anglers.

As a glance at the Table of Contents shows, this book has a very broad range of articles. There's a Story Contest winner, articles by experts on fishing, how to use a GPS to adjust your compass, what you need to know about hurricanes, how to deal with hypothermia, and much more. If the first half of the book, full of tide and current tables, is more useful, then the second half is more instructive and interesting.

A veteran reader wrote: "Your publication is a unique tool for New England sailors and coastal pilots, and serves a unique need unmet through any other resource. I hardly consider it prudent to traverse New England's coastline without a copy of ELDRIDGE on the bridge."

We would like to hear from you, our readers, about what you most rely on this book for, what articles you enjoy, and also what you'd like to see that isn't included. Our mission is to provide you with indispensable information, as well as instructive and even entertaining articles. You can contact us by email at ebb2flood@gmail.com, phone, or snail mail.

As we celebrate 140 years of continuous publication, we thank you for taking us on board. We look forward to remaining practical and useful through the changing times to come!

Robert Eldridge White, Jr. & Linda Foster White
Publishers

INLAND NAVIGATION RULES

Good Seamanship Rule (Rule 7): Every vessel shall use all available means appropriate to the prevailing circumstances and conditions to determine if risk of collision exists. If there is any doubt, such risk shall be deemed to exist.

General right-of-way (Rule 18): Vessel categories are listed in decreasing order of having the right-of-way:

- Vessel not under command (most right of way)
- Vessel restricted in ability to maneuver, in a narrow fairway or channel
- Vessel engaged in fishing with nets, lines, or trawls (but not trolling lines)
- Sailing vessel (sails only)
- Power-driven vessel (least right of way)

Vessels Under Power

Overtaking (Rule 13): A vessel overtaking another is the "give-way" vessel and must stay clear of the overtaken or "stand-on" vessel. The overtaking vessel is to sound one short blast if it intends to pass on the other vessel's starboard side, and two short blasts if it intends to pass on the other's port side. The overtaken vessel must respond with the identical sound signal if it agrees, and must maintain course and speed during the passing situation.

Meeting head-on (Rule 14): When two vessels are meeting approximately head-on, neither has right-of-way. Unless it is otherwise agreed, each vessel should turn to starboard and pass port to port.

Memorized for generations by mariners, the verse below tells what to do when power vessels meet at night.

The Rule of the Road

When all three lights I see ahead,
I turn to **Starboard** and show my **Red:**
Green to Green, Red to Red,
Perfect Safety – **Go Ahead.**

But if to **Starboard Red appear,**
It is my duty to keep clear –
To act as judgment says is proper:
To **Port** or **Starboard, Back** or **Stop** her.

And if upon my **Port** is seen
A Steamer's **Starboard** light of **Green,**
I hold my course and watch to see *
That **Green** to **Port** keeps Clear of me.

Both in safety and in doubt
Always keep a good look out.
In danger, with no room to turn,
Ease her, **Stop** her, **Go Astern.**

* "There's nought for me to do but see" is the original version

Crossing (Rule 15): When two vessels approaching each other are neither in an overtaking or meeting situation, they are deemed to be crossing. The power vessel which has the other on its starboard side is the give-way vessel and must change course, slow down, or stop. The vessel which is on the right, is in the right.

Vessels Under Sail

Port-Starboard (Rule 12): A vessel on the port tack shall keep clear of one on the starboard tack.

Windward-Leeward (Rule 12): When both vessels are on the same tack, the vessel to windward shall keep clear of a vessel to leeward.

Sail vs. Power (Rule 18): Generally, a sailboat has right of way over a powerboat. However: (1) a sailboat overtaking a powerboat must keep clear; (2) sailboats operating in a narrow channel shall keep clear of a power vessel which can safely navigate only within a narrow channel; (3) sailboats must give way to a vessel which is fishing, a vessel restricted in its ability to maneuver, and a vessel not under command.

FEDERAL SAFETY EQUIPMENT REQUIREMENTS

(These are minimum requirements. Some states require additional equipment.)

Sound Signaling Devices

Under 39.4' or 12 meters:
  Must have some means of making an efficient sound signal

Over 39.4' or 12 meters:
  Whistle or horn, audible for 1/2 mile, and a bell at least 8" dia.

Visual Distress Signals (all with approval number)

Under 16' or 5 meters:
  Night: 1 electric SOS flashlight or 3 day/night red flares

Over 16' or 5 meters:
  Day only: 1 orange flag, 3 floating or hand-held orange smoke signals
  Day and night: 3 hand-held, or 3 pistol, or 3 hand-held rocket, or 3 pyrotechnic red flares

The following signals indicate distress or need of assistance:

see p. 239, Marine Emergency and Distress Calls

- A gun or other explosive signal fired at intervals of about 1 minute
- A continuous sounding with any fog-signaling apparatus
- Rockets or shells, fired one at a time or at short intervals
- SOS transmitted by any signaling method
- "Mayday" on the radiotelephone (channel 16)
- International Code Signal flags "NC"
- An orange square flag with a black square over a black ball
- Flames on the vessel
- Rocket parachute flare or hand-held flare
- Orange colored smoke
- Slowly and repeatedly raising and lowering outstretched arms
- Signals transmitted by EPIRB
- High intensity white light flashing 50-70 times per minute

Personal Flotation Devices (must be USCG approved)

Under 16' or 5 meters:
  1 Type I, II, III, or V per person, USCG approved

Over 16' or 5 meters:
  1 Type I, II, III, or V per person, and 1 Type IV per boat, USCG approved

Portable Fire Extinguishers (approved)

Under 26'	1 B-I, if no fixed extinguisher system in machinery space. (Not required on out-boards built so that vapor entrapment cannot occur.)
26-39'	2 B-I or 1 B-II if no fixed exting. system; 1 B-I, with a fixed exting. system.
40-65'	3 B-I or 1 B-II & 1 B-I, if no fixed exting. system; 2 B-I or 1 B-II with a fixed exting. system.

Back-Fire Flame Arrestor

One approved device per carburetor of all inboard gasoline engines.

At least 2 ventilator ducts fitted with cowls or their equivalent to ventilate efficiently the bilges of every engine and fuel tank compartment of boats using gasoline or other fuel with a flashpoint less than 110°F.

NAVIGATION LIGHTS

Definition of Lights

Masthead Light — a white light fixed over the centerline showing an unbroken light over an arc of 225°, from dead ahead to 22.5° abaft the beam on either side.
Sidelights — a green light on the starboard side and a red light on the port side showing an unbroken light over an arc of the horizon of 112.5°, from dead ahead to 22.5° abaft the beam on either side.
Sternlight — a white light placed as nearly as practicable at the stern showing an unbroken light over an arc of the horizon of 135°, 67.5° from dead aft to each side of the vessel.
All-round Light — an unbroken light over an arc of the horizon of 360°.
Towing Light — a yellow light with same characteristics as the sternlight.

Note: R . and Y. Flashing Lights are now authorized for vessels assigned to Traffic Control, Medical Emergencies, Search and Rescue, Fire-Fighting, Salvage and Disabled Vessels.

When under way, in all weathers from sunset to sunrise, every vessel shall carry and exhibit the following lights

When Under Power Alone or When Under Power and Sail Combined

Under 39.4' or 12 meters:
Masthead light visible 2 miles
Sidelights visible 1 mile
Stern light visible 2 miles (may be combined with masthead light)

Over 39.4' or 12 meters to 65' or 20 meters::
Masthead light visible 3 miles
Sidelights visible 2 miles
Stern light visible 2 miles (or in lieu of separate masthead and stern lights, an all-round white light visible 2 miles)

Sailing Vessels Under Way (Sail Only)

Under 22' or 7 meters:
Either the lights listed below for sailing vessels under 65'; or a white light to be exhibited (for example, by shining it on the sail) in sufficient time to prevent collision
Under 65' or 20 meters: *may be combined in one tricolor light carried near top of mast*
Sidelights visible 2 miles
Stern light visible 2 miles

At Anchor

Vessels under 50 meters (165') must show an all-round white light visible 2 miles.
Vessels under 7 meters (22') need no light unless they are near a channel, a fairway, an anchorage or area where other vessels navigate.

Fishing

Vessels Trawling shall show, in addition to the appropriate lights above, 2 all-round lights in a vertical line, the upper green and the lower white.
Vessels Fishing (other than trawling) shall show, in addition to the appropriate lights above, 2 all-round lights, the upper red and the lower white.

When Towing or Being Towed

Towing Vessel: 2 masthead lights (if tow is less than 200 meters); 3 masthead lights in a vertical line forward (if tow exceeds 200 meters); sidelights; sternlight; a yellow tow light in vertical line above sternlight; a diamond shape where it can best be seen (if tow exceeds 200 meters).
Vessel Being Towed: sidelights; sternlight; a diamond shape where it can best be seen (if tow exceeds 200 meters).

SOUND SIGNALS FOR FOG

Ask your Chart Dealer for the latest Navigation Rules—Inland/International

Frequently, in fog, small sail or power boats cannot be heard or picked up by other vessels' radar. The Coast Guard strongly recommends that, to avoid collisions, all vessels carry Radar Reflectors mounted as high as possible.

All signals prescribed by this article for vessels under way shall be given:

First: By Power-driven Vessels – On the Whistle or Horn.
Second: By Sailing Vessels or Vessels being Towed – On the Fog Horn.

A prolonged blast shall mean a blast of 4 to 6 seconds' duration.
A short blast shall mean a blast of about one second's duration

A power-driven vessel having way upon her shall sound at intervals of no more than 2 minutes, a prolonged blast.

A power-driven vessel under way, but stopped and having no way upon her, shall sound at intervals of no more than 2 minutes, 2 prolonged blasts with about 2 seconds between them.

A sailing vessel under way, shall sound at intervals of not more than 2 minutes, 1 prolonged blast followed by 2 short blasts, regardless of tack.

A fishing vessel or a power-driven vessel towing or pushing another vessel, shall sound every 2 minutes, 1 prolonged blast followed by 2 short blasts. A vessel being towed shall sound 1 prolonged blast followed by 3 short blasts.

A vessel at anchor shall ring a bell rapidly for about 5 seconds at intervals of not more than 1 minute and may in addition sound 3 blasts, 1 short, 1 prolonged, 1 short, to give warning of her position to an approaching vessel. Vessels under 20 meters (65') shall not be required to sound these signals when anchored in a special anchorage area.

A vessel aground shall give the bell signal and shall, in addition, give 3 separate and distinct strokes of the bell.

MANEUVERING AND WARNING SIGNALS

Inland Rules:

1 short blast: I intend to leave you on my Port side.
2 short blasts: I intend to leave you on my Starboard side.
3 short blasts: I am backing
5 or more short and rapid blasts: **DANGER**

Response: If in agreement, upon hearing the 1 or 2 blast signal, a vessal shall sound the same signal and take the steps necessary to effect a safe passing. If in doubt sound the danger signal.

International Rules:

1 short blast: I intend to leave you on my Port side.
2 short blasts: I intend to leave you on my Starboard side.
2 long, 1 short blast: I am overtaking you on your Starboard side.
2 long, 2 short blasts: I am overtaking you on your Port side.

Response in overtaking: long blast, short blast, long blast, short blast if agreeable.

3 short blasts: I am backing
5 or more short and rapid blasts: **DANGER**

Why Tides and Currents Often Behave Differently

Frequently Asked Questions

We are often asked such questions as, **"Why are the times of high water and current change not the same?"** Shouldn't an ebb current begin right after a high tide? Although tides (vertical height of water) and currents (horizontal movement) are inextricably related, they often behave rather differently.

If the Earth had a uniform seabed and no land masses, it is likely that a high tide at one point would occur simultaneously with a change in the current direction. However, the existence of continents, a sea bottom which is anything but uniform, and the great ocean currents and different prevailing winds around the world, make the picture extremely complex.

As one example of how a time of high tide can differ greatly from the time of a current change, see the Relationship of High Water and Ebb Current, p. 155. Picture a fjord or long indentation into the coastline, with a narrow opening to the ocean. When a flood current is reaching its peak, or the tide is high outside the mouth of this fjord, the fjord is still filling, unable to keep pace with conditions on the outer coast.

Why do the heights of tides differ so much from one place to the next? Turn to Time of High Water at various ports, pp. 12-20, and compare the Rise in Feet of tides for Nova Scotia's outer coast (2.6 to 4.8 feet) to those for the Bay of Fundy (just below), with a range of up to 35.6 feet. Why the difference? The answer is geography, both above and below water. Tidal ranges of points out on the edge of an outer coast (Nantucket, for instance) tend to be moderate, while estuaries and deep bays with narrowing contours often experience a funneling effect which exaggerates the tidal range. Another explanation is proximity to the continental shelf: the closer a port is to the shelf, the more likely it is to experience a lower tidal range; the farther from the shelf, the more likely it is that a harbor is subject to surges, as when a wave crest hits the shallow water at a beach.

There are other anomalies between tides and currents. **Do stronger currents indicate higher tides?** Woods Hole, MA often has very strong currents through its narrow passage, sometimes as much as 7 knots, but the tidal range is less than 2 feet. Conversely, Boston Harbor has a mean tidal range of about 9.6 feet, but the average currents at the opening, between Deer Island and Hull, do not exceed 2 knots. There is no necessary correlation between current strength and range of tide.

Why did the tidal or current prediction in ELDRIDGE differ from what I saw? Unless there was an error in the Government tables we take our data from, the answer is either (1) weather-related, as when a storm either retards or advances a tidal event, or (2) the discrepancy is small enough to be explained by the approximate nature of tide and current predictions, and figures are sometimes rounded off. We appreciate hearing from readers of any observed discrepancies or errors. Call us at 617-482-8460, Monday to Friday, 9 a.m. to 5 p.m.

HOW TO USE THE TIDE AND CURRENT TABLES AND CURRENT CHARTS

High and Low Water Tide Tables

In addition to presenting tide tables for nine reference ports, from Portland to Miami, we show the approximate time of High Water and the mean (average) height of high at some 350 substations.

- On pp. 12-20, find your harbor, or the nearest one to it, and note the time difference between it and the reference port.
- Apply this time difference to the reference table for that date. On average the Low Water will follow by about 6 hours, 12 minutes.
- When the height of High Water in the reference table is higher or lower than the average, it will be correspondingly higher or lower at your harbor.

Current Tables

There are eight current tables covering from Massachusetts to the Chesapeake. At over 300 other points, on pp. 22-29, we show the approximate time of current change, the directions of ebb and flood, and the average maximum velocities.

- Find the place you are concerned with, or the listed position nearest to it, and note the time difference between it and the reference location.
- Apply this time difference to the reference table for that date. On average, the current will change approximately every 6 hours, 12 minutes.
- When the velocity of the current in the reference table exceeds the average maximum, the current in your area will also exceed the average maximum.

Naming Currents

While it is traditional to name currents as Ebb or Flood, these terms can easily confuse. We recommend using the direction as the name of the current. It is more helpful to refer to an Easterly current, which means it is Eastbound or runs toward the East, than it is to name it as an Ebb or Flood Current, which leaves the listener guessing its direction.

Current Charts and Diagrams

- Find the appropriate current chart and note the table to which it is referenced. For instance, the Long Island Sound charts (pp. 92-97) reference the Race tables.
- Turn to this table, which shows the time of start of Flood and start of Ebb, and find the time of the start of the advantageous current for that day.
- The difference between having a fair current or a head current means hours and dollars to the slower moving vessel such as a trawler or auxiliary sailboat. See Smarter Boating, p. 36.

Effect of the Moon

It is wise to pay particular attention to the phase or position of the Moon. "Astronomical" tides and currents occur around the times of full and new moons, especially when the Moon is at Perigee, or closest to the Earth. Tides will be both higher and lower than average, and currents will run stronger than average. See pp. 234-236.

TIME OF HIGH WATER

Time figures shown are the *average* differences throughout the year. Rise in feet is mean range.
(Low Water times are given *only* when they vary more than 20 min. from High Water times.)

NOTE: *Asterisk indicates that NOAA has recently dropped these substations from its listing because the data are judged to be of questionable accuracy. We have published NOAA's most recently available figures with this warning: Mariners are cautioned that the starred information is only approximate and not supported by NOAA or the Publishers of Eldridge.

*For **Canadian Ports**, if your watch is set for Atlantic Time, use the time differences listed here; if your watch is set for Eastern Time, subtract one hour from these time differences.*

	Hr.	Min.			Rise in feet
NOVA SCOTIA, Outer Coast					
Guysborough	*3*	*05*	*before*	*PORTLAND*	*3.8*
Whitehaven Harbour	*3*	*20*	"	"	*3.7*
Liscomb Harbour	*3*	*20*	"	"	*4.2*
Sheet Harbour	*3*	*20*	"	"	*4.2*
Ship Harbour	*3*	*20*	"	"	*4.2*
Jeddore Harbour	*3*	*15*	"	"	*4.3*
Halifax	*3*	*10*	"	"	*4.4*
Sable Island, north side	*3*	*15*	"	"	*2.6*
Sable Island, south side	*3*	*10*	"	"	*3.9*
Chester, Mahone Bay	*3*	*10*	"	"	*4.4*
Mahone Harbour, Mahone Bay	*3*	*10*	"	"	*4.5*
Lunenburg	*3*	*05*	"	"	*4.2*
Riverport, La Have River	*3*	*00*	"	"	*4.5*
Liverpool Bay	*2*	*55*	"	"	*4.3*
Lockeport	*2*	*40*	"	"	*4.6*
Shelburne	*2*	*40*	"	"	*4.8*
NOVA SCOTIA & NEW BRUNSWICK, Bay of Fundy					
Lower E. Pubnico	*1*	*10*	*before*	*PORTLAND*	*8.7*
Yarmouth Harbour	*0*	*20*	"	"	*11.5*
Annapolis Royal, Annapolis R.	*0*	*50*	*after*	"	*22.6*
Parrsboro, Minas Basin, Partridge Is.	*1*	*35*	"	"	*34.4*
Burntcoat Head, Minas Basin	*1*	*50*	"	"	*38.4*
Amherst Point, Cumberland Basin	*1*	*20*	"	"	*35.6*
Grindstone Is, Petitcodiac River	*1*	*05*	"	"	*31.1*
Hopewell Cape, Petitcodiac River	*1*	*00*	"	"	*33.2*
Saint John	*0*	*45*	"	"	*20.8*
Indiantown, Saint John River	*2*	*15*	"	"	*1.2*
L'Etang Harbor	*0*	*45*	"	"	*18.4*

REVERSING FALLS, SAINT JOHN, N.B.

The most turbulence in the gorge occurs on days when the tides are largest. On largest tides the outward fall is between 15 and 16 1/2 feet and is accompanied by a greater turbulence than the inward fall which is between 11 and 12 1/2 feet. The outward fall is at its greatest between two hours before and one hour after low water at St. John; the inward fall is greater just before the time of high water. For complete tidal information of Canadian ports see Tide Tables of the Atlantic Coast of Canada. (Purchase tables from nautical dealers in Canadian ports or from the Queen's Printer, Department of Public Printing, Ottawa).

PORTLAND Tables, pp. 30-35

TIME OF HIGH WATER

Time figures shown are the *average* differences throughout the year. Rise in feet is mean range. (Low Water times are given *only* when they vary more than 20 min. from High Water times.)

U.S. ATLANTIC COAST, from Maine southward

	Hr.	Min.			Rise in feet
MAINE					
Eastport	*0*	*15*	*before*	*PORTLAND*	*18.4*
Cutler, Little River	*0*	*25*	"	"	*13.5*
Shoppee Pt., Englishman Bay	*0*	*20*	"	"	*12.1*
Steele Harbor Island	*0*	*30*	"	"	*11.6*
**Jonesport*	*0*	*20*	"	"	*11.5*
Green Island, Petit Manan Bar	*0*	*30*	"	"	*10.6*
Prospect Harbor	*0*	*25*	"	"	*10.5*
Winter Harbor, Frenchman Bay	*0*	*20*	"	"	*10.1*
Bar Harbor, Mt. Desert Island	*0*	*20*	"	"	*10.6*
Southwest Harbor, Mt. Desert Island	*0*	*20*	"	"	*10.2*
Bass Harbor **high 0 15** *before, low*	*0*	*45*	"	"	*9.9*
Blue Hill Harbor, Blue Hill Bay	*0*	*10*	"	"	*10.1*
Burnt Coat Harbor, Swans Island	*0*	*20*	"	"	*9.5*
Penobscot Bay					
Center Harbor, Eggemoggin Reach	*0*	*10*	"	"	*10.1*
Little Deer Isle, Eggemoggin Reach	*0*	*05*	"	"	*10.0*
Isle Au Haut	*0*	*20*	"	"	*9.3*
Stonington, Deer Isle	*0*	*10*	"	"	*9.7*
Matinicus Harbor, Wheaton Is.	*0*	*15*	"	"	*9.0*
Vinalhaven	*0*	*10*	"	"	*9.3*
North Haven	*0*	*05*	"	"	*9.7*
Pulpit Harbor, North Haven Is.	*0*	*10*	"	"	*9.9*
Castine	*0*	*05*	"	"	*10.1*
Bucksport, Penobscott River	*0*	*25*	"	"	*10.8*
Bangor, Penobscot River **high 0 25** *before, low*	*same as*			"	*13.4*
Belfast	*0*	*10*	*before*	"	*10.2*
**Camden*	*0*	*10*	"	"	*9.6*
Rockland	*0*	*10*	"	"	*9.8*
MAINE, Outer Coast					
Tenants Harbor	*0*	*10*	*before*	*PORTLAND*	*9.3*
Monhegan Island	*0*	*15*	"	"	*8.8*
Port Clyde, St. George River	*0*	*10*	"	"	*8.9*
Thomaston, St. George River	*0*	*05*	"	"	*9.4*
New Harbor, Muscongus Bay	*0*	*10*	"	"	*8.8*
Friendship Harbor	*0*	*20*	"	"	*9.0*
Waldoboro, Medomak River	*0*	*15*	"	"	*9.5*
East Boothbay, Damariscotta River	*same as*			"	*8.9*
Boothbay Harbor	*0*	*05*	*before*	"	*8.8*
Wiscasset, Sheepscot River	*0*	*15*	*after*	"	*9.4*
Robinhood, Sasanoa River	*0*	*15*	"	"	*8.8*
Phippsburg, Kennebec River	*0*	*25*	"	"	*8.0*
Bath, Kennebec River	*1*	*00*	"	"	*6.4*
Casco Bay					
**Small Point Harbor*	*0*	*10*	*before*	"	*8.8*
Cundy Harbor, New Meadows River	*same as*			"	*8.9*
South Harpswell, Potts Harbor	*same as*			"	*8.9*
South Freeport	*0*	*10*	*after*	"	*9.0*

PORTLAND Tables, pp. 30-35

When a high tide exceeds avg. ht., the *following* low tide will be lower than avg.

*Times and Hts. are approximate. *Important*: See NOTE, top p. 12.

TIDE STATIONS

TIME OF HIGH WATER

Time figures shown are the *average* differences throughout the year. Rise in feet is mean range. (Low Water times are given *only* when they vary more than 20 min. from High Water times.)

	Hr. Min.			Rise in feet
MAINE, Cont.				
Falmouth Foreside	*same as*		*PORTLAND*	*9.2*
Great Chebeague Island	*same as*		"	*9.1*
Portland Head Light	*same as*		"	*8.9*
Cape Porpoise	*0 10*	*after*	"	*8.7*
Kennebunkport	*0 05*	"	"	*8.8*
York Harbor	*0 05*	"	"	*8.6*
NEW HAMPSHIRE				
Portsmouth	*0 20*	*after*	*PORTLAND*	*7.8*
Gosport Harbor, Isles of Shoals	*same as*		"	*8.5*
Hampton Harbor	*0 15*	*after*	"	*8.3*
MASSACHUSETTS, Outer Coast				
Newburyport, Merrimack River	**high 0 30** *after, low 1 10*	*after*	*PORTLAND*	*7.8*
Plum Island Sound, S. End	**high 0 10** *after, low 0 35*	"	"	*8.6*
Annisquam, Lobster Cove	*0 10*	"	"	*8.8*
Rockport	*0 05*	"	"	*8.7*
Gloucester Harbor	*same as*		*BOSTON*	*8.8*
**Manchester*	*same as*		"	*8.8*
Salem	*same as*		"	*8.9*
**Marblehead*	*same as*		"	*9.1*
Lynn, Lynn Harbor	*same as*		"	*9.2*
Neponset, Neponset R.	*same as*		"	*9.5*
Weymouth, Fore River Bridge	*0 10*	*after*	"	*9.5*
Hingham	*0 10*	"	"	*9.5*
Hull	*0 05*	"	"	*9.3*
Cohasset Harbor (White Head)	*0 05*	"	"	*8.8*
Scituate, Scituate Harbor	*0 05*	"	"	*8.9*
Cape Cod Bay				
Duxbury Harbor	**high 0 05** *after, low 0 35*	"	"	*9.9*
Plymouth	*0 05*	"	"	*9.8*
Cape Cod Canal, East Entrance	*same as*		"	*8.7*
Barnstable Harbor, Beach Point	*0 10*	*after*	"	*9.5*
Wellfleet	*0 15*	"	"	*10.0*
Provincetown	*0 15*	"	"	*9.1*
Cape Cod				
Stage Harbor, Chatham	*0 45*	"	"	*4.0*
Chatham Hbr, Aunt Lydias Cove	**high 0 56** *after, low 1 10*	"	"	*5.8*
Pleasant Bay, Chatham	**high 2 30** *after, low 3 25*	"	"	*3.2*
Nantucket Sound				
Wychmere Harbor	**high 0 50** *after, low 0 25*	"	"	*3.7*
Dennisport	**high 1 05** *after, low 0 40*	"	"	*3.4*
South Yarmouth, Bass River	*1 50*	"	"	*2.8*
Hyannis Port	**high 1 00** *after, low 0 25*	"	"	*3.2*
Cotuit Highlands	**high 1 15** *after, low 0 45*	"	"	*2.5*
Falmouth Heights	*0 15*	*before*	"	*1.3*
Nantucket Island				
Great Point	*0 45*	*after*	"	*3.1*
Nantucket	*1 05*	"	"	*3.0*
Muskeget Island, North side	*0 25*	"	"	*2.0*

PORTLAND Tables, pp. 30-35, BOSTON Tables, pp. 38-43

When a high tide exceeds avg. ht., the *following* low tide will be lower than avg.

*Times and Hts. are approximate. *Important*: See NOTE, top p. 12.

TIME OF HIGH WATER

Time figures shown are the *average* differences throughout the year. Rise in feet is mean range.
(Low Water times are given *only* when they vary more than 20 min. from High Water times.)

Station	High / Low	Hr.	Min.			Rise in feet
MASSACHUSETTS, Martha's Vineyard						
Edgartown	**high 1 00** *after, low*	*0*	*20*	*after*	*BOSTON*	*1.9*
Oak Bluffs	**high 0 30** *after, low*	*0*	*10*	*before*	*BOSTON*	*1.7*
Vineyard Haven	**high 0 10** *before, low*	*0*	*35*	"	"	*1.6*
**Lake Tashmoo (inside)*		*2*	*30*	"	"	*2.0*
Cedar Tree Neck	**high 0 10** *after, low*	*1*	*30*	*after*	*NEWPORT*	*2.2*
**Menemsha Bight*	**high** *same as, low*	*0*	*35*	"	"	*2.7*
Gay Head	**high 0 05** *before, low*	*0*	*45*	"	"	*2.9*
Squibnocket Point	**high 0 45** *before, low*	*same as*			"	*2.9*
Wasque Point, Chappaquiddick	**high 2 00** *after, low*	*3*	*20*	*after*	"	*1.1*
Nomans Land	**high 0 20** *before, low*	*0*	*20*	"	"	*3.0*
Vineyard Sound						
Little Hbr., Woods Hole	**high 0 30** *after, low*	*2*	*20*	"	"	*1.4*
Quick's Hole, N. side		*0*	*10*	*before*	"	*3.5*
Cuttyhunk		*1*	*20*	*after*	"	*3.4*
Buzzards Bay						
**Cuttyhunk Pond Entr.*		*same as*			"	*3.4*
W. Falmouth Harbor, Chappaquoit Pt.		*0*	*05*	*after*	"	*3.8*
**Pocasset Hbr., Barlows Landing*		*0*	*25*	"	"	*4.0*
Monument Beach		*0*	*15*	"	"	*4.0*
**Wareham River*		*0*	*20*	"	"	*4.1*
Great Hill		*0*	*10*	"	"	*4.0*
Marion, Sippican Harbor		*0*	*10*	"	"	*4.0*
Mattapoisett Harbor		*0*	*10*	"	"	*3.9*
Clarks Point		*0*	*15*	"	"	*3.6*
New Bedford		*0*	*05*	"	"	*3.7*
**South Dartmouth*		*0*	*25*	"	"	*3.7*
Westport Harbor, Westport River	**high 0 10** *after, low*	*0*	*35*	*after*	"	*3.0*
RHODE ISLAND & MASS, Narragansett Bay						
Sakonnet, Sakonnet River		*0*	*10*	*before*	*NEWPORT*	*3.2*
Beavertail Point, Conanicut Island		*0*	*05*	"	"	*3.3*
Conanicut Point, Conanicut Island		*0*	*05*	*after*	"	*3.8*
Prudence Island (south end)		*0*	*10*	"	"	*3.8*
Bristol Harbor		*0*	*15*	"	"	*4.1*
Fall River, MA		*0*	*20*	"	"	*4.4*
Bay Spring, Bullock Cove		*0*	*10*	"	"	*4.3*
Providence, State Pier no. 1		*0*	*15*	"	"	*4.4*
Pawtucket, Seekonk River		*0*	*20*	"	"	*4.6*
East Greenwich		*0*	*15*	"	"	*4.0*
Wickford		*0*	*05*	"	"	*3.7*
Narragansett Pier	**high 0 10** *before, low*	*0*	*10*	*after*	"	*3.2*
RHODE ISLAND, Outer Coast						
Pt. Judith, Harbor of Refuge	**high** *same as, low*	*0*	*35*	*after*	*NEWPORT*	*3.0*
Block Island, Old Harbor	**high 0 15** *before, low*	*0*	*15*	"	"	*2.9*
Watch Hill Pt.	**high 0 40** *after, low*	*1*	*15*	"	"	*2.6*
CONNECTICUT, L.I. Sound						
**Stonington*		*2*	*15*	*before*	*BRIDGEPORT*	*2.7*
**Noank*		*2*	*05*	"	"	*2.3*
New London, Thames River (State Pier)		*1*	*45*	"	"	*2.6*
Norwich, Thames River		*1*	*20*	"	"	*3.0*

BOSTON Tables, pp. 38-43, NEWPORT Tables, pp. 78-83, BRIDGEPORT Tables, pp. 98-103

When a high tide exceeds avg. ht., the *following* low tide will be lower than avg.
*Times and Hts. are approximate. *Important*: See NOTE, top p. 12.

TIME OF HIGH WATER

Time figures shown are the *average* differences throughout the year. Rise in feet is mean range. (Low Water times are given *only* when they vary more than 20 min. from High Water times.)

	Hr. Min.			Rise in feet
CONNECTICUT, L.I. Sound, Cont.				
Saybrook Jetty, Connecticut River	*0 35*	*before*	*BRIDGEPORT*	*3.5*
Essex, Connecticut River	*0 05*	"	"	*3.0*
Madison	*0 20*	"	"	*4.9*
Branford, Branford River	*0 05*	"	"	*5.9*
New Haven Harbor, New Haven Reach	*same as*		"	*6.2*
Milford Harbor	*same as*		"	*6.3*
Sniffens Point, Housatonic River	*0 10*	*after*	"	*6.4*
South Norwalk	*0 10*	"	"	*7.1*
Stamford	*0 05*	"	"	*7.2*
Cos Cob Harbor	*0 05*	"	"	*7.2*
**Greenwich*	*same as*		"	*7.4*
NEW YORK, Long Island Sound, North Side				
Rye Beach	*0 20*	*before*	*KINGS POINT*	*7.3*
New Rochelle	*0 15*	"	"	*7.3*
Throgs Neck	*0 10*	*after*	"	*7.0*
Whitestone, East River	*0 05*	"	"	*7.1*
College Point, Flushing Bay	*0 15*	"	"	*6.8*
Hunts Point, East River	*0 15*	"	"	*7.0*
North Brother Island, East River	*0 20*	"	"	*6.6*
Port Morris, Stony Pt., East River	*0 05*	"	"	*6.2*
NEW YORK, Long Island, North Shore				
Willets Point	*same as*		*KINGS POINT*	*7.2*
Port Washington, Manhasset Bay	*0 10*	*before*	"	*7.3*
Glen Cove, Hempstead Harbor	*0 20*	"	"	*7.3*
Oyster Bay Harbor, Oyster Bay	*0 05*	*after*	*BRIDGEPORT*	*7.3*
Cold Spring Harbor, Oyster Bay	*0 05*	*before*	"	*7.3*
Eatons Neck Point	*same as*		"	*7.1*
Lloyd Harbor, Huntington Bay	*same as*		"	*7.0*
Northport, Northport Bay	*0 05*	*before*	"	*7.3*
Port Jefferson Harbor Entrance	*same as*		"	*6.6*
Mattituck Inlet	*0 05*	*after*	"	*5.2*
Shelter Island Sound				
Orient	*1 10*	*before*	"	*2.5*
Greenport	*0 40*	"	"	*2.4*
Southold	*same as*		"	*2.3*
Sag Harbor	*0 45*	*before*	"	*2.5*
New Suffolk, Peconic Bay	*0 40*	*after*	"	*2.6*
South Jamesport, Peconic Bay	*0 40*	"	"	*2.8*
Threemile Harbor, Entr., Gardiners Bay	*1 05*	*before*	"	*2.5*
Montauk Harbor Entr.	*2 10*	"	"	*2.0*
Long Island, South Shore				
Shinnecock Inlet, Ocean **high 0 40** *before, low*	*1 05*	*before*	*SANDY HOOK*	*3.3*
Moriches Inlet	*1 00*	"	"	*2.9*
Democrat Point, Fire Island Inlet	*0 40*	"	"	*2.6*
Patchogue, Great South Bay	*3 15*	*after*	"	*1.1*
Bay Shore, Watchogue Creek Entrance	*2 15*	"	"	*1.0*
Jones Inlet (Point Lookout)	*0 20*	*before*	"	*3.6*
Bellmore, Hempstead Bay **high 1 30** *after, low*	*2 00*	*after*	"	*2.0*

BRIDGEPORT Tables, pp. 98-103, KINGS POINT Tables, pp. 104-109, SANDY HOOK Tables, pp. 134-139

When a high tide exceeds avg. ht., the *following* low tide will be lower than avg.

*Times and Hts. are approximate. *Important*: See NOTE, top p. 12.

TIME OF HIGH WATER

Time figures shown are the *average* differences throughout the year. Rise in feet is mean range.
(Low Water times are given *only* when they vary more than 20 min. from High Water times.)

	Hr. Min.			Rise in feet
NEW YORK, Long Island, South Shore, Cont.				
Freeport, Baldwin Bay	*0 40*	*after*	*SANDY HOOK*	*3.0*
E. Rockaway Inlet	*0 10*	*before*	"	*4.1*
Barren Is., Rockaway Inlet, Jamaica Bay	*same as*		"	*5.0*
NEW YORK & NEW JERSEY				
New York Harbor				
Coney Island	*0 05*	*before*	*SANDY HOOK*	*4.7*
Fort Hamilton, The Narrows	*same as*		"	*4.7*
Tarrytown, Hudson River	*1 50*	*after*	*BATTERY*	*3.2*
Poughkeepsie, Hudson River	*4 35*	"	"	*3.1*
Kingston, Hudson River	*5 20*	"	"	*3.7*
NY & NJ, the Kills and Newark Bay				
Constable Hook, Kill Van Kull	*0 20*	*before*	"	*4.6*
Port Elizabeth	*same as*		"	*5.1*
Bellville, Passaic River **high 0 10** *after, low*	*0 50*	*after*	"	*5.6*
Kearny Pt., Hackensack River.	*0 10*	"	"	*5.2*
Hackensack, Hackensack River	*1 05*	"	"	*6.0*
Lower NY Bay, Raritan Bay				
Great Kills Harbor	*0 05*	*after*	*SANDY HOOK*	*4.7*
South Amboy, Raritan River	*0 05*	*before*	"	*5.1*
New Brunswick, Raritan River	*0 30*	*after*	"	*5.7*
Keyport	*0 05*	*before*	"	*5.0*
Atlantic Highlands, Sandy Hook Bay	*0 10*	"	"	*4.7*
Highlands, Shrewsbury R., Rte. 36 bridge, Sandy Hook	*0 15*	*after*	"	*4.2*
Red Bank, Navesink River, Sandy Hook Bay	*1 20*	"	"	*3.5*
Sea Bright, Shrewsbury River, Sandy Hook Bay	*1 15*	"	"	*3.2*
NEW JERSEY, Outer Coast				
Shark River, R.R. Bridge, Shark River Island	*0 15*	*before*	"	*4.3*
Manasquan Inlet, USCG Station	*0 10*	"	"	*4.0*
Brielle, Rte. 35 bridge, Manasquan River	*0 05*	"	"	*3.9*
Barnegat Inlet,USCG Station, Barnegat Bay	*0 10*	"	"	*2.2*
Manahawkin Drawbridge **high 2 50** *after, low*	*3 40*	*after*	"	*1.3*
Beach Haven, USCG Station, Little Egg Harbor	*1 20*	"	"	*2.2*
Absecon Creek, Rte. 30 bridge	*1 05*	"	"	*3.9*
Atlantic City, Ocean	*0 25*	*before*	"	*4.0*
Beesleys Pt., Great Egg Hbr. Bay **high 0 30** *after, low*	*1 05*	*after*	"	*3.6*
Townsends Inlet, Ocean Dr. bridge	*0 10*	"	"	*3.9*
Stone Harbor, Great Channel, Hereford Inlet	*0 35*	"	"	*4.0*
Cape May Harbor, Cape May Inlet	*0 10*	"	"	*4.5*
NEW JERSEY & DELAWARE BAY				
Delaware Bay, Eastern Shore				
Brandywine Shoal Light **high 0 30** *after, low*	*1 00*	*after*	*BATTERY*	*4.9*
Cape May Point, Sunset Beach	*0 15*	"	"	*4.8*
Dennis Creek, 2.5 mi. above Entr. **high 1 15** *after, low*	*2 05*	"	"	*5.2*
Mauricetown, Maurice R.	*2 40*	"	"	*4.4*
Millville, Maurice R.	*3 55*	"	"	*5.0*

SANDY HOOK Tables, pp. 134-139, BATTERY Tables, pp. 122-127

When a high tide exceeds avg. ht., the *following* low tide will be lower than avg.

*Times and Hts. are approximate. *Important*: See NOTE, top p. 12.

TIME OF HIGH WATER

Time figures shown are the *average* differences throughout the year. Rise in feet is mean range.
(Low Water times are given *only* when they vary more than 20 min. from High Water times.)

	Hr. Min.			Rise in feet
NEW JERSEY & DELAWARE BAY, Cont.				
Delaware Bay, Western Shore				
**Cape Henlopen*	*0 10*	*after*	*BATTERY*	*4.1*
Lewes (Breakwater Harbor) **high 0 20** *after, low*	*0 45*	"	"	*4.1*
St. Jones River Ent.* **high 1 10 *after, low*	*1 55*	"	"	*4.8*
Delaware River				
**Liston Point, Delaware*	*2 05*	"	"	*5.7*
Salem, Salem River, NJ	*3 55*	"	"	*4.2*
Reedy Point, Delaware **high 3 05** *after, low*	*3 25*	"	"	*5.3*
C&D Summit Bridge, Delaware	*2 35*	"	"	*3.5*
Chesapeake City, MD	*2 20*	"	"	*2.9*
New Castle, Delaware **high 3 35** *after, low*	*4 05*	"	"	*5.2*
Wilmington Marine Terminal **high 3 55** *after, low*	*4 30*	"	"	*5.3*
Philadelphia, PA, USCG Station **high 5 30** *after, low*	*5 50*	"	"	*6.0*
Burlington, NJ **high 6 25** *after, low*	*7 05*	"	"	*7.2*
Trenton, NJ **high 6 45** *after, low*	*7 45*	"	"	*8.2*
DELAWARE, MARYLAND & VIRGINIA				
Indian River Inlet, USCG Station, Delaware	*0 55*	*after*	*SANDY HOOK*	*2.5*
Ocean City Fishing Pier	*0 20*	*before*	"	*3.4*
Harbor of Refuge, Chincoteague Bay	*0 10*	*after*	"	*2.4*
Chincoteague Channel, south end	*0 20*	"	"	*2.2*
Chincoteague Island, USCG Station	*0 40*	"	"	*1.6*
Metompkin Inlet	*0 40*	"	"	*3.6*
Wachapreague, Wachapreague Channel	*0 50*	"	"	*4.0*
**Quinby Inlet Entrance*	*0 05*	"	"	*4.0*
Great Machipongo Inlet, inside	*0 45*	"	"	*3.9*
Chesapeake Bay, Eastern Shore				
Cape Charles Harbor	*0 40*	*after*	*BATTERY*	*2.3*
Crisfield, Little Annemessex River	*4 30*	"	"	*1.9*
Salisbury, Wicomico River	*7 15*	"	"	*3.0*
Middle Hooper Island	*4 40*	*before*	*BALTIMORE*	*1.5*
Taylors Island, Little Choptank River, Slaughter Creek	*3 15*	"	"	*1.3*
**Sharps Is. Lt.*	*3 50*	"	"	*1.3*
Cambridge, Choptank River	*2 40*	"	"	*1.6*
Dover Bridge, Choptank River	*0 20*	"	"	*1.7*
Oxford, Tred Avon River	*2 50*	"	"	*1.4*
Easton Pt., Tred Avon River	*2 45*	"	"	*1.6*
St. Michaels, Miles River	*2 10*	"	"	*1.4*
Kent Island Narrows	*1 30*	"	"	*1.2*
**Bloody Pt. Bar Lt.*	*2 40*	"	"	*1.1*
Worton Creek Entrance	*1 20*	*after*	"	*1.3*
Town Point Wharf, Elk River	*3 20*	"	"	*2.2*
Chesapeake Bay, Western Shore				
Havre de Grace, Susquehanna River	*3 15*	*after*	"	*1.9*
**Pooles Is.*	*0 55*	"	"	*1.2*
Annapolis, Severn River (US Naval Academy)	*1 30*	*before*	"	*1.0*
**Sandy Point*	*1 20*	"	"	*0.8*
Thomas Pt. Shoal Lt.	*1 55*	"	"	*0.9*
**Drum Point, Pawtuxent River*	*4 50*	"	"	*1.2*
Solomons Island, Pawtuxent River	*4 40*	"	"	*1.2*
Point Lookout	*5 30*	"	"	*1.2*

BATTERY Tables, pp. 122-127, SANDY HOOK Tables, pp. 134-139, BALTIMORE Tables, pp. 156-159

When a high tide exceeds avg. ht., the *following* low tide will be lower than avg.

*Times and Hts. are approximate. *Important*: See NOTE, top p. 12.

TIME OF HIGH WATER

Time figures shown are the *average* differences throughout the year. Rise in feet is mean range. (Low Water times are given *only* when they vary more than 20 min. from High Water times.)

Station	Hr.	Min.			Rise in feet
DELAWARE, MARYLAND & VIRGINIA, Cont.					
Sunnybank, Little Wicomico River	*6*	*35*	*after*	*BATTERY*	*0.8*
Glebe Point, Great Wicomico River	*4*	*10*	"	"	*1.2*
Windmill Point, Rappahannock River	*2*	*45*	"	"	*1.2*
**Orchard Point, Rappahannock River*	*3*	*20*	"	"	*1.4*
**New Point Comfort, Mobjack Bay*	*0*	*45*	"	"	*2.3*
Tue Marshes Light, York River	*0*	*55*	"	"	*2.2*
**Perrin River, York River*	*1*	*05*	"	"	*2.3*
Yorktown, Goodwin Neck, York River	*1*	*10*	"	"	*2.2*
Hampton Roads, Sewells Pt. **high 0 50** *after, low*	*0*	*40*	"	"	*2.4*
Norfolk, Elizabeth River	*1*	*15*	"	"	*2.8*
Newport News, James River	*1*	*20*	"	"	*2.6*
Jamestown Is., James River **high 3 55** *after, low*	*4*	*15*	"	"	*2.0*
Windmill Pt., James River* **high 6 15 *after, low*	*6*	*30*	"	"	*2.3*
Chesapeake Bay Br. Tunnel **high 0 05** *before, low*	*0*	*20*	*before*	"	*2.6*
Cape Henry	*0*	*05*	"	"	*3.1*
NORTH CAROLINA					
Roanoke Sound Channel	*1*	*10*	*after*	*BATTERY*	*0.5*
Oregon Inlet Marina **high 0 20** *before, low*	*0*	*10*	*before*	"	*0.9*
Oregon Inlet, USCG Station **high 0 40** *before, low*	*1*	*00*	"	"	*1.9*
Oregon Inlet Channel	*0*	*30*	"	"	*1.2*
Cape Hatteras Fishing Pier	*1*	*00*	"	"	*3.0*
Hatteras Inlet	*0*	*50*	"	"	*2.0*
Ocracoke Inlet	*0*	*50*	"	"	*2.0*
Beaufort Inlet Channel Range	*0*	*55*	"	"	*3.2*
Morehead City	*0*	*35*	"	"	*3.1*
Bogue Inlet	*0*	*45*	"	"	*2.2*
New River Inlet	*0*	*45*	"	"	*3.0*
New Topsail Inlet **high 0 40** *before, low*	*0*	*10*	"	"	*3.0*
Bald Head, Cape Fear River	*1*	*15*	"	"	*4.5*
Wilmington **high 1 20** *after, low*	*1*	*45*	*after*	"	*4.3*
Lockwoods Folly Inlet	*0*	*55*	"	"	*4.2*
SOUTH CAROLINA					
Little River Neck, north end	*1*	*20*	*after*	*BATTERY*	*4.7*
Hog Inlet Pier	*0*	*45*	"	"	*5.0*
Myrtle Beach, Springmaid Pier	*0*	*50*	"	"	*5.0*
Pawleys Island Pier (ocean)	*0*	*55*	"	"	*4.9*
Winyah Bay Entrance, south jetty	*0*	*45*	*before*	"	*4.6*
South Island Plantation, C.G. Station	*0*	*10*	*after*	"	*3.8*
Georgetown, Sampit River **high 1 00** *after, low*	*1*	*40*	"	"	*3.7*
North Santee River Inlet	*0*	*35*	*before*	"	*4.5*
Charleston (Custom House Wharf)	*0*	*25*	"	"	*5.2*
Folly River, north, Folly Island **high** *same as, low*	*0*	*35*	"	"	*5.4*
Rockville, Bohicket Creek, North Edisto River	*0*	*05*	"	"	*5.8*
Edisto Marina, Big Bay Creek entr., South Edisto River	*0*	*20*	"	"	*6.0*
Harbor River Bridge, St. Helena Sound	*0*	*10*	"	"	*6.0*
Hutchinson Island, Ashepoo River, St. Helena Sound	*0*	*15*	*after*	"	*6.0*
Fripps Inlet, Hunting Island Bridge, St. Helena Sound	*0*	*25*	*before*	"	*6.1*
Port Royal Plantation, Hilton Head Is.	*0*	*15*	"	"	*6.1*
Battery Creek, Beaufort River Port Royal Sd, 4 mi. above entr. **high 1 00** *after, low*	*0*	*15*	*after*	"	*7.6*

BATTERY Tables, pp. 122-127

When a high tide exceeds avg. ht., the *following* low tide will be lower than avg.

*Times and Hts. are approximate. *Important*: See NOTE, top p. 12.

TIME OF HIGH WATER

Time figures shown are the *average* differences throughout the year. Rise in feet is mean range. (Low Water times are given *only* when they vary more than 20 min. from High Water times.)

	Hr. Min.			Rise in feet
SOUTH CAROLINA, Cont.				
Beaufort, Beaufort River **high 0 55** *after, low*	*0 30*	*after*	*BATTERY*	*7.4*
Braddock Point, Hilton Head Island, Calibogue Sd.	*0 10*	*before*	"	*6.7*
GEORGIA				
Savannah River Entrance, Fort Pulaski	*0 15*	*before*	*BATTERY*	*6.9*
Tybee Creek Entrance	*0 25*	"	"	*6.8*
Wilmington River, north entrance	*0 25*	*after*	"	*7.6*
Isle of Hope, Skidaway River **high 0 35** *after, low*	*0 10*	"	"	*7.8*
Egg Islands, Ossabaw Sound	*0 10*	*before*	"	*7.2*
Walburg Creek Entr., St. Catherines Sd.	*same as*		"	*7.1*
Blackbeard Island	*0 05*	*after*	"	*6.9*
Blackbeard Creek, Blackbeard Island	*0 05*	"	"	*6.5*
Old Tower, Sapelo Island, Doboy Sound	*same as*		"	*6.8*
Threemile Cut Entrance, Darien River	*0 30*	*after*	"	*7.1*
St. Simons Sound Bar	*0 15*	*before*	"	*6.5*
Frederica River, St. Simons Sound	*0 35*	*after*	"	*7.2*
Brunswick, East River, St. Simons Sound	*0 45*	"	"	*7.2*
Jekyll Is. Marina, Jekyll Creek, St. Andrew Sound	*0 30*	"	"	*6.8*
Cumberland Wharf, Cumberland River	*0 30*	*after*	"	*6.8*
FLORIDA, East Coast				
St. Marys Entrance, north jetty, Cumberland Sd.	*0 10*	*before*	*BATTERY*	*5.8*
Fernandina Beach, Amelia R. **high 0 30** *after, low*	*0 05*	*after*	"	*6.0*
Amelia City, South Amelia River	*0 50*	"	"	*5.4*
Nassau River Entrance **high 0 10** *after, low*	*0 50*	"	"	*5.1*
Mayport, (Bar Pilot Dock) **high 0 15** *after, low*	*0 15*	*before*	"	*4.6*
St. Augustine, City Dock.	*0 10*	*after*	"	*4.5*
Ponce Inlet, Halifax River **high 0 05** *after, low*	*0 30*	*after*	*MIAMI*	*2.8*
Cape Canaveral **high 1 05** *before, low*	*0 45*	*before*	"	*3.5*
Port Canaveral, Trident Pier	*same as*		"	*3.5*
Sebastian Inlet bridge **high 0 50** *before, low*	*0 25*	*before*	"	*2.2*
St. Lucie, Indian River **high 0 40** *after, low*	*1 45*	*after*	"	*1.0*
Vero Beach, ocean	*0 55*	*before*	"	*3.4*
Fort Pierce Inlet, south jetty	*0 30*	"	"	*2.6*
Stuart, St. Lucie River **high 2 15** *after, low*	*3 30*	*after*	"	*0.9*
Jupiter Inlet, south jetty	*0 10*	*before*	"	*2.5*
North Palm Beach, Lake Worth **high 0 15** *before, low*	*0 15*	*after*	"	*2.8*
Port of Palm Beach, Lake Worth **high 0 20** *before, low*	*0 05*	"	"	*2.7*
Lake Worth Pier, ocean **high 0 45** *before, low*	*0 20*	*before*	"	*2.7*
Hillsboro Inlet, C.G. Light Station	*0 15*	"	"	*2.5*
Hillsboro Inlet Marina **high 0 05** *before, low*	*0 25*	*after*	"	*2.5*
Lauderdale-by-the-Sea, fish pier. **high 0 35** *before, low*	*0 15*	*before*	"	*2.6*
Bahia Mar Yacht Club **high 0 05** *before, low*	*0 35*	*after*	"	*2.4*
Port Everglades, Turning Basin.. **high 0 30** *before, low*	*0 10*	*before*	"	*2.5*
North Miami Beach, fishing pier **high 0 20** *before, low*	*same as*		"	*2.5*
Miami, Miamarina, Biscayne Bays **high 0 20** *after, low*	*0 50*	*after*	"	*2.2*
Dinner Key Marina, Biscayne Bay **high 0 55** *after, low*	*1 50*	"	"	*1.9*
Key Biscayne Yt. Club, Biscayne B **high 0 45** *after low*	*1 30*	"	"	*2.0*
Ocean Reef Hbr., Key Largo **high 0 10** *before, low*	*0 15*	"	"	*2.3*
Tavernier Harbor, Hawk Ch. **high 0 05** *after, low*	*0 25*	"	"	*2.0*
Key West	*0 50*	*before*	*BOSTON*	*1.3*

BATTERY Tables, pp. 122-127, MIAMI Tables, pp. 160-163, BOSTON Tables, pp. 38-43

When a high tide exceeds avg. ht., the *following* low tide will be lower than avg.

*Times and Hts. are approximate. *Important*: See NOTE, top p. 12.

Piloting in a Cross Current

See also p. 222, Coping with Currents

When we are piloting in a body of water with an active current from ahead or astern, our course is not affected and the arithmetic for speed is easy. (See p. 36.) When the current comes at an angle to the bow or stern, unless our speed is far greater than the current, we need to alter course to compensate.

First, what not to do. When in a cross current it is a major mistake simply to steer toward our destination. The current will carry us more and more off course, with the heading or bearing to our destination changing all the time. We may finally get there, but we will have traveled considerably farther, on what is termed a hooked course, and possibly have entered dangerous water while doing so.

By GPS: With GPS it's all too easy to find the new heading. We enter our destination waypoint and press GoTo. There are several screens to choose from. First, carefully check the Map screen to see if there are any hazards or obstructions between us and our destination. The Highway screen, considered perhaps the most useful display, will show if we are on course by displaying the highway as straight ahead. The screen will also indicate how far to the left or right of our course we are. This is crosstrack error. We steer to that side which brings us back onto the center of the highway, and then continue to steer in such a way that we stay in the middle. We have changed our heading to achieve the desired COG, course over ground. Now the Course and Bearing numbers should be the same, and we have compensated for cross current.

By eye: Without the help of electronics but with good visibility, we know we need to alter course toward the current until a foreground object, let's say a point on the shore, remains steady in relation to an object farther away, perhaps a distant steeple. This alignment is called a range. Once we find the corrected heading, we can use our compass to maintain it, checking those objects periodically in case current or wind conditions change.

By a chart: With compromised visibility and again without electronics, the problem is solved the traditional way with a paper chart. First, consult the proper current table to determine the speed and direction of the current for the hour(s) in question. (Keep in mind that speeds and times are predictions only. They are approximate and can be altered by weather.) Plot the course, let's say 090°, as if there is no current. Then construct a one-hour vector diagram. From the departure point, construct a line in the direction of the current, let's say 180°, whose length is the distance the current would carry an object in one hour. If the predicted current is 2 knots, that's 2 n.m. Now we set our dividers for a distance which represents how far our boat speed will take us through the water in one hour, let's say 8 n.m. We will put one point of the dividers on the far end of the line representing current, and then swing the dividers until the second point intercepts the line of our intended course. The direction of that third line represents what our boat's heading needs to be (the course to steer) to maintain the original course we drew. The intercept point represents about where our boat will be along the intended course line (COG) at the end of one hour. If this leg is longer or shorter than one hour, it doesn't matter. The course to steer is the same as we determined in our one-hour vector plot, until conditions change.

TIME OF CURRENT CHANGE

(See Note at bottom of Boston Tables, pp. 38-43: Rule-of-Thumb for Current Velocities.)

NOTE: NOAA has recently dropped many substations from its listing because the data are judged to be of questionable accuracy.

CURRENTS IN THE GULF OF MAINE - In the Gulf of Maine, on the western side, the Flood Current splits at Cape Ann, Mass., and floods north and east along the shore towards the Bay of Fundy. At the same time, on the eastern side of the Gulf, at the southern tip of Nova Scotia, the Flood Current runs to the west and then north and eastwards along the shore into the Bay of Fundy. The Ebb Current is just the reverse. In addition to these large principal currents, along the Maine Coast, at least at the mouths of principal bays, there is a shoreward set during the Flood and an offshore set during the Ebb, although this set is of considerably less velocity.

West of Mount Desert, the average along-shore current is rarely more than a knot but the farther east one goes, the greater are the average velocities to be expected, up to 2 knots or more. When heading west, therefore, start off at the time shown for High Water in your area (see p. 13) and have a fair Ebb current for 6 hours. Headed east, start at the time for Low Water in your area (about 6 ½ hours after High Water) and carry the beneficial Flood current. East of Schoodic Point, the average currents are up to 2 knots and taking advantage of them will save considerable time and fuel.

Off shore, in the Gulf of Maine, unlike the along-shore currents that come to dead slack and *reverse*, there are so-called *rotary* currents. These currents constantly change direction in a clockwise flow completing the circle in about 12 ½ hours. The maximum currents are when it is flooding in the northeasterly direction or ebbing in a southwesterly direction; minimum currents occur halfway between. There is no slack water.

Entering the Bay of Fundy through Grand Manan Channel, one finds that the average velocities are from 1-2 ½ knots, although in the narrower channels off the Bay, velocities are higher (Friar Roads at Eastport has average velocities of 3 knots of more). The Current in the Bay Floods to the Northeast and Ebbs to the Southwest.

In using this table, bear in mind that **actual times of Slack or Maximum occasionally differ from the predicted times** by as much as half an hour and in rare instances as much as an hour. Referring the Time of Current Change at the subordinate stations listed below, to the predicted Current Change at the reference station gives the *approximate* time only. Therefore, to make make sure of getting the full advantage of a favorable current or slack water, the navigator should reach the entrance or strait at least half an hour before the predicted time. (This is basically the same precautionary note found in the U.S. Tidal Currents Table Book.)

Figures shown below are **average maximum** velocities in knots. We have omitted places having an average maximum velocity of less than 1 knot. To find the Time of Current Change (Start of Flood and Start of Ebb) at a selected point, refer to the table heading that particular section (in bold type) and add or subtract the time listed.

	TIME DIFFERENCES Flood Starts; Ebb Starts Hr. Min.	MAXIMUM FLOOD Dir.(true) in degrees	MAXIMUM FLOOD Avg. Max. in knots	MAXIMUM EBB Dir.(true) in degrees	MAXIMUM EBB Avg. Max. in knots
MAINE COAST – based on Boston, pp. 38-43 (Flood starts at Low Water; Ebb starts at High Water)					
Isle Au Haut, 0.8 mi. E of Richs Pt.	-0 05	336	1.4	139	1.5
Damariscotta R., off Cavis Pt.	F+1 30, E+0 00	350	0.6	215	1.0
Sheepscot R., off Barter Is.	F+1 25, E+0 05	005	0.8	200	1.1
Lowe Pt., NE of, Sasanoa R.	F+1 25, E+0 35	327	1.7	152	1.8
Lower Hell Gate, Knubble Bay*	F+1 50, E+0 35	290	3.0	155	3.5

*Velocities up to 9.0 kts. have been observed in the vicinity of the Boilers.

Important: **See NOTE, bottom p. 29.**

TIME OF CURRENT CHANGE

(See Note at bottom of Boston Tables, pp. 38-43: Rule-of-Thumb for Current Velocities.)

KENNEBEC RIVER – based on Boston, pp. 38-43

(Fl. starts at Low Water; Ebb starts at High Water)

Station	TIME DIFFERENCES Flood Starts; Ebb Starts Hr. Min.	MAXIMUM FLOOD Dir.(true) in degrees	MAXIMUM FLOOD Avg. Max. in knots	MAXIMUM EBB Dir.(true) in degrees	MAXIMUM EBB Avg. Max. in knots
Hunniwell Pt., NE of	F+2 20, E+1 25	332	2.4	151	2.9
Bald Head, 0.3 mi. SW of	F+2 40, E+1 20	321	1.6	153	2.3
Bluff Head, W of	F+2 50, E+1 50	014	2.3	184	3.4
Fiddler Ledge, N of	F+3 00, E+1 45	267	1.9	113	2.6
Doubling Pt., S of	F+1 45, E+1 45	300	2.6	127	3.0
Lincoln Ledge, E. of	F+2 45, E+1 45	359	1.9	174	2.8
Bath, 0.2 mi. S of bridge	F+2 45, E+2 05	003	1.0	177	1.5

CASCO BAY – based on Boston, pp. 38-43

(Flood starts at Low Water; Ebb starts at High Water)

Station	TIME DIFFERENCES Flood Starts; Ebb Starts Hr. Min.	MAXIMUM FLOOD Dir.(true) in degrees	MAXIMUM FLOOD Avg. Max. in knots	MAXIMUM EBB Dir.(true) in degrees	MAXIMUM EBB Avg. Max. in knots
Broad Sound, W. of Eagle Is.	F+1 00, E-0 05	010	0.9	168	1.3
Hussey Sound, SW of Overset Is.	F+0 45, E+0 25	316	1.1	153	1.2
Portland Hbr. entr., SW of Cushing Is.	F+0 30, E-0 00	322	1.0	154	1.1
Portland Bridge, Center of draw	F+1 10, E+0 45	225	0.9	050	1.0

PORTSMOUTH HARBOR – based on Boston, pp. 38-43

(Flood starts at Low Water; Ebb starts at High Water)

Station	TIME DIFFERENCES Flood Starts; Ebb Starts Hr. Min.	MAXIMUM FLOOD Dir.(true) in degrees	MAXIMUM FLOOD Avg. Max. in knots	MAXIMUM EBB Dir.(true) in degrees	MAXIMUM EBB Avg. Max. in knots
Kitts Rocks, 0.2 mi. WSW of	F+2 05, E+1 30	314	0.7	133	0.8
Portsmouth Hbr. entr.	F+2 05, E+1 30	342	1.2	194	1.5
Fort Point	F+2 10, E+1 35	328	1.6	098	2.0
Clark Is., S of	+2 15	270	1.6	085	2.3
Henderson Pt., W of	F+2 35, E+2 00	285	2.4	138	2.8

MASSACHUSETTS COAST – based on Boston, pp. 38-43

(Flood starts at Low Water; Ebb starts at High Water)

Station	TIME DIFFERENCES Flood Starts; Ebb Starts Hr. Min.	MAXIMUM FLOOD Dir.(true) in degrees	MAXIMUM FLOOD Avg. Max. in knots	MAXIMUM EBB Dir.(true) in degrees	MAXIMUM EBB Avg. Max. in knots
Merrimack River entr.	+1 35	285	2.2	105	1.4
Newburyport, Merrimack R.	F+1 25, E+2 05	288	1.5	098	1.4
Plum Is. Sound entr.	F+0 30, E+1 10	316	1.6	184	1.5
Gloucester Hbr., Blynman Canal entr.	-0 05	310	3.0	130	3.3
Hypocrite Channel	F+0 10, E+1 10	262	0.9	070	1.0

BOSTON HARBOR – based on Boston, pp. 38-43

(Flood starts at Low Water; Ebb starts at High Water)

Station	TIME DIFFERENCES Flood Starts; Ebb Starts Hr. Min.	MAXIMUM FLOOD Dir.(true) in degrees	MAXIMUM FLOOD Avg. Max. in knots	MAXIMUM EBB Dir.(true) in degrees	MAXIMUM EBB Avg. Max. in knots
Pt. Allerton, 0.4 mi. NW.	F-0 15, E+0 35	265	0.7	080	0.8
Deer Island Lt	F-0 05, E+0 18	254	1.1	111	1.2
Nantasket Rds					
Hull Gut	+0 15	163	1.2	350	1.8
West Head (W. Gut) 0.2mi. SW	F-0 10, E+1 25	167	1.4	322	1.4
Weir R. entr., Worlds End, N of	F+0 15, E+1 05	076	0.7	272	0.8
Bumkin Is., 0.4mi. W. of	F-0 20, E+0 45	195	0.5	303	0.3
Weymouth Back R., betw. Grape I. and Lower Neck	F-0 20, E+0 30	094	0.7	281	0.9

CAPE COD BAY – based on Boston, pp. 38-43

(Flood starts at Low Water; Ebb starts at High Water)

Station	TIME DIFFERENCES Flood Starts; Ebb Starts Hr. Min.	MAXIMUM FLOOD Dir.(true) in degrees	MAXIMUM FLOOD Avg. Max. in knots	MAXIMUM EBB Dir.(true) in degrees	MAXIMUM EBB Avg. Max. in knots
Race Point, 7 mi. N of	F-0 05, E+0 20	290	1.5	–	1.5
Race Point, 1 mi. NW of	F-0 10, E+0 15	226	1.0	061	0.9
Barnstable Harbor	F+0 15, E+0 40	192	1.2	004	1.4
Manomet Point	F+0 00, E+0 25	155	1.1	010	0.9
Gurnet Point, 1 mi. E of	F-0 10, E+0 15	250	1.4	–	1.0
Farnham Rock, 1 mi. E of	F-0 25, E-0 00	180	1.1	010	0.9

Important: **See NOTE, bottom p. 29.**

TIME OF CURRENT CHANGE

(See Note at bottom of Boston Tables, pp. 38-43: Rule-of-Thumb for Current Velocities.)

	TIME DIFFERENCES Flood Starts; Ebb Starts Hr. Min.	MAXIMUM FLOOD Dir.(true) in degrees	Avg. Max. in knots	MAXIMUM EBB Dir.(true) in degrees	Avg. Max. in knots
NANTUCKET SOUND – based on Pollock Rip Channel, pp. 60-65					
Pollock Rip Channel, E end	-0 20	053	2.0	212	1.8
***POLLOCK RIP CHANNEL at Butler Hole - See table, pp. 60-65**					
Monomoy Point, 0.2 mi. W of	+0 10	170	1.7	346	2.0
Halfmoon Shoal, 3.5 mi. E of	+1 10	088	1.1	295	1.0
Great Point, 0.5 mi. W of	F+0 25, E+1 15	029	1.1	195	1.2
Tuckernuck Shoal, off E end	+1 15	113	0.9	287	0.9
Nantucket Hbr. entr. chan.	F+3 20, E+2 45	171	1.2	350	1.5
Muskeget Is. chan., 1 mi. NE of	F+1 30, E+1 00	108	1.1	295	1.5
Muskeget Rock, 1.3 mi. SW of	+1 05	024	1.3	192	1.0
Muskeget Channel	+1 35	021	3.8	200	3.3
Betw. Long Shoal-Norton Shoal	+1 30	100	1.4	260	1.1
Cape Poge Lt., 1.7 mi. SSE of	+0 55	025	1.6	215	1.3
Cross Rip Channel	+1 50	091	1.3	272	0.9
Cape Poge, 3.2 mi. NE of	+2 35	095	1.6	300	1.2
Betw. Broken Gr.-Horseshoe Sh.	F+1 45, E+1 15	107	1.1	276	0.9
Point Gammon, 1.2 mi. S of	+1 10	105	1.1	260	1.0
Lewis Bay entr. chan.	+2 45	004	0.9	184	1.3
Betw. Wreck Shoal-Eldridge Shoal	+1 45	062	1.7	245	1.4
Hedge Fence Lighted Gong Buoy 22	+2 45	108	1.4	268	1.2
Betw. E. Chop-Squash Meadow	F+2 10, E+1 45	131	1.4	329	1.8
East Chop, 1 mi. N of	F+2 40, E+2 20	116	2.2	297	2.2
West Chop, 0.8 mi. N of	F+2 50, E+2 20	096	3.1	282	3.0
Betw. Hedge Fence-L'hommedieu Shoal	+2 15	106	2.1	276	2.2
Waquoit Bay entr.	F+3 20, E+3 40	348	1.5	203	1.4
L'hommedieu Shoal, N of W end	+2 20	080	2.3	268	2.3
Nobska Point, 1.8 mi. E of	+2 05	063	2.3	240	1.7
VINEYARD SOUND – based on Pollock Rip Channel, pp. 60-65					
West Chop, 0.2 mi. W of	F+1 20, E+1 50	059	2.7	241	1.4
Nobska Point, 1 mi. SE of	+2 30	071	2.6	259	2.4
Norton Point, 0.5 mi. N of	+2 00	050	3.4	240	2.4
Tarpaulin Cove, 1.5 mi. E of	F+2 50, E+2 10	055	1.9	232	2.3
Robinsons Hole, 1.2 mi. SE of	F+2 30, E+2 10	060	1.9	240	2.1
Gay Head, 3 mi. N of	+2 05	074	1.1	255	1.2
Gay Head, 1.5 mi. NW of	+1 35	012	2.0	249	2.0
VINEYARD SOUND-BUZZARDS BAY – based on Woods Hole, pp. 52-57					
Robinsons Hole, Naushon Pt.	+0 40	151	3.0	332	2.9
Quicks Hole, S end	F+1 20, E+0 30	140	1.9	300	2.0
Quicks Hole, Middle	F+1 30, E+1 00	157	2.3	327	1.8
Quicks Hole, N end	F+1 40, E+0 55	165	2.0	002	2.6
Canapitsit Channel	F+1 00, E+0 14	131	1.7	312	1.7
BUZZARDS BAY – based on Woods Hole, pp. 52-57					
Westport River entr.	-1 20	290	2.2	108	2.5
Gooseberry Nk., 2 mi. SSE of (41°27'N- 71°01'W) *rotary current, no slack water. Avg. max. 0.6 kts, approx. dir. 52° true at 3 1/2 hrs. after Flood starts at Poll. Rip. Avg. max. 0.5 kts, approx. dir. 232° true 2 1/2 hrs. after Ebb starts at Poll. Rip.*					
Betw. Ribbon Reef-Sow &Pigs Rf.	F-1 45, E-3 45	062	0.8	237	1.2
Penikese Is., 0.8 mi. NW of	F-3 00, E-1 55	050	1.2	254	1.1
Betw. Gull Is.-Nashawena Is.	F-3 39, E-3 00	091	0.9	247	1.1
Dumpling Rocks, 0.2 mi. SE of	F-3 10, E-2 30	066	0.8	190	1.1
BUZZARDS BAY – based on Cape Cod Canal, pp. 46-51					
Abiels Ledge, 0.4 mi. S of	F+0 13, E-0 15	069	1.3	236	1.8
CAPE COD CANAL - table, pp. 46-51		070	4.0	250	4.5

**See Tidal Current Chart Buzzards Bay, Vineyard and Nantucket Sounds, pp. 66-77*

Important: **See NOTE, bottom p. 29.**

TIME OF CURRENT CHANGE

(See Note at bottom of Boston Tables, pp. 38-43: Rule-of-Thumb for Current Velocities.)

	TIME DIFFERENCES Flood Starts; Ebb Starts Hr. Min.	MAXIMUM FLOOD Dir.(true) in degrees	Avg. Max. in knots	MAXIMUM EBB Dir.(true) in degrees	Avg. Max. in knots
****NARRAGANSETT BAY – based on Pollock Rip Channel, pp. 60-65**					
Tiverton, Stone Bridge, Sakonnet	F-3 00, E-2 25	010	2.7	190	2.7
Tiverton, RR Bridge, Sakonnet R.	F-3 25, E-2 50	000	2.3	180	2.4
Castle Hill, W of East Passage	F-0 05, E-1 05	013	0.7	237	1.2
Bull Point, E of	-1 10	001	1.2	206	1.5
Rose Is., NE of	F-1 55, E-1 15	310	0.8	124	1.0
Rose Is., W of	F-0 40, E-1 20	001	0.7	172	1.0
Dyer Is., W of	-1 00	023	0.8	216	1.0
Mount Hope Bridge	-1 15	047	1.1	230	1.4
Kickamuit R., Mt. Hope Bay	F-2 05, E-1 20	000	1.4	191	1.7
Warren R., Warren	-0 20	358	1.0	171	0.9
Beavertail Point, 0.8 mi NW of	F-0 10, E-1 30	003	0.5	188	1.0
Betw. Dutch Is.-Beaver Head	-1 55	030	1.0	233	1.0
Dutch Is., W of	-1 25	014	1.3	206	1.2
India Pt. RR Bridge, Seekonk R.	-1 40	020	1.0	180	1.4
BLOCK ISLAND SOUND – based on The Race, pp. 86-91					
Pt. Judith Pond entr.	-3 10	351	1.8	186	1.5
Sandy Pt., Block Is. 1.5 mi N of	F-0 25, E-1 05	315	1.9	063	2.1
Lewis Pt., 1.0 mi. SW of	F-1 30, E-0 25	298	1.9	136	1.8
Lewis Pt., 1.5 mi. W of	F-1 35, E-0 50	318	1.4	170	1.7
Southwest Ledge	-0 25	321	1.5	141	2.1
Watch Hill Pt., 2.2 mi. E of	F-0 30, E+0 45	260	1.2	086	0.7
Montauk Pt., 1.2 mi. E of	F-1 20, E-0 40	346	2.8	162	2.8
Montauk Pt., 1 mi. NE of	F-2 05, E-1 15	356	2.4	145	1.9
Betw. Shagwong Reef-Cerberus Shoal	-0 30	241	1.9	056	1.8
Betw. Cerberus Sh.-Fishers Is.	F-1 00, E+0 05	264	1.3	096	1.3
Gardiners Is., 3 mi. NE of	-0 35	305	0.9	138	1.0
GARDINERS BAY etc. – based on The Race, pp. 86-91					
Goff Point, 0.4 mi. NW of	-1 35	225	1.2	010	1.6
Acabonack Hbr. entr., 0.6 mi. ESE of	F-1 35, E-1 05	345	1.4	140	1.2
Gardiners Pt. Ruins, 1.1 mi. N of	-0 10	270	1.2	066	1.8
Betw. Gardiners Point-Plum Is.	-0 25	288	1.4	100	1.6
Jennings Pt., 0.2 mi. NNW of	+0 35	290	1.6	055	1.5
Cedar Pt., 0.2 mi. W of	F-0 10, E+0 30	195	1.8	005	1.6
North Haven Peninsula, N of	F+0 10, E+0 40	230	2.4	035	2.1
Paradise Pt., 0.4 mi. E of	+0 35	145	1.5	345	1.5
Little Peconic Bay entr.	+0 45	240	1.6	015	1.5
Robins Is., 0.5 mi. S of	F+0 30, E+0 55	245	1.7	065	0.6
FISHERS ISLAND SOUND – based on The Race, pp. 86-91					
Napatree Point, 0.7 mi. SW of	-0 50	284	1.7	113	2.2
Little Narragansett Bay entr.	-2 05	092	1.3	268	1.3
Ram Island Reef, S of	-0 50	255	1.3	088	1.6
LONG ISLAND SOUND – based on The Race, pp. 86-91					
***THE RACE (near Valiant Rock) – See pp. 86-91**		290	3.3	106	4.2
Race Point, 0.4 mi. SW of	-0 25	288	2.6	135	3.5
Little Gull Is., 1.1 mi. ENE of	+0 05	301	4.0	130	4.7
Little Gull Is., 0.8 mi. NNW of	F+0 25, E-2 20	258	1.9	043	2.9
Great Gull Is., SW of	-0 40	320	2.3	147	3.3
New London St. Pier, Thames R.	-1 30	358	0.4	178	0.4
Goshen Pt., 1.9 mi. SSE of	-0 55	285	1.2	062	1.6
Bartlett Reef, 0.2 mi. S of	F-2 05, E-1 05	255	1.4	090	1.3
Twotree Is. Channel	F-1 00, E-0 35	267	1.2	099	1.6

**See Tidal Current Chart Long Is. and Block Is. Sounds, pp. 92-97*
***Floods somewhat unstable. Flood currents differing from predicted should be expected.*

Important: **See NOTE, bottom p. 20.**

CURRENT STATIONS

TIME OF CURRENT CHANGE

(See Note at bottom of Boston Tables, pp. 38-43: Rule-of-Thumb for Current Velocities.)

	TIME DIFFERENCES Flood Starts; Ebb Starts Hr. Min.	MAXIMUM FLOOD Dir.(true) in degrees	Avg. Max. in knots	MAXIMUM EBB Dir.(true) in degrees	Avg. Max. in knots
LONG ISLAND SOUND – based on The Race, pp. 86-91					
Black Point, 0.8 mi. S of	F-0 40, E-0 15	260	1.3	073	1.4
Betw. Black Pt.-Plum Is.	+0 35	236	2.1	076	2.4
Plum Is., 0.8 mi. NNW of	F+0 10,E-1 05	247	1.7	065	2.4
Plum Gut	-1 00	306	1.9	116	3.0
Hatchett Pt., 1.1 mi. WSW of	F-2 30, E-0 40	240	1.3	045	1.2
Saybrook Bkwtr., 1.5 mi. SE of	F-1 20, E-0 45	260	1.9	070	2.0
Conn. River I-95 Bridge	F+1 15, E+0 20	356	0.9	166	1.8
Mulford Pt., 3.1 mi. NW of	+0 05	269	1.9	066	2.3
Cornfield Point, 2.8 mi. SE of	F-1 30, E-0 30	249	1.9	085	1.4
Cornfield Point, 1.1 mi. S of	-0 50	293	1.4	108	1.6
Kelsey Point, 1 mi. S of	F-1 35, E-1 05	249	2.0	118	1.5
Six Mile Reef, 2 mi. E of	F-0 30, E+0 05	235	1.6	040	2.1
Sachem Head, 1 mi. SSE of	-0 30	255	1.1	065	1.0
New Haven Harbor entr.	-0 05	277	0.7	122	0.5
Housatonic R., Milford Pt., 0.2 mi. W of	+0 15	330	1.2	135	1.2
Point No Point, 2.1 mi. S of	-0 10	251	1.3	074	1.2
Port Jefferson Harbor entr.	-0 10	150	1.6	336	1.0
Crane Neck Point, 0.5 mi. NW of	F-0 45, E-1 40	256	1.3	016	1.5
Eatons Neck Pt., 1.3 mi. N of	+0 20	283	1.4	075	1.4
Lloyd Point, 1.3 mi. NNW of	+1 30	255	1.0	055	0.9
EAST RIVER – based on Hell Gate, pp. 110-115					
Cryders Pt., 0.4 mi. NNW of	-0 30	110	1.3	285	1.1
College Pt. Rf., .25 mi. NW of	-0 30	074	1.5	261	1.4
Rikers Is. Chann. off La Guardia Field	+0 05	088	1.1	261	1.3
Hunts Point, SW of	0 00	108	1.7	280	1.3
S. Brother Is. NW of	- 0 10	054	1.5	252	1.2
Off Winthrop Ave., Astoria	0 00	040	3.4	220	2.5
Mill Rock, NE of	-0 25	103	2.3	288	0.6
Mill Rock, W of	F-0 25, E-0 00	000	1.2	180	1.0
HELL GATE (off Mill Rock) – table, pp. 110-115		050	3.4	230	4.6
Roosevelt Is., W of, off 75th St.	-0 05	037	3.8	215	4.7
Roosevelt Is., E of, off 36th Ave.	-0 10	030	3.5	210	3.4
Roosevelt Is., W of, off 67th St.	+0 10	011	3.6	230	4.0
Off 19th St. (Pier 67)	-0 10	355	1.8	179	1.9
Williamsburg Br., 0.3 mi. N of	-0 05	020	2.7	220	2.9
Brooklyn Bridge, 0.1 mi. SW of	-0 10	046	2.9	222	3.5
Buttermilk Channel * **Caution**	F-0 30, E+0 05	050	1.8	221	2.6
LONG ISLAND, South Coast – based on The Narrows, pp. 116-121					
Shinnecock Inlet	F+0 05, E-0 40	350	2.5	170	2.3
Fire Is. Inlet, 0.5 mi. S. of Oak Bch.	+0 15	082	2.4	244	2.4
Jones Inlet	F-1 15, E-0 50	035	3.1	217	2.6
East Rockaway Inlet	F-1 35, E-1 10	042	2.2	227	2.3
JAMAICA BAY – based on The Narrows, pp. 116-121					
Rockaway Inlet entr.	-1 45	085	1.8	244	2.7
Barren Is., E of	F-1 50, E-0 10	004	1.2	192	1.7
Beach Channel (bridge)	F-1 40, E-0 05	062	1.9	225	2.0
Grass Hassock Channel	-1 10	052	1.0	228	1.0

* **Caution-** During the first two hours of flood in the channel north of Governers Island, the current in the Hudson River is still ebbing while during the first 1 1/2 hours of ebb in this channel, the current in the Hudson River is still flooding.

Important: **See NOTE, bottom p. 29.**

TIME OF CURRENT CHANGE

(See Note at bottom of Boston Tables, pp. 38-43: Rule-of-Thumb for Current Velocities.)

	TIME DIFFERENCES Flood Starts; Ebb Starts Hr. Min.	MAXIMUM FLOOD Dir.(true) in degrees	Avg. Max. in knots	MAXIMUM EBB Dir.(true) in degrees	Avg. Max. in knots
NEW YORK HARBOR ENTRANCE – based on The Narrows, pp. 116-121					
Ambrose Channel	-0 40	303	1.6	123	1.7
Norton Pt., WSW of	+0 10	341	1.0	166	1.2
THE NARROWS (mid-ch.) – table, pp. 116-121		336	1.6	164	1.9
NEW YORK HARBOR, Upper Bay – based on The Narrows, pp. 116-121					
Bay Ridge, W of	F+0 00, E+0 35	354	1.4	185	1.5
Red Hook Channel	F-0 55, E-0 15	353	1.0	170	0.7
Robbins Reef Light, E of	F+0 25, E-0 05	016	1.3	204	1.6
Red Hook, 1 mi. W of	+0 45	024	1.3	206	2.3
Statue of Liberty, E of	+0 55	031	1.4	205	1.9
HUDSON RIVER, Midchannel – based on The Narrows, pp. 116-121					
George Washington Bridge	+1 35	010	1.8	203	2.5
Spuyten Duyvil	+1 35	020	1.6	–	2.1
Riverdale	F+2 25, E+1 50	015	1.4	200	2.0
Dobbs Ferry	F+2 45, E+2 10	010	1.3	–	1.7
Tarrytown	+2 40	000	1.1	–	1.5
West Point, off Duck Is	+3 40	010	1.0	–	1.1
NEW YORK HARBOR, Lower Bay – based on The Narrows, pp. 116-121					
Sandy Hook Channel	-1 20	286	1.6	094	1.9
Sandy Hook Channel, 0.4 mi. W of N. tip	-1 40	235	2.0	050	1.6
Coney Is. Lt., 1.5 mi. SSE of	-1 10	310	1.1	125	1.3
Rockaway Inlet Jetty, 1 mi. SW of	F-2 05, E-1 35	287	1.2	142	1.4
Coney Is. Channel, W end	F-1 15, E-0 30	293	1.1	102	1.2
SANDY HOOK BAY – based on The Narrows, pp. 116-121					
Highlands Bridge, Shrewsbury R.	+0 25	170	2.6	–	2.5
Sea Bright Br., Shrewsbury R.	F+1 05, E+0 45	185	1.4	–	1.7
RARITAN RIVER – based on The Narrows, pp. 116-121					
Washington Canal, N entr.	F-1 00, E-1 40	240	1.5	060	1.5
South River entr.	F-1 45, E-0 35	180	1.1	000	1.0
***ARTHUR KILL & KILL VAN KULL – based on The Narrows, pp. 116-121**					
Tottenville, Arthur Kill River	-0 50	023	1.0	211	1.1
Tufts Pt.-Smoking Pt.	-0 35	109	1.2	267	1.2
Elizabethport	+0 20	090	1.4	262	1.1
Bergen Pt., East Reach	-1 35	274	1.1	094	1.2
New Brighton	-1 35	262	1.3	072	1.9
NEW JERSEY COAST – based on Del. Bay Entr., pp. 140-145					
Manasquan Inlet	F-0 45, E-1 10	300	1.7	120	1.8
Manasquan R. Hwy. Br. Main Ch.	F-0 40, E-1 15	230	2.2	050	2.1
**Pt. Pleasant Canal, north bridge	F+1 45, E+0 50	170	1.8	350	2.0
Barnegat Inlet	F+1 00, E+0 15	270	2.2	090	2.5
Manahawkin Drawbridge	+2 30	030	1.1	210	0.9
McCrie Shoal	-0 35	280	1.3	100	1.4
Cape May Harbor entr.	-1 35	324	1.6	142	1.7
Cape May Canal, E end	-1 50	310	1.9	130	1.9

* *Tidal flow erratic due to dredging.*

***Waters are extremely turbulent. Currents of 6 to 7 knots have been reported near the bridges.*

Important: **See NOTE, bottom p. 29.**

TIME OF CURRENT CHANGE

(See Note at bottom of Boston Tables, pp. 38-43: Rule-of-Thumb for Current Velocities.)

	TIME DIFFERENCES Flood Starts; Ebb Starts Hr. Min.	MAXIMUM FLOOD Dir.(true) in degrees	MAXIMUM FLOOD Avg. Max. in knots	MAXIMUM EBB Dir.(true) in degrees	MAXIMUM EBB Avg. Max. in knots

DELAWARE BAY & RIVER – based on Del. Bay Entr., pp. 140-145

Location	Time Differences	Flood Dir.	Flood Avg. Max.	Ebb Dir.	Ebb Avg. Max.
Cape May Channel	-1 10	306	1.5	150	2.3
DELAWARE BAY ENTR. – table, pp. 140-145		327	1.4	147	1.3
Cape Henlopen, 0.7 mi. ESE of	F- 0 05, E- 040	331	1.8	139	2.4
Cape Henlopen, 2 mi. NE of	F+0 20, E-0 05	315	2.0	145	2.3
Cape Henlopen, 5 mi. N of	+0 30	344	2.0	173	1.9
Mispillion River Mouth	F+2 35, E+1 50	025	1.5	190	1.0
Bay Shore chan., City of Town Bank	- 0 40	006	0.9	183	1.0
Fourteen Ft. Bk., Lt., 1.2 mi. E of	+0 10	339	1.3	174	1.5
Maurice River entr.	+1 00	012	1.1	192	1.0
Kelly Island, 1.5 mi. E of	+0 50	348	0.9	164	1.2
Miah Maull rge. at Cross Ledge rge	+1 25	335	1.5	160	1.8
False Egg Is. Pt., 2 mi. off	+0 20	342	1.1	158	1.3
Ben Davis Pt. Shoal., SW of	+1 40	321	1.8	147	1.9
Cohansey R., 0.5 mi. above entr.	+1 30	074	1.2	254	1.4
Arnold Point, 2.2 mi. WSW of	+2 25	324	2.1	145	1.9
Smyrna River entr.	+1 55	250	1.2	070	1.5
Stony Point chan., W of	F+3 25, E+2 30	324	1.5	151	1.9
Appoquinimink R. entr.	+2 25	231	1.0	048	1.2
Reedy Is., off end of pier	+2 55	027	2.4	194	2.6
Alloway Creek entr., 0.2 mi. above	+2 15	129	2.1	325	2.1
Reedy Point, 0.85 mi. NE of	F+3 35, E+2 50	341	1.6	163	2.2
Salem River entr.	+3 40	062	1.5	245	1.6
Bulkhead Sh. chan., off Del. City	F+3 15, E+2 55	308	2.1	138	2.1
Pea Patch Is., chan., E of	+3 30	319	2.3	148	2.3
New Castle, chan., abreast of	F+3 35, E+3 00	051	1.9	230	2.4

CHESAPEAKE BAY – based on The Race, pp. 86-91

(over 90% correlation within 15 min. throughout year)

Location	Time Differences	Flood Dir.	Flood Avg. Max.	Ebb Dir.	Ebb Avg. Max.
Cape Henry Light, 2.0 mi. N of	+0 40	289	1.2	110	1.1
Chesapeake Bay entr.	F+0 20, E-0 20	300	0.8	129	1.2
Cape Henry Light, 4.6 mi. N of	F-0 25, E-0 05	294	1.3	104	1.3
Cape Henry Light, 8.3 mi. NW of	+0 30	329	1.0	133	1.1
Tail of the Horseshoe	F+0 20, E-0 05	300	0.9	110	1.0
Chesapeake Channel (Bridge Tunnel)	+0 15	335	1.8	145	1.5
Fisherman Is., 1.7 mi. S of	F-0 00, E-0 35	297	1.0	126	1.4
York Spit Channel N buoy "26"	F+1 50, E+1 05	010	0.8	195	1.1
Old Plantation Flats Lt., 0.5 mi. W of	+1 40	005	1.2	175	1.3
Wolf Trap Lt., 0.5 mi. W of	F+2 00, E+1 15	015	1.0	190	1.2
Stingray Point, 5.5 mi. E of	+2 50	343	1.0	179	0.9
Stingray Point, 12.5 mi. E of	F+2 35, E+1 50	030	1.0	175	0.8
Smith Point Lt., 6.0 mi. N of	+4 45	350	0.4	135	1.0

Cove Point - See Chesapeake Bay Current Diagram, p. 154
Pooles Island - See Chesapeake Bay Current Diagram, p. 154
Worton Point - See Chesapeake Bay Current Diagram, p. 154
CHESAPEAKE & DELAWARE CANAL - table, pp. 148-153

HAMPTON ROADS – based on The Race, pp. 86-91

(over 90% correlation within 15 min. throughout year)

Location	Time Differences	Flood Dir.	Flood Avg. Max.	Ebb Dir.	Ebb Avg. Max.
Thimble Shoal Channel (West End)	+0 05	293	0.9	116	1.2
Old Point Comfort, 0.2 mi. S of	F-0 20, E-1 15	240	1.7	075	1.4
Willoughby Spit, 0.8 mi. NW of	F-1 15, E-2 00	260	0.7	040	1.0
Sewells Point, chan., W of	F-0 25, E-1 50	195	0.9	000	1.2
Newport News, chan., middle	F-0 25, E -0 32	244	1.1	076	1.1

C&D CANAL POINTS – based on C&D Canal, pp. 148-153

Location	Time Differences	Flood Dir.	Flood Avg. Max.	Ebb Dir.	Ebb Avg. Max.
Back Creek, 0.3 mi. W of Sandy Pt.	-0 05	057	1.2	244	1.4
Reedy Point Radio Tower, S of	F-1 00, E-0 05	078	1.9	263	1.3

Important: **See NOTE, bottom p. 29.**

TIME OF CURRENT CHANGE

(See Note at bottom of Boston Tables, pp. 38-43: Rule-of-Thumb for Current Velocities.)

	TIME DIFFERENCES Flood Starts; Ebb Starts Hr. Min.	MAXIMUM FLOOD Dir.(true) in degrees	Avg. Max. in knots	MAXIMUM EBB Dir.(true) in degrees	Avg. Max. in knots
VA, NC, SC, GA & FL, outer coast – based on Hell Gate, pp. 110-115 (over 90% correlation within 15 min. throughout year)					
Hatteras Inlet	F+1 00, E+0 40	307	2.1	148	2.0
Ocracoke Inlet chan. entr.	F+1 10, E+0 45	000	1.7	145	2.4
Beaufort Inlet Approach	F+0 25, E-1 00	358	0.3	161	1.4
Cape Fear R. Bald Head	F-1 25, E-1 30	034	2.2	190	2.9
Winyah Bay entr.	+0 05	320	1.9	140	2.0
North Santee R. entr.	F-0 40, E-1 35	010	1.5	165	1.8
South Santee R. entr.	F-1 20, E-1 15	045	1.5	240	1.6
Charleston Hbr. entr., betw. jetties	-1 40	320	1.8	121	1.8
Charleston Hbr., off Ft. Sumter	-1 40	313	1.7	127	2.0
Charleston Hbr. S. ch. 0.8 mi. ENE of Ft. Johnson	F-0 55, E-1 40	275	0.8	115	2.6
Charleston Hbr., Drum Is., E of (bridge)	-1 20	020	1.2	183	2.0
North Edisto River entr.	-0 35	332	2.9	142	3.7
South Edisto River entr.	F-1 20, E-1 50	350	1.8	146	2.2
Ashepoo R. off Jefford Cr. entr.	-0 40	016	1.5	197	1.6
Port Royal Sd., SE chan. entr.	F-2 10, E-1 50	310	1.3	150	1.6
Hilton Head	-1 20	324	1.8	146	1.8
Beaufort River entr.	-1 20	010	1.3	195	1.4
Savannah River entr.	F-1 00, E-0 50	286	2.0	110	2.0
Vernon R. 1.2 mi. S of Possum Pt.	F-1 25, E-1 00	324	1.1	166	1.7
Raccoon Key & Egg Is. Shoal bet.	F-0 40, E -1 15	254	1.6	129	2.0
St. Catherines Sound entr.	F-1 40, E-0 35	291	1.8	126	1.7
Sapelo Sound entr.	F-1 30, E-0 55	290	1.7	118	2.2
Doboy Sound entr.	-1 25	289	1.6	106	1.8
Altamaha Sd., 1 mi. SE of Onemile Cut	F-0 15, E-2 00	272	1.0	092	1.9
St. Simons Sound Bar Channel	F-1 15, E-0 40	308	0.8	119	1.7
St. Andrews Sound entr.	F-1 20, E-0 50	268	2.1	103	2.2
Cumberland Sd., St. Mary's River, Ft. Clinch, 0.3 mi. N	F-1 25, E-1 00	275	1.4	087	1.6
Drum Point Is., rge. D chan	-0 50	350	1.1	170	1.5
Nassau Sd., midsound, 1 mi. N of Sawpit Cr. entr.	F-0 15, E-0 40	312	1.7	135	1.7
FLORIDA EAST COAST – based on The Narrows, pp. 116-121 (over 90% correlation within 15 min. throughout year)					
St. Johns R. entr. betw. jetties	+0 20	262	2.0	081	2.0
Mayport	+0 30	211	2.2	026	3.3
St. Johns Bluff	F+0 50, E+0 05	244	1.6	059	2.4
FLORIDA EAST COAST – based on Hell Gate, pp. 110-115 (over 90% correlation within 15 min. throughout year)					
Fort Pierce Inlet entr.	+0 40	258	2.7	080	2.8
Lake Worth Inlet, entr.	F-1 05, E-0 45	267	1.6	086	1.3
Miami Hbr., Bakers Haulover Cut	-0 05	270	2.9	090	2.5
Miami Hbr. entr.	-0 15	293	2.2	113	2.4

CURRENT STATIONS

NOTE: Velocities shown are from U.S. Gov't. figures. It is obvious, however, to local mariners and other observers, that coastal inlets may have far greater velocities than indicated here. Strong winds and opposing tides can cause even more dangerous conditions, and great caution should be used. Separate times for Flood and Ebb are given only when the times are more than 20 minutes apart.

Important: **See NOTE, bottom p. 29.**

2014 HIGH & LOW WATER
PORTLAND, ME
43°39.6'N, 70°14.8'W

		Standard Time								Standard Time					
DAY OF MONTH	DAY OF WEEK	JANUARY						DAY OF MONTH	DAY OF WEEK	FEBRUARY					
		HIGH				LOW				HIGH				LOW	
		a.m.	Ht.	p.m.	Ht.	a.m.	p.m.			a.m.	Ht.	p.m.	Ht.	a.m.	p.m.
1	W	10 17	11.6	10 57	10.2	4 01	4 43	1	S	11 49	11.6	...	...	5 34	6 07
2	T	11 10	11.8	11 49	10.4	4 54	5 34	2	S	12 23	10.8	12 42	11.3	6 27	6 57
3	F	...	...	12 03	11.7	5 48	6 26	3	M	1 13	10.8	1 35	10.8	7 21	7 48
4	S	12 43	10.5	12 58	11.4	6 44	7 20	4	T	2 06	10.5	2 31	10.1	8 17	8 41
5	S	1 36	10.5	1 53	10.9	7 39	8 12	5	W	2 59	10.2	3 27	9.4	9 14	9 35
6	M	2 32	10.3	2 52	10.3	8 38	9 08	6	T	3 54	9.7	4 28	8.8	10 14	10 32
7	T	3 29	10.1	3 53	9.7	9 40	10 06	7	F	4 53	9.3	5 32	8.3	11 18	11 33
8	W	4 28	9.8	4 57	9.1	10 44	11 06	8	S	5 54	9.1	6 36	8.0	...	12 22
9	T	5 29	9.6	6 03	8.7	11 50	...	9	S	6 55	8.9	7 37	8.0	12 35	1 23
10	F	6 30	9.5	7 07	8.4	12 07	12 54	10	M	7 52	9.0	8 31	8.1	1 34	2 19
11	S	7 27	9.5	8 06	8.4	1 07	1 53	11	T	8 42	9.1	9 18	8.3	2 27	3 07
12	S	8 21	9.5	8 58	8.4	2 02	2 46	12	W	9 27	9.3	10 00	8.6	3 13	3 49
13	M	9 09	9.6	9 45	8.5	2 52	3 33	13	T	10 07	9.5	10 37	8.8	3 55	4 26
14	T	9 52	9.7	10 26	8.6	3 37	4 15	14	F	10 44	9.6	11 11	8.9	4 32	5 00
15	W	10 31	9.7	11 04	8.7	4 18	4 53	15	S	11 19	9.6	11 43	9.1	5 08	5 32
16	T	11 07	9.7	11 39	8.8	4 56	5 28	16	S	11 53	9.6	...	...	5 42	6 03
17	F	11 42	9.7	...	...	5 31	6 01	17	M	12 15	9.2	12 27	9.5	6 16	6 36
18	S	12 12	8.8	12 16	9.6	6 06	6 33	18	T	12 47	9.3	1 03	9.3	6 53	7 10
19	S	12 46	8.9	12 52	9.4	6 42	7 06	19	W	1 23	9.4	1 42	9.1	7 32	7 49
20	M	1 20	8.9	1 29	9.2	7 19	7 42	20	T	2 02	9.5	2 26	8.9	8 15	8 32
21	T	1 56	8.9	2 09	9.0	7 59	8 20	21	F	2 46	9.5	3 16	8.7	9 04	9 20
22	W	2 36	9.0	2 53	8.7	8 43	9 02	22	S	3 37	9.5	4 12	8.5	10 00	10 16
23	T	3 20	9.0	3 42	8.5	9 33	9 50	23	S	4 34	9.5	5 16	8.4	11 02	11 19
24	F	4 09	9.1	4 38	8.3	10 28	10 44	24	M	5 38	9.7	6 24	8.5	...	12 09
25	S	5 04	9.3	5 40	8.3	11 29	11 44	25	T	6 45	9.9	7 31	8.9	12 26	1 16
26	S	6 04	9.6	6 46	8.5	...	12 33	26	W	7 51	10.3	8 34	9.5	1 32	2 18
27	M	7 06	10.0	7 50	8.8	12 46	1 37	27	T	8 53	10.8	9 30	10.1	2 35	3 15
28	T	8 08	10.5	8 51	9.4	1 49	2 37	28	F	9 50	11.2	10 22	10.6	3 33	4 08
29	W	9 06	11.1	9 47	9.9	2 49	3 33								
30	T	10 02	11.5	10 40	10.4	3 46	4 26								
31	F	10 56	11.7	11 32	10.7	4 40	5 17								

Dates when Ht. of **Low** Water is below Mean Lower Low with Ht. of lowest given for each period and Date of lowest in ( ):

1st - 7th: -2.1' (2nd)
27th - 31st: -2.1' (31st)

1st - 5th: -2.1' (1st)
25th - 28th: -1.6' (28th)

Average Rise and Fall 9.1 ft.

When a high tide exceeds avg. ht., the *following* low tide will be lower than avg.

2014 HIGH & LOW WATER
PORTLAND, ME

43°39.6'N, 70°14.8'W

***Daylight Time starts March 9 at 2 a.m.** **Daylight Saving Time**

MARCH

DATE OF MONTH	DAY OF WEEK	HIGH a.m.	Ht.	HIGH p.m.	Ht.	LOW a.m.	LOW p.m.
1	S	10 43	11.4	**11 12**	10.9	4 27	**4 58**
2	S	11 34	11.3	...	...	5 19	**5 46**
3	M	12 01	11.1	**12 25**	11.0	6 10	**6 34**
4	T	12 49	10.9	**1 16**	10.5	7 02	**7 23**
5	W	1 36	10.6	**2 06**	9.9	7 52	**8 10**
6	T	2 26	10.1	**2 59**	9.2	8 45	**9 02**
7	F	3 18	9.6	**3 56**	8.6	9 41	**9 56**
8	S	4 13	9.1	**4 56**	8.1	10 40	**10 55**
9	S	*6 13	8.8	***6 59**	7.9	...	***12 43**
10	M	7 15	8.6	**7 59**	7.9	12 58	**1 45**
11	T	8 14	8.6	**8 54**	8.0	1 58	**2 41**
12	W	9 08	8.8	**9 43**	8.3	2 54	**3 31**
13	T	9 56	9.0	**10 25**	8.6	3 43	**4 14**
14	F	10 38	9.2	**11 02**	9.0	4 26	**4 52**
15	S	11 16	9.4	**11 37**	9.3	5 04	**5 26**
16	S	11 52	9.5	...	...	5 40	**5 59**
17	M	12 09	9.5	**12 27**	9.6	6 15	**6 31**
18	T	12 42	9.7	**1 02**	9.6	6 51	**7 05**
19	W	1 16	9.9	**1 40**	9.5	7 28	**7 42**
20	T	1 53	10.0	**2 21**	9.3	8 09	**8 23**
21	F	2 35	10.0	**3 07**	9.1	8 54	**9 08**
22	S	3 22	10.0	**3 58**	8.9	9 44	**10 00**
23	S	4 14	9.9	**4 56**	8.7	10 41	**10 58**
24	M	5 14	9.8	**6 01**	8.7	11 43	...
25	T	6 21	9.7	**7 09**	8.9	12 03	**12 50**
26	W	7 30	9.9	**8 16**	9.3	1 12	**1 57**
27	T	8 38	10.1	**9 17**	9.8	2 21	**2 59**
28	F	9 40	10.5	**10 13**	10.4	3 24	**3 56**
29	S	10 37	10.7	**11 04**	10.8	4 21	**4 48**
30	S	11 29	10.9	**11 51**	11.1	5 14	**5 37**
31	M	...	...	**12 19**	10.8	6 04	**6 24**

APRIL

DAY OF MONTH	DAY OF WEEK	HIGH a.m.	Ht.	HIGH p.m.	Ht.	LOW a.m.	LOW p.m.
1	T	12 37	11.1	**1 07**	10.5	6 53	**7 10**
2	W	1 22	10.9	**1 54**	10.1	7 40	**7 55**
3	T	2 07	10.6	**2 42**	9.6	8 28	**8 41**
4	F	2 54	10.1	**3 32**	9.1	9 17	**9 30**
5	S	3 41	9.6	**4 23**	8.6	10 07	**10 20**
6	S	4 33	9.1	**5 18**	8.2	11 00	**11 15**
7	M	5 28	8.7	**6 16**	8.0	11 58	...
8	T	6 28	8.5	**7 14**	8.0	12 15	**12 56**
9	W	7 28	8.4	**8 09**	8.2	1 15	**1 52**
10	T	8 24	8.5	**8 58**	8.5	2 13	**2 43**
11	F	9 15	8.7	**9 42**	8.8	3 04	**3 28**
12	S	10 00	9.0	**10 21**	9.2	3 49	**4 08**
13	S	10 42	9.2	**10 58**	9.6	4 30	**4 45**
14	M	11 21	9.4	**11 33**	10.0	5 09	**5 21**
15	T	11 59	9.5	...	...	5 47	**5 58**
16	W	12 09	10.3	**12 38**	9.6	6 26	**6 36**
17	T	12 47	10.5	**1 20**	9.6	7 06	**7 17**
18	F	1 29	10.6	**2 05**	9.5	7 50	**8 02**
19	S	2 14	10.6	**2 53**	9.4	8 38	**8 51**
20	S	3 04	10.4	**3 47**	9.3	9 30	**9 46**
21	M	4 00	10.2	**4 46**	9.2	10 27	**10 47**
22	T	5 02	10.0	**5 50**	9.2	11 29	**11 53**
23	W	6 09	9.8	**6 56**	9.4	...	**12 34**
24	T	7 18	9.8	**8 00**	9.7	1 03	**1 39**
25	F	8 25	9.9	**9 00**	10.2	2 10	**2 40**
26	S	9 27	10.1	**9 54**	10.6	3 12	**3 36**
27	S	10 23	10.2	**10 44**	10.9	4 09	**4 28**
28	M	11 15	10.2	**11 31**	11.0	5 01	**5 16**
29	T	...	...	**12 03**	10.1	5 49	**6 02**
30	W	12 15	10.9	**12 49**	9.9	6 35	**6 46**

Dates when Ht. of **Low** Water is below Mean Lower Low with Ht. of lowest given for each period and Date of lowest in ( ):

March:
- 1st - 6th: -1.8' (1st)
- 18th - 22nd: -0.3' (19th - 21st)
- 26th - 31st: -1.5' (31st)

April:
- 1st - 4th: -1.4' (1st)
- 15th - 21st: -0.7' (18th - 19th)
- 25th - 30th: -1.1' (29th)

Average Rise and Fall 9.1 ft.

When a high tide exceeds avg. ht., the *following* low tide will be lower than avg.

2014 HIGH & LOW WATER
PORTLAND, ME
43°39.6'N, 70°14.8'W

		Daylight Saving Time								Daylight Saving Time					
DAY OF MONTH	DAY OF WEEK	MAY						DAY OF MONTH	DAY OF WEEK	JUNE					
		HIGH				LOW				HIGH				LOW	
		a.m.	Ht.	**p.m.**	Ht.	a.m.	**p.m.**			a.m.	Ht.	**p.m.**	Ht.	a.m.	**p.m.**
1	**T**	12 58	10.7	**1 34**	9.6	7 20	**7 29**	1	**S**	1 56	9.9	**2 37**	8.9	8 22	**8 28**
2	**F**	1 40	10.4	**2 19**	9.3	8 04	**8 13**	2	**M**	2 38	9.6	**3 19**	8.7	9 02	**9 12**
3	**S**	2 23	10.0	**3 04**	8.9	8 48	**8 58**	3	**T**	3 21	9.3	**4 02**	8.6	9 44	**9 58**
4	**S**	3 09	9.6	**3 52**	8.6	9 35	**9 46**	4	**W**	4 07	9.0	**4 48**	8.6	10 28	**10 48**
5	**M**	3 55	9.2	**4 40**	8.4	10 21	**10 36**	5	**T**	4 55	8.7	**5 34**	8.6	11 13	**11 39**
6	**T**	4 46	8.8	**5 31**	8.3	11 11	**11 30**	6	**F**	5 46	8.5	**6 22**	8.7	...	**12 01**
7	**W**	5 40	8.6	**6 24**	8.3	...	**12 03**	7	**S**	6 40	8.4	**7 11**	9.0	12 33	**12 49**
8	**T**	6 36	8.4	**7 16**	8.4	12 27	**12 56**	8	**S**	7 35	8.4	**8 00**	9.3	1 27	**1 39**
9	**F**	7 32	8.4	**8 06**	8.7	1 24	**1 46**	9	**M**	8 29	8.5	**8 48**	9.7	2 21	**2 29**
10	**S**	8 26	8.5	**8 52**	9.1	2 17	**2 34**	10	**T**	9 22	8.8	**9 36**	10.2	3 12	**3 18**
11	**S**	9 16	8.7	**9 35**	9.5	3 06	**3 18**	11	**W**	10 12	9.1	**10 24**	10.7	4 01	**4 07**
12	**M**	10 02	9.0	**10 16**	10.0	3 52	**4 01**	12	**T**	11 02	9.4	**11 12**	11.1	4 50	**4 56**
13	**T**	10 46	9.2	**10 57**	10.4	4 35	**4 43**	13	**F**	11 52	9.7	**...**	...	5 38	**5 46**
14	**W**	11 30	9.5	**11 39**	10.7	5 18	**5 25**	14	**S**	12 01	11.4	**12 42**	10.0	6 27	**6 37**
15	**T**	...	...	**12 15**	9.6	6 01	**6 09**	15	**S**	12 52	11.5	**1 34**	10.1	7 18	**7 30**
16	**F**	12 22	11.0	**1 01**	9.8	6 46	**6 56**	16	**M**	1 45	11.4	**2 28**	10.2	8 10	**8 26**
17	**S**	1 09	11.1	**1 50**	9.8	7 34	**7 45**	17	**T**	2 40	11.2	**3 23**	10.3	9 04	**9 24**
18	**S**	1 58	11.1	**2 42**	9.8	8 24	**8 38**	18	**W**	3 38	10.8	**4 21**	10.2	9 59	**10 26**
19	**M**	2 52	10.9	**3 37**	9.7	9 18	**9 36**	19	**T**	4 39	10.3	**5 20**	10.2	10 57	**11 30**
20	**T**	3 50	10.6	**4 36**	9.7	10 15	**10 38**	20	**F**	5 43	9.9	**6 21**	10.2	11 57	**...**
21	**W**	4 52	10.2	**5 38**	9.8	11 15	**11 44**	21	**S**	6 48	9.5	**7 21**	10.2	12 36	**12 57**
22	**T**	5 57	9.9	**6 41**	9.9	...	**12 17**	22	**S**	7 53	9.2	**8 19**	10.2	1 41	**1 57**
23	**F**	7 05	9.7	**7 42**	10.1	12 51	**1 19**	23	**M**	8 55	9.1	**9 14**	10.3	2 43	**2 54**
24	**S**	8 10	9.6	**8 40**	10.3	1 57	**2 19**	24	**T**	9 52	9.0	**10 05**	10.3	3 39	**3 47**
25	**S**	9 12	9.6	**9 35**	10.5	2 59	**3 15**	25	**W**	10 43	9.0	**10 52**	10.3	4 31	**4 36**
26	**M**	10 08	9.6	**10 25**	10.7	3 55	**4 07**	26	**T**	11 30	9.0	**11 35**	10.2	5 17	**5 21**
27	**T**	11 00	9.6	**11 11**	10.7	4 46	**4 55**	27	**F**	...	...	**12 13**	9.0	6 00	**6 03**
28	**W**	11 47	9.5	**11 54**	10.6	5 34	**5 41**	28	**S**	12 15	10.1	**12 52**	9.0	6 40	**6 43**
29	**T**	...	...	**12 32**	9.4	6 18	**6 24**	29	**S**	12 54	10.0	**1 31**	8.9	7 18	**7 21**
30	**F**	12 35	10.4	**1 14**	9.2	7 00	**7 05**	30	**M**	1 31	9.8	**2 08**	8.9	7 54	**8 00**
31	**S**	1 16	10.2	**1 55**	9.0	7 41	**7 46**								

Dates when Ht. of **Low** Water is below Mean Lower Low with Ht. of lowest given for each period and Date of lowest in ( ):

1st - 2nd: -0.8' (1st)
14th - 22nd: -1.1' (17th - 18th)
26th - 30th: -0.6' (28th)

12th - 20th: -1.4' (15th - 16th)
26th: -0.2'

Average Rise and Fall 9.1 ft.

When a high tide exceeds avg. ht., the *following* low tide will be lower than avg.

2014 HIGH & LOW WATER
PORTLAND, ME

43°39.6'N, 70°14.8'W

DATE OF MONTH	DAY OF WEEK	JULY (Daylight Saving Time)						DAY OF MONTH	DAY OF WEEK	AUGUST (Daylight Saving Time)					
		HIGH				LOW				HIGH				LOW	
		a.m.	Ht.	**p.m.**	Ht.	a.m.	**p.m.**			a.m.	Ht.	**p.m.**	Ht.	a.m.	**p.m.**
1	**T**	2 10	9.6	**2 46**	8.8	8 31	**8 41**	1	**F**	2 58	9.1	**3 24**	9.1	9 09	**9 31**
2	**W**	2 49	9.4	**3 24**	8.8	9 08	**9 22**	2	**S**	3 40	8.9	**4 05**	9.2	9 48	**10 17**
3	**T**	3 31	9.1	**4 05**	8.9	9 47	**10 07**	3	**S**	4 26	8.7	**4 51**	9.3	10 32	**11 07**
4	**F**	4 16	8.8	**4 49**	8.9	10 29	**10 56**	4	**M**	5 17	8.5	**5 42**	9.4	11 22	**...**
5	**S**	5 02	8.6	**5 33**	9.0	11 12	**11 46**	5	**T**	6 13	8.4	**6 36**	9.6	12 03	**12 16**
6	**S**	5 54	8.4	**6 22**	9.2	...	**12 01**	6	**W**	7 13	8.5	**7 35**	10.0	1 03	**1 14**
7	**M**	6 49	8.4	**7 14**	9.5	12 41	**12 52**	7	**T**	8 16	8.7	**8 35**	10.4	2 04	**2 15**
8	**T**	7 47	8.4	**8 07**	9.9	1 38	**1 47**	8	**F**	9 17	9.2	**9 34**	10.9	3 04	**3 15**
9	**W**	8 45	8.7	**9 02**	10.4	2 35	**2 42**	9	**S**	10 14	9.7	**10 31**	11.3	4 02	**4 14**
10	**T**	9 42	9.0	**9 56**	10.9	3 31	**3 38**	10	**S**	11 09	10.2	**11 26**	11.6	4 56	**5 10**
11	**F**	10 37	9.5	**10 49**	11.3	4 24	**4 32**	11	**M**	...	...	**12 02**	10.7	5 48	**6 05**
12	**S**	11 30	9.9	**11 43**	11.6	5 17	**5 26**	12	**T**	12 21	11.7	**12 54**	10.9	6 39	**6 59**
13	**S**	...	...	**12 23**	10.3	6 08	**6 20**	13	**W**	1 14	11.6	**1 46**	11.0	7 30	**7 54**
14	**M**	12 36	11.7	**1 16**	10.5	7 00	**7 15**	14	**T**	2 09	11.2	**2 39**	11.0	8 22	**8 50**
15	**T**	1 30	11.6	**2 09**	10.7	7 52	**8 11**	15	**F**	3 04	10.7	**3 33**	10.7	9 14	**9 48**
16	**W**	2 26	11.3	**3 03**	10.7	8 45	**9 09**	16	**S**	4 02	10.0	**4 29**	10.4	10 09	**10 48**
17	**T**	3 23	10.8	**3 59**	10.6	9 39	**10 09**	17	**S**	5 02	9.4	**5 27**	10.0	11 06	**11 51**
18	**F**	4 22	10.3	**4 57**	10.4	10 35	**11 11**	18	**M**	6 05	8.9	**6 28**	9.7	...	**12 07**
19	**S**	5 24	9.7	**5 56**	10.2	11 33	**...**	19	**T**	7 08	8.6	**7 29**	9.5	12 55	**1 08**
20	**S**	6 28	9.2	**6 56**	10.0	12 16	**12 33**	20	**W**	8 10	8.4	**8 28**	9.4	1 57	**2 08**
21	**M**	7 33	8.8	**7 56**	9.9	1 20	**1 33**	21	**T**	9 07	8.4	**9 21**	9.5	2 54	**3 04**
22	**T**	8 35	8.7	**8 53**	9.9	2 22	**2 32**	22	**F**	9 57	8.6	**10 09**	9.5	3 46	**3 53**
23	**W**	9 32	8.6	**9 45**	9.9	3 20	**3 26**	23	**S**	10 41	8.7	**10 51**	9.6	4 31	**4 38**
24	**T**	10 23	8.7	**10 32**	9.9	4 11	**4 16**	24	**S**	11 21	8.9	**11 30**	9.7	5 11	**5 18**
25	**F**	11 09	8.8	**11 15**	9.9	4 57	**5 01**	25	**M**	11 56	9.0	**...**	...	5 47	**5 55**
26	**S**	11 49	8.8	**11 54**	9.9	5 38	**5 41**	26	**T**	12 06	9.7	**12 30**	9.2	6 20	**6 30**
27	**S**	...	...	**12 27**	8.9	6 16	**6 20**	27	**W**	12 41	9.6	**1 02**	9.3	6 52	**7 05**
28	**M**	12 31	9.8	**1 02**	8.9	6 51	**6 56**	28	**T**	1 16	9.5	**1 35**	9.3	7 24	**7 40**
29	**T**	1 07	9.7	**1 37**	9.0	7 25	**7 33**	29	**F**	1 51	9.3	**2 10**	9.4	7 57	**8 18**
30	**W**	1 43	9.6	**2 11**	9.0	7 58	**8 10**	30	**S**	2 29	9.2	**2 47**	9.4	8 34	**8 59**
31	**T**	2 20	9.4	**2 47**	9.1	8 32	**8 49**	31	**S**	3 10	8.9	**3 29**	9.5	9 14	**9 45**

Dates when Ht. of **Low** Water is below Mean Lower Low with Ht. of lowest given for each period and Date of lowest in ( ):

11th - 18th: -1.7' (14th - 15th)

8th - 16th: -1.8' (12th)

Average Rise and Fall 9.1 ft.

When a high tide exceeds avg. ht., the *following* low tide will be lower than avg.

2014 HIGH & LOW WATER
PORTLAND, ME

43°39.6'N, 70°14.8'W

Daylight Saving Time | **Daylight Saving Time**

DAY OF MONTH	DAY OF WEEK	SEPTEMBER						DAY OF MONTH	DAY OF WEEK	OCTOBER					
		HIGH				LOW				HIGH				LOW	
		a.m.	Ht.	p.m.	Ht.	a.m.	p.m.			a.m.	Ht.	p.m.	Ht.	a.m.	p.m.
1	**M**	3 56	8.7	**4 16**	9.5	10 00	**10 36**	1	**W**	4 28	8.8	**4 45**	9.8	10 30	**11 12**
2	**T**	4 48	8.6	**5 09**	9.6	10 51	**11 34**	2	**T**	5 29	8.8	**5 48**	9.8	11 31	**...**
3	**W**	5 46	8.5	**6 08**	9.7	11 49	**...**	3	**F**	6 33	8.9	**6 54**	9.9	12 15	**12 37**
4	**T**	6 51	8.6	**7 12**	10.0	12 37	**12 53**	4	**S**	7 39	9.3	**8 02**	10.2	1 21	**1 44**
5	**F**	7 54	8.9	**8 15**	10.3	1 40	**1 56**	5	**S**	8 40	9.8	**9 04**	10.5	2 23	**2 47**
6	**S**	8 57	9.4	**9 17**	10.8	2 42	**2 59**	6	**M**	9 37	10.4	**10 02**	10.8	3 21	**3 47**
7	**S**	9 55	10.0	**10 16**	11.2	3 41	**3 59**	7	**T**	10 30	10.9	**10 57**	11.0	4 15	**4 42**
8	**M**	10 49	10.6	**11 11**	11.4	4 35	**4 55**	8	**W**	11 20	11.3	**11 49**	11.0	5 06	**5 34**
9	**T**	11 41	11.0	**...**	...	5 27	**5 49**	9	**T**	...	...	**12 09**	11.4	5 55	**6 25**
10	**W**	12 05	11.5	**12 31**	11.3	6 17	**6 42**	10	**F**	12 40	10.8	**12 57**	11.3	6 43	**7 15**
11	**T**	12 57	11.3	**1 21**	11.3	7 07	**7 35**	11	**S**	1 30	10.4	**1 45**	11.0	7 31	**8 05**
12	**F**	1 50	10.9	**2 12**	11.1	7 57	**8 28**	12	**S**	2 21	9.9	**2 34**	10.5	8 20	**8 56**
13	**S**	2 43	10.3	**3 04**	10.7	8 48	**9 23**	13	**M**	3 13	9.4	**3 25**	10.0	9 11	**9 50**
14	**S**	3 38	9.7	**3 58**	10.2	9 41	**10 20**	14	**T**	4 07	8.9	**4 18**	9.5	10 05	**10 45**
15	**M**	4 36	9.2	**4 55**	9.7	10 37	**11 20**	15	**W**	5 03	8.6	**5 16**	9.0	11 02	**11 44**
16	**T**	5 37	8.7	**5 55**	9.3	11 37	**...**	16	**T**	6 02	8.3	**6 16**	8.8	...	**12 02**
17	**W**	6 39	8.4	**6 56**	9.1	12 23	**12 38**	17	**F**	6 59	8.3	**7 15**	8.7	12 42	**1 02**
18	**T**	7 39	8.3	**7 56**	9.0	1 24	**1 39**	18	**S**	7 54	8.4	**8 11**	8.7	1 38	**1 59**
19	**F**	8 34	8.4	**8 50**	9.1	2 21	**2 35**	19	**S**	8 43	8.6	**9 01**	8.8	2 29	**2 50**
20	**S**	9 24	8.6	**9 39**	9.2	3 12	**3 25**	20	**M**	9 27	8.9	**9 47**	9.0	3 14	**3 36**
21	**S**	10 07	8.8	**10 22**	9.4	3 56	**4 10**	21	**T**	10 07	9.3	**10 28**	9.2	3 55	**4 18**
22	**M**	10 46	9.1	**11 02**	9.5	4 36	**4 50**	22	**W**	10 43	9.6	**11 07**	9.3	4 32	**4 56**
23	**T**	11 22	9.3	**11 38**	9.5	5 12	**5 27**	23	**T**	11 18	9.8	**11 44**	9.4	5 07	**5 33**
24	**W**	11 55	9.5	**...**	...	5 45	**6 02**	24	**F**	11 53	10.0	**...**	...	5 42	**6 09**
25	**T**	12 13	9.5	**12 27**	9.6	6 17	**6 37**	25	**S**	12 22	9.4	**12 29**	10.2	6 18	**6 47**
26	**F**	12 48	9.4	**1 00**	9.8	6 50	**7 13**	26	**S**	1 00	9.4	**1 07**	10.3	6 56	**7 28**
27	**S**	1 25	9.3	**1 36**	9.8	7 25	**7 51**	27	**M**	1 42	9.3	**1 50**	10.3	7 38	**8 13**
28	**S**	2 03	9.2	**2 15**	9.9	8 03	**8 34**	28	**T**	2 28	9.2	**2 37**	10.2	8 25	**9 02**
29	**M**	2 46	9.0	**2 59**	9.8	8 46	**9 21**	29	**W**	3 18	9.1	**3 30**	10.1	9 17	**9 56**
30	**T**	3 34	8.9	**3 49**	9.8	9 35	**10 13**	30	**T**	4 14	9.1	**4 29**	9.9	10 14	**10 55**
								31	**F**	5 15	9.1	**5 33**	9.8	11 18	**11 58**

Dates when Ht. of **Low** Water is below Mean Lower Low with Ht. of lowest given for each period and Date of lowest in ( ):

6th - 13th: -1.5' (9th - 10th)

5th - 12th: -1.5' (8th - 9th)
25th - 28th: -0.3'

Average Rise and Fall 9.1 ft.

When a high tide exceeds avg. ht., the *following* low tide will be lower than avg.

2014 HIGH & LOW WATER PORTLAND, ME

43°39.6'N, 70°14.8'W

*Standard Time starts Nov. 2 at 2 a.m.

DATE OF MONTH	DAY OF WEEK	NOVEMBER HIGH a.m.	Ht.	p.m.	Ht.	LOW a.m.	p.m.
1	**S**	6 19	9.3	**6 40**	9.8	...	**12 25**
2	**S**	*6 23	9.7	***6 48**	9.9	1 02	***12 33**
3	**M**	7 23	10.1	**7 51**	10.1	1 03	**1 37**
4	**T**	8 21	10.6	**8 51**	10.2	2 03	**2 37**
5	**W**	9 12	11.0	**9 44**	10.3	2 56	**3 30**
6	**T**	10 02	11.2	**10 35**	10.3	3 46	**4 21**
7	**F**	10 49	11.2	**11 24**	10.2	4 35	**5 10**
8	**S**	11 34	11.1	**...**	...	5 21	**5 57**
9	**S**	12 11	9.9	**12 20**	10.7	6 07	**6 43**
10	**M**	12 58	9.5	**1 05**	10.3	6 54	**7 30**
11	**T**	1 46	9.2	**1 52**	9.8	7 41	**8 18**
12	**W**	2 35	8.8	**2 41**	9.3	8 30	**9 07**
13	**T**	3 25	8.5	**3 33**	8.9	9 23	**9 58**
14	**F**	4 18	8.4	**4 28**	8.6	10 18	**10 52**
15	**S**	5 11	8.3	**5 26**	8.4	11 16	**11 45**
16	**S**	6 04	8.4	**6 22**	8.3	...	**12 13**
17	**M**	6 54	8.6	**7 16**	8.4	12 36	**1 07**
18	**T**	7 40	9.0	**8 05**	8.6	1 23	**1 56**
19	**W**	8 23	9.3	**8 51**	8.8	2 07	**2 41**
20	**T**	9 03	9.7	**9 34**	9.0	2 49	**3 23**
21	**F**	9 42	10.1	**10 15**	9.2	3 29	**4 03**
22	**S**	10 21	10.4	**10 56**	9.3	4 08	**4 43**
23	**S**	11 02	10.6	**11 39**	9.5	4 49	**5 25**
24	**M**	11 45	10.8	**...**	...	5 33	**6 09**
25	**T**	12 24	9.5	**12 31**	10.8	6 19	**6 56**
26	**W**	1 12	9.5	**1 21**	10.6	7 08	**7 46**
27	**T**	2 04	9.5	**2 16**	10.4	8 03	**8 40**
28	**F**	3 00	9.5	**3 15**	10.1	9 02	**9 38**
29	**S**	4 00	9.6	**4 19**	9.8	10 06	**10 39**
30	**S**	5 02	9.7	**5 27**	9.6	11 13	**11 42**

Standard Time

DAY OF MONTH	DAY OF WEEK	DECEMBER HIGH a.m.	Ht.	p.m.	Ht.	LOW a.m.	p.m.
1	**M**	6 05	9.9	**6 34**	9.5	...	**12 21**
2	**T**	7 06	10.2	**7 39**	9.5	12 44	**1 25**
3	**W**	8 03	10.5	**8 38**	9.6	1 43	**2 25**
4	**T**	8 57	10.7	**9 34**	9.6	2 39	**3 20**
5	**F**	9 46	10.8	**10 23**	9.7	3 30	**4 09**
6	**S**	10 32	10.8	**11 09**	9.6	4 17	**4 56**
7	**S**	11 16	10.7	**11 54**	9.4	5 03	**5 40**
8	**M**	11 58	10.4	**...**	...	5 46	**6 22**
9	**T**	12 36	9.2	**12 40**	10.1	6 29	**7 04**
10	**W**	1 18	9.0	**1 22**	9.7	7 12	**7 46**
11	**T**	2 01	8.8	**2 06**	9.3	7 56	**8 28**
12	**F**	2 45	8.6	**2 52**	8.9	8 43	**9 12**
13	**S**	3 30	8.5	**3 41**	8.5	9 32	**9 58**
14	**S**	4 18	8.4	**4 33**	8.3	10 25	**10 46**
15	**M**	5 08	8.5	**5 28**	8.1	11 20	**11 37**
16	**T**	5 58	8.6	**6 24**	8.0	...	**12 16**
17	**W**	6 48	8.9	**7 19**	8.1	12 27	**1 10**
18	**T**	7 37	9.2	**8 11**	8.4	1 17	**2 00**
19	**F**	8 23	9.7	**8 59**	8.7	2 06	**2 48**
20	**S**	9 08	10.1	**9 46**	9.0	2 53	**3 34**
21	**S**	9 54	10.6	**10 32**	9.4	3 39	**4 19**
22	**M**	10 39	10.9	**11 18**	9.7	4 25	**5 04**
23	**T**	11 27	11.1	**...**	...	5 13	**5 51**
24	**W**	12 06	9.9	**12 16**	11.2	6 02	**6 39**
25	**T**	12 55	10.0	**1 08**	11.0	6 54	**7 30**
26	**F**	1 47	10.1	**2 03**	10.7	7 49	**8 23**
27	**S**	2 43	10.0	**3 02**	10.2	8 48	**9 19**
28	**S**	3 41	10.0	**4 04**	9.7	9 51	**10 19**
29	**M**	4 42	10.0	**5 11**	9.3	10 58	**11 21**
30	**T**	5 45	10.0	**6 20**	9.0	...	**12 06**
31	**W**	6 47	10.0	**7 26**	8.9	12 24	**1 12**

Dates when Ht. of **Low** Water is below Mean Lower Low with Ht. of lowest given for each period and Date of lowest in ( ):

3rd - 10th: -1.3' (6th - 7th)
21st - 29th: -0.9' (24th - 25th)

2nd - 9th: -1.0' (5th - 6th)
20th - 28th: -1.5' (24th)

Average Rise and Fall 9.1 ft.

When a high tide exceeds avg. ht., the *following* low tide will be lower than avg.

Smarter Boating in Currents

If your vessel is a sailboat or a displacement powerboat your normal cruising speed is probably under 10 knots. In this range, current can become a significant factor. (See the Current Tables for the Cape Cod Canal and the Race, and the Current Diagrams for Vineyard Sound, showing some currents of 4 to 5 knots.) You can save a remarkable amount of time and, if under power, a great deal of fuel expense by using the current for maximum efficiency.

SAIL: Slow vs. Flow

The arithmetic is simple. If your 35' sailboat has a boat speed (BS) through the water of 5 knots under power or sail, then a 2-knot current directly against you means your speed made good (SMG) is 3 knots, and the same current going with you boosts that to 7 knots. Tacking into or with a current changes the simple arithmetic shown here. (See Coping With Currents, p. 21) The time difference can be great: a destination 10 miles away is 3 hours 20 minutes against the current, but only 1 hour 26 minutes with the current. Leaving earlier or later to go with the current leaves more time (almost 2 hours) to relax either at your departure point or destination. Of course if you're just out for a sail on a beautiful day, the arithmetic may not matter! If your sailboat is under power, keep reading.

POWER: Ego vs. Eco

As long as speed thrills, as we know it does, some boaters will demand it. But the trend is headed the other way. Today it is more about being economical, not egomaniacal. By far the most dramatic saving in fuel cost, or nautical miles per gallon (NMPG), comes from cutting back on the throttle; however, there are further savings from using the current to your advantage, especially with slower vessels.

Consider a trawler that burns 10 gallons of fuel per hour at a speed of 8 knots. If the cost of fuel is, say, $4 per gallon, that's $40 per hour. For a destination 24 nautical miles away, going directly against a current of 2 knots, her SMG is only 6 knots, requiring 4 hours for the trip, and costing her owner $160. If the skipper had gone with a current of 2 knots, then her SMG would be 10 knots, her transit time 2 hours 24 minutes, with a fuel expense of only $96. The time saved, 1 hour 36 minutes, allows more time for relaxation (TFR) either before departure or after arrival, and the $64 saved could buy a nice meal ashore. That's smarter boating!

Consult the table below for SMG and time/fuel consequences in currents.

SMG *WITH* CURRENT, and Time/Fuel GAINS

Current Speed Kts *With* +		**+1 kt**	**+2 kts**	**+3 kts**	**+4 kts**
Boat Speed: **4 kts**	SMG =	**5 kts**	**6 kts**	**7 kts**	**8 kts**
Time/Fuel **Gain**		**20%**	**33%**	**43%**	**50%**
Boat Speed: **6 kts**	SMG =	**7 kts**	**8 kts**	**9 kts**	**10 kts**
Time/Fuel **Gain**		**14%**	**25%**	**33%**	**40%**
Boat Speed: **8 kts**	SMG =	**9 kts**	**10 kts**	**11 kts**	**12 kts**
Time/Fuel **Gain**		**11%**	**20%**	**28%**	**33%**
Boat Speed: **10 kts**	SMG =	**11 kts**	**12 kts**	**13 kts**	**14 kts**
Time/Fuel **Gain**		**9%**	**17%**	**24%**	**29%**

SMG *AGAINST* CURRENT, and Time/Fuel LOSSES

Current Speed Kts *Against* -		**-1 kt**	**-2 kts**	**-3 kts**	**-4 kts**
Boat Speed: **4 kts**	SMG =	**3 kts**	**2 kts**	**1 kts**	**0 kts**
Time/Fuel **Loss**		**33%**	**100%**	**300%**	**---**
Boat Speed: **6 kts**	SMG =	**5 kts**	**4 kts**	**3 kts**	**2 kts**
Time/Fuel **Loss**		**20%**	**50%**	**100%**	**200%**
Boat Speed: **8 kts**	SMG =	**7 kts**	**6 kts**	**5 kts**	**4 kts**
Time/Fuel **Loss**		**14%**	**33%**	**60%**	**100%**
Boat Speed: **10 kts**	SMG =	**9 kts**	**8 kts**	**7 kts**	**6 kts**
Time/Fuel **Loss**		**11%**	**25%**	**43%**	**67%**

Boston Harbor Currents

This diagram shows the direction of the Flood Currents in Boston Harbor at the Maximum* Flood velocity, generally 3.5 hours after Low Water at Boston.

The Ebb Currents flow in precisely the opposite direction (note one exception, shown by dotted arrow east of Winthrop), and reach these maximum velocities about 4 hours after High Water at Boston.

The velocities of the Ebb Currents are about the same as those of the Flood Currents. Where the Ebb Current differs by .2 kts., the velocity of the Ebb is shown in parentheses.

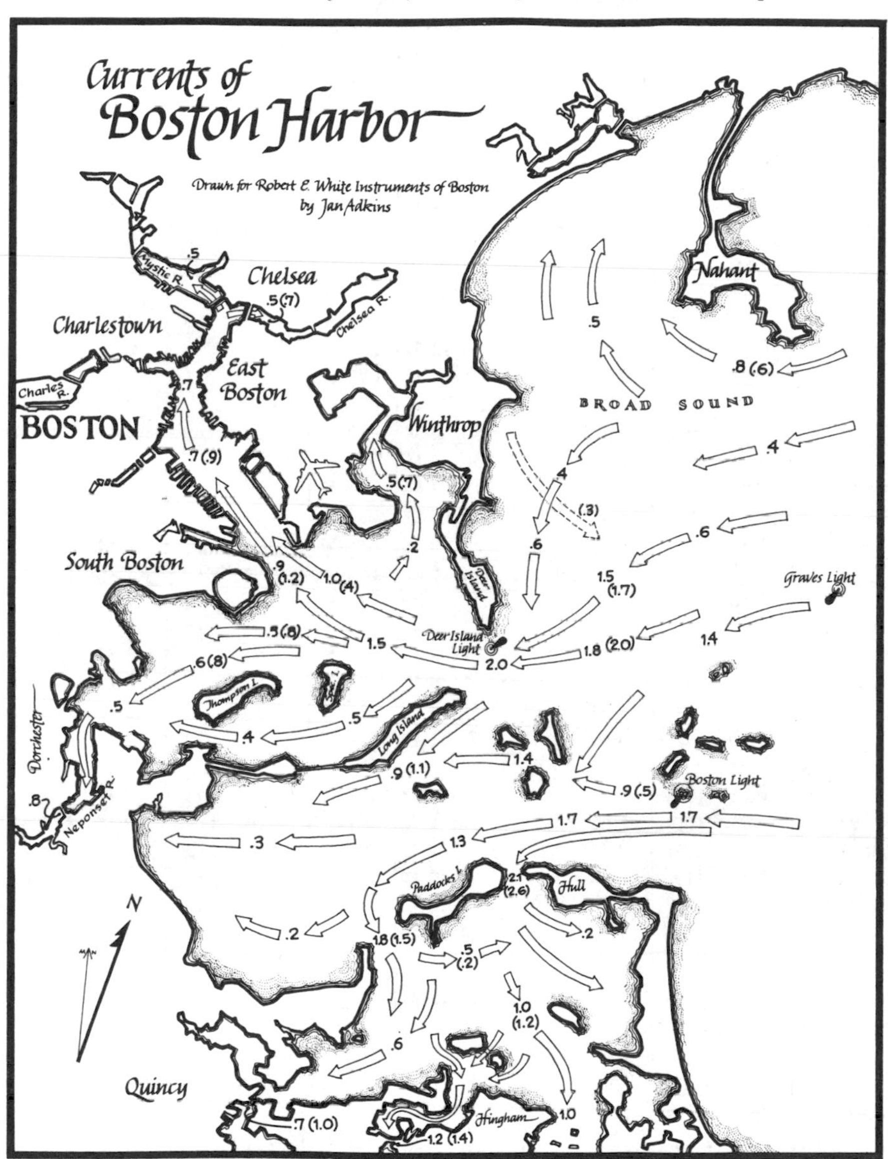

*The Velocities shown on this Current Diagram are the **maximums** normally encountered each month at Full Moon and at New Moon. At other times the velocities will be lower. As a rule of thumb, the velocities shown are those found on days when High Water at Boston is 11.0' to 11.5' (see Boston High Water Tables pp. 38-43). When the height of High Water is 10.5', subtract 10% from the velocities shown; at 10.0', subtract 20%; at 9.0', 30%; at 8.0', 40%; below 7.5', 50%.

2014 HIGH & LOW WATER BOSTON, MA

42°21.3'N, 71°03.1'W

Standard Time | **Standard Time**

DAY OF MONTH	DAY OF WEEK	JANUARY						DAY OF MONTH	DAY OF WEEK	FEBRUARY					
		HIGH				LOW				HIGH				LOW	
		a.m.	Ht.	**p.m.**	Ht.	a.m.	**p.m.**			a.m.	Ht.	**p.m.**	Ht.	a.m.	**p.m.**
1	**W**	10 29	12.0	**11 09**	10.6	4 17	**4 56**	1	**S**	...	...	**12 01**	12.1	5 45	**6 17**
2	**T**	11 22	12.2	**...**	...	5 10	**5 47**	2	**S**	12 33	11.3	**12 52**	11.8	6 38	**7 07**
3	**F**	12 01	10.9	**12 15**	12.2	6 02	**6 38**	3	**M**	1 23	11.2	**1 44**	11.3	7 30	**7 57**
4	**S**	12 54	11.0	**1 10**	11.9	6 57	**7 30**	4	**T**	2 15	11.0	**2 38**	10.6	8 24	**8 48**
5	**S**	1 46	10.9	**2 04**	11.4	7 50	**8 21**	5	**W**	3 06	10.6	**3 33**	9.9	9 18	**9 39**
6	**M**	2 40	10.8	**3 00**	10.8	8 46	**9 15**	6	**T**	4 00	10.2	**4 31**	9.2	10 15	**10 34**
7	**T**	3 36	10.5	**3 59**	10.1	9 44	**10 10**	7	**F**	4 56	9.7	**5 32**	8.6	11 16	**11 32**
8	**W**	4 33	10.2	**5 00**	9.4	10 45	**11 07**	8	**S**	5 55	9.4	**6 35**	8.3	...	**12 18**
9	**T**	5 32	10.0	**6 04**	9.0	11 48	**...**	9	**S**	6 55	9.3	**7 36**	8.2	12 31	**1 20**
10	**F**	6 31	9.8	**7 07**	8.7	12 06	**12 51**	10	**M**	7 52	9.3	**8 31**	8.3	1 28	**2 17**
11	**S**	7 28	9.8	**8 07**	8.6	1 04	**1 52**	11	**T**	8 43	9.5	**9 20**	8.6	2 22	**3 06**
12	**S**	8 22	9.8	**9 00**	8.6	1 59	**2 46**	12	**W**	9 30	9.7	**10 02**	8.8	3 10	**3 48**
13	**M**	9 11	9.9	**9 47**	8.8	2 50	**3 34**	13	**T**	10 12	9.9	**10 41**	9.1	3 54	**4 27**
14	**T**	9 54	10.0	**10 29**	8.9	3 36	**4 15**	14	**F**	10 51	10.0	**11 18**	9.3	4 35	**5 04**
15	**W**	10 35	10.1	**11 08**	9.0	4 18	**4 54**	15	**S**	11 28	10.1	**11 53**	9.5	5 14	**5 40**
16	**T**	11 14	10.1	**11 45**	9.1	4 59	**5 31**	16	**S**	...	...	**12 05**	10.1	5 53	**6 16**
17	**F**	11 51	10.1	**...**	...	5 38	**6 08**	17	**M**	12 28	9.7	**12 42**	10.0	6 32	**6 53**
18	**S**	12 22	9.2	**12 29**	10.1	6 18	**6 45**	18	**T**	1 04	9.8	**1 21**	9.8	7 12	**7 32**
19	**S**	12 58	9.3	**1 07**	9.9	6 57	**7 23**	19	**W**	1 42	9.9	**2 02**	9.6	7 54	**8 13**
20	**M**	1 35	9.3	**1 47**	9.7	7 38	**8 02**	20	**T**	2 22	9.9	**2 47**	9.4	8 39	**8 57**
21	**T**	2 14	9.4	**2 29**	9.4	8 21	**8 43**	21	**F**	3 07	9.9	**3 37**	9.1	9 29	**9 46**
22	**W**	2 55	9.4	**3 14**	9.2	9 07	**9 27**	22	**S**	3 58	9.9	**4 32**	8.9	10 24	**10 42**
23	**T**	3 40	9.4	**4 04**	8.9	9 57	**10 16**	23	**S**	4 54	10.0	**5 33**	8.8	11 23	**11 42**
24	**F**	4 30	9.6	**4 59**	8.7	10 52	**11 10**	24	**M**	5 55	10.1	**6 37**	9.0	...	**12 26**
25	**S**	5 24	9.8	**5 58**	8.7	11 50	**...**	25	**T**	6 58	10.4	**7 40**	9.4	12 44	**1 28**
26	**S**	6 21	10.1	**7 00**	8.9	12 07	**12 51**	26	**W**	8 01	10.9	**8 41**	9.9	1 46	**2 28**
27	**M**	7 21	10.5	**8 01**	9.3	1 07	**1 51**	27	**T**	9 01	11.3	**9 38**	10.5	2 46	**3 24**
28	**T**	8 20	11.0	**9 00**	9.8	2 06	**2 48**	28	**F**	9 58	11.6	**10 31**	11.0	3 42	**4 16**
29	**W**	9 17	11.5	**9 56**	10.3	3 03	**3 44**								
30	**T**	10 13	11.9	**10 50**	10.8	3 59	**4 36**								
31	**F**	11 07	12.1	**11 42**	11.1	4 53	**5 28**								

Dates when Ht. of **Low** Water is below Mean Lower Low with Ht. of lowest given for each period and Date of lowest in ( ):

1st - 7th: -2.2' (2nd - 3rd)
27th - 31st: -2.2' (31st)

1st - 5th: -2.2' (1st)
25th - 28th: -1.7' (28th)

Average Rise and Fall 9.5 ft.

When a high tide exceeds avg. ht., the *following* low tide will be lower than avg. Since there is a high degree of correlation between the height of High Water and the velocities of the Flood and Ebb Currents for that same day, we offer a rough rule of thumb for estimating the current velocities, for ALL the Current Charts and Diagrams in this book. **Rule of Thumb:** Refer to Boston High Water. If the height of High Water is 11.0' or over, use the Current Chart velocities as shown. When the height is 10.5', subtract 10%; at 10.0', subtract 20%; at 9.0', 30%; at 8.0', 40%; below 7.5', 50%.

2014 HIGH & LOW WATER BOSTON, MA

42°21.3'N, 71°03.1'W

***Daylight Time starts March 9 at 2 a.m.** **Daylight Saving Time**

DAY OF MONTH	DAY OF WEEK	MARCH HIGH a.m.	Ht.	p.m.	Ht.	LOW a.m.	p.m.	DAY OF MONTH	DAY OF WEEK	APRIL HIGH a.m.	Ht.	p.m.	Ht.	LOW a.m.	p.m.
1	**S**	10 52	11.8	**11 21**	11.4	4 36	**5 07**	1	**T**	12 45	11.5	**1 14**	11.0	6 59	**7 18**
2	**S**	11 43	11.8	**...**	...	5 28	**5 55**	2	**W**	1 31	11.4	**2 01**	10.6	7 47	**8 03**
3	**M**	12 10	11.5	**12 33**	11.5	6 19	**6 43**	3	**T**	2 16	11.0	**2 48**	10.1	8 34	**8 49**
4	**T**	12 58	11.4	**1 24**	11.0	7 10	**7 31**	4	**F**	3 03	10.6	**3 38**	9.5	9 22	**9 38**
5	**W**	1 45	11.1	**2 13**	10.4	7 59	**8 18**	5	**S**	3 50	10.0	**4 27**	9.0	10 11	**10 27**
6	**T**	2 34	10.6	**3 05**	9.7	8 50	**9 08**	6	**S**	4 41	9.5	**5 20**	8.6	11 03	**11 20**
7	**F**	3 24	10.1	**3 59**	9.0	9 43	**10 00**	7	**M**	5 35	9.2	**6 16**	8.4	11 58	**...**
8	**S**	4 18	9.6	**4 56**	8.5	10 40	**10 56**	8	**T**	6 32	8.9	**7 13**	8.4	12 16	**12 54**
9	**S**	*6 15	9.2	***6 57**	8.2	...	***12 39**	9	**W**	7 31	8.9	**8 08**	8.5	1 13	**1 50**
10	**M**	7 15	9.0	**7 57**	8.2	12 54	**1 40**	10	**T**	8 26	9.0	**8 59**	8.8	2 09	**2 41**
11	**T**	8 14	9.0	**8 53**	8.3	1 53	**2 37**	11	**F**	9 18	9.2	**9 45**	9.2	3 02	**3 28**
12	**W**	9 09	9.2	**9 43**	8.6	2 48	**3 28**	12	**S**	10 05	9.4	**10 27**	9.6	3 50	**4 12**
13	**T**	9 58	9.4	**10 27**	9.0	3 39	**4 12**	13	**S**	10 49	9.7	**11 07**	10.0	4 34	**4 53**
14	**F**	10 42	9.7	**11 07**	9.3	4 24	**4 53**	14	**M**	11 31	9.9	**11 45**	10.4	5 17	**5 34**
15	**S**	11 23	9.9	**11 44**	9.7	5 07	**5 31**	15	**T**	...	...	**12 12**	10.1	5 59	**6 14**
16	**S**	...	...	**12 01**	10.0	5 47	**6 08**	16	**W**	12 24	10.7	**12 53**	10.1	6 41	**6 56**
17	**M**	12 21	9.9	**12 39**	10.1	6 27	**6 46**	17	**T**	1 04	10.9	**1 36**	10.1	7 25	**7 39**
18	**T**	12 56	10.2	**1 18**	10.1	7 07	**7 24**	18	**F**	1 47	11.0	**2 21**	10.0	8 10	**8 25**
19	**W**	1 33	10.3	**1 58**	10.0	7 48	**8 04**	19	**S**	2 33	11.0	**3 10**	9.9	8 58	**9 14**
20	**T**	2 12	10.4	**2 40**	9.8	8 31	**8 47**	20	**S**	3 23	10.9	**4 03**	9.7	9 50	**10 08**
21	**F**	2 55	10.5	**3 26**	9.6	9 17	**9 33**	21	**M**	4 18	10.7	**5 00**	9.6	10 45	**11 06**
22	**S**	3 42	10.4	**4 17**	9.4	10 07	**10 24**	22	**T**	5 17	10.5	**6 01**	9.6	11 44	**...**
23	**S**	4 34	10.3	**5 14**	9.2	11 02	**11 21**	23	**W**	6 21	10.3	**7 04**	9.8	12 08	**12 45**
24	**M**	5 33	10.2	**6 15**	9.1	...	**12 02**	24	**T**	7 27	10.3	**8 05**	10.1	1 12	**1 46**
25	**T**	6 36	10.2	**7 19**	9.3	12 23	**1 05**	25	**F**	8 31	10.3	**9 04**	10.5	2 15	**2 45**
26	**W**	7 41	10.4	**8 23**	9.7	1 27	**2 07**	26	**S**	9 31	10.5	**9 58**	10.9	3 15	**3 40**
27	**T**	8 46	10.6	**9 23**	10.2	2 30	**3 07**	27	**S**	10 28	10.6	**10 49**	11.2	4 12	**4 32**
28	**F**	9 46	10.9	**10 18**	10.8	3 30	**4 03**	28	**M**	11 20	10.6	**11 36**	11.3	5 04	**5 21**
29	**S**	10 43	11.2	**11 10**	11.2	4 27	**4 55**	29	**T**	...	...	**12 08**	10.5	5 53	**6 08**
30	**S**	11 36	11.3	**11 59**	11.5	5 20	**5 44**	30	**W**	12 21	11.3	**12 55**	10.3	6 39	**6 53**
31	**M**	...	...	**12 26**	11.2	6 11	**6 31**								

Dates when Ht. of **Low** Water is below Mean Lower Low with Ht. of lowest given for each period and Date of lowest in ( ):

March	April
1st - 6th: -1.9' (1st)	1st - 4th: -1.5' (1st)
18th - 22nd: -0.4' (19th - 20th)	15th - 22nd: -0.8' (17th - 18th)
26th - 31st: -1.5' (31st)	25th - 30th: -1.1' (29th)

Average Rise and Fall 9.5 ft.

When a high tide exceeds avg. ht., the *following* low tide will be lower than avg. Since there is a high degree of correlation between the height of High Water and the velocities of the Flood and Ebb Currents for that same day, we offer a rough rule of thumb for estimating the current velocities, for ALL the Current Charts and Diagrams in this book. **Rule of Thumb:** Refer to Boston High Water. If the height of High Water is 11.0' or over, use the Current Chart velocities as shown. When the height is 10.5', subtract 10%; at 10.0', subtract 20%; at 9.0', 30%; at 8.0', 40%; below 7.5', 50%.

2014 HIGH & LOW WATER BOSTON, MA

42°21.3'N, 71°03.1'W

DAY OF MONTH	DAY OF WEEK	MAY						DAY OF MONTH	DAY OF WEEK	JUNE					
		Daylight Saving Time								Daylight Saving Time					
		HIGH				LOW				HIGH				LOW	
		a.m.	Ht.	**p.m.**	Ht.	a.m.	**p.m.**			a.m.	Ht.	**p.m.**	Ht.	a.m.	**p.m.**
1	**T**	1 05	11.1	**1 40**	10.1	7 25	**7 37**	1	**S**	2 06	10.3	**2 43**	9.3	8 27	**8 39**
2	**F**	1 49	10.8	**2 24**	9.7	8 09	**8 21**	2	**M**	2 49	10.0	**3 26**	9.2	9 10	**9 24**
3	**S**	2 32	10.4	**3 10**	9.4	8 54	**9 07**	3	**T**	3 34	9.7	**4 11**	9.0	9 54	**10 11**
4	**S**	3 19	10.0	**3 57**	9.1	9 41	**9 55**	4	**W**	4 22	9.4	**4 59**	9.0	10 40	**11 02**
5	**M**	4 06	9.6	**4 45**	8.8	10 27	**10 45**	5	**T**	5 10	9.1	**5 45**	9.0	11 27	**11 52**
6	**T**	4 57	9.3	**5 37**	8.7	11 17	**11 38**	6	**F**	6 02	8.9	**6 34**	9.2	...	**12 16**
7	**W**	5 50	9.0	**6 29**	8.7	...	**12 09**	7	**S**	6 55	8.8	**7 24**	9.4	12 46	**1 06**
8	**T**	6 45	8.9	**7 21**	8.8	12 33	**1 01**	8	**S**	7 49	8.9	**8 12**	9.8	1 39	**1 56**
9	**F**	7 40	8.9	**8 11**	9.1	1 28	**1 53**	9	**M**	8 42	9.0	**9 01**	10.2	2 32	**2 47**
10	**S**	8 33	9.0	**8 59**	9.5	2 21	**2 42**	10	**T**	9 34	9.3	**9 49**	10.7	3 24	**3 37**
11	**S**	9 24	9.2	**9 44**	9.9	3 11	**3 29**	11	**W**	10 25	9.6	**10 37**	11.1	4 14	**4 26**
12	**M**	10 12	9.5	**10 27**	10.4	3 59	**4 14**	12	**T**	11 15	9.9	**11 26**	11.5	5 03	**5 15**
13	**T**	10 58	9.7	**11 10**	10.8	4 45	**4 59**	13	**F**	...	...	**12 05**	10.2	5 53	**6 05**
14	**W**	11 43	10.0	**11 53**	11.2	5 31	**5 44**	14	**S**	12 16	11.8	**12 55**	10.5	6 42	**6 56**
15	**T**	...	...	**12 29**	10.2	6 16	**6 29**	15	**S**	1 06	11.9	**1 47**	10.6	7 32	**7 48**
16	**F**	12 38	11.4	**1 16**	10.3	7 03	**7 16**	16	**M**	1 59	11.9	**2 39**	10.7	8 23	**8 41**
17	**S**	1 25	11.5	**2 04**	10.3	7 51	**8 06**	17	**T**	2 53	11.6	**3 34**	10.7	9 16	**9 37**
18	**S**	2 15	11.5	**2 56**	10.3	8 41	**8 58**	18	**W**	3 50	11.2	**4 30**	10.7	10 10	**10 36**
19	**M**	3 08	11.3	**3 50**	10.2	9 33	**9 53**	19	**T**	4 49	10.7	**5 27**	10.6	11 05	**11 36**
20	**T**	4 04	11.0	**4 47**	10.2	10 28	**10 52**	20	**F**	5 51	10.2	**6 26**	10.6	...	**12 03**
21	**W**	5 04	10.7	**5 46**	10.2	11 26	**11 53**	21	**S**	6 54	9.8	**7 24**	10.6	12 39	**1 01**
22	**T**	6 07	10.3	**6 47**	10.3	...	**12 25**	22	**S**	7 57	9.6	**8 22**	10.6	1 41	**1 59**
23	**F**	7 11	10.1	**7 46**	10.5	12 57	**1 24**	23	**M**	8 57	9.4	**9 16**	10.6	2 42	**2 54**
24	**S**	8 15	10.0	**8 44**	10.7	1 59	**2 22**	24	**T**	9 54	9.3	**10 07**	10.6	3 39	**3 47**
25	**S**	9 15	9.9	**9 38**	10.9	3 00	**3 18**	25	**W**	10 46	9.3	**10 55**	10.6	4 30	**4 36**
26	**M**	10 11	9.9	**10 28**	11.0	3 56	**4 10**	26	**T**	11 33	9.3	**11 38**	10.5	5 17	**5 22**
27	**T**	11 03	9.9	**11 15**	11.0	4 47	**4 58**	27	**F**	...	...	**12 16**	9.3	6 00	**6 06**
28	**W**	11 51	9.8	**11 59**	10.9	5 35	**5 44**	28	**S**	12 20	10.5	**12 56**	9.3	6 41	**6 48**
29	**T**	...	...	**12 36**	9.8	6 20	**6 29**	29	**S**	1 01	10.4	**1 36**	9.3	7 21	**7 29**
30	**F**	12 42	10.8	**1 19**	9.6	7 03	**7 12**	30	**M**	1 41	10.2	**2 15**	9.3	8 00	**8 11**
31	**S**	1 23	10.6	**2 01**	9.5	7 45	**7 55**								

Dates when Ht. of **Low** Water is below Mean Lower Low with Ht. of lowest given for each period and Date of lowest in ( ):

1st - 2nd: -0.8' (1st)
14th - 21st: -1.2' (17th - 18th)
26th - 30th: -0.6' (28th)

11th - 20th: -1.6' (15th - 16th)

Average Rise and Fall 9.5 ft.

When a high tide exceeds avg. ht., the *following* low tide will be lower than avg. Since there is a high degree of correlation between the height of High Water and the velocities of the Flood and Ebb Currents for that same day, we offer a rough rule of thumb for estimating the current velocities, for ALL the Current Charts and Diagrams in this book. **Rule of Thumb:** Refer to Boston High Water. If the height of High Water is 11.0' or over, use the Current Chart velocities as shown. When the height is 10.5', subtract 10%; at 10.0', subtract 20%; at 9.0', 30%; at 8.0', 40%; below 7.5', 50%.

2014 HIGH & LOW WATER
BOSTON, MA
42°21.3'N, 71°03.1'W

Day of Month	Day of Week	JULY HIGH a.m.	Ht.	p.m.	Ht.	LOW a.m.	p.m.	Day of Month	Day of Week	AUGUST HIGH a.m.	Ht.	p.m.	Ht.	LOW a.m.	p.m.
		Daylight Saving Time								**Daylight Saving Time**					
1	**T**	2 22	10.0	**2 55**	9.3	8 40	**8 54**	1	**F**	3 16	9.5	**3 41**	9.6	9 28	**9 52**
2	**W**	3 04	9.8	**3 36**	9.3	9 20	**9 39**	2	**S**	4 00	9.3	**4 24**	9.6	10 11	**10 39**
3	**T**	3 47	9.5	**4 19**	9.3	10 03	**10 25**	3	**S**	4 47	9.1	**5 10**	9.7	10 57	**11 31**
4	**F**	4 34	9.2	**5 05**	9.3	10 48	**11 15**	4	**M**	5 39	8.9	**6 02**	9.8	11 48	**...**
5	**S**	5 22	9.0	**5 50**	9.5	11 34	**...**	5	**T**	6 34	8.8	**6 55**	10.1	12 26	**12 41**
6	**S**	6 14	8.8	**6 39**	9.7	12 06	**12 23**	6	**W**	7 32	8.9	**7 51**	10.4	1 23	**1 38**
7	**M**	7 08	8.8	**7 30**	10.0	1 00	**1 16**	7	**T**	8 31	9.2	**8 49**	10.9	2 21	**2 36**
8	**T**	8 03	8.9	**8 23**	10.4	1 55	**2 09**	8	**F**	9 29	9.7	**9 46**	11.4	3 18	**3 33**
9	**W**	8 59	9.2	**9 16**	10.8	2 50	**3 03**	9	**S**	10 25	10.2	**10 43**	11.8	4 14	**4 29**
10	**T**	9 55	9.5	**10 09**	11.3	3 45	**3 57**	10	**S**	11 20	10.7	**11 38**	12.1	5 07	**5 24**
11	**F**	10 49	10.0	**11 02**	11.7	4 38	**4 51**	11	**M**	...	...	**12 13**	11.2	5 59	**6 18**
12	**S**	11 42	10.4	**11 56**	12.0	5 30	**5 44**	12	**T**	12 32	12.2	**1 04**	11.5	6 50	**7 11**
13	**S**	...	...	**12 35**	10.8	6 21	**6 37**	13	**W**	1 25	12.0	**1 56**	11.6	7 40	**8 05**
14	**M**	12 49	12.2	**1 27**	11.0	7 12	**7 30**	14	**T**	2 19	11.6	**2 48**	11.5	8 31	**8 59**
15	**T**	1 43	12.1	**2 19**	11.2	8 03	**8 24**	15	**F**	3 13	11.1	**3 41**	11.2	9 22	**9 54**
16	**W**	2 37	11.7	**3 13**	11.2	8 55	**9 19**	16	**S**	4 09	10.4	**4 35**	10.8	10 15	**10 51**
17	**T**	3 33	11.2	**4 07**	11.1	9 47	**10 17**	17	**S**	5 07	9.8	**5 32**	10.4	11 10	**11 51**
18	**F**	4 31	10.7	**5 03**	10.9	10 42	**11 16**	18	**M**	6 08	9.2	**6 30**	10.1	...	**12 08**
19	**S**	5 31	10.1	**6 01**	10.6	11 38	**...**	19	**T**	7 10	8.9	**7 30**	9.8	12 53	**1 07**
20	**S**	6 33	9.5	**6 59**	10.4	12 17	**12 35**	20	**W**	8 11	8.7	**8 28**	9.7	1 55	**2 05**
21	**M**	7 35	9.2	**7 58**	10.2	1 20	**1 34**	21	**T**	9 08	8.7	**9 22**	9.8	2 53	**3 00**
22	**T**	8 37	9.0	**8 54**	10.2	2 21	**2 31**	22	**F**	9 59	8.9	**10 10**	9.9	3 45	**3 51**
23	**W**	9 34	8.9	**9 46**	10.2	3 19	**3 25**	23	**S**	10 44	9.0	**10 54**	10.0	4 29	**4 36**
24	**T**	10 25	9.0	**10 34**	10.2	4 11	**4 15**	24	**S**	11 24	9.3	**11 34**	10.1	5 10	**5 18**
25	**F**	11 11	9.1	**11 18**	10.2	4 56	**5 00**	25	**M**	...	...	**12 01**	9.5	5 47	**5 59**
26	**S**	11 52	9.2	**11 59**	10.2	5 38	**5 43**	26	**T**	12 13	10.1	**12 37**	9.6	6 24	**6 38**
27	**S**	...	...	**12 31**	9.3	6 16	**6 24**	27	**W**	12 51	10.1	**1 13**	9.8	7 00	**7 17**
28	**M**	12 38	10.2	**1 08**	9.4	6 54	**7 04**	28	**T**	1 28	10.0	**1 49**	9.8	7 37	**7 57**
29	**T**	1 16	10.1	**1 45**	9.5	7 31	**7 44**	29	**F**	2 07	9.8	**2 26**	9.9	8 16	**8 38**
30	**W**	1 55	10.0	**2 23**	9.5	8 09	**8 25**	30	**S**	2 47	9.6	**3 06**	9.9	8 56	**9 22**
31	**T**	2 35	9.8	**3 01**	9.5	8 48	**9 08**	31	**S**	3 31	9.4	**3 49**	9.9	9 38	**10 09**

Dates when Ht. of **Low** Water is below Mean Lower Low with Ht. of lowest given for each period and Date of lowest in ( ):

10th - 18th: -1.9' (14th)

8th - 16th: -1.9' (12th)

Average Rise and Fall 9.5 ft.

When a high tide exceeds avg. ht., the *following* **low tide will be lower than avg.** Since there is a high degree of correlation between the height of High Water and the velocities of the Flood and Ebb Currents for that same day, we offer a rough rule of thumb for estimating the current velocities, for ALL the Current Charts and Diagrams in this book. **Rule of Thumb:** Refer to Boston High Water. If the height of High Water is 11.0' or over, use the Current Chart velocities as shown. When the height is 10.5', subtract 10%; at 10.0', subtract 20%; at 9.0', 30%; at 8.0', 40%; below 7.5', 50%.

2014 HIGH & LOW WATER BOSTON, MA

42°21.3'N, 71°03.1'W

Daylight Saving Time

DAY OF MONTH	DAY OF WEEK	SEPTEMBER					
		HIGH				LOW	
		a.m.	Ht.	**p.m.**	Ht.	a.m.	**p.m.**
1	**M**	4 18	9.1	**4 36**	9.9	10 25	**11 01**
2	**T**	5 10	9.0	**5 29**	10.0	11 17	**11 57**
3	**W**	6 07	8.9	**6 27**	10.2	...	**12 14**
4	**T**	7 08	9.1	**7 28**	10.5	12 57	**1 15**
5	**F**	8 08	9.4	**8 28**	10.8	1 56	**2 15**
6	**S**	9 07	9.9	**9 28**	11.3	2 55	**3 14**
7	**S**	10 04	10.5	**10 25**	11.6	3 51	**4 12**
8	**M**	10 58	11.1	**11 20**	11.9	4 45	**5 07**
9	**T**	11 50	11.5	**...**	...	5 37	**6 00**
10	**W**	12 14	11.9	**12 40**	11.8	6 27	**6 52**
11	**T**	1 06	11.7	**1 30**	11.8	7 16	**7 44**
12	**F**	1 58	11.3	**2 20**	11.6	8 05	**8 36**
13	**S**	2 51	10.8	**3 11**	11.2	8 55	**9 29**
14	**S**	3 45	10.1	**4 04**	10.7	9 47	**10 23**
15	**M**	4 41	9.5	**4 59**	10.1	10 41	**11 21**
16	**T**	5 39	9.0	**5 57**	9.7	11 37	**...**
17	**W**	6 40	8.7	**6 57**	9.4	12 20	**12 36**
18	**T**	7 39	8.6	**7 56**	9.4	1 21	**1 35**
19	**F**	8 35	8.7	**8 50**	9.4	2 18	**2 31**
20	**S**	9 25	8.9	**9 40**	9.6	3 09	**3 22**
21	**S**	10 09	9.2	**10 25**	9.7	3 54	**4 08**
22	**M**	10 50	9.5	**11 06**	9.9	4 35	**4 51**
23	**T**	11 27	9.8	**11 45**	10.0	5 13	**5 31**
24	**W**	...	...	**12 03**	10.0	5 51	**6 11**
25	**T**	12 23	10.0	**12 39**	10.1	6 28	**6 50**
26	**F**	1 02	9.9	**1 16**	10.3	7 06	**7 30**
27	**S**	1 41	9.8	**1 54**	10.3	7 45	**8 12**
28	**S**	2 22	9.6	**2 34**	10.3	8 26	**8 56**
29	**M**	3 06	9.5	**3 19**	10.3	9 11	**9 44**
30	**T**	3 55	9.3	**4 09**	10.2	10 00	**10 37**

Daylight Saving Time

DAY OF MONTH	DAY OF WEEK	OCTOBER					
		HIGH				LOW	
		a.m.	Ht.	**p.m.**	Ht.	a.m.	**p.m.**
1	**W**	4 48	9.2	**5 05**	10.2	10 54	**11 34**
2	**T**	5 47	9.2	**6 05**	10.2	11 53	**...**
3	**F**	6 48	9.4	**7 08**	10.4	12 33	**12 55**
4	**S**	7 50	9.8	**8 12**	10.6	1 35	**1 59**
5	**S**	8 48	10.3	**9 12**	11.0	2 33	**2 58**
6	**M**	9 44	10.9	**10 09**	11.2	3 30	**3 56**
7	**T**	10 37	11.4	**11 04**	11.4	4 23	**4 50**
8	**W**	11 28	11.7	**11 57**	11.4	5 14	**5 43**
9	**T**	...	...	**12 17**	11.9	6 03	**6 33**
10	**F**	12 48	11.2	**1 05**	11.8	6 52	**7 23**
11	**S**	1 38	10.8	**1 53**	11.5	7 40	**8 12**
12	**S**	2 28	10.3	**2 42**	11.0	8 28	**9 02**
13	**M**	3 19	9.8	**3 32**	10.4	9 18	**9 53**
14	**T**	4 11	9.3	**4 24**	9.9	10 09	**10 47**
15	**W**	5 06	8.9	**5 20**	9.5	11 03	**11 42**
16	**T**	6 03	8.7	**6 18**	9.2	...	**12 01**
17	**F**	7 00	8.6	**7 16**	9.0	12 39	**12 59**
18	**S**	7 54	8.7	**8 11**	9.1	1 34	**1 55**
19	**S**	8 44	9.0	**9 03**	9.2	2 26	**2 47**
20	**M**	9 29	9.3	**9 50**	9.4	3 12	**3 35**
21	**T**	10 11	9.7	**10 33**	9.6	3 55	**4 19**
22	**W**	10 51	10.0	**11 15**	9.7	4 36	**5 01**
23	**T**	11 29	10.3	**11 55**	9.8	5 16	**5 42**
24	**F**	...	...	**12 06**	10.5	5 56	**6 23**
25	**S**	12 35	9.8	**12 45**	10.7	6 36	**7 05**
26	**S**	1 16	9.8	**1 25**	10.8	7 18	**7 48**
27	**M**	2 00	9.7	**2 09**	10.8	8 01	**8 34**
28	**T**	2 46	9.6	**2 56**	10.7	8 49	**9 24**
29	**W**	3 36	9.5	**3 49**	10.5	9 40	**10 17**
30	**T**	4 31	9.5	**4 46**	10.4	10 36	**11 13**
31	**F**	5 29	9.5	**5 48**	10.3	11 36	**...**

Dates when Ht. of **Low** Water is below Mean Lower Low with Ht. of lowest given for each period and Date of lowest in ( ):

6th - 13th: -1.7' (10th)

5th - 12th: -1.6' (8th - 9th)
24th - 29th: -0.5' (26th - 27th)

Average Rise and Fall 9.5 ft.

When a high tide exceeds avg. ht., the *following* low tide will be lower than avg. Since there is a high degree of correlation between the height of High Water and the velocities of the Flood and Ebb Currents for that same day, we offer a rough rule of thumb for estimating the current velocities, for ALL the Current Charts and Diagrams in this book. **Rule of Thumb:** Refer to Boston High Water. If the height of High Water is 11.0' or over, use the Current Chart velocities as shown. When the height is 10.5', subtract 10%; at 10.0', subtract 20%; at 9.0', 30%; at 8.0', 40%; below 7.5', 50%.

2014 HIGH & LOW WATER
BOSTON, MA
42°21.3'N, 71°03.1'W

***Standard Time starts Nov. 2 at 2 a.m.** **Standard Time**

DAY OF MONTH	DAY OF WEEK	NOVEMBER						DAY OF MONTH	DAY OF WEEK	DECEMBER					
		HIGH				LOW				HIGH				LOW	
		a.m.	Ht.	**p.m.**	Ht.	a.m.	**p.m.**			a.m.	Ht.	**p.m.**	Ht.	a.m.	**p.m.**
1	**S**	6 30	9.7	**6 51**	10.2	12 13	**12 39**	1	**M**	6 12	10.3	**6 40**	9.8	...	**12 26**
2	**S**	*6 31	10.1	***6 55**	10.3	1 13	***12 42**	2	**T**	7 11	10.6	**7 42**	9.8	12 50	**1 28**
3	**M**	7 30	10.6	**7 57**	10.4	1 12	**1 43**	3	**W**	8 07	10.9	**8 42**	9.9	1 47	**2 27**
4	**T**	8 26	11.0	**8 56**	10.6	2 09	**2 42**	4	**T**	9 02	11.1	**9 37**	9.9	2 42	**3 22**
5	**W**	9 18	11.4	**9 50**	10.7	3 02	**3 35**	5	**F**	9 50	11.2	**10 27**	9.9	3 33	**4 12**
6	**T**	10 08	11.6	**10 41**	10.7	3 52	**4 26**	6	**S**	10 37	11.2	**11 14**	9.9	4 21	**4 59**
7	**F**	10 56	11.7	**11 30**	10.5	4 41	**5 15**	7	**S**	11 22	11.1	**11 59**	9.7	5 08	**5 44**
8	**S**	11 42	11.5	**...**	...	5 29	**6 03**	8	**M**	...	...	**12 05**	10.8	5 53	**6 27**
9	**S**	12 18	10.3	**12 28**	11.2	6 15	**6 49**	9	**T**	12 42	9.6	**12 48**	10.5	6 37	**7 10**
10	**M**	1 05	9.9	**1 13**	10.8	7 01	**7 36**	10	**W**	1 25	9.4	**1 32**	10.1	7 21	**7 53**
11	**T**	1 52	9.5	**2 00**	10.3	7 49	**8 23**	11	**T**	2 08	9.2	**2 16**	9.8	8 06	**8 36**
12	**W**	2 40	9.2	**2 49**	9.8	8 37	**9 11**	12	**F**	2 53	9.0	**3 03**	9.4	8 53	**9 22**
13	**T**	3 30	8.9	**3 41**	9.4	9 28	**10 02**	13	**S**	3 40	8.9	**3 53**	9.0	9 43	**10 09**
14	**F**	4 22	8.7	**4 35**	9.0	10 22	**10 54**	14	**S**	4 28	8.8	**4 45**	8.7	10 35	**10 58**
15	**S**	5 15	8.7	**5 31**	8.8	11 17	**11 46**	15	**M**	5 18	8.9	**5 39**	8.5	11 29	**11 49**
16	**S**	6 07	8.8	**6 26**	8.8	...	**12 13**	16	**T**	6 08	9.0	**6 34**	8.5	...	**12 23**
17	**M**	6 58	9.0	**7 20**	8.8	12 37	**1 07**	17	**W**	6 58	9.3	**7 27**	8.6	12 40	**1 17**
18	**T**	7 45	9.4	**8 10**	9.0	1 26	**1 58**	18	**T**	7 47	9.7	**8 19**	8.8	1 30	**2 08**
19	**W**	8 30	9.8	**8 57**	9.2	2 13	**2 45**	19	**F**	8 34	10.1	**9 09**	9.1	2 20	**2 58**
20	**T**	9 12	10.1	**9 42**	9.4	2 58	**3 30**	20	**S**	9 21	10.6	**9 57**	9.4	3 08	**3 46**
21	**F**	9 54	10.5	**10 26**	9.6	3 41	**4 14**	21	**S**	10 08	11.0	**10 45**	9.8	3 56	**4 33**
22	**S**	10 35	10.8	**11 10**	9.8	4 25	**4 58**	22	**M**	10 54	11.4	**11 32**	10.1	4 44	**5 20**
23	**S**	11 18	11.1	**11 54**	9.9	5 08	**5 42**	23	**T**	11 42	11.6	**...**	...	5 32	**6 07**
24	**M**	...	...	**12 02**	11.2	5 53	**6 27**	24	**W**	12 20	10.3	**12 32**	11.7	6 21	**6 56**
25	**T**	12 40	9.9	**12 49**	11.2	6 40	**7 15**	25	**T**	1 10	10.4	**1 23**	11.5	7 12	**7 46**
26	**W**	1 28	10.0	**1 39**	11.1	7 30	**8 05**	26	**F**	2 01	10.5	**2 17**	11.1	8 06	**8 38**
27	**T**	2 19	10.0	**2 32**	10.9	8 23	**8 58**	27	**S**	2 55	10.5	**3 14**	10.7	9 02	**9 32**
28	**F**	3 14	9.9	**3 30**	10.6	9 19	**9 53**	28	**S**	3 51	10.4	**4 14**	10.2	10 01	**10 28**
29	**S**	4 12	10.0	**4 31**	10.2	10 20	**10 51**	29	**M**	4 50	10.4	**5 18**	9.7	11 04	**11 27**
30	**S**	5 11	10.1	**5 35**	10.0	11 22	**11 50**	30	**T**	5 50	10.4	**6 23**	9.4	...	**12 08**
								31	**W**	6 51	10.4	**7 27**	9.2	12 27	**1 11**

Dates when Ht. of **Low** Water is below Mean Lower Low with Ht. of lowest given for each period and Date of lowest in ( ):

3rd - 10th: -1.4' (7th)
21st - 29th: -1.1' (24th - 25th)

2nd - 9th: -1.0' (5th - 6th)
20th - 28th: -1.7' (24th)

Average Rise and Fall 9.5 ft.

When a high tide exceeds avg. ht., the *following* low tide will be lower than avg. Since there is a high degree of correlation between the height of High Water and the velocities of the Flood and Ebb Currents for that same day, we offer a rough rule of thumb for estimating the current velocities, for ALL the Current Charts and Diagrams in this book. **Rule of Thumb:** Refer to Boston High Water. If the height of High Water is 11.0' or over, use the Current Chart velocities as shown. When the height is 10.5', subtract 10%; at 10.0', subtract 20%; at 9.0', 30%; at 8.0', 40%; below 7.5', 50%.

CAPE COD CANAL

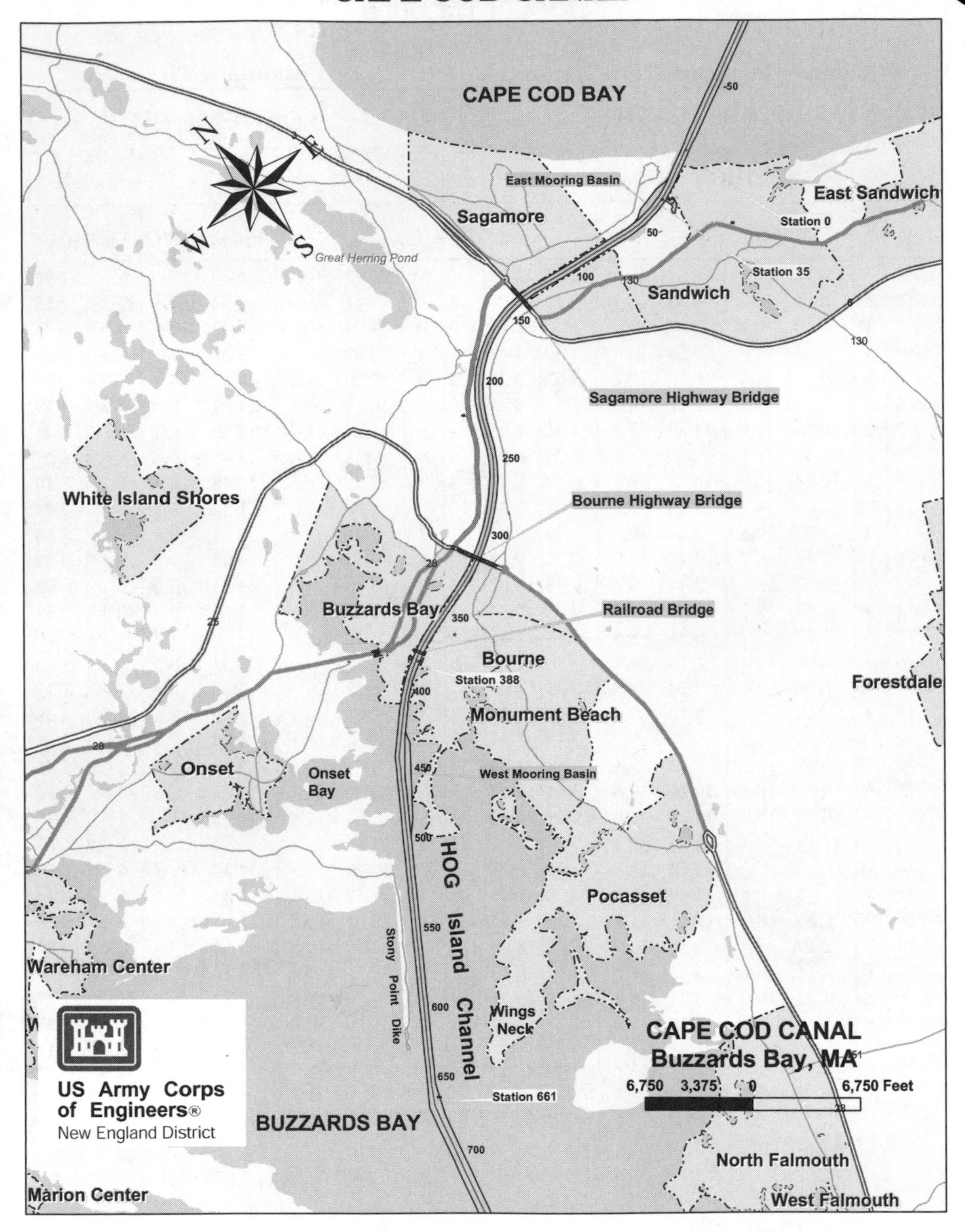

SMALL BOAT BASINS ON EITHER END OF THE CANAL: On E. end, 13-ft. mean low water, on S. side of Sandwich, available for mooring small boat traffic; On W. end, channel 13-ft. at mean low water, 100 ft. wide leads from NE side of Hog Is. Ch. abreast of Hog Is. to harbor in Onset Bay. Fuel, supplies and phone services at both locations.

See Cape Cod Canal Currents pp. 46-51.

CAPE COD CANAL REGULATIONS

For complete regulations see 33 CFR, Part 207 and 36 CFR, Part 327

No excessive wake – Speed Limit 10 m.p.h. (8.7 kts.)

Vessels going *with* the current have right of way over those going *against* it.

Clearance under all bridges: 135 feet at mean high water. Available clearance can be reduced by construction work on the bridges so mariners are advised to contact the Marine Traffic Controller for current clearance dimensions prior to transit. Buzzards Bay Railroad Bridge is maintained in up, or open position, except when lowered for trains or maintenance.

Obtaining Clearance

Vessels 65 feet and over shall not enter the Canal until clearance has been given by radio from the Marine Traffic Controller. These vessels shall request clearance at least 15 minutes prior to entering the Canal at any point.

Vessels of any kind unable to make a through transit of the Canal against a head current of 6 kts. within a time limit of 2-1/2 hrs. are required to obtain helper tug assistance or wait for a fair current prior to receiving clearance from the Controller.

Two-way traffic through the Canal for all vessels is allowed when Controller on duty considers conditions suitable.

Communications

Direct communications are available at all hours by VHF radio or by phoning 978-318-8500. Call on Channel 13 to establish contact. Transmissions may then be switched to Channel 14 as the working channel. Channel 16 is also available but should be limited to emergency situations. Vessels shall maintain a radio guard on Channel 13 during the entire passage.

Traffic Lights

Traffic Lights are at Eastern End at Sandwich (Cape Cod Bay entrance) and at Western End near Wings Neck (Buzzards Bay entrance). When traffic lights are extinguished: all vessels over 65 feet are cautioned not to enter Canal until clearance given, as above.

Entering From EASTERN END: (Lights on South side of entrance to Canal.)

RED LIGHT: Any type of vessel 65 feet in length and over must stop clear of the Cape Cod Bay entrance channel.

YELLOW LIGHT: Vessels 65 feet in length and over and drawing less than 25 feet may proceed as far as the East Mooring Basin where they must stop.

GREEN LIGHT: Vessels may proceed westward through the Canal.

Entering From WESTERN END: (Lights near Wings Neck at West Entrance to Hog Is. Channel)

RED LIGHT: Vessels 65 feet and over in length and drawing less than 25 feet must keep southerly of Hog Island Channel Entrance Buoys Nos. 1 and 2 and utilize the general anchorage areas adjacent to the improved channel. Vessel traffic drawing 25 feet and over are directed not to enter the Canal channel at the Cleveland Ledge Light entrance and shall lay to or anchor in Buzzards Bay until clearance is granted by the Marine Traffic Controller or a green traffic light at Wings Neck is displayed.

YELLOW LIGHT: Vessels may proceed through Hog Island Channel as far as the West Mooring Basin where they must stop.

GREEN LIGHT: Vessels may proceed eastward through the Canal.

Prohibited Activities

Jet skis, sea planes, paddle-driven craft and sailing vessels not under power are prohibited from transiting the Canal.

Fishing from a vessel within the channel limits of the Canal is prohibited.

Anchoring within the channel limits of the Canal, except in emergencies with notice given to the Traffic Controller, is prohibited.

2014 CURRENT TABLE
CAPE COD CANAL
41°44.5'N, 70°36.8'W at R.R. Bridge

Standard Time

JANUARY

DAY OF MONTH	DAY OF WEEK	CURRENT TURNS TO EAST Flood Starts a.m.	p.m.	Kts.	CURRENT TURNS TO WEST Ebb Starts a.m.	p.m.	Kts.
1	W	2 35	3 03	p4.9	8 50	9 36	a5.1
2	T	3 26	3 53	p5.0	9 39	10 26	a5.2
3	F	4 16	4 43	p5.0	10 29	11 16	a5.3
4	S	5 08	5 34	p5.0	11 20	...	5.2
5	S	5 59	6 25	p4.8	12 07	12 13	p5.1
6	M	6 52	7 18	p4.6	12 59	1 09	p4.9
7	T	7 49	8 15	p4.4	1 53	2 09	4.6
8	W	8 49	9 14	4.1	2 49	3 12	4.4
9	T	9 51	10 16	3.9	3 48	4 18	a4.3
10	F	10 55	11 19	a3.9	4 46	5 23	a4.2
11	S	11 56	...	3.9	5 44	6 24	a4.2
12	S	12 18	12 51	p4.0	6 37	7 19	a4.2
13	M	1 11	1 41	p4.1	7 27	8 09	a4.3
14	T	1 59	2 25	p4.2	8 13	8 54	a4.4
15	W	2 42	3 06	p4.2	8 55	9 35	a4.5
16	T	3 21	3 43	p4.3	9 34	10 13	a4.5
17	F	3 58	4 19	p4.3	10 12	10 50	a4.6
18	S	4 34	4 55	p4.3	10 48	11 26	a4.6
19	S	5 10	5 31	p4.2	11 24	...	4.5
20	M	5 47	6 08	p4.2	12 02	12 02	p4.4
21	T	6 26	6 48	p4.1	12 39	12 42	p4.3
22	W	7 10	7 32	p4.0	1 18	1 26	p4.2
23	T	7 58	8 21	p3.9	2 02	2 18	4.0
24	F	8 52	9 17	3.8	2 51	3 19	4.0
25	S	9 52	10 18	3.8	3 47	4 26	a4.0
26	S	10 56	11 22	3.9	4 47	5 33	a4.2
27	M	11 59	...	4.1	5 48	6 37	a4.4
28	T	12 25	12 58	p4.4	6 46	7 34	a4.6
29	W	1 23	1 54	p4.7	7 41	8 28	a4.9
30	T	2 18	2 47	p4.9	8 34	9 19	a5.1
31	F	3 10	3 37	p5.0	9 25	10 08	a5.2

Standard Time

FEBRUARY

DAY OF MONTH	DAY OF WEEK	CURRENT TURNS TO EAST Flood Starts a.m.	p.m.	Kts.	CURRENT TURNS TO WEST Ebb Starts a.m.	p.m.	Kts.
1	S	4 00	4 26	p5.1	10 15	10 56	a5.3
2	S	4 49	5 15	p5.0	11 05	11 44	a5.3
3	M	5 38	6 04	4.8	11 56	...	5.1
4	T	6 29	6 55	4.6	12 32	12 49	4.9
5	W	7 21	7 46	a4.4	1 22	1 45	a4.7
6	T	8 16	8 42	a4.1	2 15	2 44	a4.4
7	F	9 16	9 42	a3.9	3 11	3 48	a4.2
8	S	10 19	10 45	a3.7	4 10	4 53	a4.1
9	S	11 22	11 47	a3.7	5 10	5 56	a4.0
10	M	...	12 21	3.8	6 08	6 53	a4.0
11	T	12 44	1 14	p3.9	7 00	7 43	a4.1
12	W	1 34	2 00	p4.1	7 48	8 28	a4.3
13	T	2 17	2 41	p4.2	8 32	9 09	a4.4
14	F	2 57	3 19	p4.3	9 12	9 46	a4.5
15	S	3 33	3 54	p4.3	9 49	10 22	a4.6
16	S	4 09	4 29	p4.3	10 26	10 56	a4.6
17	M	4 43	5 04	p4.3	11 01	11 30	a4.6
18	T	5 19	5 40	p4.3	11 37	...	4.5
19	W	5 57	6 19	4.2	12 05	12 16	4.4
20	T	6 39	7 02	4.1	12 42	1 00	4.3
21	F	7 26	7 51	a4.0	1 23	1 51	a4.2
22	S	8 20	8 47	a4.0	2 12	2 52	a4.2
23	S	9 21	9 51	a3.9	3 10	4 00	a4.1
24	M	10 28	10 59	a4.0	4 15	5 11	a4.2
25	T	11 35	...	4.1	5 22	6 16	a4.4
26	W	12 05	12 39	p4.4	6 26	7 15	a4.6
27	T	1 06	1 36	p4.7	7 24	8 09	a4.8
28	F	2 02	2 30	p4.9	8 19	8 59	a5.1

The Kts. (knots) columns show the **maximum** predicted velocities of the stronger one of the Flood Currents and the stronger one of the Ebb Currents for each day.
The letter "a" means the velocity shown should occur **after** the **a.m.** Current Change. The letter "p" means the velocity shown should occur **after** the **p.m.** Current Change (even if next morning). No "a" or "p" means a.m. and p.m. velocities are the same for that day.
Avg. Max. Velocity: Flood 4.0 Kts., Ebb 4.5 Kts.
Max. Flood 3 hrs. after Flood Starts, ±20 min.
Max. Ebb 3 hrs. after Ebb Starts, ±20 min.
Average rise and fall: canal east end, 8.7 ft. (time of high water same as Boston); west end, at Monument Beach, 4.0 ft. (time of high water 15 min. after Newport).

See pp. 22-29 for Current Change at other points.

2014 CURRENT TABLE
CAPE COD CANAL
41°44.5'N, 70°36.8'W at R.R. Bridge

***Daylight Time starts Mar. 9 at 2 a.m.** — **Daylight Saving Time**

MARCH

DAY OF MONTH	DAY OF WEEK	CURRENT TURNS TO EAST Flood Starts a.m.	**p.m.**	Kts.	CURRENT TURNS TO WEST Ebb Starts a.m.	**p.m.**	Kts.
1	**S**	2 53	**3 20**	p5.0	9 10	**9 47**	a5.2
2	**S**	3 42	**4 08**	p5.0	10 00	**10 33**	a5.2
3	**M**	4 29	**4 54**	4.9	10 49	**11 18**	a5.2
4	**T**	5 17	**5 42**	a4.8	11 38	**...**	5.0
5	**W**	6 03	**6 28**	a4.6	12 03	**12 28**	a4.9
6	**T**	6 51	**7 16**	a4.4	12 50	**1 20**	a4.7
7	**F**	7 43	**8 08**	a4.1	1 39	**2 16**	a4.4
8	**S**	8 39	**9 06**	a3.8	2 33	**3 17**	a4.2
9	**S**	*10 40	***11 09**	a3.6	*4 31	***5 20**	a4.0
10	**M**	11 43	**...**	3.6	5 33	**6 23**	a3.9
11	**T**	12 12	**12 44**	p3.7	6 33	**7 20**	a3.9
12	**W**	1 11	**1 39**	p3.8	7 29	**8 11**	a4.0
13	**T**	2 03	**2 27**	p4.0	8 19	**8 56**	a4.2
14	**F**	2 48	**3 10**	p4.1	9 04	**9 37**	a4.3
15	**S**	3 28	**3 49**	p4.2	9 45	**10 14**	a4.5
16	**S**	4 05	**4 25**	p4.3	10 24	**10 50**	a4.6
17	**M**	4 41	**5 01**	p4.4	11 01	**11 24**	4.6
18	**T**	5 16	**5 37**	4.4	11 38	**11 58**	4.6
19	**W**	5 53	**6 14**	a4.4	...	**12 15**	4.6
20	**T**	6 31	**6 54**	a4.4	12 33	**12 56**	a4.6
21	**F**	7 14	**7 38**	a4.3	1 10	**1 41**	a4.5
22	**S**	8 02	**8 29**	a4.2	1 53	**2 34**	a4.4
23	**S**	8 57	**9 27**	a4.1	2 44	**3 35**	a4.3
24	**M**	9 59	**10 32**	a4.0	3 45	**4 43**	a4.3
25	**T**	11 07	**11 42**	a4.1	4 53	**5 52**	a4.3
26	**W**	...	**12 15**	4.2	6 03	**6 57**	a4.4
27	**T**	12 49	**1 19**	p4.4	7 09	**7 55**	a4.6
28	**F**	1 50	**2 18**	p4.6	8 08	**8 48**	a4.8
29	**S**	2 45	**3 11**	p4.8	9 03	**9 37**	4.9
30	**S**	3 36	**4 00**	p4.8	9 55	**10 23**	5.0
31	**M**	4 23	**4 47**	4.8	10 44	**11 08**	a5.1

APRIL

DAY OF MONTH	DAY OF WEEK	CURRENT TURNS TO EAST Flood Starts a.m.	**p.m.**	Kts.	CURRENT TURNS TO WEST Ebb Starts a.m.	**p.m.**	Kts.
1	**T**	5 09	**5 32**	a4.8	11 32	**11 51**	5.0
2	**W**	5 53	**6 16**	a4.7	...	**12 19**	4.8
3	**T**	6 37	**7 01**	a4.6	12 34	**1 07**	a4.8
4	**F**	7 24	**7 47**	a4.3	1 18	**1 56**	a4.6
5	**S**	8 11	**8 36**	a4.1	2 05	**2 48**	a4.4
6	**S**	9 02	**9 30**	a3.8	2 55	**3 44**	a4.2
7	**M**	9 59	**10 30**	a3.6	3 52	**4 44**	a4.0
8	**T**	10 59	**11 32**	a3.5	4 53	**5 44**	a3.8
9	**W**	...	**12 01**	3.6	5 54	**6 40**	a3.8
10	**T**	12 32	**12 56**	p3.7	6 52	**7 31**	3.9
11	**F**	1 25	**1 47**	p3.9	7 44	**8 17**	4.1
12	**S**	2 12	**2 32**	p4.0	8 31	**8 59**	p4.3
13	**S**	2 54	**3 13**	p4.2	9 15	**9 38**	4.4
14	**M**	3 33	**3 53**	p4.3	9 55	**10 15**	p4.6
15	**T**	4 11	**4 31**	4.4	10 35	**10 51**	p4.7
16	**W**	4 49	**5 10**	a4.5	11 15	**11 27**	p4.7
17	**T**	5 28	**5 50**	a4.5	11 56	**...**	4.6
18	**F**	6 10	**6 33**	a4.6	12 04	**12 40**	a4.7
19	**S**	6 55	**7 20**	a4.5	12 46	**1 28**	a4.7
20	**S**	7 45	**8 13**	a4.4	1 32	**2 22**	a4.6
21	**M**	8 40	**9 13**	a4.3	2 26	**3 23**	a4.5
22	**T**	9 42	**10 18**	a4.2	3 28	**4 28**	a4.4
23	**W**	10 48	**11 26**	a4.2	4 36	**5 33**	a4.4
24	**T**	11 55	**...**	4.3	5 46	**6 35**	a4.4
25	**F**	12 32	**12 59**	p4.4	6 52	**7 32**	4.5
26	**S**	1 32	**1 57**	p4.5	7 52	**8 25**	4.7
27	**S**	2 27	**2 50**	p4.6	8 48	**9 13**	4.8
28	**M**	3 17	**3 40**	4.6	9 39	**9 59**	p4.9
29	**T**	4 04	**4 26**	a4.7	10 28	**10 43**	p4.9
30	**W**	4 48	**5 10**	a4.7	11 15	**11 25**	p4.8

The Kts. (knots) columns show the **maximum** predicted velocities of the stronger one of the Flood Currents and the stronger one of the Ebb Currents for each day.
The letter "a" means the velocity shown should occur **after** the **a.m.** Current Change. The letter "p" means the velocity shown should occur **after** the **p.m.** Current Change (even if next morning). No "a" or "p" means a.m. and p.m. velocities are the same for that day.
Avg. Max. Velocity: Flood 4.0 Kts., Ebb 4.5 Kts.
Max. Flood 3 hrs. after Flood Starts, ±20 min.
Max. Ebb 3 hrs. after Ebb Starts, ±20 min.
Average rise and fall: canal east end, 8.7 ft. (time of high water same as Boston); west end, at Monument Beach, 4.0 ft. (time of high water 15 min. after Newport).

See pp. 22-29 for Current Change at other points.

2014 CURRENT TABLE
CAPE COD CANAL
41°44.5'N, 70°36.8'W at R.R. Bridge

Daylight Saving Time

MAY

DAY OF MONTH	DAY OF WEEK	CURRENT TURNS TO EAST Flood Starts			CURRENT TURNS TO WEST Ebb Starts		
		a.m.	**p.m.**	Kts.	a.m.	**p.m.**	Kts.
1	**T**	5 31	**5 52**	a4.6	...	**12 01**	4.6
2	**F**	6 14	**6 35**	a4.5	12 07	**12 46**	a4.7
3	**S**	6 56	**7 18**	a4.3	12 49	**1 33**	a4.6
4	**S**	7 42	**8 05**	a4.1	1 33	**2 21**	a4.4
5	**M**	8 28	**8 55**	a3.9	2 20	**3 11**	a4.2
6	**T**	9 19	**9 50**	a3.7	3 13	**4 05**	a4.0
7	**W**	10 15	**10 49**	a3.6	4 10	**5 01**	a3.9
8	**T**	11 12	**11 47**	a3.6	5 10	**5 56**	3.8
9	**F**	...	**12 08**	3.7	6 09	**6 47**	3.9
10	**S**	12 41	**1 01**	p3.8	7 04	**7 34**	p4.1
11	**S**	1 31	**1 50**	p4.0	7 54	**8 18**	p4.3
12	**M**	2 16	**2 35**	p4.1	8 41	**9 00**	p4.5
13	**T**	2 59	**3 19**	4.2	9 26	**9 39**	p4.6
14	**W**	3 41	**4 01**	4.4	10 09	**10 19**	p4.8
15	**T**	4 23	**4 44**	a4.6	10 53	**10 59**	p4.9
16	**F**	5 06	**5 28**	a4.7	11 38	**11 41**	p4.9
17	**S**	5 51	**6 15**	a4.7	...	**12 25**	4.6
18	**S**	6 39	**7 05**	a4.7	12 26	**1 16**	a4.9
19	**M**	7 31	**8 00**	a4.6	1 16	**2 10**	a4.8
20	**T**	8 26	**8 59**	a4.5	2 12	**3 08**	a4.7
21	**W**	9 26	**10 03**	a4.4	3 14	**4 09**	a4.6
22	**T**	10 30	**11 08**	a4.3	4 20	**5 11**	a4.5
23	**F**	11 34	**...**	4.3	5 28	**6 11**	a4.5
24	**S**	12 12	**12 37**	p4.3	6 33	**7 08**	4.5
25	**S**	1 12	**1 35**	p4.3	7 35	**8 00**	p4.6
26	**M**	2 08	**2 29**	4.4	8 31	**8 50**	p4.7
27	**T**	2 58	**3 19**	a4.5	9 24	**9 36**	p4.7
28	**W**	3 45	**4 05**	a4.5	10 12	**10 19**	p4.8
29	**T**	4 29	**4 48**	a4.5	10 59	**11 01**	p4.7
30	**F**	5 11	**5 29**	a4.4	11 43	**11 42**	p4.6
31	**S**	5 51	**6 10**	a4.3	...	**12 26**	4.3

Daylight Saving Time

JUNE

DAY OF MONTH	DAY OF WEEK	CURRENT TURNS TO EAST Flood Starts			CURRENT TURNS TO WEST Ebb Starts		
		a.m.	**p.m.**	Kts.	a.m.	**p.m.**	Kts.
1	**S**	6 32	**6 51**	a4.2	12 23	**1 08**	a4.5
2	**M**	7 13	**7 34**	a4.1	1 04	**1 52**	a4.4
3	**T**	7 56	**8 20**	a4.0	1 48	**2 37**	a4.3
4	**W**	8 43	**9 11**	a3.9	2 36	**3 25**	a4.1
5	**T**	9 32	**10 03**	a3.8	3 28	**4 16**	a4.0
6	**F**	10 24	**10 59**	a3.7	4 24	**5 08**	a3.9
7	**S**	11 19	**11 54**	a3.7	5 23	**5 59**	3.9
8	**S**	...	**12 14**	3.8	6 20	**6 49**	p4.1
9	**M**	12 48	**1 06**	p3.9	7 15	**7 36**	p4.3
10	**T**	1 38	**1 57**	4.0	8 07	**8 21**	p4.5
11	**W**	2 26	**2 46**	4.2	8 57	**9 06**	p4.7
12	**T**	3 13	**3 34**	a4.4	9 45	**9 50**	p4.8
13	**F**	4 00	**4 21**	a4.6	10 32	**10 35**	p5.0
14	**S**	4 47	**5 09**	a4.8	11 20	**11 21**	p5.0
15	**S**	5 35	**5 58**	a4.8	...	**12 09**	4.7
16	**M**	6 24	**6 50**	a4.8	12 10	**1 00**	a5.1
17	**T**	7 16	**7 44**	a4.8	1 02	**1 53**	a5.0
18	**W**	8 10	**8 42**	a4.7	1 58	**2 49**	a4.9
19	**T**	9 08	**9 42**	a4.5	2 58	**3 46**	a4.7
20	**F**	10 08	**10 46**	a4.3	4 02	**4 45**	a4.6
21	**S**	11 11	**11 49**	a4.2	5 08	**5 44**	4.4
22	**S**	...	**12 13**	4.1	6 14	**6 42**	p4.4
23	**M**	12 50	**1 13**	4.1	7 16	**7 36**	p4.5
24	**T**	1 47	**2 08**	a4.2	8 14	**8 26**	p4.5
25	**W**	2 39	**2 59**	a4.3	9 07	**9 14**	p4.6
26	**T**	3 26	**3 45**	a4.3	9 55	**9 58**	p4.6
27	**F**	4 10	**4 27**	a4.3	10 40	**10 40**	p4.6
28	**S**	4 51	**5 07**	a4.3	11 22	**11 20**	p4.6
29	**S**	5 30	**5 45**	a4.3	...	**12 03**	4.2
30	**M**	6 08	**6 24**	a4.2	12 01	**12 42**	a4.5

The Kts. (knots) columns show the **maximum** predicted velocities of the stronger one of the Flood Currents and the stronger one of the Ebb Currents for each day.
The letter "a" means the velocity shown should occur **after** the **a.m.** Current Change. The letter "p" means the velocity shown should occur **after** the **p.m.** Current Change (even if next morning). No "a" or "p" means a.m. and p.m. velocities are the same for that day.
Avg. Max. Velocity: Flood 4.0 Kts., Ebb 4.5 Kts.
Max. Flood 3 hrs. after Flood Starts, ±20 min.
Max. Ebb 3 hrs. after Ebb Starts, ±20 min.
Average rise and fall: canal east end, 8.7 ft. (time of high water same as Boston); west end, at Monument Beach, 4.0 ft. (time of high water 15 min. after Newport).

See pp. 22-29 for Current Change at other points.

2014 CURRENT TABLE
CAPE COD CANAL

41°44.5'N, 70°36.8'W at R.R. Bridge

Daylight Saving Time — **Daylight Saving Time**

JULY								AUGUST							
DAY OF MONTH	DAY OF WEEK	CURRENT TURNS TO						DAY OF MONTH	DAY OF WEEK	CURRENT TURNS TO					
		EAST Flood Starts			WEST Ebb Starts					EAST Flood Starts			WEST Ebb Starts		
		a.m.	**p.m.**	Kts.	a.m.	**p.m.**	Kts.			a.m.	**p.m.**	Kts.	a.m.	**p.m.**	Kts.
1	**T**	6 46	**7 04**	a4.2	12 38	**1 21**	a4.5	1	**F**	7 33	**7 53**	a4.1	1 29	**2 03**	a4.3
2	**W**	7 25	**7 46**	a4.1	1 18	**2 02**	a4.3	2	**S**	8 15	**8 39**	a4.0	2 12	**2 44**	a4.2
3	**T**	8 07	**8 31**	a4.0	2 01	**2 44**	a4.2	3	**S**	9 02	**9 30**	a3.8	3 01	**3 31**	4.0
4	**F**	8 53	**9 20**	a3.9	2 48	**3 30**	a4.1	4	**M**	9 55	**10 28**	3.7	3 57	**4 23**	p4.0
5	**S**	9 41	**10 12**	a3.8	3 40	**4 18**	a4.0	5	**T**	10 52	**11 28**	p3.8	5 00	**5 21**	p4.1
6	**S**	10 33	**11 08**	a3.7	4 37	**5 10**	3.9	6	**W**	11 54	**...**	3.8	6 06	**6 20**	p4.2
7	**M**	11 29	**...**	3.7	5 37	**6 03**	p4.1	7	**T**	12 30	**12 56**	a4.0	7 09	**7 18**	p4.5
8	**T**	12 05	**12 27**	3.8	6 38	**6 55**	p4.2	8	**F**	1 30	**1 54**	4.2	8 07	**8 13**	p4.7
9	**W**	1 02	**1 23**	4.0	7 35	**7 47**	p4.5	9	**S**	2 26	**2 50**	a4.5	9 01	**9 06**	p5.0
10	**T**	1 56	**2 17**	4.2	8 30	**8 37**	p4.7	10	**S**	3 19	**3 42**	a4.8	9 51	**9 58**	p5.2
11	**F**	2 48	**3 10**	a4.5	9 22	**9 26**	p4.9	11	**M**	4 10	**4 32**	a4.9	10 40	**10 48**	p5.3
12	**S**	3 39	**4 01**	a4.7	10 12	**10 15**	p5.1	12	**T**	4 59	**5 22**	a5.0	11 28	**11 39**	p5.3
13	**S**	4 28	**4 51**	a4.9	11 01	**11 04**	p5.2	13	**W**	5 48	**6 11**	a5.0	...	**12 16**	5.0
14	**M**	5 18	**5 41**	a5.0	11 51	**11 55**	p5.2	14	**T**	6 38	**7 02**	a4.9	12 30	**1 05**	a5.2
15	**T**	6 08	**6 32**	a5.0	...	**12 40**	4.9	15	**F**	7 28	**7 54**	a4.7	1 23	**1 55**	a5.0
16	**W**	6 58	**7 25**	a4.9	12 47	**1 31**	a5.1	16	**S**	8 21	**8 50**	a4.4	2 19	**2 48**	a4.7
17	**T**	7 51	**8 20**	a4.7	1 41	**2 24**	a5.0	17	**S**	9 17	**9 50**	4.1	3 19	**3 44**	4.4
18	**F**	8 46	**9 18**	a4.5	2 39	**3 19**	a4.8	18	**M**	10 18	**10 53**	3.9	4 22	**4 44**	p4.2
19	**S**	9 44	**10 19**	a4.3	3 41	**4 16**	a4.5	19	**T**	11 21	**11 57**	p3.8	5 28	**5 45**	p4.1
20	**S**	10 45	**11 23**	a4.1	4 46	**5 15**	4.3	20	**W**	...	**12 25**	3.6	6 32	**6 44**	p4.1
21	**M**	11 48	**...**	3.9	5 52	**6 14**	p4.3	21	**T**	12 58	**1 24**	a3.9	7 31	**7 39**	p4.2
22	**T**	12 26	**12 50**	a4.0	6 56	**7 11**	p4.3	22	**F**	1 53	**2 15**	a4.0	8 23	**8 29**	p4.3
23	**W**	1 25	**1 47**	a4.1	7 54	**8 04**	p4.3	23	**S**	2 41	**3 00**	a4.1	9 09	**9 14**	p4.4
24	**T**	2 18	**2 38**	a4.1	8 47	**8 52**	p4.4	24	**S**	3 24	**3 41**	a4.2	9 51	**9 55**	p4.5
25	**F**	3 06	**3 24**	a4.2	9 35	**9 37**	p4.5	25	**M**	4 03	**4 18**	a4.2	10 29	**10 34**	p4.6
26	**S**	3 49	**4 05**	a4.2	10 18	**10 19**	p4.5	26	**T**	4 39	**4 53**	a4.3	11 05	**11 11**	p4.6
27	**S**	4 29	**4 44**	a4.3	10 58	**10 58**	p4.6	27	**W**	5 14	**5 28**	a4.3	11 40	**11 47**	p4.6
28	**M**	5 06	**5 20**	a4.3	11 36	**11 36**	p4.6	28	**T**	5 49	**6 03**	a4.3	...	**12 14**	4.4
29	**T**	5 42	**5 57**	a4.3	...	**12 12**	4.3	29	**F**	6 24	**6 40**	4.2	12 24	**12 48**	a4.5
30	**W**	6 18	**6 33**	a4.2	12 13	**12 48**	a4.5	30	**S**	7 02	**7 20**	4.1	1 01	**1 24**	a4.4
31	**T**	6 54	**7 12**	a4.2	12 50	**1 25**	a4.4	31	**S**	7 43	**8 05**	4.0	1 43	**2 03**	4.2

The Kts. (knots) columns show the **maximum** predicted velocities of the stronger one of the Flood Currents and the stronger one of the Ebb Currents for each day.
The letter "a" means the velocity shown should occur **after** the **a.m.** Current Change. The letter "p" means the velocity shown should occur **after** the **p.m.** Current Change (even if next morning). No "a" or "p" means a.m. and p.m. velocities are the same for that day.
Avg. Max. Velocity: Flood 4.0 Kts., Ebb 4.5 Kts.
Max. Flood 3 hrs. after Flood Starts, ±20 min.
Max. Ebb 3 hrs. after Ebb Starts, ±20 min.
Average rise and fall: canal east end, 8.7 ft. (time of high water same as Boston); west end, at Monument Beach, 4.0 ft. (time of high water 15 min. after Newport).

See pp. 22-29 for Current Change at other points.

2014 CURRENT TABLE
CAPE COD CANAL

41°44.5'N, 70°36.8'W at R.R. Bridge

Daylight Saving Time

SEPTEMBER

DAY OF MONTH	DAY OF WEEK	CURRENT TURNS TO EAST Flood Starts a.m.	p.m.	Kts.	CURRENT TURNS TO WEST Ebb Starts a.m.	p.m.	Kts.
1	M	8 29	**8 56**	3.9	2 31	**2 49**	4.1
2	T	9 22	**9 54**	p3.8	3 28	**3 43**	p4.1
3	W	10 23	**10 58**	p3.9	4 33	**4 46**	p4.1
4	T	11 30	**...**	3.7	5 41	**5 52**	p4.3
5	F	12 04	**12 34**	a4.0	6 46	**6 55**	p4.5
6	S	1 07	**1 36**	a4.3	7 45	**7 54**	p4.7
7	S	2 05	**2 32**	a4.6	8 39	**8 49**	p5.0
8	M	3 00	**3 24**	a4.8	9 30	**9 42**	p5.1
9	T	3 51	**4 13**	4.9	10 18	**10 32**	p5.2
10	W	4 40	**5 02**	a5.0	11 04	**11 22**	p5.2
11	T	5 28	**5 50**	a5.0	11 50	**...**	5.1
12	F	6 15	**6 38**	4.8	12 12	**12 37**	a5.1
13	S	7 04	**7 28**	a4.6	1 04	**1 25**	a4.9
14	S	7 54	**8 21**	4.3	1 57	**2 16**	4.6
15	M	8 48	**9 18**	4.0	2 55	**3 10**	4.3
16	T	9 47	**10 19**	p3.8	3 56	**4 10**	p4.1
17	W	10 51	**11 23**	p3.7	5 00	**5 12**	p4.0
18	T	11 55	**...**	3.4	6 03	**6 14**	p4.0
19	F	12 25	**12 54**	a3.8	7 00	**7 10**	p4.0
20	S	1 21	**1 46**	a3.9	7 52	**8 01**	p4.2
21	S	2 10	**2 32**	a4.0	8 37	**8 47**	p4.3
22	M	2 53	**3 12**	a4.1	9 19	**9 29**	p4.4
23	T	3 32	**3 49**	4.2	9 57	**10 08**	p4.5
24	W	4 09	**4 25**	4.3	10 32	**10 46**	p4.6
25	T	4 44	**5 00**	4.3	11 07	**11 22**	p4.6
26	F	5 20	**5 35**	4.3	11 40	**11 59**	4.5
27	S	5 56	**6 12**	4.3	...	**12 14**	4.5
28	S	6 34	**6 53**	p4.3	12 38	**12 49**	4.4
29	M	7 16	**7 38**	p4.2	1 21	**1 30**	4.3
30	T	8 04	**8 30**	p4.1	2 10	**2 17**	p4.2

Daylight Saving Time

OCTOBER

DAY OF MONTH	DAY OF WEEK	CURRENT TURNS TO EAST Flood Starts a.m.	p.m.	Kts.	CURRENT TURNS TO WEST Ebb Starts a.m.	p.m.	Kts.
1	W	8 59	**9 29**	p4.0	3 08	**3 14**	p4.2
2	T	10 02	**10 34**	p4.0	4 13	**4 20**	p4.2
3	F	11 09	**11 41**	p4.1	5 21	**5 30**	p4.3
4	S	...	**12 17**	3.9	6 25	**6 37**	p4.5
5	S	12 46	**1 18**	a4.3	7 24	**7 38**	p4.7
6	M	1 45	**2 14**	a4.6	8 17	**8 34**	p4.9
7	T	2 40	**3 06**	4.7	9 07	**9 26**	p5.0
8	W	3 31	**3 55**	p4.9	9 54	**10 17**	p5.1
9	T	4 19	**4 42**	4.9	10 40	**11 06**	5.1
10	F	5 06	**5 28**	p4.9	11 25	**11 55**	a5.1
11	S	5 53	**6 15**	p4.7	...	**12 10**	5.0
12	S	6 39	**7 02**	p4.5	12 45	**12 56**	p4.8
13	M	7 27	**7 51**	p4.2	1 36	**1 44**	p4.5
14	T	8 18	**8 44**	p4.0	2 29	**2 36**	p4.3
15	W	9 14	**9 42**	p3.8	3 27	**3 33**	p4.1
16	T	10 14	**10 43**	p3.6	4 27	**4 35**	p3.9
17	F	11 17	**11 43**	p3.7	5 27	**5 37**	p3.9
18	S	...	**12 17**	3.4	6 23	**6 35**	p3.9
19	S	12 40	**1 10**	a3.8	7 14	**7 28**	p4.1
20	M	1 31	**1 57**	a3.9	8 00	**8 16**	p4.2
21	T	2 16	**2 39**	4.0	8 42	**9 00**	4.3
22	W	2 58	**3 18**	p4.2	9 21	**9 41**	p4.5
23	T	3 36	**3 55**	p4.3	9 58	**10 20**	4.5
24	F	4 14	**4 32**	p4.4	10 33	**10 59**	a4.6
25	S	4 51	**5 09**	p4.5	11 08	**11 38**	a4.7
26	S	5 30	**5 49**	p4.5	11 44	**...**	4.7
27	M	6 10	**6 31**	p4.4	12 19	**12 22**	p4.6
28	T	6 55	**7 18**	p4.4	1 04	**1 05**	p4.5
29	W	7 45	**8 11**	p4.3	1 55	**1 55**	p4.4
30	T	8 41	**9 09**	p4.2	2 52	**2 54**	p4.4
31	F	9 44	**10 14**	p4.1	3 55	**4 01**	p4.3

The Kts. (knots) columns show the **maximum** predicted velocities of the stronger one of the Flood Currents and the stronger one of the Ebb Currents for each day.
The letter "a" means the velocity shown should occur **after** the **a.m.** Current Change. The letter "p" means the velocity shown should occur **after** the **p.m.** Current Change (even if next morning). No "a" or "p" means a.m. and p.m. velocities are the same for that day.
Avg. Max. Velocity: Flood 4.0 Kts., Ebb 4.5 Kts.
Max. Flood 3 hrs. after Flood Starts, ±20 min.
Max. Ebb 3 hrs. after Ebb Starts, ±20 min.
Average rise and fall: canal east end, 8.7 ft. (time of high water same as Boston); west end, at Monument Beach, 4.0 ft. (time of high water 15 min. after Newport).

See pp. 22-29 for Current Change at other points.

2014 CURRENT TABLE
CAPE COD CANAL

41°44.5'N, 70°36.8'W at R.R. Bridge

***Standard Time starts Nov. 2 at 2 a.m.** **Standard Time**

NOVEMBER

Day of Month	Day of Week	CURRENT TURNS TO: EAST Flood Starts a.m.	p.m.	Kts.	WEST Ebb Starts a.m.	p.m.	Kts.
1	S	10 51	11 20	p4.2	5 00	5 11	p4.4
2	S	*10 58	*11 25	p4.3	*5 02	*5 18	p4.5
3	M	...	12 01	4.2	6 01	6 21	p4.6
4	T	12 26	12 58	4.5	6 54	7 18	p4.8
5	W	1 20	1 49	p4.7	7 45	8 12	a4.9
6	T	2 12	2 38	p4.8	8 32	9 03	a5.0
7	F	3 00	3 24	p4.8	9 17	9 52	a5.0
8	S	3 46	4 09	p4.7	10 01	10 39	a5.0
9	S	4 31	4 54	p4.6	10 45	11 26	a4.9
10	M	5 15	5 38	p4.4	11 29	...	4.7
11	T	6 00	6 24	p4.2	12 14	12 14	p4.5
12	W	6 47	7 11	p4.0	1 03	1 02	p4.3
13	T	7 38	8 03	p3.8	1 54	1 55	p4.1
14	F	8 33	8 58	p3.7	2 48	2 53	p3.9
15	S	9 32	9 55	p3.6	3 44	3 53	p3.8
16	S	10 31	10 52	p3.7	4 38	4 53	p3.9
17	M	11 27	11 45	p3.8	5 30	5 49	3.9
18	T	...	12 17	3.7	6 19	6 40	p4.1
19	W	12 35	1 03	3.9	7 03	7 27	4.2
20	T	1 20	1 45	p4.1	7 45	8 12	a4.4
21	F	2 03	2 26	p4.3	8 24	8 54	a4.6
22	S	2 44	3 06	p4.5	9 02	9 36	a4.7
23	S	3 25	3 47	p4.6	9 40	10 19	a4.8
24	M	4 07	4 29	p4.6	10 19	11 03	a4.8
25	T	4 51	5 14	p4.6	11 02	11 50	a4.8
26	W	5 38	6 02	p4.6	11 48	...	4.8
27	T	6 29	6 55	p4.5	12 40	12 40	p4.7
28	F	7 25	7 52	p4.4	1 35	1 38	p4.6
29	S	8 26	8 53	p4.3	2 34	2 43	p4.5
30	S	9 31	9 58	p4.2	3 36	3 52	p4.4

DECEMBER

Day of Month	Day of Week	CURRENT TURNS TO: EAST Flood Starts a.m.	p.m.	Kts.	WEST Ebb Starts a.m.	p.m.	Kts.
1	M	10 37	11 02	p4.3	4 37	5 00	4.4
2	T	11 40	...	4.2	5 36	6 04	4.5
3	W	12 04	12 38	p4.4	6 31	7 03	a4.6
4	T	1 02	1 33	p4.5	7 23	7 58	a4.7
5	F	1 54	2 22	p4.6	8 11	8 50	a4.8
6	S	2 43	3 08	p4.6	8 57	9 38	a4.9
7	S	3 28	3 52	p4.6	9 41	10 23	a4.9
8	M	4 11	4 34	p4.5	10 23	11 08	a4.8
9	T	4 53	5 16	p4.4	11 05	11 51	a4.7
10	W	5 35	5 57	p4.2	11 47	...	4.5
11	T	6 17	6 40	p4.1	12 34	12 31	p4.4
12	F	7 02	7 24	p3.9	1 19	1 18	p4.2
13	S	7 51	8 13	p3.8	2 06	2 09	p4.0
14	S	8 44	9 05	p3.7	2 56	3 05	p3.9
15	M	9 40	10 00	p3.6	3 48	4 04	3.8
16	T	10 37	10 56	p3.7	4 41	5 04	3.8
17	W	11 32	11 50	p3.8	5 32	6 00	a4.0
18	T	...	12 23	3.8	6 21	6 53	4.1
19	F	12 42	1 12	p4.1	7 07	7 42	a4.3
20	S	1 30	1 58	p4.3	7 51	8 29	a4.5
21	S	2 16	2 42	p4.5	8 33	9 15	a4.7
22	M	3 02	3 27	p4.7	9 16	10 00	a4.9
23	T	3 47	4 12	p4.8	10 00	10 46	a5.0
24	W	4 33	4 58	p4.8	10 45	11 33	a5.0
25	T	5 22	5 47	p4.8	11 33	...	5.0
26	F	6 12	6 39	p4.7	12 22	12 26	p4.9
27	S	7 07	7 33	p4.5	1 15	1 23	p4.8
28	S	8 05	8 32	p4.4	2 10	2 25	p4.6
29	M	9 08	9 35	p4.2	3 09	3 32	4.4
30	T	10 13	10 39	4.1	4 10	4 40	a4.4
31	W	11 18	11 43	a4.1	5 10	5 46	a4.4

The Kts. (knots) columns show the **maximum** predicted velocities of the stronger one of the Flood Currents and the stronger one of the Ebb Currents for each day.
The letter "a" means the velocity shown should occur **after** the **a.m.** Current Change. The letter "p" means the velocity shown should occur **after** the **p.m.** Current Change (even if next morning). No "a" or "p" means a.m. and p.m. velocities are the same for that day.
Avg. Max. Velocity: Flood 4.0 Kts., Ebb 4.5 Kts.
Max. Flood 3 hrs. after Flood Starts, ±20 min.
Max. Ebb 3 hrs. after Ebb Starts, ±20 min.
Average rise and fall: canal east end, 8.7 ft. (time of high water same as Boston); west end, at Monument Beach, 4.0 ft. (time of high water 15 min. after Newport).

See pp. 22-29 for Current Change at other points.

2014 CURRENT TABLE

WOODS HOLE, MA, The Strait

41°31.16'N, 70°40.97'W

JANUARY

Standard Time

Day of Month	Day of Week	SOUTHEAST Flood Starts a.m.	p.m.	Kts.	NORTHWEST Ebb Starts a.m.	p.m.	Kts.
1	W	3 28	4 11	a2.6	9 51	10 30	a3.9
2	T	4 22	5 03	2.6	10 44	11 23	a4.0
3	F	5 17	5 55	2.6	11 37	...	4.0
4	S	6 13	6 49	2.6	12 16	12 31	p3.9
5	S	7 09	7 41	p2.6	1 09	1 24	p3.7
6	M	8 08	8 38	p2.5	2 02	2 19	p3.5
7	T	9 11	9 36	p2.4	2 55	3 14	3.1
8	W	10 14	10 33	p2.3	3 49	4 10	a2.9
9	T	11 14	11 28	2.2	4 44	5 08	a2.6
10	F	...	12 11	2.2	5 39	6 06	a2.6
11	S	12 20	1 05	p2.2	6 34	7 03	a2.7
12	S	1 11	1 57	p2.3	7 28	7 57	a2.8
13	M	2 00	2 45	p2.4	8 18	8 47	a2.8
14	T	2 46	3 30	p2.4	9 04	9 34	a2.8
15	W	3 31	4 12	p2.3	9 48	10 18	a2.8
16	T	4 14	4 53	p2.2	10 30	11 01	a2.9
17	F	4 56	5 32	2.0	11 11	11 43	a3.0
18	S	5 38	6 10	2.1	11 51	...	3.1
19	S	6 19	6 47	2.1	12 23	12 31	p3.2
20	M	7 00	7 25	2.1	1 03	1 11	p3.2
21	T	7 44	8 06	p2.2	1 44	1 53	p3.2
22	W	8 32	8 51	p2.2	2 25	2 38	p3.2
23	T	9 26	9 41	p2.2	3 09	3 27	3.1
24	F	10 23	10 35	p2.3	3 56	4 22	a3.1
25	S	11 20	11 30	p2.3	4 48	5 22	a3.1
26	S	...	12 17	2.1	5 45	6 25	a3.1
27	M	12 26	1 13	a2.3	6 45	7 26	a3.3
28	T	1 23	2 09	a2.4	7 44	8 24	a3.5
29	W	2 19	3 04	2.4	8 41	9 19	a3.6
30	T	3 15	3 56	p2.6	9 35	10 13	a3.8
31	F	4 09	4 48	p2.7	10 28	11 05	a3.9

FEBRUARY

Standard Time

Day of Month	Day of Week	SOUTHEAST Flood Starts a.m.	p.m.	Kts.	NORTHWEST Ebb Starts a.m.	p.m.	Kts.
1	S	5 04	5 39	p2.8	11 21	11 57	a3.9
2	S	5 58	6 29	p2.7	...	12 13	3.8
3	M	6 52	7 19	p2.6	12 48	1 06	p3.6
4	T	7 49	8 12	p2.5	1 39	1 58	p3.4
5	W	8 48	9 07	2.3	2 30	2 51	a3.1
6	T	9 49	10 04	a2.2	3 22	3 46	a2.8
7	F	10 48	10 59	a2.1	4 15	4 42	a2.6
8	S	11 45	11 51	a2.1	5 09	5 38	a2.3
9	S	...	12 38	2.1	6 04	6 34	a2.5
10	M	12 43	1 29	p2.1	6 57	7 28	a2.5
11	T	1 32	2 16	p2.1	7 48	8 18	a2.6
12	W	2 18	3 00	p2.2	8 35	9 04	a2.6
13	T	3 02	3 41	p2.1	9 18	9 47	a2.7
14	F	3 45	4 19	p2.1	9 59	10 28	2.8
15	S	4 26	4 57	2.1	10 39	11 08	3.0
16	S	5 06	5 33	2.2	11 19	11 48	a3.2
17	M	5 46	6 09	2.3	11 59	...	3.4
18	T	6 26	6 46	2.3	12 27	12 40	p3.4
19	W	7 09	7 26	p2.3	1 07	1 23	p3.5
20	T	7 56	8 11	p2.3	1 49	2 10	3.4
21	F	8 50	9 05	p2.3	2 34	3 02	a3.4
22	S	9 51	10 05	p2.3	3 25	4 00	a3.4
23	S	10 54	11 07	p2.3	4 21	5 01	a3.3
24	M	11 54	...	2.1	5 21	6 04	a3.2
25	T	12 08	12 54	a2.3	6 24	7 06	a3.2
26	W	1 08	1 52	a2.3	7 26	8 05	a3.3
27	T	2 06	2 47	p2.5	8 24	9 01	a3.5
28	F	3 02	3 39	p2.8	9 19	9 53	a3.6

See the Woods Hole Current Diagram inset on pp. 66-77.

Mariners should exercise great caution when transiting Woods Hole Passage as velocities have been reported to exceed NOAA's predictions.

To hold longest fair current from Buzzards Bay headed East through Vineyard and Nantucket Sounds go through Woods Hole 2 1/2 hrs. after flood starts SE in Woods Hole. (Any earlier means adverse currents in the Sounds.)

2014 CURRENT TABLE
WOODS HOLE, MA, The Strait

41°31.16'N, 70°40.97'W

***Daylight Time starts Mar. 9 at 2 a.m.** **Daylight Saving Time**

MARCH								APRIL							
DAY OF MONTH	DAY OF WEEK	CURRENT TURNS TO						DAY OF MONTH	DAY OF WEEK	CURRENT TURNS TO					
		SOUTHEAST Flood Starts			NORTHWEST Ebb Starts					SOUTHEAST Flood Starts			NORTHWEST Ebb Starts		
		a.m.	**p.m.**	Kts.	a.m.	**p.m.**	Kts.			a.m.	**p.m.**	Kts.	a.m.	**p.m.**	Kts.
1	S	3 56	**4 29**	p2.8	10 12	**10 45**	a3.7	1	T	6 25	**6 42**	a2.8	12 12	**12 35**	a3.5
2	S	4 50	**5 18**	2.8	11 04	**11 35**	a3.7	2	W	7 15	**7 29**	a2.6	1 00	**1 25**	a3.4
3	M	5 42	**6 06**	2.7	11 55	**...**	3.6	3	T	8 05	**8 17**	a2.5	1 48	**2 15**	a3.2
4	T	6 35	**6 56**	a2.6	12 25	**12 46**	3.5	4	F	8 57	**9 09**	a2.3	2 36	**3 05**	a3.0
5	W	7 27	**7 44**	a2.5	1 14	**1 37**	a3.3	5	S	9 51	**10 03**	a2.1	3 25	**3 55**	a2.8
6	T	8 23	**8 37**	a2.3	2 04	**2 28**	a3.1	6	S	10 47	**10 58**	a2.0	4 14	**4 46**	a2.5
7	F	9 21	**9 34**	a2.2	2 54	**3 21**	a2.8	7	M	11 41	**11 52**	a1.8	5 04	**5 37**	a2.3
8	S	10 19	**10 29**	a2.0	3 45	**4 14**	a2.5	8	T	...	**12 23**	1.8	5 55	**6 25**	a2.1
9	S	...	***12 15**	1.9	*5 38	***6 09**	a2.3	9	W	12 40	**1 15**	1.8	6 47	**7 17**	a2.1
10	M	12 23	**1 07**	p1.9	6 31	**7 03**	a2.1	10	T	1 31	**2 03**	1.6	7 37	**8 08**	p2.2
11	T	1 14	**1 56**	p1.9	7 24	**7 56**	a2.3	11	F	2 17	**2 45**	p1.8	8 25	**8 54**	p2.5
12	W	2 02	**2 42**	p1.9	8 15	**8 45**	a2.4	12	S	3 01	**3 25**	p1.9	9 11	**9 36**	p2.8
13	T	2 49	**3 25**	p1.9	9 02	**9 31**	a2.4	13	S	3 43	**4 03**	p2.1	9 54	**10 17**	p3.0
14	F	3 33	**4 05**	p1.9	9 46	**10 14**	a2.6	14	M	4 24	**4 41**	p2.3	10 36	**10 57**	p3.3
15	S	4 14	**4 42**	2.0	10 27	**10 53**	p2.9	15	T	5 05	**5 19**	p2.5	11 19	**11 37**	p3.6
16	S	4 55	**5 19**	p2.2	11 07	**11 32**	3.1	16	W	5 47	**5 59**	p2.6	...	**12 03**	3.5
17	M	5 35	**5 55**	p2.4	11 48	**...**	3.3	17	T	6 30	**6 42**	p2.6	12 19	**12 50**	a3.8
18	T	6 15	**6 32**	p2.5	12 12	**12 29**	p3.5	18	F	7 15	**7 28**	2.5	1 04	**1 40**	a3.9
19	W	6 56	**7 11**	p2.5	12 51	**1 13**	3.6	19	S	8 05	**8 20**	2.4	1 53	**2 32**	a3.9
20	T	7 39	**7 53**	2.4	1 33	**1 59**	a3.7	20	S	9 01	**9 20**	2.3	2 45	**3 27**	a3.8
21	F	8 27	**8 41**	p2.4	2 18	**2 49**	a3.7	21	M	10 04	**10 27**	p2.3	3 41	**4 24**	a3.6
22	S	9 22	**9 39**	p2.3	3 06	**3 43**	a3.7	22	T	11 10	**11 34**	2.2	4 41	**5 24**	a3.4
23	S	10 25	**10 44**	p2.3	4 00	**4 41**	a3.5	23	W	...	**12 14**	2.2	5 43	**6 24**	a3.2
24	M	11 31	**11 49**	p2.2	4 59	**5 42**	a3.4	24	T	12 39	**1 13**	p2.3	6 45	**7 25**	a3.0
25	T	...	**12 34**	2.1	6 01	**6 44**	a3.2	25	F	1 40	**2 10**	p2.5	7 47	**8 23**	p3.1
26	W	12 53	**1 35**	2.2	7 05	**7 46**	a3.1	26	S	2 38	**3 04**	p2.6	8 47	**9 18**	p3.2
27	T	1 54	**2 32**	p2.4	8 07	**8 45**	a3.2	27	S	3 34	**3 55**	p2.8	9 42	**10 10**	p3.3
28	F	2 53	**3 27**	p2.6	9 06	**9 40**	3.3	28	M	4 26	**4 44**	2.8	10 34	**10 59**	p3.4
29	S	3 49	**4 18**	p2.8	10 01	**10 32**	3.4	29	T	5 17	**5 31**	a2.9	11 25	**11 48**	p3.4
30	S	4 42	**5 07**	p2.9	10 54	**11 23**	3.5	30	W	6 06	**6 18**	a2.8	...	**12 15**	3.1
31	M	5 34	**5 55**	a2.9	11 45	**...**	3.5								

See the Woods Hole Current Diagram inset on pp. 66-77.

Mariners should exercise great caution when transiting Woods Hole Passage as velocities have been reported to exceed NOAA's predictions.

CAUTION: Going *from* Buzzards Bay *into* Vineyard Sound, whether through Woods Hole, or Robinsons Hole or Quicks Hole, *Red* Buoys must be kept on the LEFT or PORT hand, *Green* Buoys kept on the RIGHT or STARBOARD hand. You are considered to be proceeding seaward and should thus follow the rules for LEAVING a harbor.

See pp. 22-29 for Current Change at other points.

2014 CURRENT TABLE

WOODS HOLE, MA, The Strait

41°31.16'N, 70°40.97'W

Daylight Saving Time

MAY

DAY OF MONTH	DAY OF WEEK	CURRENT TURNS TO SOUTHEAST Flood Starts a.m.	p.m.	Kts.	CURRENT TURNS TO NORTHWEST Ebb Starts a.m.	p.m.	Kts.
1	T	6 54	7 04	a2.6	12 35	1 04	a3.3
2	F	7 41	7 50	a2.5	1 22	1 52	a3.1
3	S	8 30	8 39	a2.3	2 08	2 40	a3.0
4	S	9 21	9 31	a2.1	2 55	3 27	a2.8
5	M	10 12	10 24	a1.9	3 42	4 15	a2.6
6	T	11 03	11 18	a1.7	4 29	5 03	a2.4
7	W	11 52	...	1.6	5 17	5 51	a2.3
8	T	12 08	12 37	p1.7	6 06	6 39	2.2
9	F	12 56	1 21	p1.8	6 55	7 27	p2.3
10	S	1 43	2 03	p1.9	7 45	8 14	p2.6
11	S	2 28	2 44	p2.0	8 34	8 58	p2.9
12	M	3 12	3 25	p2.2	9 21	9 41	p3.2
13	T	3 55	4 05	p2.4	10 07	10 24	p3.5
14	W	4 38	4 48	p2.5	10 53	11 07	p3.7
15	T	5 22	5 32	p2.6	11 41	11 53	p3.9
16	F	6 08	6 19	p2.6	...	12 31	3.6
17	S	6 57	7 09	p2.6	12 42	1 23	a4.0
18	S	7 48	8 04	p2.5	1 34	2 16	a4.0
19	M	8 44	9 05	a2.4	2 28	3 10	a3.9
20	T	9 45	10 11	2.3	3 24	4 06	a3.7
21	W	10 49	11 19	a2.3	4 23	5 04	a3.5
22	T	11 51	...	2.3	5 23	6 03	a3.2
23	F	12 23	12 50	p2.4	6 24	7 02	a3.0
24	S	1 24	1 45	p2.4	7 26	8 00	p2.9
25	S	2 21	2 39	2.5	8 25	8 55	p3.1
26	M	3 17	3 30	2.6	9 21	9 47	p3.2
27	T	4 09	4 19	a2.8	10 14	10 36	p3.2
28	W	4 58	5 06	a2.8	11 04	11 23	p3.2
29	T	5 46	5 52	a2.7	11 53	...	2.9
30	F	6 32	6 38	a2.6	12 09	12 40	a3.1
31	S	7 17	7 23	a2.4	12 55	1 27	a3.1

Daylight Saving Time

JUNE

DAY OF MONTH	DAY OF WEEK	CURRENT TURNS TO SOUTHEAST Flood Starts a.m.	p.m.	Kts.	CURRENT TURNS TO NORTHWEST Ebb Starts a.m.	p.m.	Kts.
1	S	8 01	8 09	a2.2	1 40	2 12	a3.0
2	M	8 47	8 57	a2.0	2 24	2 57	a2.8
3	T	9 33	9 47	a1.8	3 08	3 42	a2.7
4	W	10 22	10 40	a1.8	3 52	4 27	a2.6
5	T	11 07	11 30	a1.8	4 37	5 12	a2.5
6	F	11 53	...	1.8	5 24	5 58	a2.5
7	S	12 20	12 37	p1.9	6 13	6 44	p2.5
8	S	1 07	1 20	p2.0	7 04	7 32	p2.7
9	M	1 54	2 04	p2.2	7 57	8 20	p3.0
10	T	2 41	2 49	p2.3	8 50	9 08	p3.3
11	W	3 28	3 35	p2.4	9 41	9 55	p3.6
12	T	4 14	4 22	p2.5	10 31	10 43	p3.8
13	F	5 02	5 11	p2.6	11 22	11 32	p4.0
14	S	5 51	6 02	p2.7	...	12 13	3.6
15	S	6 41	6 55	2.6	12 24	1 06	a4.1
16	M	7 33	7 51	2.5	1 17	1 59	a4.1
17	T	8 27	8 50	a2.5	2 12	2 52	a3.9
18	W	9 25	9 55	a2.4	3 07	3 47	a3.7
19	T	10 26	11 01	a2.4	4 04	4 43	a3.5
20	F	11 27	...	2.3	5 03	5 40	a3.1
21	S	12 05	12 24	p2.3	6 02	6 37	2.8
22	S	1 05	1 20	2.3	7 03	7 34	p2.8
23	M	2 03	2 13	2.4	8 02	8 30	p3.0
24	T	2 58	3 05	a2.5	8 59	9 22	p3.0
25	W	3 49	3 54	a2.6	9 52	10 11	p3.1
26	T	4 38	4 41	a2.7	10 41	10 58	p3.0
27	F	5 24	5 27	a2.6	11 29	11 43	p3.0
28	S	6 08	6 11	a2.5	...	12 14	2.8
29	S	6 50	6 55	a2.3	12 27	12 59	a3.0
30	M	7 31	7 38	a2.1	1 10	1 42	a3.0

See the Woods Hole Current Diagram inset on pp. 66-77.

Mariners should exercise great caution when transiting Woods Hole Passage as velocities have been reported to exceed NOAA's predictions.

To hold longest fair current from Buzzards Bay headed East through Vineyard and Nantucket Sounds go through Woods Hole 2 1/2 hrs. after flood starts SE in Woods Hole. (Any earlier means adverse currents in the Sounds.)

2014 CURRENT TABLE
WOODS HOLE, MA, The Strait

41°31.16'N, 70°40.97'W

JULY

Daylight Saving Time

DAY OF MONTH	DAY OF WEEK	CURRENT TURNS TO SOUTHEAST Flood Starts			CURRENT TURNS TO NORTHWEST Ebb Starts		
		a.m.	**p.m.**	Kts.	a.m.	**p.m.**	Kts.
1	T	8 11	**8 22**	a1.9	1 52	**2 25**	a3.0
2	W	8 53	**9 09**	a1.9	2 34	**3 07**	a2.9
3	T	9 36	**9 58**	a1.9	3 15	**3 49**	a2.9
4	F	10 21	**10 50**	a2.0	3 58	**4 31**	a2.8
5	S	11 06	**11 41**	a2.0	4 44	**5 15**	2.7
6	S	11 53	**...**	2.1	5 33	**6 02**	2.7
7	M	12 31	**12 40**	p2.1	6 27	**6 52**	p2.8
8	T	1 22	**1 29**	p2.2	7 24	**7 45**	p3.1
9	W	2 12	**2 19**	p2.3	8 22	**8 39**	p3.3
10	T	3 03	**3 10**	p2.4	9 17	**9 31**	p3.6
11	F	3 54	**4 02**	p2.5	10 11	**10 24**	p3.8
12	S	4 44	**4 54**	p2.6	11 03	**11 16**	p4.0
13	S	5 35	**5 47**	p2.7	11 55	**...**	3.6
14	M	6 26	**6 42**	p2.7	12 08	**12 48**	a4.1
15	T	7 17	**7 37**	2.6	1 02	**1 40**	a4.0
16	W	8 09	**8 35**	a2.5	1 55	**2 32**	a3.9
17	T	9 03	**9 37**	a2.5	2 49	**3 26**	a3.7
18	F	10 01	**10 41**	a2.4	3 45	**4 20**	a3.4
19	S	11 01	**11 44**	a2.3	4 42	**5 15**	a3.1
20	S	11 59	**...**	2.2	5 40	**6 11**	a2.7
21	M	12 44	**12 55**	2.2	6 39	**7 08**	p2.6
22	T	1 41	**1 48**	a2.3	7 38	**8 04**	p2.8
23	W	2 36	**2 40**	a2.4	8 35	**8 57**	p2.9
24	T	3 26	**3 29**	a2.5	9 28	**9 46**	p2.9
25	F	4 13	**4 16**	a2.5	10 16	**10 32**	p2.9
26	S	4 58	**5 00**	a2.5	11 02	**11 16**	p2.9
27	S	5 39	**5 43**	a2.4	11 46	**11 58**	p3.0
28	M	6 19	**6 25**	a2.2	...	**12 28**	2.8
29	T	6 57	**7 06**	a2.1	12 39	**1 09**	a3.0
30	W	7 34	**7 48**	a2.1	1 19	**1 50**	a3.1
31	T	8 12	**8 31**	a2.1	1 59	**2 29**	a3.1

AUGUST

Daylight Saving Time

DAY OF MONTH	DAY OF WEEK	CURRENT TURNS TO SOUTHEAST Flood Starts			CURRENT TURNS TO NORTHWEST Ebb Starts		
		a.m.	**p.m.**	Kts.	a.m.	**p.m.**	Kts.
1	F	8 51	**9 17**	a2.1	2 40	**3 09**	a3.1
2	S	9 34	**10 07**	a2.1	3 22	**3 51**	a3.1
3	S	10 21	**11 01**	a2.1	4 08	**4 35**	3.0
4	M	11 13	**11 57**	a2.2	5 00	**5 23**	p3.0
5	T	...	**12 06**	2.2	5 56	**6 17**	p3.0
6	W	12 51	**1 00**	p2.2	6 56	**7 15**	p3.1
7	T	1 47	**1 56**	p2.3	7 57	**8 14**	p3.3
8	F	2 41	**2 51**	p2.4	8 56	**9 12**	p3.5
9	S	3 35	**3 46**	p2.5	9 51	**10 07**	p3.7
10	S	4 27	**4 40**	p2.6	10 44	**11 00**	p3.9
11	M	5 18	**5 34**	p2.7	11 36	**11 53**	p4.0
12	T	6 08	**6 28**	2.7	...	**12 28**	3.7
13	W	6 58	**7 22**	a2.7	12 45	**1 20**	a3.9
14	T	7 48	**8 18**	a2.6	1 38	**2 11**	a3.8
15	F	8 40	**9 17**	a2.5	2 31	**3 03**	a3.6
16	S	9 36	**10 19**	2.3	3 25	**3 55**	a3.3
17	S	10 35	**11 21**	2.2	4 20	**4 49**	a2.9
18	M	11 33	**...**	2.1	5 16	**5 44**	a2.6
19	T	12 20	**12 29**	a2.2	6 14	**6 40**	p2.5
20	W	1 17	**1 23**	a2.2	7 11	**7 35**	p2.6
21	T	2 09	**2 14**	a2.2	8 07	**8 29**	p2.7
22	F	2 59	**3 03**	a2.3	9 00	**9 18**	p2.7
23	S	3 44	**3 48**	a2.3	9 47	**10 04**	p2.7
24	S	4 26	**4 32**	a2.3	10 32	**10 46**	p2.8
25	M	5 06	**5 13**	a2.2	11 14	**11 27**	p2.9
26	T	5 44	**5 54**	2.1	11 54	**...**	2.9
27	W	6 20	**6 34**	p2.2	12 07	**12 34**	a3.1
28	T	6 56	**7 14**	2.2	12 47	**1 13**	3.2
29	F	7 32	**7 55**	2.2	1 27	**1 51**	3.3
30	S	8 10	**8 38**	a2.2	2 08	**2 31**	3.3
31	S	8 52	**9 28**	a2.2	2 52	**3 13**	3.3

See the Woods Hole Current Diagram inset on pp. 66-77.

Mariners should exercise great caution when transiting Woods Hole Passage as velocities have been reported to exceed NOAA's predictions.

CAUTION: Going *from* Buzzards Bay *into* Vineyard Sound, whether through Woods Hole, or Robinsons Hole or Quicks Hole, *Red* Buoys must be kept on the LEFT or PORT hand, *Green* Buoys kept on the RIGHT or STARBOARD hand. You are considered to be proceeding seaward and should thus follow the rules for LEAVING a harbor.

See pp. 22-29 for Current Change at other points.

2014 CURRENT TABLE
WOODS HOLE, MA, The Strait
41°31.16'N, 70°40.97'W

Daylight Saving Time

SEPTEMBER

Day of Month	Day of Week	Southeast Flood Starts a.m.	p.m.	Kts.	Northwest Ebb Starts a.m.	p.m.	Kts.
1	M	9 41	10 25	a2.2	3 40	3 59	p3.3
2	T	10 38	11 25	a2.2	4 34	4 52	p3.2
3	W	11 38	...	2.2	5 32	5 50	p3.1
4	T	12 26	12 39	p2.2	6 33	6 51	p3.1
5	F	1 23	1 38	p2.3	7 35	7 54	p3.3
6	S	2 20	2 36	p2.4	8 35	8 54	p3.4
7	S	3 15	3 32	p2.5	9 31	9 50	p3.6
8	M	4 08	4 27	2.6	10 24	10 43	p3.7
9	T	4 59	5 21	p2.8	11 16	11 36	p3.8
10	W	5 48	6 14	2.8	...	12 07	3.7
11	T	6 38	7 07	2.7	12 28	12 58	a3.8
12	F	7 27	8 00	p2.6	1 20	1 48	a3.6
13	S	8 17	8 55	2.4	2 12	2 38	a3.4
14	S	9 11	9 54	p2.3	3 04	3 29	a3.1
15	M	10 08	10 55	p2.2	3 58	4 22	a2.8
16	T	11 07	11 53	p2.1	4 52	5 15	a2.5
17	W	...	12 03	1.8	5 47	6 10	2.2
18	T	12 30	12 56	p1.8	6 42	7 04	p2.4
19	F	1 38	1 46	a2.0	7 36	7 57	p2.4
20	S	2 26	2 34	a2.1	8 28	8 46	p2.5
21	S	3 10	3 19	a2.1	9 15	9 32	p2.5
22	M	3 50	4 02	a2.0	9 58	10 14	p2.7
23	T	4 29	4 43	2.0	10 39	10 55	p2.9
24	W	5 06	5 23	p2.2	11 18	11 35	p3.1
25	T	5 42	6 02	p2.3	11 57	...	3.2
26	F	6 18	6 42	2.3	12 15	12 35	p3.4
27	S	6 56	7 23	a2.4	12 57	1 15	p3.5
28	S	7 35	8 07	a2.4	1 41	1 56	p3.6
29	M	8 19	8 56	a2.3	2 28	2 41	p3.6
30	T	9 11	9 54	a2.3	3 18	3 32	p3.5

Daylight Saving Time

OCTOBER

Day of Month	Day of Week	Southeast Flood Starts a.m.	p.m.	Kts.	Northwest Ebb Starts a.m.	p.m.	Kts.
1	W	10 12	10 58	a2.3	4 13	4 27	p3.4
2	T	11 17	...	2.2	5 11	5 28	p3.3
3	F	12 01	12 21	p2.2	6 12	6 31	p3.2
4	S	1 02	1 23	p2.3	7 13	7 34	p3.2
5	S	1 59	2 21	p2.4	8 13	8 35	p3.3
6	M	2 54	3 19	p2.5	9 10	9 32	p3.4
7	T	3 47	4 13	p2.7	10 04	10 26	3.5
8	W	4 37	5 07	2.8	10 55	11 19	a3.6
9	T	5 27	5 59	p2.8	11 45	...	3.6
10	F	6 16	6 50	p2.7	12 10	12 35	p3.6
11	S	7 04	7 41	p2.6	1 02	1 24	3.4
12	S	7 54	8 33	p2.5	1 52	2 13	3.2
13	M	8 46	9 28	p2.3	2 43	3 03	2.9
14	T	9 41	10 25	p2.2	3 34	3 53	2.7
15	W	10 38	11 21	p2.0	4 26	4 45	2.4
16	T	11 34	...	1.7	5 18	5 37	2.2
17	F	12 13	12 26	a1.9	6 10	6 29	a2.1
18	S	1 02	1 16	a1.9	7 01	7 19	a2.1
19	S	1 47	2 03	a1.8	7 51	8 07	a2.2
20	M	2 30	2 48	a1.8	8 38	8 57	2.4
21	T	3 10	3 31	p1.9	9 22	9 40	p2.7
22	W	3 49	4 12	2.0	10 02	10 22	2.9
23	T	4 27	4 52	2.2	10 42	11 04	3.1
24	F	5 05	5 32	2.3	11 21	11 47	a3.4
25	S	5 43	6 13	2.4	...	12 01	3.6
26	S	6 23	6 56	a2.5	12 32	12 43	p3.8
27	M	7 07	7 42	a2.5	1 18	1 28	p3.8
28	T	7 54	8 33	a2.4	2 08	2 17	p3.8
29	W	8 49	9 31	2.3	3 00	3 11	p3.7
30	T	9 52	10 34	a2.3	3 55	4 08	p3.6
31	F	10 59	11 38	a2.3	4 52	5 09	p3.4

See the Woods Hole Current Diagram inset on pp. 66-77.

Mariners should exercise great caution when transiting Woods Hole Passage as velocities have been reported to exceed NOAA's predictions.

To hold longest fair current from Buzzards Bay headed East through Vineyard and Nantucket Sounds go through Woods Hole 2 1/2 hrs. after flood starts SE in Woods Hole. (Any earlier means adverse currents in the Sounds.)

2014 CURRENT TABLE
WOODS HOLE, MA, The Strait

41°31.16'N, 70°40.97'W

***Standard Time starts Nov. 2 at 2 a.m.** **Standard Time**

NOVEMBER

DAY OF MONTH	DAY OF WEEK	CURRENT TURNS TO SOUTHEAST Flood Starts a.m.	p.m.	Kts.	CURRENT TURNS TO NORTHWEST Ebb Starts a.m.	p.m.	Kts.
1	S	...	12 05	2.3	5 51	6 11	p3.2
2	S	12 38	*12 07	p2.3	*5 51	*6 14	P3.1
3	M	12 36	1 07	2.3	6 51	7 15	3.1
4	T	1 32	2 05	2.5	7 48	8 13	3.2
5	W	2 24	2 59	p2.7	8 42	9 08	a3.4
6	T	3 15	3 52	p2.9	9 33	10 00	a3.4
7	F	4 05	4 43	p2.9	10 23	10 51	a3.5
8	S	4 54	5 33	p2.8	11 12	11 42	a3.4
9	S	5 42	6 21	p2.6	...	12 01	3.3
10	M	6 31	7 10	p2.5	12 31	12 48	p3.1
11	T	7 20	8 00	p2.3	1 20	1 36	p2.9
12	W	8 12	8 52	p2.1	2 09	2 24	p2.7
13	T	9 06	9 45	p1.9	2 58	3 12	2.5
14	F	10 00	10 34	p1.8	3 46	4 01	2.3
15	S	10 52	11 21	p1.7	4 35	4 50	2.2
16	S	11 42	...	1.6	5 23	5 40	2.2
17	M	12 05	12 29	a1.7	6 11	6 30	2.2
18	T	12 47	1 14	a1.8	6 58	7 18	2.4
19	W	1 29	1 58	1.9	7 42	8 05	a2.7
20	T	2 09	2 41	a2.1	8 25	8 51	2.9
21	F	2 50	3 23	2.2	9 06	9 36	a3.2
22	S	3 30	4 05	a2.4	9 48	10 22	a3.5
23	S	4 13	4 49	a2.5	10 31	11 09	a3.7
24	M	4 57	5 34	a2.6	11 17	11 58	a3.9
25	T	5 45	6 22	a2.6	...	12 07	4.0
26	W	6 36	7 14	a2.5	12 49	12 59	p4.0
27	T	7 32	8 10	2.4	1 42	1 53	p3.8
28	F	8 35	9 12	2.3	2 36	2 50	p3.7
29	S	9 42	10 14	2.3	3 32	3 50	p3.4
30	S	10 48	11 14	p2.3	4 30	4 51	p3.2

DECEMBER

DAY OF MONTH	DAY OF WEEK	CURRENT TURNS TO SOUTHEAST Flood Starts a.m.	p.m.	Kts.	CURRENT TURNS TO NORTHWEST Ebb Starts a.m.	p.m.	Kts.
1	M	11 51	...	2.2	5 29	5 53	3.0
2	T	12 12	12 51	2.3	6 28	6 54	a3.0
3	W	1 07	1 48	p2.5	7 25	7 53	a3.1
4	T	2 02	2 44	p2.7	8 20	8 49	a3.2
5	F	2 53	3 36	p2.8	9 11	9 41	a3.3
6	S	3 43	4 25	p2.9	10 00	10 31	a3.3
7	S	4 32	5 13	p2.8	10 49	11 20	a3.2
8	M	5 19	6 00	p2.6	11 36	...	3.2
9	T	6 06	6 45	p2.4	12 08	12 22	p3.1
10	W	6 53	7 30	p2.2	12 55	1 08	p2.9
11	T	7 41	8 16	p2.0	1 41	1 53	p2.8
12	F	8 31	9 04	p1.8	2 26	2 38	p2.7
13	S	9 23	9 51	p1.8	3 11	3 23	2.5
14	S	10 15	10 37	p1.8	3 57	4 10	2.4
15	M	11 05	11 21	p1.8	4 42	4 58	2.3
16	T	11 53	...	1.7	5 28	5 48	a2.4
17	W	12 04	12 39	a1.9	6 15	6 39	a2.5
18	T	12 48	1 25	a2.0	7 02	7 31	a2.7
19	F	1 32	2 11	a2.1	7 48	8 21	a3.0
20	S	2 16	2 56	a2.3	8 34	9 10	a3.3
21	S	3 02	3 41	a2.4	9 21	9 59	a3.6
22	M	3 48	4 28	2.5	10 08	10 48	a3.8
23	T	4 37	5 16	a2.6	10 58	11 39	a4.0
24	W	5 28	6 05	a2.7	11 49	...	4.1
25	T	6 21	6 56	a2.6	12 31	12 42	p4.1
26	F	7 17	7 50	2.5	1 23	1 37	p3.9
27	S	8 19	8 49	2.4	2 16	2 33	p3.7
28	S	9 25	9 50	2.3	3 11	3 31	p3.4
29	M	10 30	10 50	p2.3	4 08	4 30	3.1
30	T	11 33	11 48	p2.3	5 05	5 31	a2.9
31	W	...	12 34	2.3	6 04	6 33	a2.8

See the Woods Hole Current Diagram inset on pp. 66-77.

Mariners should exercise great caution when transiting Woods Hole Passage as velocities have been reported to exceed NOAA's predictions.

CAUTION: Going *from* Buzzards Bay *into* Vineyard Sound, whether through Woods Hole, or Robinsons Hole or Quicks Hole, *Red* Buoys must be kept on the LEFT or PORT hand, *Green* Buoys kept on the RIGHT or STARBOARD hand. You are considered to be proceeding seaward and should thus follow the rules for LEAVING a harbor.

See pp. 22-29 for Current Change at other points.

My dear Captain and Mr. Mate,

As I cannot talk with you, I will do the next best thing. I will write you a letter.

Do you know, Captain and Mr. Mate of a place on the Atlantic Coast that is called "The Graveyard"? I propose to tell you something about it, and do what I can to keep vessels out of it. "The Graveyard" so called, is that part of the coast which lies between Sow and Pigs Rocks and Naushon Island. This place has been called "The Graveyard" for many years, because many a good craft has laid her bones there, and many a captain has lost his reputation there also. If a vessel gets into this graveyard, there must be a cause for it. Did it ever occur to you that seldom does a vessel go ashore on Gay Head, or on the south side of the Sound? but that hundreds of them have been piled up in "The Graveyard", or on the north side of the Sound? I will explain why this is so: if you are bound into Vineyard Sound in thick weather, you will probably refer to the "Gay Head and Cross Rip" table in this book, to see when the tide turns in or out. You will notice at the head of each table that it says, "This table shows the time that the current turns Easterly and Westerly, off Gay Head in ship channel." That means off Gay Head when it bears about South. Now, as a rule, captains figure on the currents after they leave the Lightship, as running Easterly into the Sound, when as a matter of fact the first of the flood between the Lightship and Gay Head runs nearly North; and the current does not begin to run to the eastward until you are well into the Sound, as shewn by the chart on the opposite page. Vessels bound into Vineyard Sound from the Westward will have the current of ebb on the starboard bow. (see arrows on the hulls in the chart on the opposite page)

I have explained this matter, and I leave the rest to your judgment and careful consideration; and thus you will undoubtedly keep your vessel out of "The Graveyard".

Yours for a fair tide,

Geo. W. Eldridge.

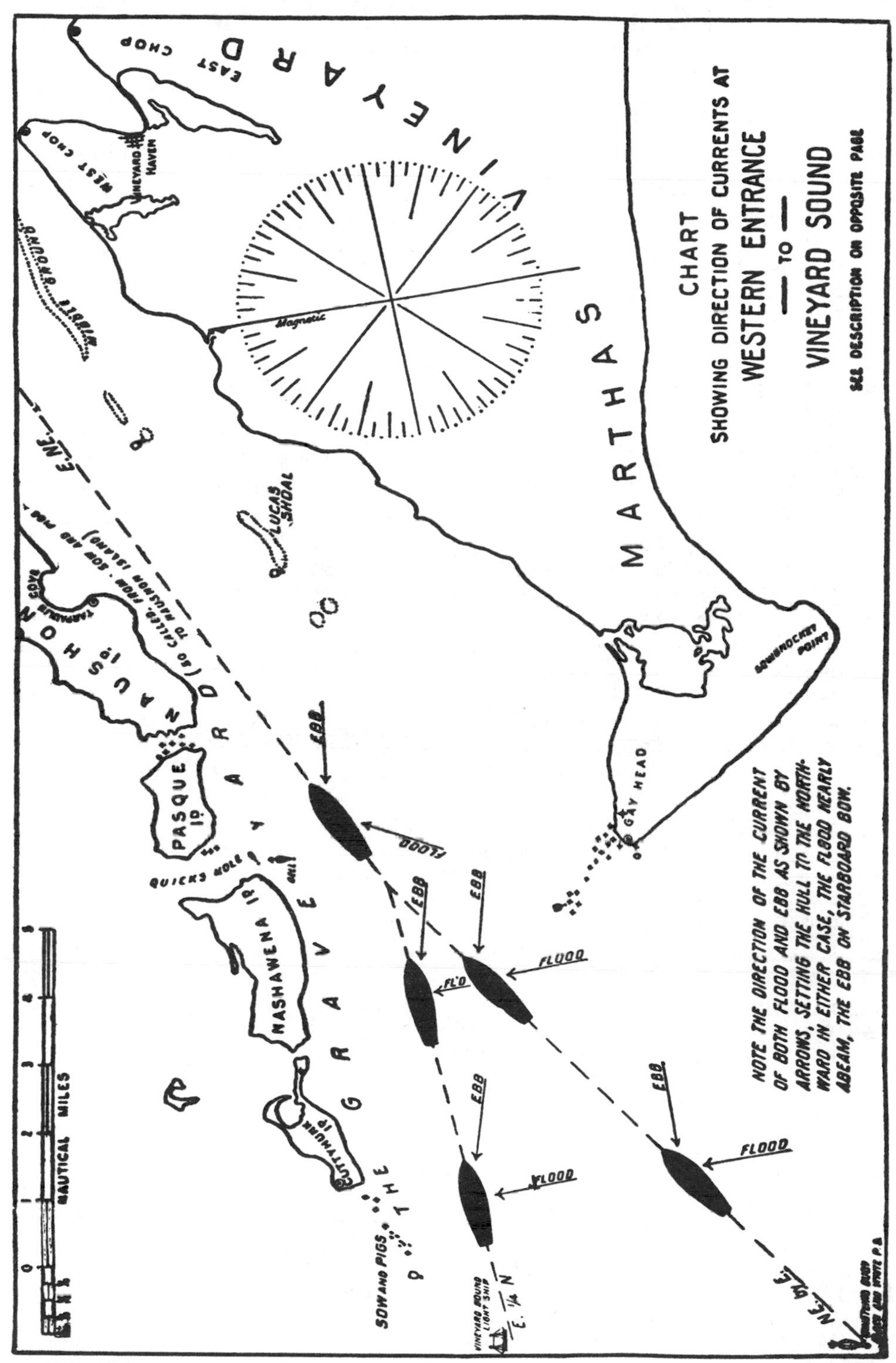

This lightship, shown on Capt. Eldridge's chart, on the western edge, was replaced many years ago by a buoy.

2014 CURRENT TABLE
POLLOCK RIP CHANNEL, MA

41°33'N, 69°59'W SE of Monomoy Pt. at Butler Hole

Standard Time — **Standard Time**

JANUARY

DAY OF MONTH	DAY OF WEEK	CURRENT TURNS TO NORTHEAST Flood Starts a.m.	p.m.	Kts.	CURRENT TURNS TO SOUTHWEST Ebb Starts a.m.	p.m.	Kts.
1	**W**	2 17	**2 36**	p2.4	8 40	**9 21**	a2.1
2	**T**	3 09	**3 26**	p2.5	9 29	**10 11**	a2.2
3	**F**	4 00	**4 17**	p2.5	10 20	**11 02**	a2.2
4	**S**	4 53	**5 11**	p2.4	11 13	**11 54**	a2.2
5	**S**	5 45	**6 04**	p2.4	...	**12 08**	2.1
6	**M**	6 41	**7 01**	p2.2	12 47	**1 06**	a2.0
7	**T**	7 39	**8 01**	p2.1	1 43	**2 07**	a1.9
8	**W**	8 40	**9 04**	2.0	2 41	**3 11**	a1.8
9	**T**	9 42	**10 09**	a2.0	3 41	**4 15**	a1.7
10	**F**	10 44	**11 13**	a2.0	4 40	**5 19**	a1.7
11	**S**	11 43	**...**	2.1	5 38	**6 19**	a1.7
12	**S**	12 13	**12 38**	p2.2	6 33	**7 13**	a1.7
13	**M**	1 07	**1 27**	p2.2	7 23	**8 02**	a1.7
14	**T**	1 55	**2 11**	p2.3	8 09	**8 46**	a1.8
15	**W**	2 38	**2 50**	p2.3	8 51	**9 27**	a1.8
16	**T**	3 17	**3 27**	p2.2	9 31	**10 05**	a1.8
17	**F**	3 53	**4 02**	p2.2	10 09	**10 41**	1.8
18	**S**	4 29	**4 38**	p2.2	10 47	**11 18**	a1.9
19	**S**	5 04	**5 14**	p2.2	11 26	**11 56**	a1.9
20	**M**	5 42	**5 53**	p2.1	...	**12 07**	1.9
21	**T**	6 22	**6 34**	p2.1	12 36	**12 51**	a1.9
22	**W**	7 04	**7 20**	p2.0	1 19	**1 38**	1.8
23	**T**	7 51	**8 10**	1.8	2 05	**2 30**	a1.8
24	**F**	8 42	**9 05**	a1.8	2 54	**3 26**	a1.7
25	**S**	9 38	**10 04**	a1.8	3 48	**4 25**	a1.7
26	**S**	10 36	**11 06**	a1.8	4 44	**5 25**	a1.7
27	**M**	11 35	**...**	1.9	5 41	**6 24**	a1.7
28	**T**	12 08	**12 33**	p2.1	6 37	**7 21**	a1.8
29	**W**	1 07	**1 29**	p2.2	7 31	**8 14**	a1.9
30	**T**	2 02	**2 22**	p2.4	8 24	**9 05**	2.0
31	**F**	2 55	**3 14**	p2.5	9 15	**9 55**	a2.2

FEBRUARY

DAY OF MONTH	DAY OF WEEK	CURRENT TURNS TO NORTHEAST Flood Starts a.m.	p.m.	Kts.	CURRENT TURNS TO SOUTHWEST Ebb Starts a.m.	p.m.	Kts.
1	**S**	3 45	**4 05**	p2.5	10 06	**10 44**	a2.2
2	**S**	4 35	**4 55**	p2.5	10 58	**11 33**	a2.2
3	**M**	5 25	**5 47**	p2.4	11 50	**...**	2.1
4	**T**	6 18	**6 41**	2.2	12 23	**12 45**	2.0
5	**W**	7 11	**7 36**	a2.1	1 16	**1 43**	a1.9
6	**T**	8 08	**8 36**	a2.0	2 11	**2 43**	a1.8
7	**F**	9 08	**9 39**	a1.9	3 09	**3 46**	a1.6
8	**S**	10 10	**10 43**	a1.9	4 08	**4 49**	a1.6
9	**S**	11 11	**11 44**	a2.0	5 07	**5 49**	a1.6
10	**M**	...	**12 07**	2.1	6 03	**6 45**	a1.6
11	**T**	12 39	**12 58**	p2.2	6 55	**7 34**	a1.7
12	**W**	1 28	**1 43**	p2.2	7 43	**8 18**	1.7
13	**T**	2 11	**2 24**	p2.2	8 26	**8 58**	a1.8
14	**F**	2 50	**3 01**	p2.2	9 05	**9 35**	1.8
15	**S**	3 25	**3 36**	p2.2	9 43	**10 11**	1.9
16	**S**	3 59	**4 10**	p2.2	10 20	**10 46**	1.9
17	**M**	4 33	**4 46**	p2.2	10 58	**11 23**	a2.0
18	**T**	5 08	**5 23**	p2.2	11 37	**...**	2.0
19	**W**	5 46	**6 03**	p2.1	12 01	**12 19**	a2.0
20	**T**	6 27	**6 48**	2.0	12 42	**1 05**	a1.9
21	**F**	7 13	**7 37**	a1.9	1 27	**1 57**	a1.9
22	**S**	8 05	**8 34**	a1.9	2 18	**2 54**	a1.8
23	**S**	9 03	**9 37**	a1.8	3 14	**3 56**	a1.7
24	**M**	10 07	**10 44**	a1.8	4 15	**5 00**	a1.6
25	**T**	11 12	**11 50**	a1.9	5 17	**6 03**	a1.6
26	**W**	...	**12 15**	2.1	6 18	**7 02**	1.7
27	**T**	12 52	**1 14**	p2.2	7 16	**7 56**	a1.9
28	**F**	1 48	**2 09**	p2.4	8 10	**8 47**	2.0

The Kts. (knots) columns show the **maximum** predicted velocities of the stronger one of the Flood Currents and the stronger one of the Ebb Currents for each day.
The letter "a" means the velocity shown should occur **after** the **a.m.** Current Change. The letter "p" means the velocity shown should occur **after** the **p.m.** Current Change (even if next morning). No "a" or "p" means a.m. and p.m. velocities are the same for that day.
Avg. Max. Velocity: Flood 2.0 Kts., Ebb 1.8 Kts.
Max. Flood 3 hrs. 20 min. after Flood Starts, ±15 min.
Max. Ebb 2 hrs. 45 min. after Ebb Starts, ±15 min.
Gay Head (1 1/2 mi. NW of): avg. max velocity, Flood 2.0 kts., Ebb 2.0 kts. Time of Flood and Ebb 1 hr. 35 min. after Pollock Rip. Cross Rip: avg. max. velocity, Flood 1.3 kts., Ebb 0.9 kts. Time of Flood and Ebb 1 hr. 50 min. after Pollock Rip. Use POLLOCK RIP tables with current charts on pp. 66-77. See pp. 22-29 for Current Change at other points.

2014 CURRENT TABLE
POLLOCK RIP CHANNEL, MA

41°33'N, 69°59'W SE of Monomoy Pt. at Butler Hole

***Daylight Time starts March 9 at 2 a.m.** **Daylight Saving Time**

MARCH

DAY OF MONTH	DAY OF WEEK	CURRENT TURNS TO NORTHEAST Flood Starts a.m.	p.m.	Kts.	CURRENT TURNS TO SOUTHWEST Ebb Starts a.m.	p.m.	Kts.
1	S	2 40	3 01	p2.4	9 02	9 36	2.1
2	S	3 29	3 50	p2.5	9 52	10 23	a2.2
3	M	4 16	4 39	p2.4	10 42	11 10	a2.2
4	T	5 04	5 28	2.3	11 32	11 58	a2.1
5	W	5 52	6 17	a2.2	...	12 24	1.9
6	T	6 42	7 10	a2.1	12 47	1 18	a1.9
7	F	7 35	8 06	a2.0	1 40	2 15	a1.7
8	S	8 32	9 07	a1.9	2 36	3 15	a1.6
9	S	*10 32	*11 09	a1.9	*4 34	*5 15	a1.5
10	M	11 33	...	1.9	5 33	6 14	a1.5
11	T	12 09	12 30	p2.0	6 31	7 09	1.5
12	W	1 05	1 23	p2.1	7 24	7 59	1.6
13	T	1 54	2 09	p2.1	8 12	8 43	1.7
14	F	2 38	2 52	p2.2	8 56	9 24	1.8
15	S	3 17	3 30	p2.2	9 37	10 02	1.8
16	S	3 53	4 06	p2.2	10 15	10 38	1.9
17	M	4 27	4 42	p2.2	10 53	11 13	2.0
18	T	5 01	5 17	p2.2	11 30	11 50	2.0
19	W	5 36	5 55	a2.2	...	12 10	2.0
20	T	6 13	6 36	a2.2	12 28	12 52	2.0
21	F	6 55	7 22	a2.1	1 10	1 39	a2.0
22	S	7 43	8 14	a2.1	1 56	2 32	a1.9
23	S	8 36	9 13	a2.0	2 49	3 30	a1.8
24	M	9 38	10 19	a1.9	3 49	4 34	a1.7
25	T	10 45	11 28	a1.9	4 53	5 39	a1.6
26	W	11 54	...	1.9	5 59	6 43	1.6
27	T	12 35	12 59	p2.1	7 03	7 42	1.7
28	F	1 37	2 00	p2.2	8 02	8 37	1.8
29	S	2 32	2 55	p2.3	8 57	9 27	2.0
30	S	3 23	3 46	p2.4	9 48	10 15	2.0
31	M	4 11	4 35	2.3	10 38	11 01	a2.1

APRIL

DAY OF MONTH	DAY OF WEEK	CURRENT TURNS TO NORTHEAST Flood Starts a.m.	p.m.	Kts.	CURRENT TURNS TO SOUTHWEST Ebb Starts a.m.	p.m.	Kts.
1	T	4 56	5 21	a2.4	11 26	11 46	2.0
2	W	5 41	6 08	a2.3	...	12 14	1.9
3	T	6 26	6 55	a2.2	12 32	1 03	a1.9
4	F	7 14	7 45	a2.1	1 19	1 53	a1.8
5	S	8 02	8 37	a2.0	2 09	2 46	a1.7
6	S	8 56	9 33	a1.9	3 02	3 42	a1.6
7	M	9 52	10 32	a1.9	3 59	4 39	a1.5
8	T	10 51	11 30	a1.9	4 57	5 35	1.4
9	W	11 48	...	1.9	5 54	6 28	1.5
10	T	12 25	12 41	p2.0	6 48	7 18	p1.6
11	F	1 15	1 30	p2.1	7 37	8 03	p1.7
12	S	1 59	2 14	p2.1	8 23	8 46	p1.8
13	S	2 40	2 55	2.1	9 05	9 25	p1.9
14	M	3 17	3 34	2.1	9 45	10 03	1.9
15	T	3 53	4 12	a2.2	10 25	10 40	p2.0
16	W	4 28	4 50	a2.2	11 04	11 18	2.0
17	T	5 06	5 30	a2.3	11 45	11 58	p2.1
18	F	5 46	6 14	a2.3	...	12 30	2.0
19	S	6 30	7 03	a2.2	12 42	1 19	a2.0
20	S	7 20	7 57	a2.1	1 32	2 13	a1.9
21	M	8 17	8 58	a2.0	2 28	3 12	a1.8
22	T	9 20	10 05	a2.0	3 30	4 15	a1.7
23	W	10 27	11 13	a1.9	4 36	5 19	1.6
24	T	11 37	...	2.0	5 43	6 22	p1.7
25	F	12 19	12 43	p2.1	6 47	7 21	p1.8
26	S	1 20	1 44	p2.2	7 47	8 16	p1.9
27	S	2 15	2 39	2.2	8 42	9 06	1.9
28	M	3 05	3 31	a2.3	9 34	9 54	p2.0
29	T	3 52	4 18	a2.3	10 23	10 39	p2.0
30	W	4 36	5 04	a2.3	11 10	11 23	1.9

The Kts. (knots) columns show the **maximum** predicted velocities of the stronger one of the Flood Currents and the stronger one of the Ebb Currents for each day.
The letter "a" means the velocity shown should occur **after** the **a.m.** Current Change. The letter "p" means the velocity shown should occur **after** the **p.m.** Current Change (even if next morning). No "a" or "p" means a.m. and p.m. velocities are the same for that day.

Avg. Max. Velocity: Flood 2.0 Kts., Ebb 1.8 Kts.

Max. Flood 3 hrs. 20 min. after Flood Starts, ±15 min.

Max. Ebb 2 hrs. 45 min. after Ebb Starts, ±15 min.

Gay Head (1 1/2 mi. NW of): avg. max velocity, Flood 2.0 kts., Ebb 2.0 kts. Time of Flood and Ebb 1 hr. 35 min. after Pollock Rip. Cross Rip: avg. max. velocity, Flood 1.3 kts., Ebb 0.9 kts. Time of Flood and Ebb 1 hr. 50 min. after Pollock Rip. Use POLLOCK RIP tables with current charts on pp. 66-77. See pp. 22-29 for Current Change at other points.

2014 CURRENT TABLE
POLLOCK RIP CHANNEL, MA

41°33'N, 69°59'W SE of Monomoy Pt. at Butler Hole

Daylight Saving Time

MAY

DAY OF MONTH	DAY OF WEEK	CURRENT TURNS TO NORTHEAST Flood Starts a.m.	p.m.	Kts.	CURRENT TURNS TO SOUTHWEST Ebb Starts a.m.	p.m.	Kts.
1	T	5 19	**5 48**	a2.3	11 56	**...**	1.8
2	F	6 01	**6 32**	a2.2	12 07	**12 42**	a1.8
3	S	6 45	**7 18**	a2.1	12 52	**1 29**	a1.8
4	S	7 32	**8 08**	a2.0	1 39	**2 17**	a1.7
5	M	8 20	**8 59**	a1.9	2 30	**3 08**	a1.6
6	T	9 13	**9 53**	a1.9	3 23	**4 00**	1.5
7	W	10 07	**10 48**	a1.9	4 18	**4 53**	1.5
8	T	11 02	**11 41**	a1.9	5 14	**5 45**	p1.6
9	F	11 55	**...**	1.9	6 08	**6 34**	p1.6
10	S	12 31	**12 46**	p2.0	6 59	**7 21**	p1.7
11	S	1 17	**1 34**	p2.0	7 47	**8 05**	p1.8
12	M	2 00	**2 18**	2.0	8 32	**8 47**	p1.9
13	T	2 40	**3 01**	a2.1	9 15	**9 28**	p1.9
14	W	3 19	**3 42**	a2.2	9 57	**10 08**	p2.0
15	T	3 59	**4 25**	a2.3	10 40	**10 49**	p2.1
16	F	4 40	**5 09**	a2.3	11 24	**11 33**	p2.1
17	S	5 24	**5 56**	a2.3	...	**12 11**	2.0
18	S	6 11	**6 48**	a2.3	12 20	**1 02**	a2.1
19	M	7 04	**7 44**	a2.2	1 13	**1 56**	a2.0
20	T	8 01	**8 45**	a2.2	2 10	**2 54**	a1.9
21	W	9 04	**9 50**	a2.1	3 13	**3 56**	a1.8
22	T	10 11	**10 56**	a2.0	4 18	**4 58**	1.7
23	F	11 18	**...**	2.0	5 25	**5 59**	1.7
24	S	12 01	**12 24**	p2.1	6 29	**6 58**	p1.8
25	S	1 00	**1 26**	2.1	7 30	**7 53**	p1.8
26	M	1 55	**2 22**	a2.2	8 27	**8 44**	p1.9
27	T	2 46	**3 14**	a2.3	9 19	**9 32**	p1.9
28	W	3 33	**4 02**	a2.3	10 08	**10 18**	p1.9
29	T	4 17	**4 46**	a2.3	10 54	**11 01**	p1.8
30	F	4 58	**5 28**	a2.2	11 38	**11 43**	p1.8
31	S	5 38	**6 10**	a2.2	...	**12 20**	1.7

Daylight Saving Time

JUNE

DAY OF MONTH	DAY OF WEEK	CURRENT TURNS TO NORTHEAST Flood Starts a.m.	p.m.	Kts.	CURRENT TURNS TO SOUTHWEST Ebb Starts a.m.	p.m.	Kts.
1	S	6 19	**6 52**	a2.1	12 26	**1 03**	a1.7
2	M	7 01	**7 36**	a2.1	1 11	**1 47**	a1.7
3	T	7 46	**8 23**	a2.0	1 57	**2 33**	a1.7
4	W	8 35	**9 14**	a2.0	2 47	**3 21**	1.6
5	T	9 24	**10 03**	a1.9	3 39	**4 10**	1.6
6	F	10 16	**10 55**	a1.9	4 32	**5 00**	p1.6
7	S	11 09	**11 45**	a1.9	5 26	**5 50**	p1.7
8	S	...	**12 01**	1.8	6 18	**6 38**	p1.7
9	M	12 33	**12 52**	1.9	7 09	**7 25**	p1.8
10	T	1 20	**1 41**	a2.0	7 58	**8 11**	p1.8
11	W	2 04	**2 29**	a2.1	8 45	**8 55**	p1.9
12	T	2 48	**3 15**	a2.2	9 31	**9 39**	p2.0
13	F	3 33	**4 02**	a2.3	10 18	**10 25**	p2.1
14	S	4 18	**4 50**	a2.3	11 05	**11 12**	p2.1
15	S	5 06	**5 40**	a2.4	11 54	**...**	2.0
16	M	5 56	**6 33**	a2.4	12 02	**12 45**	a2.1
17	T	6 50	**7 29**	a2.3	12 56	**1 38**	a2.1
18	W	7 47	**8 28**	a2.3	1 54	**2 35**	a2.0
19	T	8 48	**9 30**	a2.2	2 55	**3 34**	a1.9
20	F	9 52	**10 34**	a2.1	3 59	**4 34**	p1.8
21	S	10 58	**11 37**	2.0	5 05	**5 35**	1.7
22	S	...	**12 04**	2.0	6 10	**6 34**	p1.8
23	M	12 38	**1 06**	a2.1	7 12	**7 30**	p1.8
24	T	1 35	**2 04**	a2.2	8 10	**8 23**	p1.8
25	W	2 27	**2 57**	a2.3	9 03	**9 12**	p1.8
26	T	3 14	**3 44**	a2.3	9 51	**9 57**	p1.8
27	F	3 57	**4 27**	a2.3	10 35	**10 39**	p1.8
28	S	4 37	**5 06**	a2.2	11 16	**11 20**	p1.8
29	S	5 16	**5 45**	a2.2	11 56	**...**	1.7
30	M	5 53	**6 23**	a2.1	12 01	**12 35**	a1.8

The Kts. (knots) columns show the **maximum** predicted velocities of the stronger one of the Flood Currents and the stronger one of the Ebb Currents for each day.
The letter "a" means the velocity shown should occur **after** the **a.m.** Current Change. The letter "p" means the velocity shown should occur **after** the **p.m.** Current Change (even if next morning). No "a" or "p" means a.m. and p.m. velocities are the same for that day.
Avg. Max. Velocity: Flood 2.0 Kts., Ebb 1.8 Kts.
Max. Flood 3 hrs. 20 min. after Flood Starts, ±15 min.
Max. Ebb 2 hrs. 45 min. after Ebb Starts, ±15 min.
Gay Head (1 1/2 mi. NW of): avg. max velocity, Flood 2.0 kts., Ebb 2.0 kts. Time of Flood and Ebb 1 hr. 35 min. after Pollock Rip. Cross Rip: avg. max. velocity, Flood 1.3 kts., Ebb 0.9 kts. Time of Flood and Ebb 1 hr. 50 min. after Pollock Rip. Use POLLOCK RIP tables with current charts on pp. 66-77. See pp. 22-29 for Current Change at other points.

2014 CURRENT TABLE
POLLOCK RIP CHANNEL, MA

41°33'N, 69°59'W SE of Monomoy Pt. at Butler Hole

Daylight Saving Time — **Daylight Saving Time**

JULY

DAY OF MONTH	DAY OF WEEK	CURRENT TURNS TO NORTHEAST Flood Starts a.m.	p.m.	Kts.	CURRENT TURNS TO SOUTHWEST Ebb Starts a.m.	p.m.	Kts.
1	T	6 32	7 03	a2.1	12 42	1 16	a1.8
2	W	7 13	7 46	a2.1	1 25	1 57	a1.8
3	T	7 56	8 31	a2.0	2 11	2 42	1.7
4	F	8 43	9 19	a1.9	2 59	3 28	1.7
5	S	9 32	10 08	a1.9	3 51	4 16	p1.7
6	S	10 24	10 59	1.8	4 44	5 06	p1.7
7	M	11 18	11 50	1.8	5 39	5 57	p1.7
8	T	...	12 13	1.7	6 33	6 48	p1.7
9	W	12 42	1 08	a1.9	7 26	7 38	p1.8
10	T	1 32	2 01	a2.0	8 18	8 27	p1.9
11	F	2 22	2 52	a2.2	9 08	9 16	p2.0
12	S	3 11	3 43	a2.3	9 57	10 05	p2.1
13	S	4 01	4 33	a2.4	10 46	10 54	p2.2
14	M	4 51	5 24	a2.4	11 35	11 46	p2.2
15	T	5 42	6 16	a2.4	...	12 26	2.1
16	W	6 35	7 10	a2.4	12 39	1 18	a2.1
17	T	7 30	8 06	a2.3	1 36	2 12	2.0
18	F	8 29	9 06	a2.2	2 35	3 09	1.9
19	S	9 31	10 08	2.0	3 38	4 08	1.8
20	S	10 36	11 12	p2.0	4 42	5 09	p1.7
21	M	11 42	...	1.9	5 48	6 09	p1.7
22	T	12 14	12 45	a2.1	6 51	7 07	p1.7
23	W	1 12	1 44	a2.2	7 49	8 01	p1.7
24	T	2 05	2 36	a2.2	8 42	8 50	p1.8
25	F	2 53	3 22	a2.3	9 29	9 35	p1.8
26	S	3 36	4 04	a2.3	10 11	10 16	p1.8
27	S	4 14	4 41	a2.2	10 51	10 56	p1.8
28	M	4 51	5 17	a2.2	11 28	11 34	p1.8
29	T	5 26	5 52	a2.2	...	12 05	1.8
30	W	6 02	6 29	a2.1	12 13	12 42	a1.9
31	T	6 40	7 07	a2.1	12 53	1 21	a1.9

AUGUST

DAY OF MONTH	DAY OF WEEK	CURRENT TURNS TO NORTHEAST Flood Starts a.m.	p.m.	Kts.	CURRENT TURNS TO SOUTHWEST Ebb Starts a.m.	p.m.	Kts.
1	F	7 21	7 49	a2.0	1 36	2 02	1.8
2	S	8 04	8 34	a2.0	2 22	2 47	1.8
3	S	8 52	9 23	1.8	3 12	3 35	1.7
4	M	9 45	10 16	p1.8	4 05	4 26	p1.7
5	T	10 41	11 11	p1.8	5 02	5 20	p1.6
6	W	11 41	...	1.7	6 00	6 16	p1.7
7	T	12 09	12 40	a1.9	6 58	7 11	p1.7
8	F	1 06	1 38	a2.0	7 54	8 05	p1.9
9	S	2 01	2 33	a2.2	8 46	8 57	p2.0
10	S	2 54	3 25	a2.3	9 37	9 48	p2.1
11	M	3 45	4 16	a2.4	10 26	10 39	p2.2
12	T	4 36	5 05	a2.5	11 15	11 30	p2.2
13	W	5 27	5 56	a2.5	...	12 04	2.1
14	T	6 18	6 47	a2.4	12 22	12 55	a2.2
15	F	7 12	7 41	a2.3	1 17	1 47	a2.1
16	S	8 08	8 39	a2.1	2 14	2 42	a1.9
17	S	9 08	9 39	p2.0	3 15	3 41	1.7
18	M	10 12	10 43	p2.0	4 18	4 41	1.6
19	T	11 18	11 46	p2.0	5 23	5 42	p1.6
20	W	...	12 21	1.8	6 25	6 41	p1.6
21	T	12 45	1 19	a2.1	7 23	7 36	p1.7
22	F	1 39	2 11	a2.2	8 15	8 25	p1.7
23	S	2 27	2 56	a2.2	9 01	9 10	p1.8
24	S	3 09	3 35	a2.3	9 42	9 51	p1.8
25	M	3 47	4 12	a2.2	10 20	10 30	p1.9
26	T	4 23	4 46	a2.2	10 56	11 07	p1.9
27	W	4 58	5 19	a2.2	11 31	11 44	1.9
28	T	5 32	5 54	a2.1	...	12 07	1.9
29	F	6 09	6 30	a2.1	12 23	12 45	1.9
30	S	6 48	7 10	2.0	1 04	1 25	1.9
31	S	7 30	7 54	1.9	1 48	2 08	1.8

The Kts. (knots) columns show the **maximum** predicted velocities of the stronger one of the Flood Currents and the stronger one of the Ebb Currents for each day.
The letter "a" means the velocity shown should occur **after** the **a.m.** Current Change. The letter "p" means the velocity shown should occur **after** the **p.m.** Current Change (even if next morning). No "a" or "p" means a.m. and p.m. velocities are the same for that day.

Avg. Max. Velocity: Flood 2.0 Kts., Ebb 1.8 Kts.

Max. Flood 3 hrs. 20 min. after Flood Starts, ±15 min.

Max. Ebb 2 hrs. 45 min. after Ebb Starts, ±15 min.

Gay Head (1 1/2 mi. NW of): avg. max velocity, Flood 2.0 kts., Ebb 2.0 kts. Time of Flood and Ebb 1 hr. 35 min. after Pollock Rip. Cross Rip: avg. max. velocity, Flood 1.3 kts., Ebb 0.9 kts. Time of Flood and Ebb 1 hr. 50 min. after Pollock Rip. Use POLLOCK RIP tables with current charts on pp. 66-77. See pp. 22-29 for Current Change at other points.

2014 CURRENT TABLE
POLLOCK RIP CHANNEL, MA

41°33'N, 69°59'W SE of Monomoy Pt. at Butler Hole

Daylight Saving Time

SEPTEMBER

DAY OF MONTH	DAY OF WEEK	CURRENT TURNS TO NORTHEAST Flood Starts			CURRENT TURNS TO SOUTHWEST Ebb Starts		
		a.m.	**p.m.**	Kts.	a.m.	**p.m.**	Kts.
1	**M**	8 18	**8 43**	p1.9	2 38	**2 57**	1.7
2	**T**	9 12	**9 38**	p1.8	3 32	**3 51**	p1.7
3	**W**	10 11	**10 38**	p1.8	4 31	**4 49**	p1.6
4	**T**	11 16	**11 42**	p1.9	5 33	**5 49**	p1.6
5	**F**	...	**12 19**	1.7	6 33	**6 49**	p1.7
6	**S**	12 44	**1 20**	a2.0	7 31	**7 46**	p1.8
7	**S**	1 43	**2 16**	a2.2	8 26	**8 41**	p2.0
8	**M**	2 38	**3 08**	a2.3	9 17	**9 33**	p2.1
9	**T**	3 30	**3 58**	a2.4	10 06	**10 23**	p2.2
10	**W**	4 21	**4 46**	a2.4	10 54	**11 14**	p2.2
11	**T**	5 10	**5 34**	a2.4	11 41	**...**	2.1
12	**F**	6 00	**6 24**	2.3	12 05	**12 30**	2.1
13	**S**	6 51	**7 15**	p2.2	12 57	**1 21**	a2.0
14	**S**	7 46	**8 10**	2.0	1 52	**2 15**	1.8
15	**M**	8 44	**9 08**	p2.0	2 51	**3 12**	1.6
16	**T**	9 45	**10 10**	p1.9	3 52	**4 12**	1.5
17	**W**	10 49	**11 12**	p2.0	4 54	**5 13**	p1.5
18	**T**	11 52	**...**	1.8	5 54	**6 12**	1.5
19	**F**	12 12	**12 48**	a2.0	6 50	**7 07**	p1.6
20	**S**	1 06	**1 39**	a2.1	7 41	**7 56**	p1.7
21	**S**	1 54	**2 23**	a2.2	8 26	**8 41**	p1.8
22	**M**	2 37	**3 03**	a2.2	9 07	**9 23**	1.8
23	**T**	3 16	**3 39**	a2.2	9 46	**10 01**	1.9
24	**W**	3 53	**4 13**	a2.2	10 22	**10 39**	1.9
25	**T**	4 28	**4 46**	2.1	10 57	**11 16**	1.9
26	**F**	5 03	**5 20**	2.1	11 33	**11 54**	a2.0
27	**S**	5 39	**5 56**	2.1	...	**12 10**	2.0
28	**S**	6 18	**6 35**	p2.1	12 35	**12 50**	1.9
29	**M**	7 02	**7 20**	p2.0	1 20	**1 35**	p1.9
30	**T**	7 51	**8 11**	p1.9	2 10	**2 25**	p1.8

Daylight Saving Time

OCTOBER

DAY OF MONTH	DAY OF WEEK	CURRENT TURNS TO NORTHEAST Flood Starts			CURRENT TURNS TO SOUTHWEST Ebb Starts		
		a.m.	**p.m.**	Kts.	a.m.	**p.m.**	Kts.
1	**W**	8 47	**9 08**	p1.9	3 05	**3 21**	p1.7
2	**T**	9 49	**10 12**	p1.8	4 05	**4 23**	1.6
3	**F**	10 55	**11 19**	p1.9	5 08	**5 28**	p1.6
4	**S**	...	**12 02**	1.7	6 11	**6 31**	p1.7
5	**S**	12 24	**1 03**	a2.0	7 09	**7 30**	p1.8
6	**M**	1 26	**1 59**	a2.2	8 04	**8 26**	p1.9
7	**T**	2 22	**2 51**	2.3	8 56	**9 19**	2.0
8	**W**	3 15	**3 40**	p2.4	9 45	**10 09**	2.1
9	**T**	4 05	**4 27**	p2.4	10 32	**10 59**	2.1
10	**F**	4 54	**5 14**	p2.4	11 19	**11 48**	a2.1
11	**S**	5 42	**6 01**	p2.3	...	**12 06**	2.0
12	**S**	6 31	**6 49**	p2.2	12 39	**12 55**	1.9
13	**M**	7 23	**7 41**	p2.0	1 31	**1 47**	1.7
14	**T**	8 18	**8 36**	p1.9	2 25	**2 42**	1.6
15	**W**	9 16	**9 34**	p1.9	3 22	**3 39**	1.5
16	**T**	10 16	**10 33**	p1.9	4 20	**4 39**	p1.5
17	**F**	11 15	**11 31**	p2.0	5 17	**5 37**	1.5
18	**S**	...	**12 10**	1.8	6 11	**6 32**	1.5
19	**S**	12 26	**1 00**	a2.0	7 01	**7 22**	1.6
20	**M**	1 15	**1 45**	a2.1	7 47	**8 08**	1.7
21	**T**	2 01	**2 26**	2.1	8 30	**8 51**	1.8
22	**W**	2 42	**3 04**	2.1	9 09	**9 32**	a1.9
23	**T**	3 20	**3 39**	p2.2	9 47	**10 10**	1.9
24	**F**	3 57	**4 13**	p2.2	10 24	**10 49**	a2.0
25	**S**	4 34	**4 49**	p2.2	11 00	**11 29**	a2.0
26	**S**	5 13	**5 26**	p2.2	11 39	**...**	2.0
27	**M**	5 54	**6 08**	p2.2	12 11	**12 21**	p2.0
28	**T**	6 39	**6 54**	p2.1	12 57	**1 07**	1.9
29	**W**	7 30	**7 47**	p2.1	1 47	**1 59**	1.8
30	**T**	8 27	**8 46**	p2.0	2 43	**2 58**	1.7
31	**F**	9 31	**9 51**	p1.9	3 43	**4 02**	1.6

The Kts. (knots) columns show the **maximum** predicted velocities of the stronger one of the Flood Currents and the stronger one of the Ebb Currents for each day.
The letter "a" means the velocity shown should occur **after** the **a.m.** Current Change. The letter "p" means the velocity shown should occur **after** the **p.m.** Current Change (even if next morning). No "a" or "p" means a.m. and p.m. velocities are the same for that day.
Avg. Max. Velocity: Flood 2.0 Kts., Ebb 1.8 Kts.
Max. Flood 3 hrs. 20 min. after Flood Starts, ±15 min.
Max. Ebb 2 hrs. 45 min. after Ebb Starts, ±15 min.
Gay Head (1 1/2 mi. NW of): avg. max velocity, Flood 2.0 kts., Ebb 2.0 kts. Time of Flood and Ebb 1 hr. 35 min. after Pollock Rip. Cross Rip: avg. max. velocity, Flood 1.3 kts., Ebb 0.9 kts. Time of Flood and Ebb 1 hr. 50 min. after Pollock Rip. Use POLLOCK RIP tables with current charts on pp. 66-77. See pp. 22-29 for Current Change at other points.

2014 CURRENT TABLE
POLLOCK RIP CHANNEL, MA
41°33'N, 69°59'W SE of Monomoy Pt. at Butler Hole

***Standard Time starts Nov. 2 at 2 a.m.** **Standard Time**

NOVEMBER

DAY OF MONTH	DAY OF WEEK	CURRENT TURNS TO NORTHEAST Flood Starts a.m.	**p.m.**	Kts.	CURRENT TURNS TO SOUTHWEST Ebb Starts a.m.	**p.m.**	Kts.
1	**S**	10 37	**10 59**	p1.9	4 46	**5 08**	1.6
2	**S**	*10 43	***11 06**	p2.0	*4 48	***5 13**	1.7
3	**M**	11 45	...	2.0	5 47	**6 14**	p1.8
4	**T**	12 10	**12 43**	p2.2	6 43	**7 12**	1.9
5	**W**	1 07	**1 34**	p2.3	7 36	**8 05**	1.9
6	**T**	2 01	**2 23**	p2.4	8 25	**8 56**	a2.0
7	**F**	2 51	**3 10**	p2.4	9 12	**9 45**	a2.0
8	**S**	3 39	**3 55**	p2.3	9 58	**10 33**	a2.0
9	**S**	4 26	**4 40**	p2.3	10 44	**11 21**	a1.9
10	**M**	5 12	**5 25**	p2.2	11 31	...	1.8
11	**T**	6 00	**6 13**	p2.1	12 09	**12 19**	1.7
12	**W**	6 50	**7 02**	p2.0	12 59	**1 11**	1.6
13	**T**	7 42	**7 55**	p1.9	1 50	**2 05**	1.5
14	**F**	8 37	**8 50**	p1.9	2 42	**3 00**	1.5
15	**S**	9 32	**9 46**	p1.9	3 35	**3 57**	1.5
16	**S**	10 26	**10 40**	p1.9	4 28	**4 52**	a1.6
17	**M**	11 17	**11 32**	p2.0	5 18	**5 44**	1.6
18	**T**	...	**12 04**	2.0	6 05	**6 32**	a1.7
19	**W**	12 20	**12 47**	2.0	6 50	**7 18**	a1.8
20	**T**	1 05	**1 27**	p2.1	7 32	**8 01**	a1.9
21	**F**	1 47	**2 06**	p2.2	8 12	**8 43**	a1.9
22	**S**	2 28	**2 43**	p2.2	8 52	**9 24**	a2.0
23	**S**	3 08	**3 22**	p2.3	9 31	**10 06**	a2.0
24	**M**	3 49	**4 03**	p2.3	10 12	**10 50**	a2.1
25	**T**	4 33	**4 47**	p2.3	10 57	**11 37**	a2.1
26	**W**	5 21	**5 35**	p2.3	11 45	...	2.0
27	**T**	6 13	**6 29**	p2.2	12 27	**12 39**	1.9
28	**F**	7 10	**7 28**	p2.1	1 22	**1 38**	1.8
29	**S**	8 12	**8 32**	p2.0	2 21	**2 42**	1.7
30	**S**	9 17	**9 40**	p2.0	3 22	**3 48**	1.7

DECEMBER

DAY OF MONTH	DAY OF WEEK	CURRENT TURNS TO NORTHEAST Flood Starts a.m.	**p.m.**	Kts.	CURRENT TURNS TO SOUTHWEST Ebb Starts a.m.	**p.m.**	Kts.
1	**M**	10 22	**10 47**	p2.0	4 24	**4 54**	a1.7
2	**T**	11 25	**11 52**	2.0	5 24	**5 58**	1.7
3	**W**	...	**12 24**	2.2	6 22	**6 57**	a1.8
4	**T**	12 53	**1 19**	p2.3	7 16	**7 53**	a1.9
5	**F**	1 47	**2 08**	p2.3	8 07	**8 44**	a1.9
6	**S**	2 38	**2 55**	p2.4	8 54	**9 32**	a1.9
7	**S**	3 25	**3 39**	p2.3	9 40	**10 18**	a1.9
8	**M**	4 10	**4 21**	p2.3	10 24	**11 02**	a1.8
9	**T**	4 53	**5 03**	p2.2	11 08	**11 46**	a1.8
10	**W**	5 36	**5 45**	p2.1	11 53	...	1.7
11	**T**	6 20	**6 30**	p2.0	12 30	**12 39**	1.7
12	**F**	7 06	**7 16**	p2.0	1 15	**1 28**	1.6
13	**S**	7 55	**8 06**	p1.9	2 03	**2 20**	1.6
14	**S**	8 45	**8 58**	p1.9	2 52	**3 14**	a1.6
15	**M**	9 37	**9 52**	p1.8	3 42	**4 08**	a1.6
16	**T**	10 29	**10 45**	1.8	4 32	**5 02**	a1.6
17	**W**	11 19	**11 38**	a1.9	5 22	**5 54**	a1.7
18	**T**	...	**12 06**	1.9	6 10	**6 44**	a1.7
19	**F**	12 27	**12 51**	p2.0	6 56	**7 31**	a1.8
20	**S**	1 14	**1 34**	p2.1	7 40	**8 16**	a1.9
21	**S**	2 00	**2 16**	p2.2	8 23	**9 00**	a1.9
22	**M**	2 44	**2 59**	p2.3	9 06	**9 45**	a2.0
23	**T**	3 29	**3 44**	p2.4	9 50	**10 30**	a2.1
24	**W**	4 15	**4 30**	p2.4	10 37	**11 18**	a2.1
25	**T**	5 04	**5 20**	p2.4	11 27	...	2.1
26	**F**	5 56	**6 13**	p2.3	12 08	**12 21**	p2.1
27	**S**	6 51	**7 11**	p2.2	1 01	**1 19**	1.9
28	**S**	7 51	**8 13**	p2.1	1 58	**2 22**	a1.9
29	**M**	8 54	**9 19**	p2.0	2 58	**3 28**	a1.8
30	**T**	9 59	**10 27**	1.9	3 59	**4 35**	a1.7
31	**W**	11 04	**11 34**	a2.0	5 01	**5 40**	a1.7

The Kts. (knots) columns show the **maximum** predicted velocities of the stronger one of the Flood Currents and the stronger one of the Ebb Currents for each day.
The letter "a" means the velocity shown should occur **after** the **a.m.** Current Change. The letter "p" means the velocity shown should occur **after** the **p.m.** Current Change (even if next morning). No "a" or "p" means a.m. and p.m. velocities are the same for that day.
Avg. Max. Velocity: Flood 2.0 Kts., Ebb 1.8 Kts.
Max. Flood 3 hrs. 20 min. after Flood Starts, ±15 min.
Max. Ebb 2 hrs. 45 min. after Ebb Starts, ±15 min.
Gay Head (1 1/2 mi. NW of): avg. max velocity, Flood 2.0 kts., Ebb 2.0 kts. Time of Flood and Ebb 1 hr. 35 min. after Pollock Rip. Cross Rip: avg. max. velocity, Flood 1.3 kts., Ebb 0.9 kts. Time of Flood and Ebb 1 hr. 50 min. after Pollock Rip. Use POLLOCK RIP tables with current charts on pp. 66-77. See pp. 22-29 for Current Change at other points.

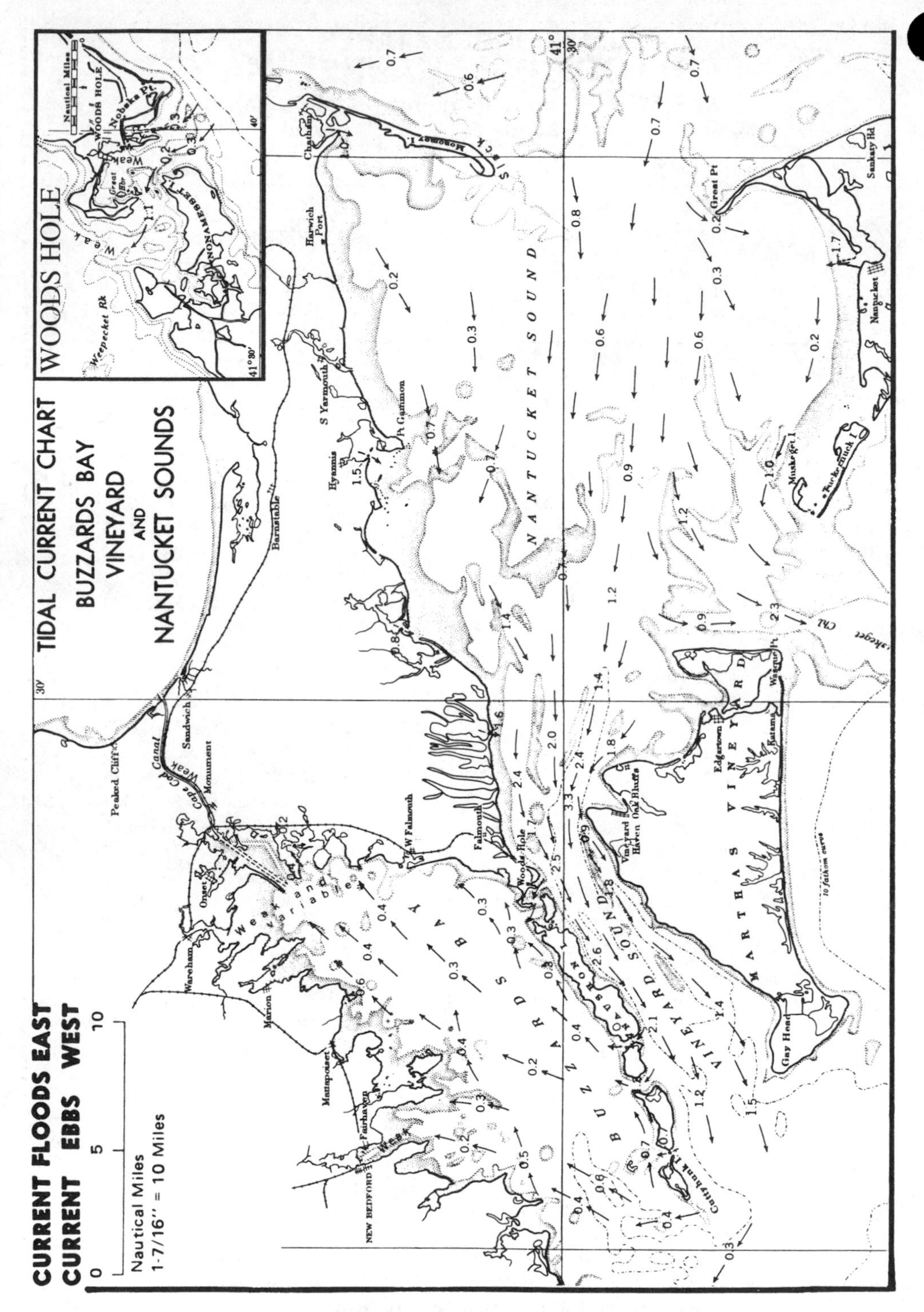

FLOOD STARTS AT POLLOCK RIP CHANNEL
OR: 4 HOURS **AFTER** HIGH WATER AT BOSTON

Velocities shown are at Spring Tides. **See Note at bottom of Boston Tables: Rule-of-Thumb for Current Velocities.** *(Pollock Rip Ch. is SE of Monomoy Pt.)*

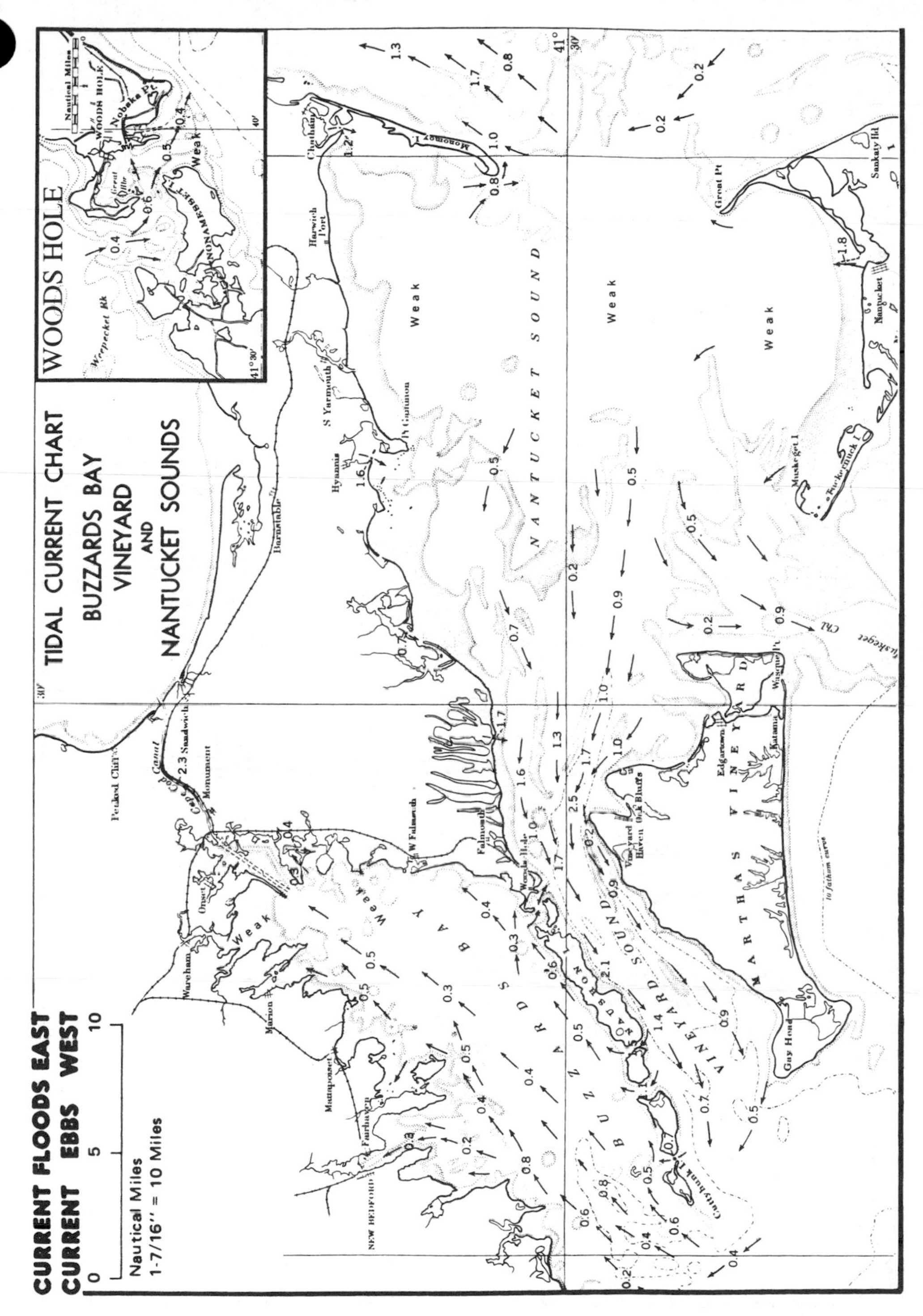

1 HOUR **AFTER** FLOOD STARTS AT POLLOCK RIP CHANNEL
OR: 5 HOURS **AFTER** HIGH WATER AT BOSTON

Velocities shown are at Spring Tides. **See Note at bottom of Boston Tables: Rule-of-Thumb for Current Velocities.** *(Pollock Rip Ch. is SE of Monomoy Pt.)*

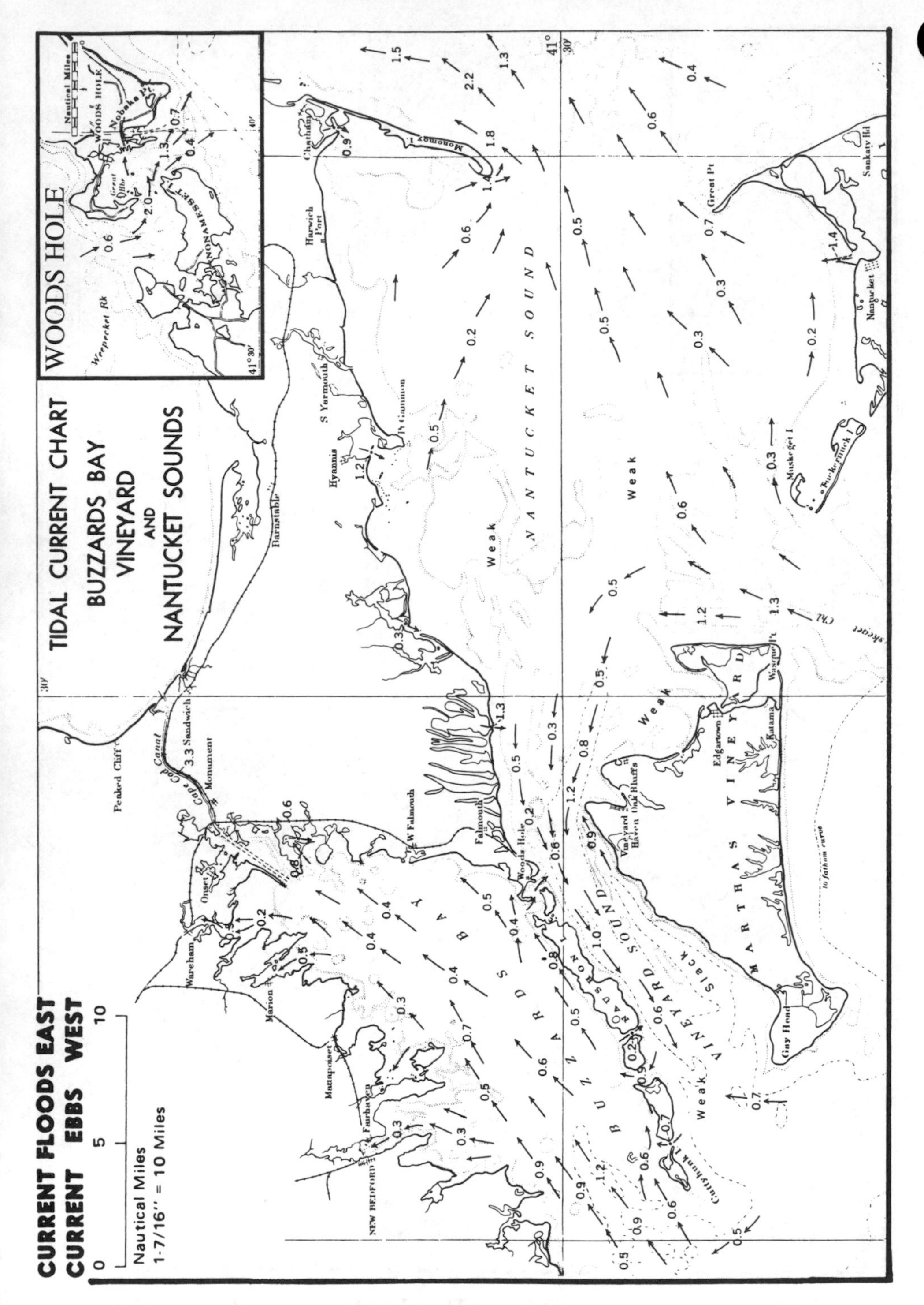

2 HOURS **AFTER** FLOOD STARTS AT POLLOCK RIP CHANNEL
OR: LOW WATER AT BOSTON

Velocities shown are at Spring Tides. **See Note at bottom of Boston Tables: Rule-of-Thumb for Current Velocities.** *(Pollock Rip Ch. is SE of Monomoy Pt.)*

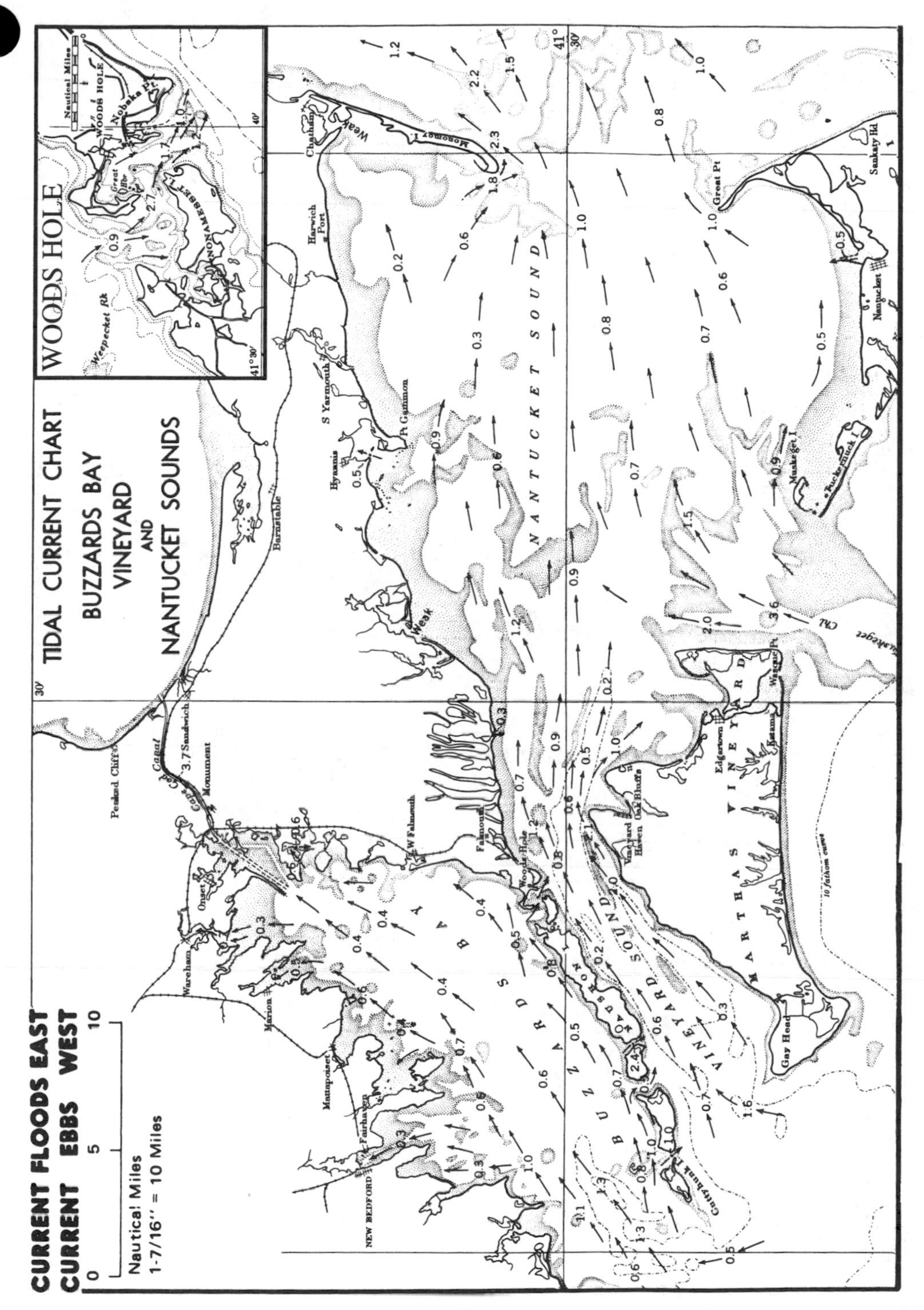

3 HOURS **AFTER** FLOOD STARTS AT POLLOCK RIP CHANNEL
OR: 1 HOUR **AFTER** LOW WATER AT BOSTON

Velocities shown are at Spring Tides. **See Note at bottom of Boston Tables: Rule-of-Thumb for Current Velocities.** *(Pollock Rip Ch. is SE of Monomoy Pt.)*

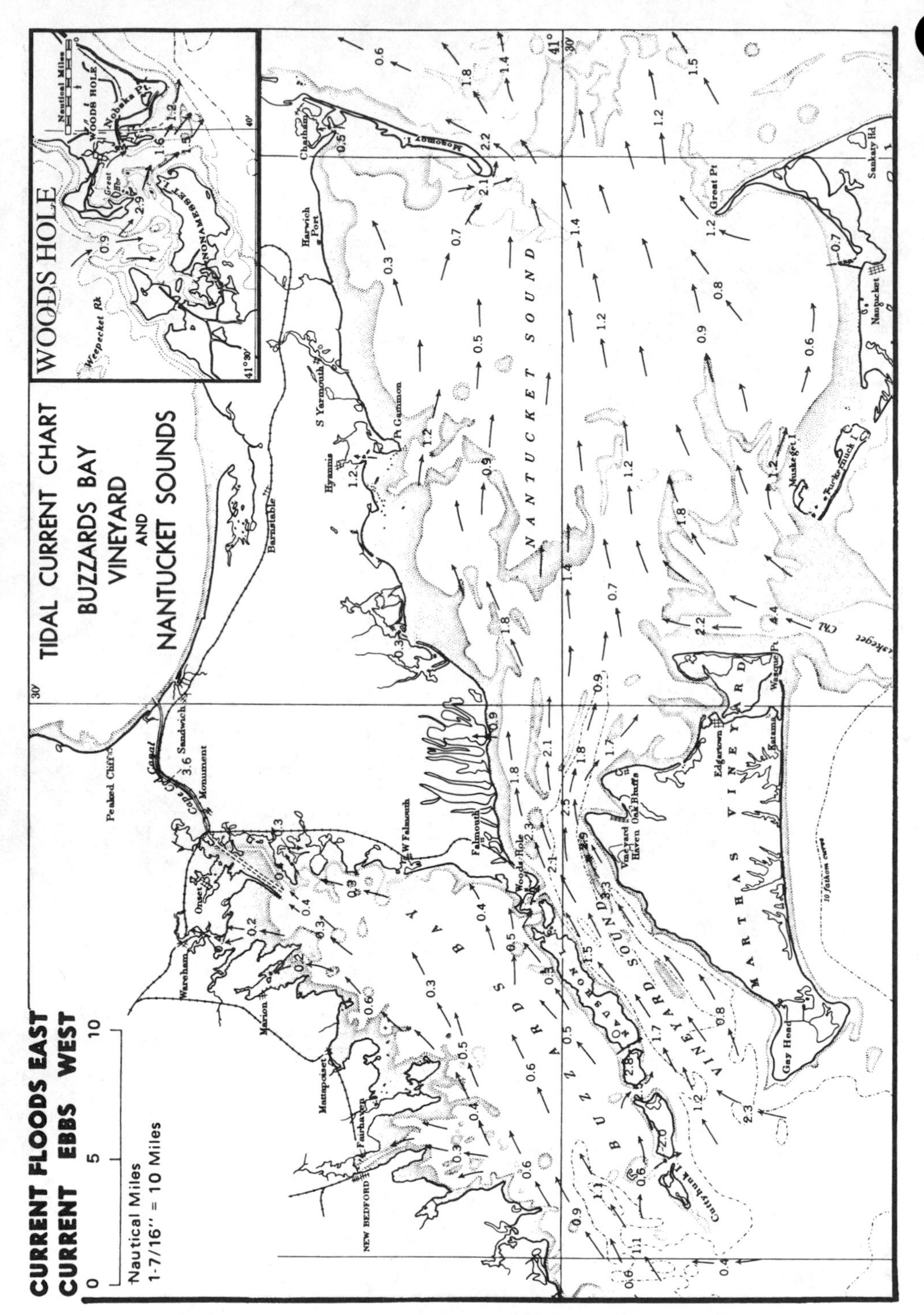

4 HOURS **AFTER** FLOOD STARTS AT POLLOCK RIP CHANNEL
OR: 2 HOURS **AFTER** LOW WATER AT BOSTON

Velocities shown are at Spring Tides. **See Note at bottom of Boston Tables: Rule-of-Thumb for Current Velocities.** *(Pollock Rip Ch. is SE of Monomoy Pt.)*

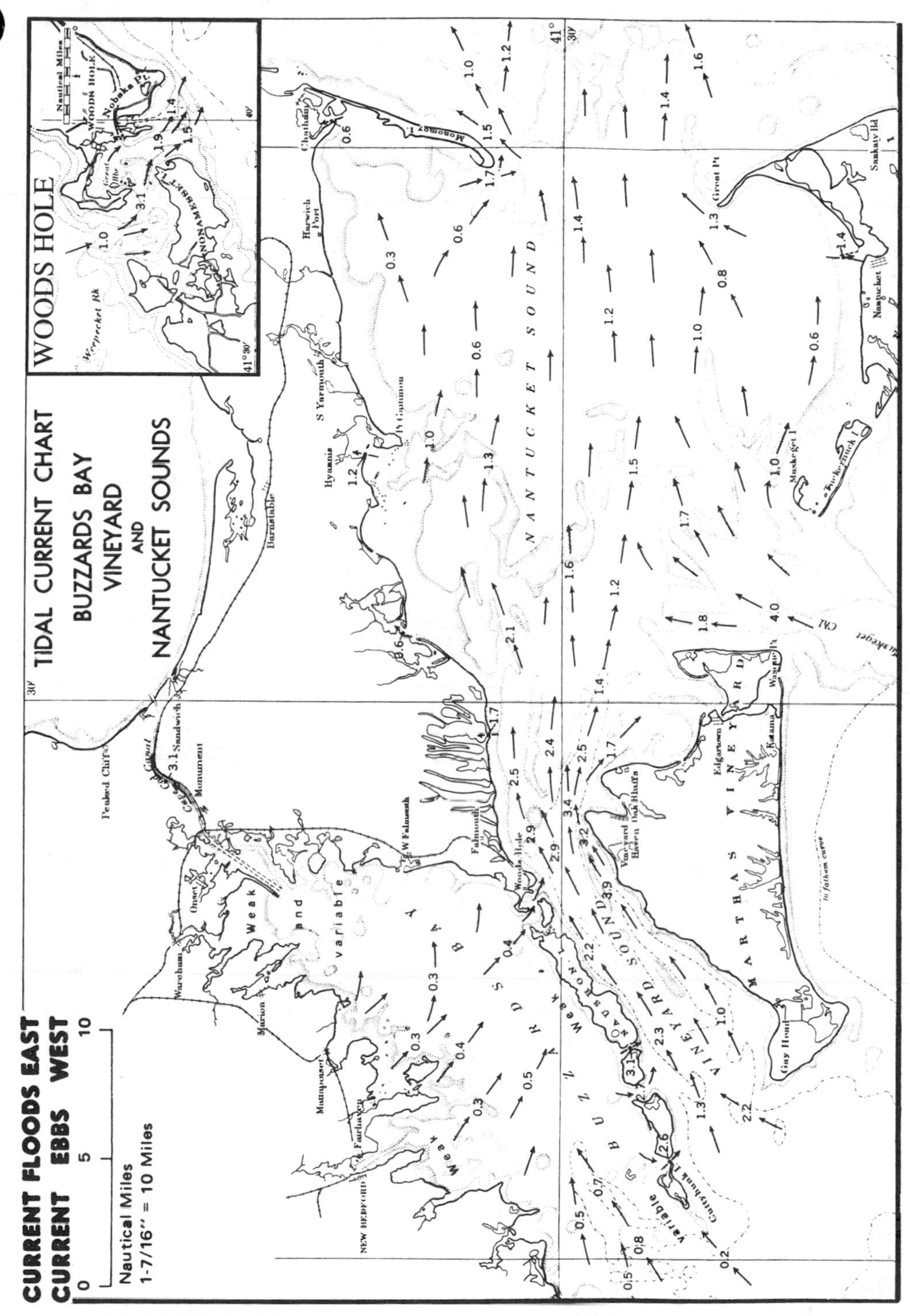

5 HOURS **AFTER** FLOOD STARTS AT POLLOCK RIP CHANNEL
OR: 3 HOURS **AFTER** LOW WATER AT BOSTON

Velocities shown are at Spring Tides. **See Note at bottom of Boston Tables: Rule-of-Thumb for Current Velocities.** *(Pollock Rip Ch. is SE of Monomoy Pt.)*

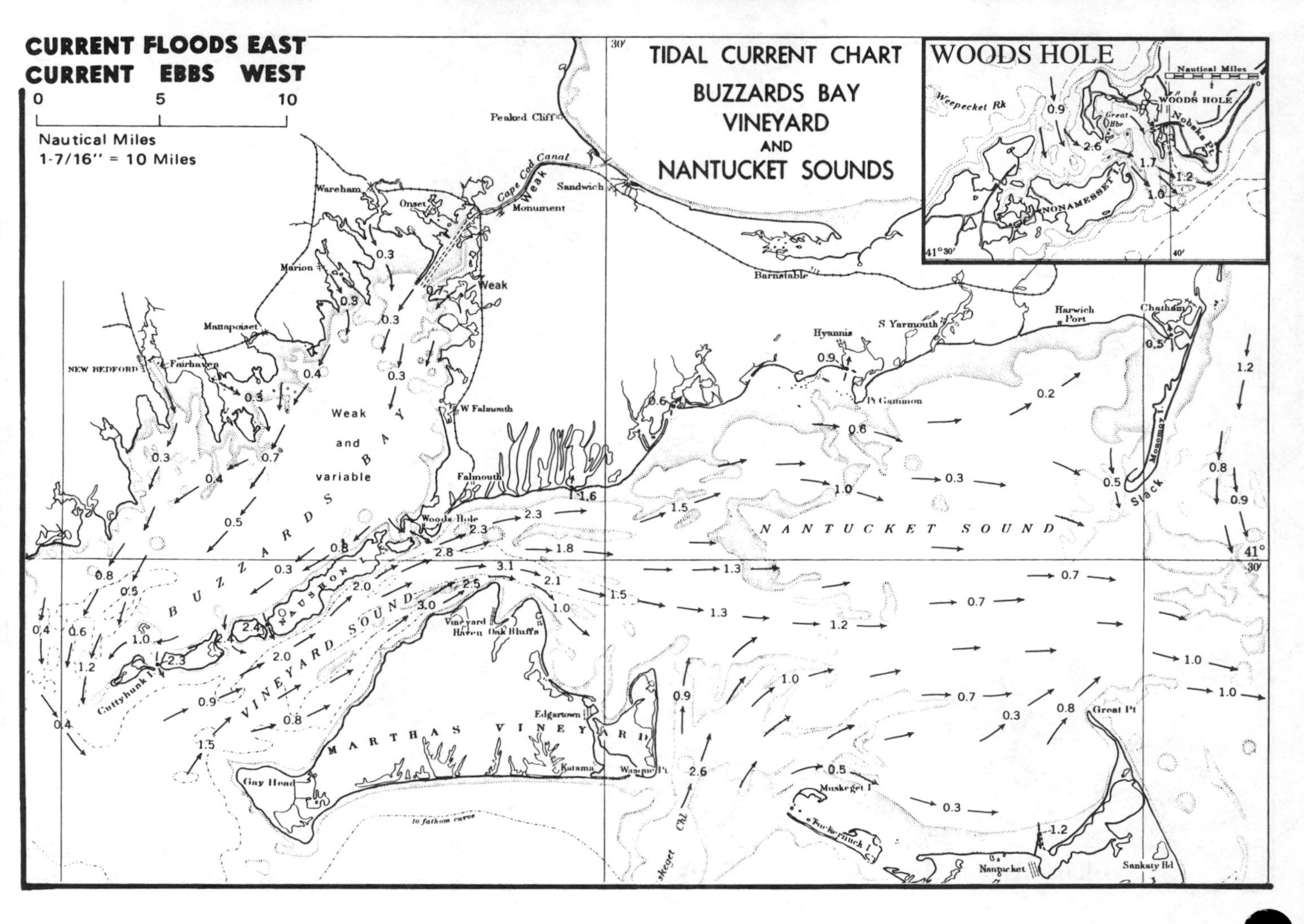

EBB STARTS AT POLLOCK RIP CHANNEL
OR: 4 HOURS **AFTER** LOW WATER AT BOSTON

Velocities shown are at Spring Tides. **See Note at bottom of Boston Tables: Rule-of-Thumb for Current Velocities.** *(Pollock Rip Ch. is SE of Monomoy Pt.)*

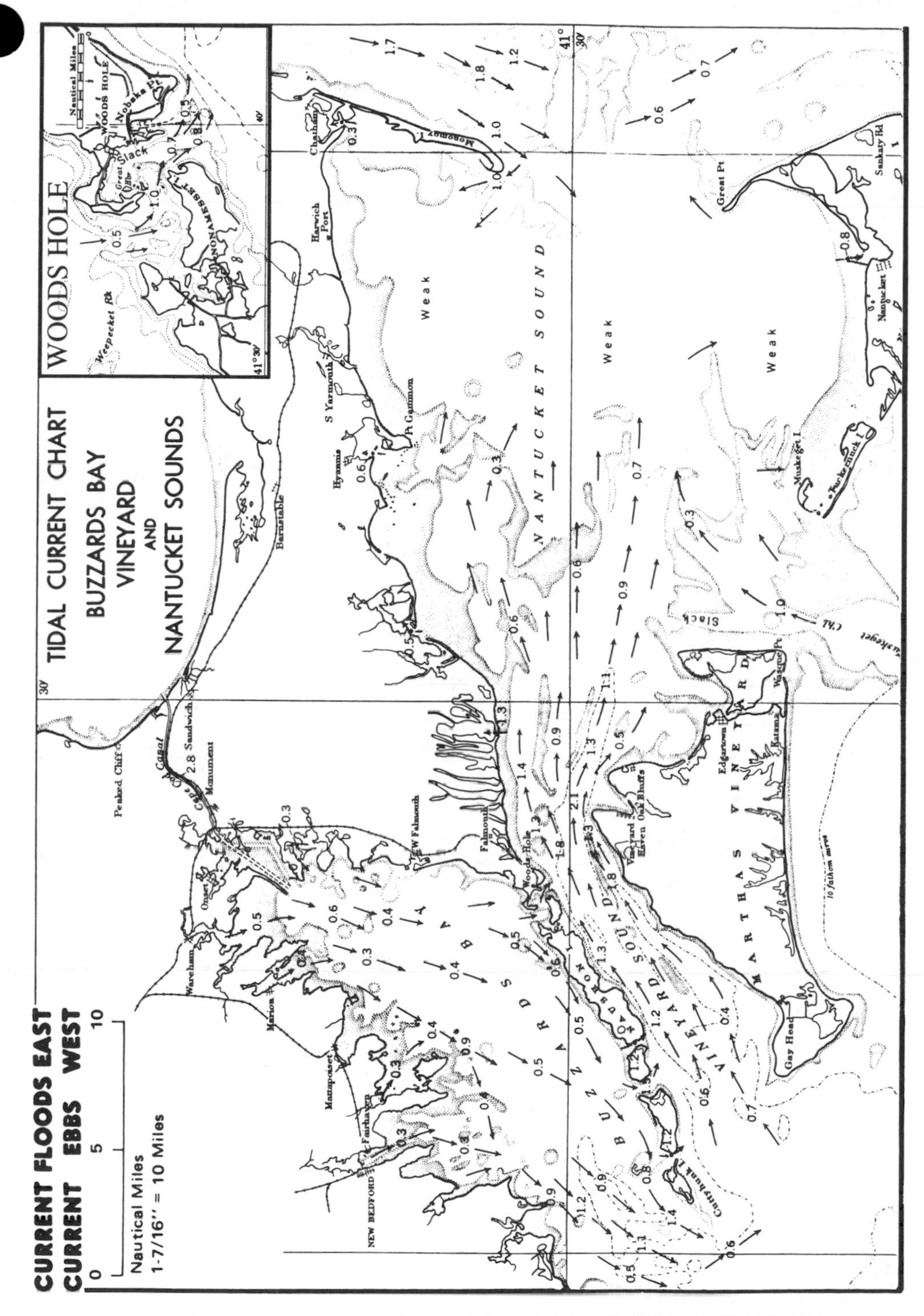

1 HOUR **AFTER** EBB STARTS AT POLLOCK RIP CHANNEL
OR: 5 HOURS **AFTER** LOW WATER AT BOSTON

Velocities shown are at Spring Tides. **See Note at bottom of Boston Tables: Rule-of-Thumb for Current Velocities.** *(Pollock Rip Ch. is SE of Monomoy Pt.)*

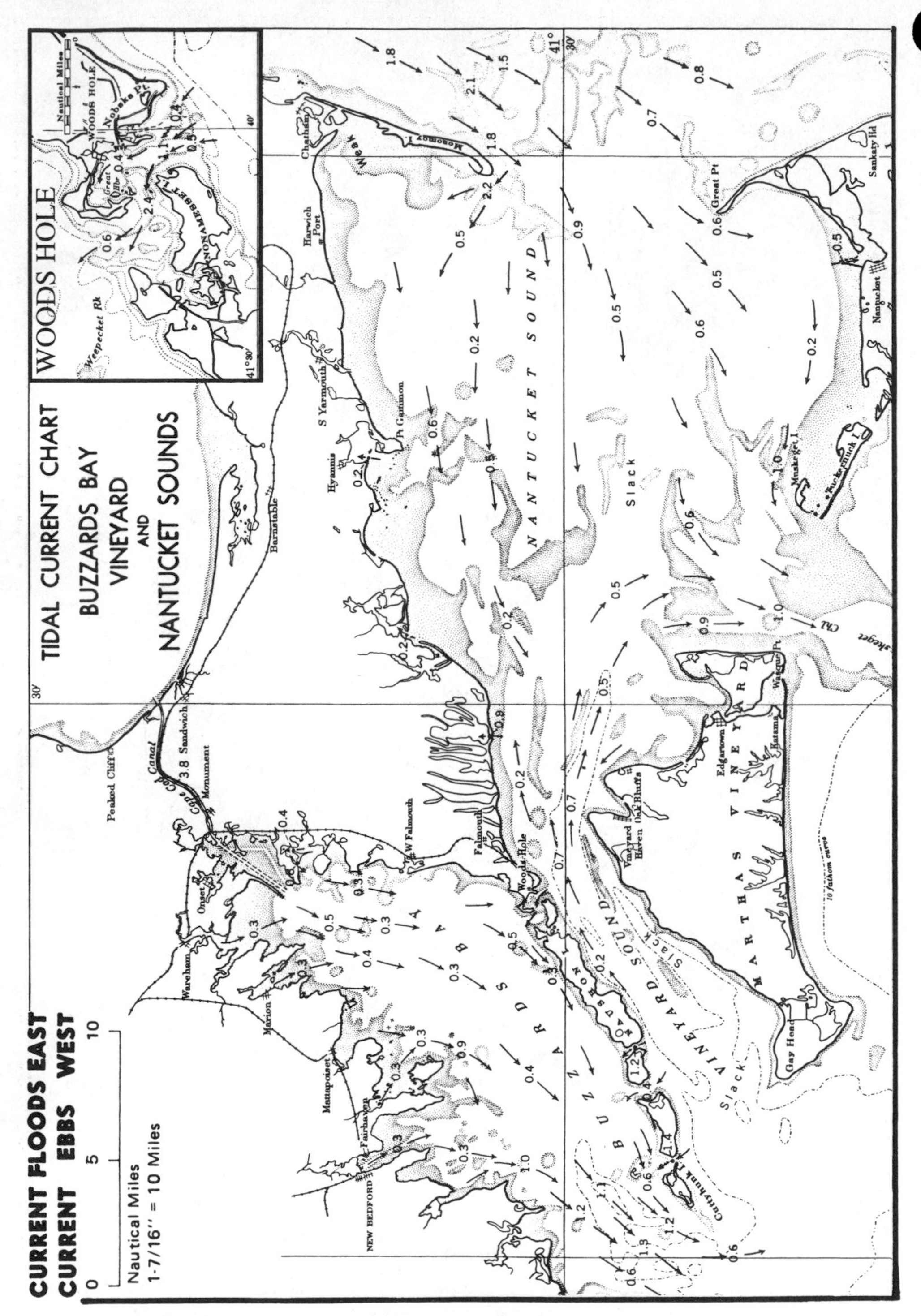

2 HOURS **AFTER** EBB STARTS AT POLLOCK RIP CHANNEL
OR: HIGH WATER AT BOSTON

Velocities shown are at Spring Tides. **See Note at bottom of Boston Tables: Rule-of-Thumb for Current Velocities.** *(Pollock Rip Ch. is SE of Monomoy Pt.)*

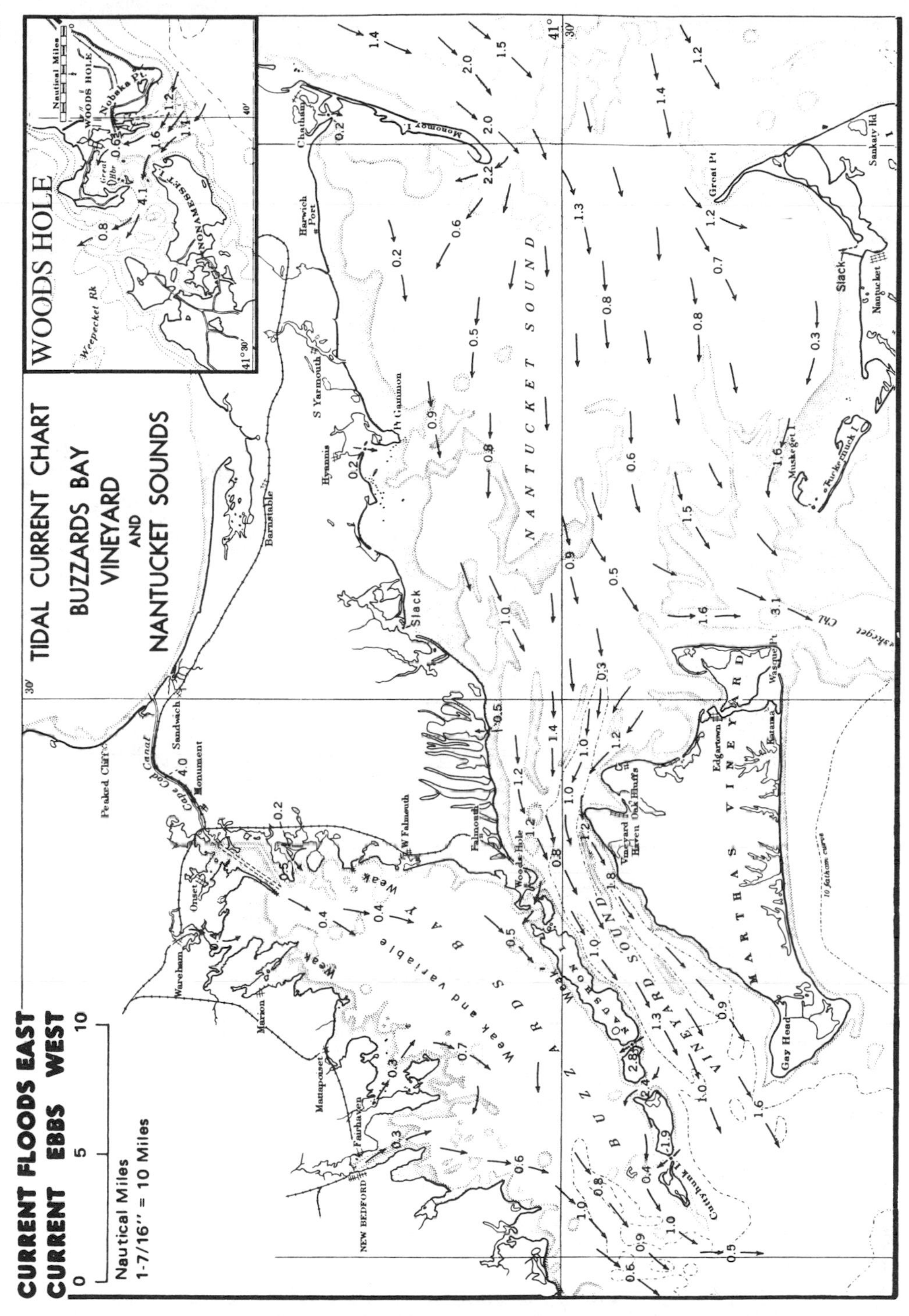

3 HOURS **AFTER** EBB STARTS AT POLLOCK RIP CHANNEL
OR: 1 HOUR **AFTER** HIGH WATER AT BOSTON

Velocities shown are at Spring Tides. **See Note at bottom of Boston Tables: Rule-of-Thumb for Current Velocities.** *(Pollock Rip Ch. is SE of Monomoy Pt.)*

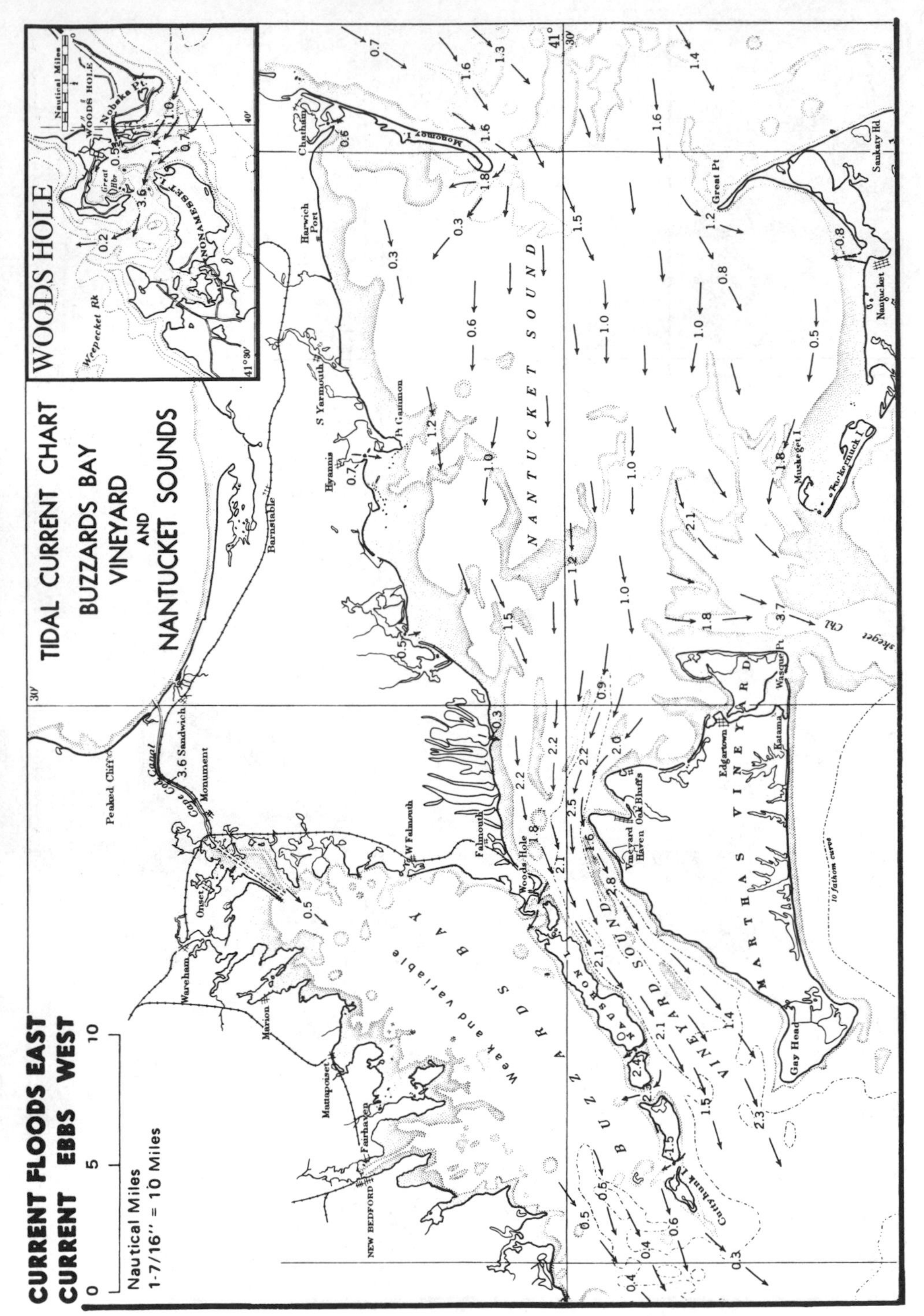

4 HOURS **AFTER** EBB STARTS AT POLLOCK RIP CHANNEL
OR: 2 HOURS **AFTER** HIGH WATER AT BOSTON

Velocities shown are at Spring Tides. **See Note at bottom of Boston Tables: Rule-of-Thumb for Current Velocities.** *(Pollock Rip Ch. is SE of Monomoy Pt.)*

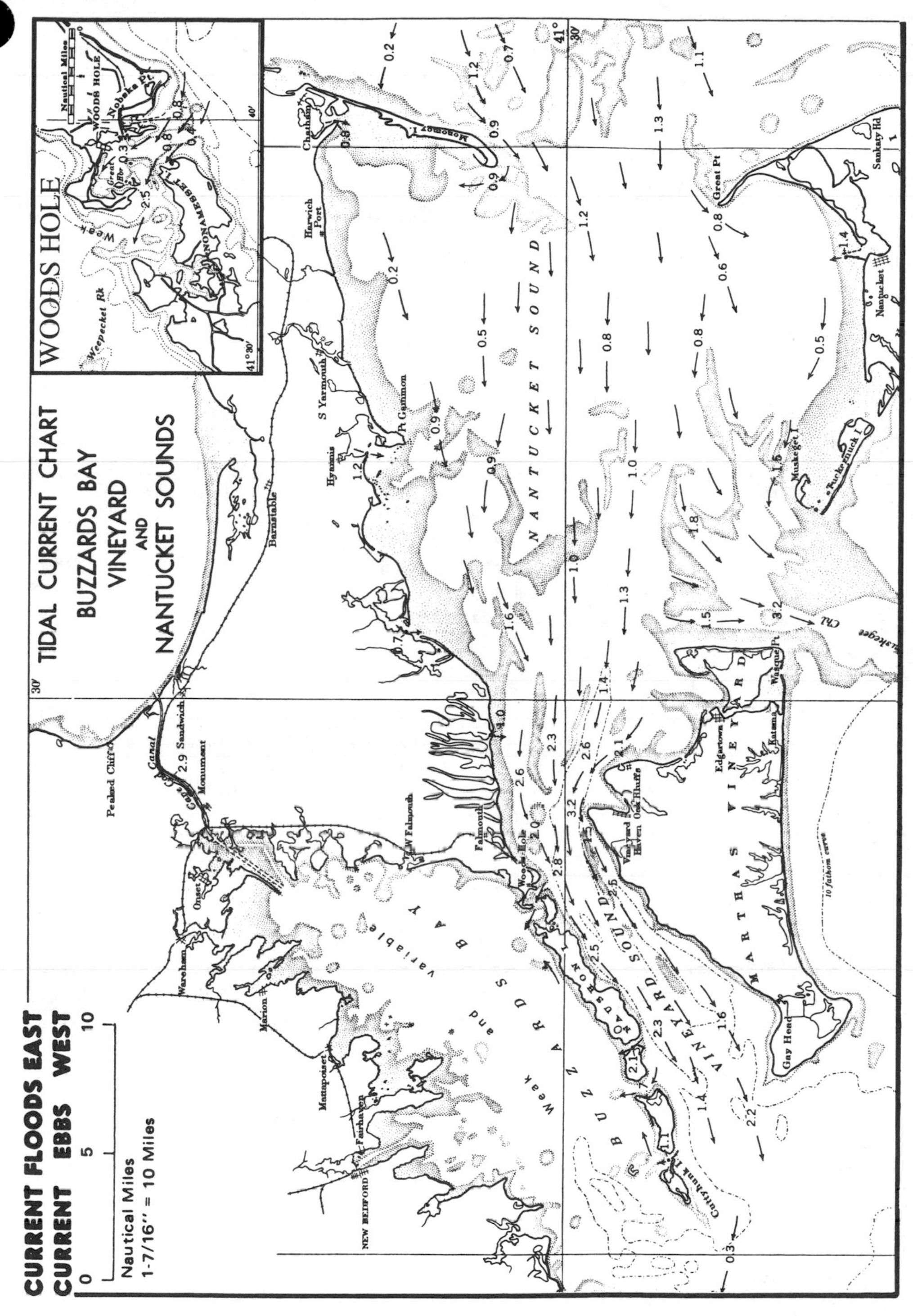

5 HOURS **AFTER** EBB STARTS AT POLLOCK RIP CHANNEL
OR: 3 HOURS **AFTER** HIGH WATER AT BOSTON

Velocities shown are at Spring Tides. **See Note at bottom of Boston Tables: Rule-of-Thumb for Current Velocities.** *(Pollock Rip Ch. is SE of Monomoy Pt.)*

2014 HIGH & LOW WATER
NEWPORT, RI
41°30.3'N, 71°19.6'W

Standard Time (January) — **Standard Time** (February)

DAY OF MONTH	DAY OF WEEK	JANUARY HIGH a.m.	Ht.	p.m.	Ht.	JANUARY LOW a.m.	p.m.	DAY OF MONTH	DAY OF WEEK	FEBRUARY HIGH a.m.	Ht.	p.m.	Ht.	FEBRUARY LOW a.m.	p.m.
1	W	7 13	4.7	**7 36**	4.1	12 13	**1 15**	1	S	8 39	4.6	**9 02**	4.4	1 53	**2 29**
2	T	8 04	4.8	**8 28**	4.2	1 07	**2 05**	2	S	9 29	4.4	**9 53**	4.3	2 45	**3 11**
3	F	8 56	4.7	**9 21**	4.2	2 02	**2 53**	3	M	10 21	4.1	**10 46**	4.1	3 34	**3 52**
4	S	9 51	4.5	**10 17**	4.1	2 56	**3 40**	4	T	11 14	3.7	**11 40**	3.8	4 24	**4 34**
5	S	10 44	4.2	**11 11**	4.0	3 48	**4 25**	5	W	...	...	**12 06**	3.3	5 16	**5 17**
6	M	11 38	3.8	**...**	...	4 45	**5 13**	6	T	12 34	3.5	**1 00**	3.0	6 28	**6 09**
7	T	12 07	3.8	**12 34**	3.5	5 53	**6 06**	7	F	1 30	3.2	**1 57**	2.8	8 27	**7 11**
8	W	1 04	3.6	**1 31**	3.2	7 47	**7 09**	8	S	2 31	3.0	**2 58**	2.6	9 34	**8 20**
9	T	2 03	3.4	**2 30**	2.9	9 08	**8 13**	9	S	3 36	2.9	**3 59**	2.7	10 22	**9 21**
10	F	3 04	3.3	**3 30**	2.8	10 04	**9 05**	10	M	4 37	3.0	**4 55**	2.8	10 58	**10 12**
11	S	4 06	3.3	**4 29**	2.8	10 50	**9 49**	11	T	5 27	3.1	**5 42**	3.0	11 30	**10 58**
12	S	5 02	3.3	**5 21**	2.9	11 26	**10 31**	12	W	6 10	3.2	**6 24**	3.2	11 59	**11 43**
13	M	5 50	3.4	**6 07**	3.0	11 56	**11 14**	13	T	6 46	3.3	**7 02**	3.3	...	**12 35**
14	T	6 32	3.5	**6 48**	3.1	...	**12 26**	14	F	7 20	3.4	**7 37**	3.4	12 27	**1 08**
15	W	7 10	3.5	**7 27**	3.2	12 01	**12 59**	15	S	7 53	3.4	**8 13**	3.5	1 09	**1 41**
16	T	7 45	3.5	**8 04**	3.3	12 41	**1 34**	16	S	8 26	3.4	**8 48**	3.5	1 49	**2 13**
17	F	8 19	3.4	**8 41**	3.3	1 24	**2 08**	17	M	9 02	3.4	**9 25**	3.5	2 26	**2 43**
18	S	8 53	3.4	**9 18**	3.2	2 05	**2 41**	18	T	9 40	3.3	**10 05**	3.5	3 02	**3 13**
19	S	9 28	3.3	**9 57**	3.2	2 44	**3 13**	19	W	10 23	3.2	**10 49**	3.4	3 39	**3 46**
20	M	10 06	3.1	**10 37**	3.1	3 22	**3 44**	20	T	11 09	3.1	**11 37**	3.4	4 19	**4 25**
21	T	10 47	3.0	**11 19**	3.1	4 01	**4 18**	21	F	...	...	**12 01**	3.0	5 06	**5 12**
22	W	11 33	2.9	**...**	...	4 43	**4 57**	22	S	12 31	3.4	**12 57**	2.9	6 06	**6 11**
23	T	12 06	3.1	**12 23**	2.9	5 34	**5 46**	23	S	1 31	3.4	**1 59**	2.9	7 27	**7 25**
24	F	12 58	3.2	**1 18**	2.8	6 39	**6 46**	24	M	2 36	3.5	**3 06**	3.1	8 54	**8 44**
25	S	1 55	3.3	**2 20**	2.8	7 59	**7 56**	25	T	3 45	3.7	**4 13**	3.4	10 02	**9 55**
26	S	2 59	3.5	**3 26**	3.0	9 15	**9 04**	26	W	4 50	4.0	**5 14**	3.8	10 57	**10 58**
27	M	4 06	3.8	**4 32**	3.2	10 18	**10 06**	27	T	5 48	4.3	**6 10**	4.1	11 46	**11 56**
28	T	5 08	4.1	**5 32**	3.6	11 14	**11 05**	28	F	6 40	4.5	**7 01**	4.4	...	**12 33**
29	W	6 04	4.4	**6 27**	3.9	...	**12 07**								
30	T	6 57	4.6	**7 19**	4.2	12 03	**12 57**								
31	F	7 48	4.7	**8 10**	4.4	12 59	**1 44**								

Dates when Ht. of **Low** Water is below Mean Lower Low with Ht. of lowest given for each period and Date of lowest in ( ):

January:
1st - 6th: -1.0' (2nd)
14th: -0.2'
16th - 19th: -0.3' (17th)
26th - 31st: -1.0' (31st)

February:
1st - 4th: -1.0' (1st)
14th - 19th: -0.3' (15th - 18th)
25th - 28th: -0.8' (28th)

Average Rise and Fall 3.5 ft.

When a high tide exceeds avg. ht., the *following* low tide will be lower than avg.

2014 HIGH & LOW WATER
NEWPORT, RI

41°30.3'N, 71°19.6'W

***Daylight Time starts March 9 at 2 a.m.** **Daylight Saving Time**

DAY OF MONTH	DAY OF WEEK	MARCH						DAY OF MONTH	DAY OF WEEK	APRIL					
		HIGH				LOW				HIGH				LOW	
		a.m.	Ht.	**p.m.**	Ht.	a.m.	**p.m.**			a.m.	Ht.	**p.m.**	Ht.	a.m.	**p.m.**
1	**S**	7 30	4.5	**7 51**	4.6	12 51	**1 17**	1	**T**	9 45	4.1	**10 04**	4.5	3 12	**3 05**
2	**S**	8 19	4.5	**8 40**	4.6	1 43	**1 59**	2	**W**	10 32	3.9	**10 51**	4.2	3 52	**3 43**
3	**M**	9 08	4.3	**9 29**	4.4	2 30	**2 39**	3	**T**	11 20	3.6	**11 38**	3.8	4 31	**4 23**
4	**T**	9 58	4.0	**10 20**	4.1	3 16	**3 18**	4	**F**	...	...	**12 10**	3.3	5 11	**5 05**
5	**W**	10 47	3.6	**11 09**	3.8	3 57	**3 56**	5	**S**	12 27	3.4	**12 58**	3.1	5 53	**5 50**
6	**T**	11 37	3.3	**...**	...	4 41	**4 38**	6	**S**	1 16	3.1	**1 50**	2.9	6 45	**6 44**
7	**F**	12 01	3.4	**12 29**	3.0	5 30	**5 24**	7	**M**	2 09	2.9	**2 44**	2.8	7 59	**7 53**
8	**S**	12 54	3.1	**1 24**	2.8	6 41	**6 21**	8	**T**	3 05	2.7	**3 42**	2.8	9 29	**9 16**
9	**S**	1 52	2.9	***3 22**	2.7	*9 41	***8 34**	9	**W**	4 07	2.7	**4 39**	3.0	10 23	**10 24**
10	**M**	3 57	2.8	**4 24**	2.7	10 37	**9 51**	10	**T**	5 05	2.8	**5 31**	3.2	11 04	**11 17**
11	**T**	5 01	2.8	**5 22**	2.8	11 17	**10 52**	11	**F**	5 54	2.9	**6 16**	3.4	11 41	**...**
12	**W**	5 55	2.9	**6 12**	3.0	11 51	**11 41**	12	**S**	6 35	3.2	**6 56**	3.7	12 02	**12 18**
13	**T**	6 39	3.0	**6 54**	3.3	...	**12 25**	13	**S**	7 14	3.4	**7 34**	3.9	12 45	**12 54**
14	**F**	7 16	3.2	**7 32**	3.5	12 27	**1 00**	14	**M**	7 52	3.5	**8 12**	4.1	1 27	**1 29**
15	**S**	7 50	3.4	**8 07**	3.7	1 10	**1 34**	15	**T**	8 32	3.6	**8 51**	4.2	2 07	**2 05**
16	**S**	8 23	3.5	**8 42**	3.8	1 50	**2 08**	16	**W**	9 13	3.7	**9 33**	4.3	2 47	**2 42**
17	**M**	8 59	3.5	**9 19**	3.9	2 29	**2 40**	17	**T**	9 58	3.7	**10 19**	4.3	3 27	**3 20**
18	**T**	9 37	3.5	**9 57**	3.9	3 07	**3 12**	18	**F**	10 46	3.6	**11 08**	4.2	4 08	**4 01**
19	**W**	10 18	3.5	**10 39**	3.9	3 43	**3 45**	19	**S**	11 38	3.5	**...**	...	4 51	**4 46**
20	**T**	11 03	3.4	**11 26**	3.8	4 21	**4 21**	20	**S**	12 02	4.1	**12 33**	3.5	5 40	**5 38**
21	**F**	11 52	3.3	**...**	...	5 02	**5 02**	21	**M**	12 59	3.9	**1 30**	3.5	6 39	**6 40**
22	**S**	12 17	3.7	**12 45**	3.2	5 49	**5 50**	22	**T**	1 59	3.8	**2 31**	3.5	7 56	**8 03**
23	**S**	1 12	3.7	**1 42**	3.1	6 47	**6 50**	23	**W**	3 02	3.7	**3 34**	3.7	9 21	**9 47**
24	**M**	2 13	3.6	**2 44**	3.2	8 07	**8 07**	24	**T**	4 07	3.7	**4 38**	3.9	10 22	**11 01**
25	**T**	3 18	3.6	**3 50**	3.3	9 39	**9 37**	25	**F**	5 11	3.8	**5 39**	4.2	11 09	**11 57**
26	**W**	4 26	3.7	**4 56**	3.6	10 44	**10 54**	26	**S**	6 09	3.9	**6 33**	4.4	11 51	**...**
27	**T**	5 31	3.9	**5 57**	4.0	11 35	**11 56**	27	**S**	7 01	4.0	**7 23**	4.6	12 46	**12 31**
28	**F**	6 29	4.1	**6 52**	4.4	...	**12 20**	28	**M**	7 50	4.0	**8 10**	4.7	1 31	**1 11**
29	**S**	7 22	4.3	**7 43**	4.6	12 51	**1 03**	29	**T**	8 36	4.0	**8 55**	4.6	2 13	**1 51**
30	**S**	8 11	4.3	**8 30**	4.7	1 42	**1 45**	30	**W**	9 22	3.9	**9 39**	4.4	2 52	**2 32**
31	**M**	8 58	4.3	**9 17**	4.7	2 29	**2 25**								

Dates when Ht. of **Low** Water is below Mean Lower Low with Ht. of lowest given for each period and Date of lowest in ( ):

1st - 5th: -0.9' (1st - 2nd)
16th - 20th: -0.4' (18th)
27th - 31st: -0.7' (30th - 31st)

1st - 3rd: -0.6' (1st)
14th - 18th: -0.3' (15th - 18th)
26th - 30th: -0.3'

Average Rise and Fall 3.5 ft.

When a high tide exceeds avg. ht., the *following* low tide will be lower than avg.

2014 HIGH & LOW WATER NEWPORT, RI

41°30.3'N, 71°19.6'W

Daylight Saving Time

DAY OF MONTH	DAY OF WEEK	MAY					
		HIGH				LOW	
		a.m.	Ht.	p.m.	Ht.	a.m.	p.m.
1	T	10 07	3.7	10 24	4.1	3 29	3 12
2	F	10 53	3.5	11 08	3.8	4 06	3 54
3	S	11 40	3.3	11 53	3.5	4 43	4 36
4	S	...	...	12 29	3.2	5 24	5 22
5	M	12 38	3.2	1 16	3.0	6 08	6 13
6	T	1 24	3.0	2 04	3.0	7 03	7 15
7	W	2 12	2.8	2 55	3.0	8 12	8 33
8	T	3 03	2.8	3 48	3.1	9 18	9 46
9	F	3 59	2.8	4 41	3.3	10 09	10 43
10	S	4 56	2.9	5 30	3.5	10 52	11 31
11	S	5 47	3.1	6 16	3.8	11 31	...
12	M	6 35	3.3	6 59	4.1	12 15	12 11
13	T	7 20	3.5	7 42	4.4	12 59	12 51
14	W	8 05	3.7	8 26	4.5	1 44	1 32
15	T	8 51	3.8	9 13	4.6	2 29	2 16
16	F	9 39	3.8	10 02	4.6	3 14	3 01
17	S	10 30	3.8	10 54	4.5	3 59	3 48
18	S	11 24	3.8	11 48	4.3	4 45	4 37
19	M	...	...	12 20	3.8	5 34	5 32
20	T	12 45	4.1	1 18	3.8	6 31	6 38
21	W	1 43	3.9	2 16	3.9	7 39	8 12
22	T	2 43	3.8	3 17	4.0	8 53	9 58
23	F	3 46	3.6	4 19	4.1	9 52	11 02
24	S	4 48	3.6	5 19	4.2	10 39	11 54
25	S	5 47	3.6	6 14	4.3	11 20	...
26	M	6 40	3.7	7 04	4.4	12 39	12 01
27	T	7 29	3.7	7 50	4.4	1 20	12 39
28	W	8 15	3.7	8 34	4.4	1 57	1 20
29	T	8 59	3.7	9 17	4.2	2 32	2 03
30	F	9 43	3.6	9 58	4.0	3 07	2 46
31	S	10 27	3.5	10 40	3.7	3 43	3 29

Daylight Saving Time

DAY OF MONTH	DAY OF WEEK	JUNE					
		HIGH				LOW	
		a.m.	Ht.	p.m.	Ht.	a.m.	p.m.
1	S	11 11	3.4	11 20	3.5	4 19	4 12
2	M	11 56	3.3	...	...	4 57	4 57
3	T	12 01	3.3	12 40	3.2	5 37	5 45
4	W	12 43	3.1	1 25	3.2	6 22	6 40
5	T	1 25	3.0	2 09	3.2	7 12	7 44
6	F	2 11	2.9	2 57	3.3	8 10	8 57
7	S	3 03	2.9	3 48	3.4	9 08	10 01
8	S	4 00	2.9	4 42	3.6	9 59	10 55
9	M	5 00	3.1	5 35	3.9	10 45	11 44
10	T	5 57	3.3	6 27	4.2	11 31	...
11	W	6 50	3.5	7 16	4.5	12 32	12 17
12	T	7 40	3.8	8 05	4.7	1 21	1 05
13	F	8 30	3.9	8 55	4.9	2 10	1 55
14	S	9 21	4.1	9 46	4.8	2 59	2 46
15	S	10 14	4.1	10 39	4.7	3 47	3 39
16	M	11 09	4.2	11 33	4.5	4 34	4 32
17	T	...	...	12 04	4.2	5 22	5 29
18	W	12 29	4.3	1 01	4.2	6 13	6 36
19	T	1 25	4.0	1 58	4.1	7 10	8 18
20	F	2 23	3.7	2 56	4.1	8 15	9 52
21	S	3 22	3.5	3 57	4.1	9 16	10 55
22	S	4 24	3.4	4 57	4.1	10 07	11 46
23	M	5 24	3.4	5 54	4.1	10 50	...
24	T	6 19	3.4	6 46	4.1	12 30	-A-
25	W	7 09	3.5	7 32	4.2	1 08	12 12
26	T	7 54	3.5	8 15	4.1	1 41	12 55
27	F	8 37	3.6	8 55	4.0	2 12	1 39
28	S	9 19	3.6	9 33	3.9	2 45	2 24
29	S	10 00	3.5	10 11	3.7	3 20	3 08
30	M	10 41	3.5	10 48	3.5	3 55	3 51

A also at 11:31 a.m.

Dates when Ht. of **Low** Water is below Mean Lower Low with Ht. of lowest given for each period and Date of lowest in ( ):

1st: -0.2'
14th - 18th: -0.4' (16th)

12th - 17th: -0.4' (13th - 16th)

Average Rise and Fall 3.5 ft.

When a high tide exceeds avg. ht., the *following* low tide will be lower than avg.

2014 HIGH & LOW WATER NEWPORT, RI

41°30.3'N, 71°19.6'W

DAY OF MONTH	DAY OF WEEK	JULY						DAY OF MONTH	DAY OF WEEK	AUGUST					
		Daylight Saving Time								Daylight Saving Time					
		HIGH				LOW				HIGH				LOW	
		a.m.	Ht.	p.m.	Ht.	a.m.	p.m.			a.m.	Ht.	p.m.	Ht.	a.m.	p.m.
1	T	11 22	3.4	11 25	3.4	4 30	4 34	1	F	...	...	12 05	3.5	5 05	5 31
2	W	...	...	12 03	3.4	5 05	5 17	2	S	12 14	3.2	12 48	3.5	5 41	6 17
3	T	12 04	3.2	12 44	3.3	5 42	6 04	3	S	1 00	3.2	1 35	3.6	6 24	7 16
4	F	12 47	3.1	1 27	3.3	6 24	6 59	4	M	1 53	3.1	2 29	3.6	7 18	8 31
5	S	1 31	3.0	2 12	3.4	7 11	8 03	5	T	2 49	3.1	3 27	3.8	8 21	9 47
6	S	2 22	3.0	3 03	3.5	8 07	9 15	6	W	3 52	3.2	4 31	4.0	9 30	10 51
7	M	3 19	3.0	4 00	3.7	9 06	10 19	7	T	4 59	3.4	5 35	4.3	10 33	11 46
8	T	4 21	3.1	5 00	4.0	10 03	11 15	8	F	6 01	3.7	6 34	4.6	11 33	...
9	W	5 24	3.3	5 58	4.3	10 57	11 59	9	S	6 58	4.1	7 28	4.9	12 37	12 31
10	T	6 23	3.6	6 54	4.6	11 51	...	10	S	7 52	4.5	8 19	5.0	1 27	1 28
11	F	7 18	3.9	7 46	4.8	12 59	12 45	11	M	8 44	4.7	9 10	5.0	2 16	2 25
12	S	8 11	4.2	8 37	5.0	1 51	1 40	12	T	9 35	4.8	10 01	4.9	3 02	3 19
13	S	9 03	4.4	9 29	5.0	2 41	2 35	13	W	10 28	4.8	10 53	4.6	3 46	4 12
14	M	9 56	4.5	10 21	4.8	3 29	3 30	14	T	11 21	4.7	11 47	4.3	4 29	5 04
15	T	10 50	4.5	11 15	4.6	4 15	4 25	15	F	...	...	12 15	4.5	5 12	6 01
16	W	11 44	4.5	...	...	5 00	5 21	16	S	12 40	3.9	1 10	4.2	5 57	7 17
17	T	12 09	4.3	12 40	4.4	5 46	6 25	17	S	1 35	3.6	2 07	3.9	6 47	9 10
18	F	1 04	4.0	1 36	4.2	6 35	7 59	18	M	2 32	3.3	3 06	3.7	7 47	10 19
19	S	2 00	3.7	2 33	4.1	7 31	9 36	19	T	3 32	3.2	4 09	3.6	8 56	11 11
20	S	2 58	3.4	3 33	3.9	8 33	10 40	20	W	4 34	3.1	5 11	3.5	10 00	11 51
21	M	3 59	3.3	4 35	3.8	9 33	11 32	21	T	5 32	3.2	6 05	3.6	10 52	...
22	T	5 00	3.2	5 34	3.8	10 23	...	22	F	6 23	3.4	6 51	3.7	12 21	-C-
23	W	5 57	3.2	6 27	3.8	12 16	-A-	23	S	7 08	3.5	7 30	3.7	12 47	12 22
24	T	6 47	3.4	7 13	3.9	12 50	-B-	24	S	7 47	3.7	8 05	3.8	1 16	1 06
25	F	7 32	3.5	7 54	3.9	1 19	12 36	25	M	8 24	3.8	8 38	3.8	1 48	1 49
26	S	8 14	3.6	8 32	3.9	1 48	1 21	26	T	9 00	3.9	9 11	3.8	2 21	2 30
27	S	8 53	3.7	9 07	3.8	2 20	2 06	27	W	9 35	3.9	9 45	3.7	2 54	3 10
28	M	9 31	3.7	9 42	3.7	2 53	2 49	28	T	10 11	3.8	10 22	3.6	3 26	3 47
29	T	10 09	3.7	10 16	3.6	3 27	3 31	29	F	10 49	3.8	11 02	3.5	3 58	4 24
30	W	10 46	3.6	10 52	3.5	4 00	4 11	30	S	11 30	3.7	11 47	3.4	4 30	5 02
31	T	11 25	3.6	11 31	3.4	4 32	4 50	31	S	...	...	12 16	3.7	5 06	5 46

A also at 11:08 a.m.   **B** also at 11:52 a.m.   **C** also at 11:38 a.m.

Dates when Ht. of **Low** Water is below Mean Lower Low with Ht. of lowest given for each period and Date of lowest in ( ):

10th - 17th: -0.6' (14th)

8th - 15th: -0.6' (11th - 13th)

Average Rise and Fall 3.5 ft.

When a high tide exceeds avg. ht., the *following* low tide will be lower than avg.

2014 HIGH & LOW WATER
NEWPORT, RI
41°30.3'N, 71°19.6'W

Daylight Saving Time — **Daylight Saving Time**

DAY OF MONTH	DAY OF WEEK	SEPTEMBER						DAY OF MONTH	DAY OF WEEK	OCTOBER					
		HIGH				LOW				HIGH				LOW	
		a.m.	Ht.	**p.m.**	Ht.	a.m.	**p.m.**			a.m.	Ht.	**p.m.**	Ht.	a.m.	**p.m.**
1	**M**	12 36	3.2	**1 07**	3.7	5 48	**6 41**	1	**W**	1 13	3.3	**1 44**	3.8	6 18	**7 32**
2	**T**	1 29	3.2	**2 02**	3.8	6 41	**7 55**	2	**T**	2 13	3.3	**2 46**	3.9	7 28	**9 04**
3	**W**	2 28	3.2	**3 04**	3.9	7 49	**9 23**	3	**F**	3 16	3.5	**3 51**	4.0	8 55	**10 13**
4	**T**	3 33	3.3	**4 11**	4.0	9 08	**10 33**	4	**S**	4 22	3.8	**4 57**	4.1	10 17	**11 05**
5	**F**	4 39	3.6	**5 15**	4.3	10 19	**11 26**	5	**S**	5 24	4.2	**5 55**	4.4	11 20	**11 49**
6	**S**	5 42	4.0	**6 15**	4.5	11 23	**...**	6	**M**	6 21	4.5	**6 50**	4.5	...	**12 16**
7	**S**	6 39	4.4	**7 09**	4.8	12 14	**12 22**	7	**T**	7 14	4.9	**7 41**	4.6	12 33	**1 09**
8	**M**	7 33	4.7	**8 00**	4.9	1 01	**1 18**	8	**W**	8 03	5.0	**8 29**	4.6	1 16	**1 59**
9	**T**	8 23	5.0	**8 50**	4.9	1 47	**2 12**	9	**T**	8 52	5.0	**9 18**	4.5	1 59	**2 47**
10	**W**	9 13	5.1	**9 40**	4.7	2 31	**3 04**	10	**F**	9 40	4.9	**10 07**	4.2	2 41	**3 32**
11	**T**	10 04	5.0	**10 30**	4.5	3 14	**3 53**	11	**S**	10 30	4.6	**10 57**	3.9	3 23	**4 15**
12	**F**	10 55	4.7	**11 22**	4.1	3 56	**4 41**	12	**S**	11 20	4.3	**11 48**	3.6	4 04	**4 58**
13	**S**	11 48	4.4	**...**	...	4 37	**5 30**	13	**M**	...	...	**12 12**	3.9	4 47	**5 44**
14	**S**	12 15	3.8	**12 42**	4.1	5 20	**6 26**	14	**T**	12 41	3.4	**1 05**	3.6	5 33	**6 42**
15	**M**	1 09	3.5	**1 38**	3.8	6 08	**8 18**	15	**W**	1 35	3.2	**2 00**	3.3	6 27	**8 29**
16	**T**	2 05	3.3	**2 36**	3.5	7 04	**9 46**	16	**T**	2 31	3.1	**2 57**	3.1	7 35	**9 42**
17	**W**	3 04	3.1	**3 38**	3.3	8 16	**10 37**	17	**F**	3 28	3.1	**3 56**	3.0	9 01	**10 22**
18	**T**	4 04	3.1	**4 40**	3.3	9 36	**11 13**	18	**S**	4 25	3.2	**4 52**	3.1	10 11	**10 55**
19	**F**	5 03	3.2	**5 35**	3.3	10 36	**11 42**	19	**S**	5 17	3.4	**5 40**	3.2	11 01	**11 28**
20	**S**	5 54	3.4	**6 21**	3.4	11 23	**...**	20	**M**	6 03	3.6	**6 20**	3.3	11 44	**...**
21	**S**	6 38	3.6	**6 59**	3.6	12 09	**12 06**	21	**T**	6 42	3.8	**6 58**	3.5	12 01	**12 25**
22	**M**	7 17	3.8	**7 34**	3.7	12 40	**12 48**	22	**W**	7 19	4.0	**7 34**	3.6	12 35	**1 06**
23	**T**	7 53	4.0	**8 07**	3.7	1 13	**1 29**	23	**T**	7 55	4.1	**8 11**	3.7	1 10	**1 46**
24	**W**	8 27	4.0	**8 41**	3.8	1 46	**2 09**	24	**F**	8 32	4.2	**8 50**	3.7	1 45	**2 26**
25	**T**	9 02	4.1	**9 16**	3.7	2 20	**2 47**	25	**S**	9 12	4.3	**9 32**	3.7	2 21	**3 05**
26	**F**	9 39	4.1	**9 55**	3.7	2 52	**3 24**	26	**S**	9 55	4.2	**10 18**	3.6	2 58	**3 45**
27	**S**	10 19	4.0	**10 38**	3.5	3 25	**4 02**	27	**M**	10 42	4.2	**11 08**	3.5	3 37	**4 26**
28	**S**	11 03	4.0	**11 26**	3.4	4 00	**4 41**	28	**T**	11 34	4.1	**...**	...	4 19	**5 12**
29	**M**	11 52	3.9	**...**	...	4 38	**5 25**	29	**W**	12 03	3.4	**12 29**	4.0	5 07	**6 06**
30	**T**	12 17	3.3	**12 46**	3.9	5 23	**6 19**	30	**T**	1 00	3.4	**1 28**	3.9	6 05	**7 15**
								31	**F**	1 59	3.5	**2 29**	3.8	7 18	**8 41**

Dates when Ht. of **Low** Water is below Mean Lower Low with Ht. of lowest given for each period and Date of lowest in ( ):

7th - 12th: -0.6' (9th - 11th)

5th - 11th: -0.6' (9th)

Average Rise and Fall 3.5 ft.

When a high tide exceeds avg. ht., the *following* low tide will be lower than avg.

2014 HIGH & LOW WATER
NEWPORT, RI
41°30.3'N, 71°19.6'W

***Standard Time starts Nov. 2 at 2 a.m.** **Standard Time**

DAY OF MONTH	DAY OF WEEK	NOVEMBER HIGH a.m.	Ht.	**p.m.**	Ht.	NOVEMBER LOW a.m.	**p.m.**	DAY OF MONTH	DAY OF WEEK	DECEMBER HIGH a.m.	Ht.	**p.m.**	Ht.	DECEMBER LOW a.m.	**p.m.**
1	**S**	3 01	3.7	**3 32**	3.8	8 55	**9 50**	1	**M**	2 45	3.9	**3 14**	3.6	9 26	**9 13**
2	**S**	*3 04	3.9	***3 35**	3.9	*9 21	***9 40**	2	**T**	3 47	4.0	**4 15**	3.6	10 23	**9 59**
3	**M**	4 06	4.2	**4 35**	4.0	10 21	**10 24**	3	**W**	4 45	4.2	**5 11**	3.7	11 12	**10 41**
4	**T**	5 04	4.5	**5 31**	4.1	11 14	**11 07**	4	**T**	5 40	4.3	**6 04**	3.7	11 57	**11 23**
5	**W**	5 55	4.7	**6 21**	4.2	11 59	**11 47**	5	**F**	6 28	4.4	**6 51**	3.8	...	**12 37**
6	**T**	6 45	4.8	**7 09**	4.2	...	**12 46**	6	**S**	7 14	4.4	**7 37**	3.7	12 04	**1 15**
7	**F**	7 32	4.8	**7 56**	4.1	12 29	**1 29**	7	**S**	7 59	4.2	**8 22**	3.6	12 46	**1 51**
8	**S**	8 18	4.6	**8 43**	3.9	1 11	**2 11**	8	**M**	8 42	4.0	**9 07**	3.5	1 29	**2 27**
9	**S**	9 05	4.3	**9 31**	3.7	1 53	**2 50**	9	**T**	9 26	3.7	**9 52**	3.3	2 12	**3 03**
10	**M**	9 52	4.0	**10 20**	3.5	2 36	**3 29**	10	**W**	10 09	3.5	**10 38**	3.2	2 55	**3 40**
11	**T**	10 41	3.7	**11 11**	3.3	3 19	**4 10**	11	**T**	10 52	3.2	**11 23**	3.1	3 39	**4 19**
12	**W**	11 30	3.4	**...**	...	4 04	**4 55**	12	**F**	11 34	3.0	**...**	...	4 26	**5 02**
13	**T**	12 01	3.1	**12 18**	3.1	4 55	**5 50**	13	**S**	12 09	3.0	**12 17**	2.8	5 18	**5 52**
14	**F**	12 52	3.0	**1 08**	2.9	5 55	**7 02**	14	**S**	12 55	2.9	**1 02**	2.7	6 20	**6 51**
15	**S**	1 44	3.0	**1 59**	2.8	7 12	**8 11**	15	**M**	1 43	2.9	**1 51**	2.6	7 34	**7 53**
16	**S**	2 37	3.1	**2 52**	2.8	8 31	**9 00**	16	**T**	2 33	3.0	**2 45**	2.7	8 44	**8 48**
17	**M**	3 30	3.2	**3 44**	2.9	9 28	**9 41**	17	**W**	3 27	3.2	**3 43**	2.8	9 40	**9 35**
18	**T**	4 19	3.4	**4 33**	3.0	10 15	**10 20**	18	**T**	4 20	3.4	**4 38**	3.0	10 29	**10 19**
19	**W**	5 03	3.6	**5 18**	3.2	10 58	**10 57**	19	**F**	5 10	3.7	**5 30**	3.2	11 15	**11 03**
20	**T**	5 44	3.9	**6 01**	3.4	11 41	**11 35**	20	**S**	5 58	4.0	**6 18**	3.4	11 59	**11 48**
21	**F**	6 25	4.1	**6 44**	3.5	...	**12 23**	21	**S**	6 44	4.2	**7 06**	3.6	...	**12 47**
22	**S**	7 07	4.3	**7 27**	3.6	12 14	**1 06**	22	**M**	7 31	4.4	**7 54**	3.7	12 34	**1 33**
23	**S**	7 50	4.4	**8 13**	3.7	12 55	**1 49**	23	**T**	8 20	4.5	**8 44**	3.8	1 22	**2 19**
24	**M**	8 36	4.4	**9 01**	3.7	1 37	**2 32**	24	**W**	9 10	4.4	**9 36**	3.9	2 12	**3 04**
25	**T**	9 26	4.3	**9 53**	3.6	2 21	**3 16**	25	**T**	10 03	4.3	**10 31**	3.9	3 02	**3 49**
26	**W**	10 19	4.2	**10 48**	3.6	3 08	**4 02**	26	**F**	10 58	4.1	**11 27**	3.9	3 55	**4 36**
27	**T**	11 14	4.0	**11 45**	3.6	3 59	**4 54**	27	**S**	11 54	3.8	**...**	...	4 53	**5 28**
28	**F**	...	...	**12 12**	3.9	4 58	**5 55**	28	**S**	12 24	3.8	**12 51**	3.6	6 05	**6 30**
29	**S**	12 43	3.7	**1 11**	3.7	6 13	**7 08**	29	**M**	1 22	3.8	**1 50**	3.4	8 00	**7 40**
30	**S**	1 43	3.8	**2 12**	3.6	8 02	**8 19**	30	**T**	2 24	3.7	**2 52**	3.2	9 22	**8 44**
								31	**W**	3 27	3.7	**3 54**	3.2	10 21	**9 35**

Dates when Ht. of **Low** Water is below Mean Lower Low with Ht. of lowest given for each period and Date of lowest in ( ):

3rd - 9th: -0.5' (5th, 7th)
22nd - 26th: -0.3' (24th - 25th)

2nd - 9th: -0.4' (4th, 6th - 7th)
19th - 27th: -0.7' (23rd - 24th)

Average Rise and Fall 3.5 ft.

When a high tide exceeds avg. ht., the *following* low tide will be lower than avg.

Narragansett Bay Currents

This Current Diagram shows average maximum currents with a normal range (3.5 ft.) of tides at Newport. See pp. 78-83.

Maximum Ebb Currents, 3 hours *after* High Water at Newport, are shown by double-headed arrows and velocities are underlined.

Maximum Flood Currents, 2 1/2 hours *before* High Water at Newport, are shown by single-headed arrows and velocities are not underlined.

When height of High Water at Newport is 3.0 ft., subtract 30% from all velocities shown. When height is 4.0 ft., add 20%; when 4.5 ft., add 40%; when 5.0 ft., add 60%.

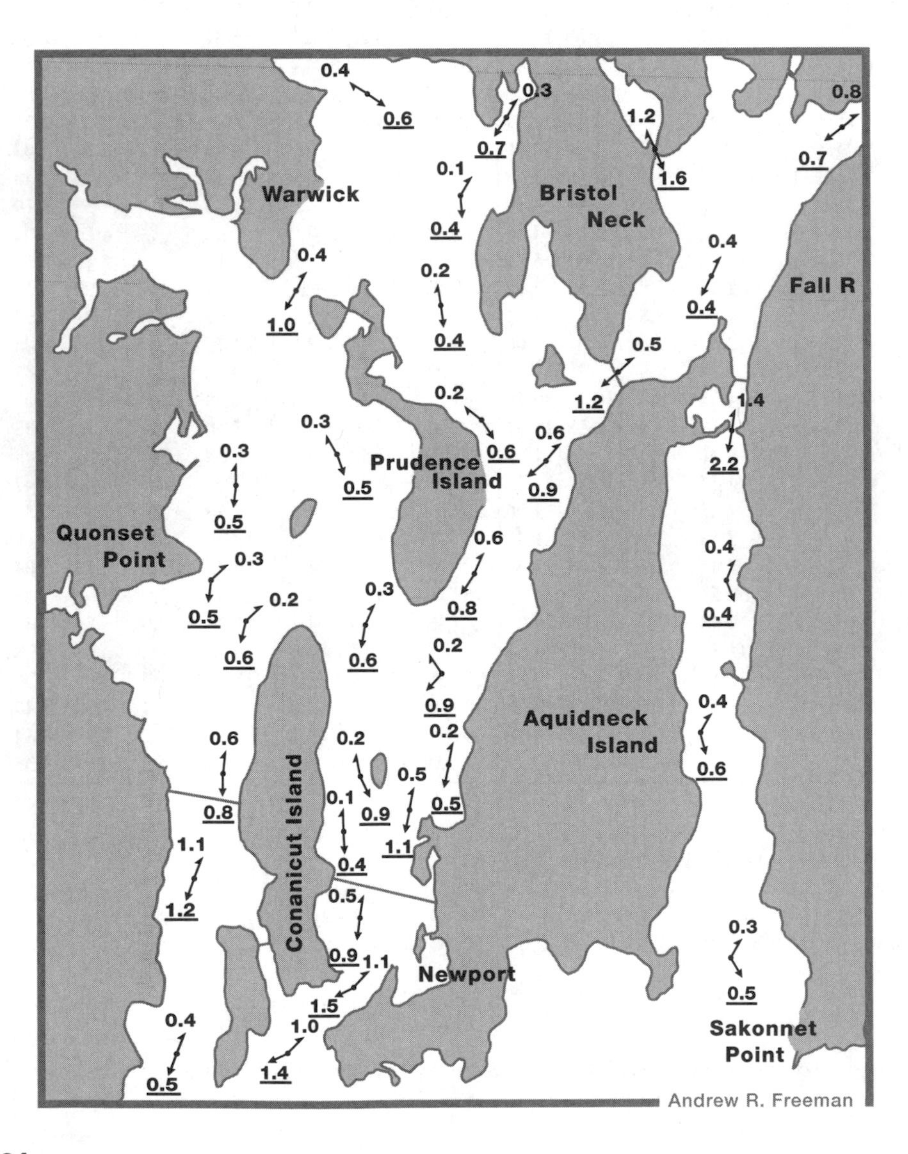

Holding a Fair Current between Eastern Long Island and Nantucket

There is a curious phenomenon which can be used to advantage by every vessel, and particularly the slower cruiser or auxiliary, in making the passage *either* way between eastern Long Island Sound, on the west, and Buzzards Bay, Vineyard and Nantucket Sounds on the east.

Note in the very simplified diagram below, that in Long Island Sound, the Ebb Current flows to the *east*, and in Buzzards Bay, Vineyard and Nantucket Sounds the Ebb Current flows to the *west*. (Off Newport, these opposed Ebb Currents merge and flow *south*.) The reverse is also true: the Flood Current flows *west* through Long Island Sound and *east* through Buzzards Bay, Vineyard and Nantucket Sounds. (Half arrow indicates Ebb Current, whole arrow indicates Flood Current.)

In making a *complete* passage through the area of the diagram, simply ride the favoring Ebb Current toward Newport from either direction and, pick up the favoring Flood Current in leaving the Newport area.

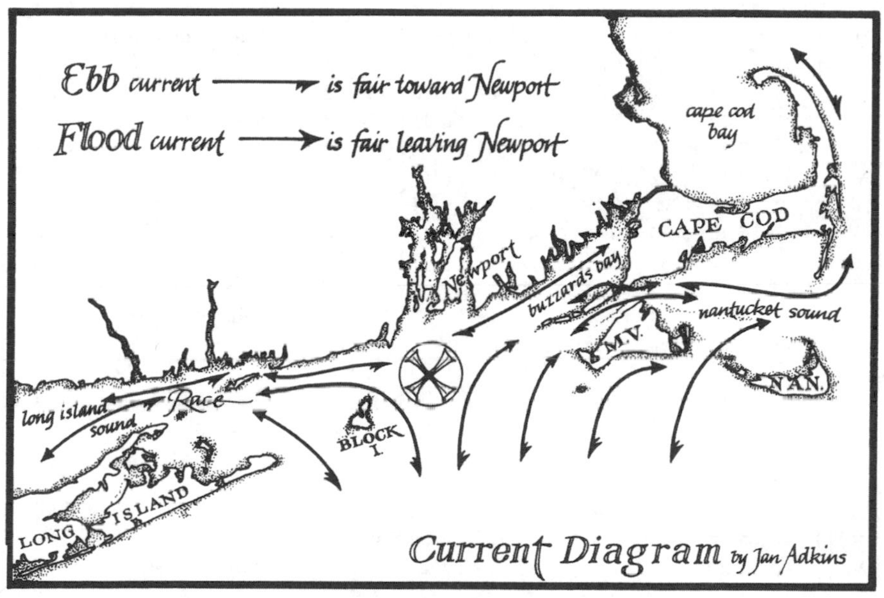

Arrive at "X" at the times shown for "Current Turns to Northwest at The Race," tables pp. 86-91.

The E-W currents between Pt. Judith and Cuttyhunk are only 1/2 to 1 kt., while those to the West of Pt. Judith and to the East of Cuttyhunk are much greater. Bearing this in mind, those making *only a partial trip* through the area may find it better even to buck a slight head current in the Pt. Judith-Cuttyhunk area so as to pick up the maximum hours of strong favoring currents beyond those points.

For example, if headed for the Cape Cod Canal, refer to the Tidal Current Chart Buzzards Bay, Vineyard & Nantucket Sound, pp. 66-77 and arrive just N. of Cuttyhunk as Flood Starts at Pollock Rip, pp. 60-65 to ensure the most favorable currents. If headed for Nantucket, refer to the same Charts and arrive just S. of Cuttyhunk at 3 hours after Flood Starts at Pollock Rip, pp. 60-65. If headed into Long Island Sound, refer to the Tidal Current Chart Long Island Sound and Block Island Sound, pp. 92-97 and arrive at Pt. Judith when Flood Current turns West at The Race, p. 95.

2014 CURRENT TABLE

THE RACE, LONG ISLAND SOUND

41°13.69'N, 72°03.75'W 0.2 nm E.N.E. of Valiant Rock

Standard Time — **Standard Time**

JANUARY

DAY OF MONTH	DAY OF WEEK	NORTHWEST Flood Starts a.m.	NORTHWEST Flood Starts **p.m.**	NORTHWEST Flood Starts Kts.	SOUTHEAST Ebb Starts a.m.	SOUTHEAST Ebb Starts **p.m.**	SOUTHEAST Ebb Starts Kts.
1	W	4 56	**5 37**	4.3	10 56	**11 34**	a5.6
2	T	5 51	**6 28**	4.4	11 49	**...**	5.6
3	F	6 47	**7 20**	4.4	12 27	**12 43**	p5.5
4	S	7 44	**8 14**	p4.3	1 21	**1 37**	p5.3
5	S	8 41	**9 08**	p4.1	2 15	**2 34**	a5.1
6	M	9 42	**10 05**	p3.8	3 11	**3 32**	a4.9
7	T	10 46	**11 04**	p3.5	4 08	**4 34**	a4.6
8	W	11 51	**...**	3.0	5 06	**5 37**	a4.4
9	T	12 04	**12 54**	a3.3	6 05	**6 41**	a4.2
10	F	1 04	**1 54**	a3.2	7 04	**7 42**	a4.2
11	S	2 01	**2 49**	a3.1	7 59	**8 38**	a4.2
12	S	2 55	**3 39**	3.1	8 51	**9 27**	a4.3
13	M	3 44	**4 23**	3.2	9 37	**10 11**	a4.3
14	T	4 29	**5 03**	3.2	10 19	**10 51**	a4.3
15	W	5 10	**5 40**	p3.3	10 58	**11 27**	a4.3
16	T	5 50	**6 15**	p3.3	11 34	**...**	4.2
17	F	6 28	**6 49**	p3.4	12 02	**12 09**	p4.2
18	S	7 05	**7 23**	p3.4	12 37	**12 45**	p4.2
19	S	7 43	**7 57**	p3.4	1 12	**1 21**	a4.2
20	M	8 22	**8 34**	p3.4	1 49	**2 01**	a4.2
21	T	9 05	**9 14**	p3.3	2 29	**2 43**	a4.2
22	W	9 52	**9 59**	p3.3	3 11	**3 30**	a4.2
23	T	10 44	**10 50**	p3.3	3 59	**4 22**	a4.2
24	F	11 40	**11 47**	p3.3	4 51	**5 19**	a4.2
25	S	...	**12 39**	2.8	5 47	**6 22**	a4.2
26	S	12 47	**1 39**	a3.4	6 48	**7 26**	a4.4
27	M	1 47	**2 37**	a3.5	7 49	**8 28**	a4.7
28	T	2 47	**3 32**	a3.8	8 49	**9 27**	a5.0
29	W	3 45	**4 26**	a4.1	9 46	**10 23**	a5.3
30	T	4 41	**5 18**	p4.4	10 41	**11 16**	a5.5
31	F	5 36	**6 09**	p4.5	11 34	**...**	5.6

FEBRUARY

DAY OF MONTH	DAY OF WEEK	NORTHWEST Flood Starts a.m.	NORTHWEST Flood Starts **p.m.**	NORTHWEST Flood Starts Kts.	SOUTHEAST Ebb Starts a.m.	SOUTHEAST Ebb Starts **p.m.**	SOUTHEAST Ebb Starts Kts.
1	S	6 31	**6 59**	p4.5	12 08	**12 27**	p5.5
2	S	7 26	**7 50**	p4.4	1 00	**1 20**	a5.4
3	M	8 22	**8 42**	p4.1	1 52	**2 14**	a5.2
4	T	9 20	**9 38**	p3.7	2 45	**3 09**	a4.9
5	W	10 20	**10 35**	p3.4	3 39	**4 07**	a4.6
6	T	11 22	**11 35**	p3.1	4 35	**5 08**	a4.2
7	F	...	**12 24**	2.7	5 33	**6 09**	a4.0
8	S	12 36	**1 24**	a2.9	6 31	**7 10**	a3.8
9	S	1 34	**2 19**	a2.8	7 28	**8 05**	a3.8
10	M	2 28	**3 08**	2.8	8 21	**8 55**	a3.8
11	T	3 17	**3 52**	2.9	9 08	**9 38**	a3.9
12	W	4 02	**4 31**	p3.1	9 50	**10 17**	a4.0
13	T	4 43	**5 08**	p3.2	10 29	**10 54**	a4.1
14	F	5 21	**5 42**	p3.4	11 06	**11 29**	4.2
15	S	5 59	**6 16**	p3.5	11 41	**...**	4.2
16	S	6 35	**6 50**	p3.6	12 04	**12 18**	a4.4
17	M	7 13	**7 25**	p3.6	12 40	**12 56**	a4.5
18	T	7 52	**8 02**	p3.6	1 17	**1 35**	a4.5
19	W	8 35	**8 44**	p3.5	1 58	**2 19**	a4.5
20	T	9 22	**9 31**	p3.4	2 42	**3 06**	a4.5
21	F	10 15	**10 25**	p3.3	3 30	**3 59**	a4.4
22	S	11 13	**11 25**	p3.3	4 25	**4 57**	a4.2
23	S	...	**12 15**	2.9	5 24	**6 01**	a4.2
24	M	12 28	**1 17**	a3.3	6 28	**7 07**	a4.3
25	T	1 32	**2 17**	a3.4	7 32	**8 10**	a4.5
26	W	2 34	**3 13**	3.7	8 34	**9 10**	4.8
27	T	3 32	**4 07**	p4.1	9 32	**10 05**	5.1
28	F	4 28	**4 58**	p4.4	10 27	**10 57**	p5.4

The Kts. (knots) columns show the **maximum** predicted velocities of the stronger one of the Flood Currents and the stronger one of the Ebb Currents for each day.
The letter "a" means the velocity shown should occur **after** the **a.m.** Current Change. The letter "p" means the velocity shown should occur **after** the **p.m.** Current Change (even if next morning). No "a" or "p" means a.m. and p.m. velocities are the same for that day.
Avg. Max. Velocity: Flood 3.3 Kts., Ebb 4.2 Kts.
Max. Flood 2 hrs. 45 min. after Flood Starts, ±15 min.
Max. Ebb 3 hrs. 25 min. after Ebb Starts, ±15 min.
Use THE RACE tables with current charts pp. 92-97

See pp. 22-29 for Current Change at other points.

2014 CURRENT TABLE

THE RACE, LONG ISLAND SOUND

41°13.69'N, 72°03.75'W 0.2 nm E.N.E. of Valiant Rock

***Daylight Time starts March 9 at 2 a.m.** **Daylight Saving Time**

MARCH

DAY OF MONTH	DAY OF WEEK	CURRENT TURNS TO NORTHWEST Flood Starts			CURRENT TURNS TO SOUTHEAST Ebb Starts		
		a.m.	**p.m.**	Kts.	a.m.	**p.m.**	Kts.
1	**S**	5 23	**5 48**	p4.5	11 19	**11 48**	p5.5
2	**S**	6 15	**6 37**	p4.5	...	**12 10**	5.3
3	**M**	7 08	**7 27**	p4.3	12 37	**1 01**	a5.5
4	**T**	8 01	**8 18**	p4.0	1 27	**1 53**	a5.3
5	**W**	8 54	**9 10**	p3.6	2 16	**2 45**	a4.9
6	**T**	9 50	**10 05**	3.1	3 08	**3 39**	a4.5
7	**F**	10 49	**11 04**	2.8	4 01	**4 35**	a4.0
8	**S**	11 48	**...**	2.5	4 56	**5 33**	a3.7
9	**S**	12 04	***1 46**	a2.6	*6 53	***7 30**	a3.5
10	**M**	2 02	**2 40**	2.5	7 49	**8 25**	a3.4
11	**T**	2 56	**3 29**	p2.6	8 42	**9 14**	3.4
12	**W**	3 45	**4 12**	p2.8	9 31	**9 58**	3.6
13	**T**	4 29	**4 52**	p3.0	10 15	**10 38**	p3.9
14	**F**	5 10	**5 30**	p3.2	10 56	**11 16**	p4.2
15	**S**	5 50	**6 06**	p3.4	11 35	**11 53**	p4.4
16	**S**	6 28	**6 41**	p3.6	...	**12 13**	4.2
17	**M**	7 06	**7 17**	p3.7	12 30	**12 51**	a4.7
18	**T**	7 45	**7 55**	p3.8	1 08	**1 31**	a4.8
19	**W**	8 26	**8 35**	p3.8	1 48	**2 13**	a4.9
20	**T**	9 10	**9 20**	p3.7	2 30	**2 59**	a4.8
21	**F**	9 58	**10 10**	p3.5	3 17	**3 48**	a4.7
22	**S**	10 53	**11 07**	p3.3	4 08	**4 42**	a4.5
23	**S**	11 52	**...**	3.0	5 05	**5 42**	a4.3
24	**M**	12 10	**12 56**	a3.2	6 07	**6 46**	a4.2
25	**T**	1 16	**1 58**	a3.2	7 12	**7 52**	a4.2
26	**W**	2 21	**2 58**	p3.4	8 18	**8 55**	p4.5
27	**T**	3 23	**3 55**	p3.8	9 21	**9 53**	p4.9
28	**F**	4 22	**4 48**	p4.1	10 19	**10 47**	p5.2
29	**S**	5 17	**5 38**	p4.3	11 13	**11 38**	p5.4
30	**S**	6 09	**6 28**	p4.4	...	**12 05**	5.0
31	**M**	7 00	**7 16**	p4.3	12 27	**12 54**	a5.5

APRIL

DAY OF MONTH	DAY OF WEEK	CURRENT TURNS TO NORTHWEST Flood Starts			CURRENT TURNS TO SOUTHEAST Ebb Starts		
		a.m.	**p.m.**	Kts.	a.m.	**p.m.**	Kts.
1	**T**	7 49	**8 04**	p4.1	1 14	**1 43**	a5.4
2	**W**	8 38	**8 52**	3.8	2 01	**2 31**	a5.1
3	**T**	9 28	**9 42**	3.4	2 48	**3 19**	a4.8
4	**F**	10 20	**10 36**	a3.1	3 35	**4 09**	a4.3
5	**S**	11 12	**11 30**	a2.7	4 24	**5 00**	a3.9
6	**S**	...	**12 06**	2.5	5 15	**5 52**	a3.5
7	**M**	12 27	**1 00**	2.4	6 08	**6 45**	a3.2
8	**T**	1 24	**1 52**	p2.4	7 02	**7 37**	3.1
9	**W**	2 17	**2 41**	p2.5	7 56	**8 26**	p3.3
10	**T**	3 06	**3 26**	p2.7	8 47	**9 12**	p3.6
11	**F**	3 52	**4 07**	p2.9	9 34	**9 55**	p4.0
12	**S**	4 35	**4 47**	p3.2	10 19	**10 36**	p4.3
13	**S**	5 16	**5 26**	p3.5	11 01	**11 16**	p4.6
14	**M**	5 56	**6 05**	p3.7	11 43	**11 56**	p4.9
15	**T**	6 36	**6 46**	p3.9	...	**12 25**	4.4
16	**W**	7 18	**7 27**	p3.9	12 38	**1 09**	a5.1
17	**T**	8 02	**8 12**	p3.9	1 21	**1 54**	a5.1
18	**F**	8 49	**9 01**	p3.8	2 07	**2 42**	a5.1
19	**S**	9 39	**9 55**	p3.6	2 57	**3 34**	a4.9
20	**S**	10 35	**10 55**	p3.4	3 51	**4 30**	a4.7
21	**M**	11 34	**11 59**	3.2	4 49	**5 30**	a4.4
22	**T**	...	**12 37**	3.2	5 52	**6 33**	a4.2
23	**W**	1 06	**1 39**	p3.3	6 58	**7 37**	p4.3
24	**T**	2 12	**2 39**	p3.5	8 04	**8 38**	p4.6
25	**F**	3 13	**3 35**	p3.8	9 07	**9 36**	p4.9
26	**S**	4 11	**4 28**	p4.0	10 06	**10 29**	p5.2
27	**S**	5 05	**5 19**	p4.1	11 00	**11 19**	p5.3
28	**M**	5 56	**6 08**	p4.2	11 50	**...**	4.7
29	**T**	6 44	**6 55**	p4.1	12 06	**12 38**	a5.3
30	**W**	7 31	**7 42**	a3.9	12 51	**1 24**	a5.2

The Kts. (knots) columns show the **maximum** predicted velocities of the stronger one of the Flood Currents and the stronger one of the Ebb Currents for each day.
The letter "a" means the velocity shown should occur **after** the **a.m.** Current Change. The letter "p" means the velocity shown should occur **after** the **p.m.** Current Change (even if next morning). No "a" or "p" means a.m. and p.m. velocities are the same for that day.
Avg. Max. Velocity: Flood 3.3 Kts., Ebb 4.2 Kts.
Max. Flood 2 hrs. 45 min. after Flood Starts, ±15 min.
Max. Ebb 3 hrs. 25 min. after Ebb Starts, ±15 min.
Use THE RACE tables with current charts pp. 92-97

See pp. 22-29 for Current Change at other points.

2014 CURRENT TABLE
THE RACE, LONG ISLAND SOUND

41°13.69'N, 72°03.75'W 0.2 nm E.N.E. of Valiant Rock

MAY — Daylight Saving Time								JUNE — Daylight Saving Time							
		CURRENT TURNS TO								CURRENT TURNS TO					
		NORTHWEST Flood Starts			SOUTHEAST Ebb Starts					NORTHWEST Flood Starts			SOUTHEAST Ebb Starts		
DAY OF MONTH	DAY OF WEEK	a.m.	**p.m.**	Kts.	a.m.	**p.m.**	Kts.	DAY OF MONTH	DAY OF WEEK	a.m.	**p.m.**	Kts.	a.m.	**p.m.**	Kts.
1	**T**	8 17	**8 29**	a3.6	1 36	**2 09**	a4.9	1	**S**	9 16	**9 33**	a3.2	2 32	**3 06**	a4.2
2	**F**	9 02	**9 16**	a3.4	2 19	**2 53**	a4.6	2	**M**	9 56	**10 19**	a3.0	3 12	**3 47**	a3.9
3	**S**	9 47	**10 04**	a3.1	3 03	**3 38**	a4.2	3	**T**	10 38	**11 06**	a2.9	3 53	**4 28**	a3.6
4	**S**	10 35	**10 56**	a2.8	3 47	**4 23**	a3.8	4	**W**	11 22	**11 57**	a2.8	4 36	**5 11**	p3.5
5	**M**	11 22	**11 48**	a2.6	4 33	**5 09**	a3.5	5	**T**	...	**12 07**	2.7	5 23	**5 57**	p3.5
6	**T**	...	**12 11**	2.5	5 21	**5 57**	3.2	6	**F**	12 47	**12 54**	p2.8	6 14	**6 45**	p3.6
7	**W**	12 41	**1 00**	p2.5	6 11	**6 46**	p3.3	7	**S**	1 38	**1 43**	p2.9	7 08	**7 35**	p3.8
8	**T**	1 33	**1 48**	p2.6	7 04	**7 34**	p3.4	8	**S**	2 28	**2 32**	p3.1	8 04	**8 26**	p4.1
9	**F**	2 23	**2 34**	p2.7	7 57	**8 22**	p3.7	9	**M**	3 18	**3 22**	p3.3	9 00	**9 18**	p4.5
10	**S**	3 11	**3 19**	p3.0	8 49	**9 09**	p4.1	10	**T**	4 06	**4 11**	p3.6	9 53	**10 08**	p4.9
11	**S**	3 56	**4 03**	p3.2	9 39	**9 55**	p4.4	11	**W**	4 55	**5 01**	p3.8	10 45	**10 59**	p5.2
12	**M**	4 40	**4 47**	p3.5	10 27	**10 40**	p4.8	12	**T**	5 43	**5 51**	p4.1	11 36	**11 49**	p5.4
13	**T**	5 24	**5 31**	p3.8	11 14	**11 26**	p5.1	13	**F**	6 32	**6 42**	p4.2	...	**12 27**	4.8
14	**W**	6 09	**6 16**	p4.0	...	**12 01**	4.5	14	**S**	7 21	**7 35**	p4.2	12 39	**1 18**	a5.5
15	**T**	6 54	**7 03**	p4.1	12 11	**12 47**	a5.3	15	**S**	8 11	**8 29**	a4.2	1 31	**2 10**	a5.5
16	**F**	7 41	**7 52**	p4.1	12 59	**1 36**	a5.3	16	**M**	9 03	**9 27**	a4.1	2 24	**3 04**	a5.3
17	**S**	8 30	**8 45**	3.9	1 48	**2 26**	a5.3	17	**T**	9 58	**10 27**	a4.0	3 20	**4 00**	a5.0
18	**S**	9 22	**9 41**	a3.8	2 40	**3 20**	a5.1	18	**W**	10 55	**11 32**	a3.8	4 19	**4 58**	a4.7
19	**M**	10 17	**10 42**	a3.7	3 35	**4 16**	a4.8	19	**T**	11 55	**...**	3.6	5 20	**5 58**	p4.5
20	**T**	11 16	**11 47**	a3.5	4 34	**5 16**	a4.6	20	**F**	12 38	**12 56**	p3.5	6 25	**6 59**	p4.5
21	**W**	...	**12 17**	3.5	5 37	**6 18**	4.3	21	**S**	1 43	**1 57**	p3.5	7 31	**7 59**	p4.5
22	**T**	12 54	**1 18**	p3.5	6 43	**7 20**	p4.4	22	**S**	2 46	**2 55**	p3.5	8 35	**8 57**	p4.6
23	**F**	2 00	**2 18**	p3.6	7 49	**8 20**	p4.6	23	**M**	3 44	**3 51**	p3.5	9 35	**9 51**	p4.7
24	**S**	3 01	**3 15**	p3.7	8 53	**9 17**	p4.8	24	**T**	4 37	**4 43**	p3.6	10 29	**10 41**	p4.7
25	**S**	3 59	**4 10**	p3.8	9 52	**10 11**	p5.0	25	**W**	5 26	**5 32**	p3.6	11 17	**11 27**	p4.7
26	**M**	4 52	**5 01**	p3.9	10 46	**11 00**	p5.1	26	**T**	6 11	**6 18**	3.5	...	**12 02**	4.1
27	**T**	5 42	**5 50**	p3.9	11 35	**11 46**	p5.0	27	**F**	6 52	**7 01**	a3.5	12 09	**12 42**	a4.6
28	**W**	6 29	**6 37**	p3.8	...	**12 21**	4.3	28	**S**	7 31	**7 42**	a3.5	12 49	**1 20**	a4.5
29	**T**	7 13	**7 22**	a3.7	12 30	**1 04**	a4.9	29	**S**	8 08	**8 22**	a3.4	1 26	**1 57**	a4.3
30	**F**	7 55	**8 06**	a3.5	1 12	**1 46**	a4.7	30	**M**	8 44	**9 02**	a3.3	2 03	**2 34**	a4.1
31	**S**	8 35	**8 49**	a3.3	1 52	**2 26**	a4.4								

The Kts. (knots) columns show the **maximum** predicted velocities of the stronger one of the Flood Currents and the stronger one of the Ebb Currents for each day.
The letter "a" means the velocity shown should occur **after** the **a.m.** Current Change. The letter "p" means the velocity shown should occur **after** the **p.m.** Current Change (even if next morning). No "a" or "p" means a.m. and p.m. velocities are the same for that day.
Avg. Max. Velocity: Flood 3.3 Kts., Ebb 4.2 Kts.
Max. Flood 2 hrs. 45 min. after Flood Starts, ±15 min.
Max. Ebb 3 hrs. 25 min. after Ebb Starts, ±15 min.
Use THE RACE tables with current charts pp. 92-97

See pp. 22-29 for Current Change at other points.

2014 CURRENT TABLE

THE RACE, LONG ISLAND SOUND

41°13.69'N, 72°03.75'W 0.2 nm E.N.E. of Valiant Rock

Daylight Saving Time | **Daylight Saving Time**

JULY

DAY OF MONTH	DAY OF WEEK	CURRENT TURNS TO NORTHWEST Flood Starts a.m.	**p.m.**	Kts.	CURRENT TURNS TO SOUTHEAST Ebb Starts a.m.	**p.m.**	Kts.
1	T	9 20	**9 43**	a3.2	2 39	**3 10**	a4.0
2	W	9 57	**10 26**	a3.1	3 18	**3 49**	3.8
3	T	10 36	**11 12**	a3.0	3 58	**4 29**	p3.8
4	F	11 20	**...**	3.0	4 43	**5 13**	p3.8
5	S	12 01	**12 06**	p3.0	5 32	**6 01**	p3.8
6	S	12 54	**12 58**	p3.0	6 26	**6 53**	p4.0
7	M	1 47	**1 51**	p3.1	7 24	**7 49**	p4.2
8	T	2 42	**2 46**	p3.3	8 24	**8 45**	p4.5
9	W	3 35	**3 41**	p3.6	9 23	**9 41**	p4.8
10	T	4 28	**4 36**	p3.9	10 19	**10 36**	p5.2
11	F	5 19	**5 30**	p4.2	11 14	**11 29**	p5.4
12	S	6 10	**6 24**	p4.3	...	**12 07**	5.0
13	S	7 01	**7 18**	p4.4	12 22	**12 59**	a5.6
14	M	7 51	**8 13**	a4.4	1 15	**1 51**	a5.6
15	T	8 43	**9 10**	a4.4	2 08	**2 45**	a5.4
16	W	9 36	**10 09**	a4.2	3 03	**3 39**	5.1
17	T	10 32	**11 12**	a4.0	4 01	**4 35**	p4.8
18	F	11 30	**...**	3.7	5 01	**5 34**	p4.6
19	S	12 17	**12 32**	p3.5	6 04	**6 34**	p4.4
20	S	1 22	**1 34**	p3.3	7 09	**7 35**	p4.3
21	M	2 25	**2 35**	p3.2	8 14	**8 34**	p4.3
22	T	3 24	**3 32**	p3.2	9 13	**9 30**	p4.3
23	W	4 17	**4 24**	p3.3	10 07	**10 20**	p4.4
24	T	5 05	**5 12**	p3.3	10 55	**11 05**	p4.4
25	F	5 48	**5 57**	3.3	11 37	**11 46**	p4.3
26	S	6 27	**6 38**	a3.4	...	**12 15**	4.0
27	S	7 03	**7 16**	a3.4	12 24	**12 51**	a4.3
28	M	7 37	**7 54**	a3.4	1 00	**1 25**	a4.2
29	T	8 11	**8 31**	a3.4	1 35	**2 00**	4.2
30	W	8 45	**9 09**	a3.4	2 10	**2 35**	p4.2
31	T	9 20	**9 50**	a3.3	2 47	**3 12**	p4.2

AUGUST

DAY OF MONTH	DAY OF WEEK	CURRENT TURNS TO NORTHWEST Flood Starts a.m.	**p.m.**	Kts.	CURRENT TURNS TO SOUTHEAST Ebb Starts a.m.	**p.m.**	Kts.
1	F	9 58	**10 33**	a3.3	3 26	**3 52**	p4.1
2	S	10 40	**11 21**	a3.2	4 10	**4 36**	p4.1
3	S	11 28	**...**	3.1	4 58	**5 25**	p4.0
4	M	12 16	**12 23**	p3.1	5 52	**6 19**	p4.0
5	T	1 12	**1 20**	p3.2	6 52	**7 18**	p4.2
6	W	2 10	**2 19**	p3.3	7 55	**8 19**	p4.4
7	T	3 08	**3 19**	p3.6	8 57	**9 19**	p4.8
8	F	4 03	**4 17**	p3.9	9 57	**10 17**	p5.1
9	S	4 57	**5 13**	p4.2	10 53	**11 12**	p5.4
10	S	5 49	**6 08**	p4.4	11 47	**...**	5.2
11	M	6 40	**7 02**	4.5	12 06	**12 39**	5.5
12	T	7 30	**7 57**	a4.6	12 59	**1 31**	5.5
13	W	8 21	**8 52**	a4.5	1 51	**2 22**	5.4
14	T	9 13	**9 49**	a4.3	2 45	**3 15**	p5.2
15	F	10 08	**10 49**	a4.0	3 41	**4 10**	p4.9
16	S	11 05	**11 52**	a3.6	4 39	**5 07**	p4.5
17	S	...	**12 07**	3.3	5 40	**6 06**	p4.2
18	M	12 56	**1 09**	p3.0	6 43	**7 07**	p4.0
19	T	1 59	**2 11**	p2.9	7 46	**8 07**	p3.9
20	W	2 57	**3 09**	p2.9	8 45	**9 03**	p3.9
21	T	3 50	**4 01**	p3.0	9 38	**9 54**	p4.0
22	F	4 36	**4 48**	p3.1	10 24	**10 39**	p4.0
23	S	5 18	**5 31**	3.2	11 05	**11 19**	p4.1
24	S	5 56	**6 11**	a3.3	11 43	**11 56**	4.1
25	M	6 31	**6 48**	a3.4	...	**12 17**	4.2
26	T	7 04	**7 24**	a3.5	12 32	**12 51**	p4.3
27	W	7 38	**8 01**	a3.5	1 07	**1 25**	p4.4
28	T	8 11	**8 38**	a3.5	1 42	**2 01**	p4.5
29	F	8 47	**9 17**	a3.5	2 20	**2 39**	p4.5
30	S	9 25	**10 01**	a3.4	3 00	**3 20**	p4.4
31	S	10 09	**10 49**	a3.3	3 44	**4 05**	p4.3

The Kts. (knots) columns show the **maximum** predicted velocities of the stronger one of the Flood Currents and the stronger one of the Ebb Currents for each day.
The letter "a" means the velocity shown should occur **after** the **a.m.** Current Change. The letter "p" means the velocity shown should occur **after** the **p.m.** Current Change (even if next morning). No "a" or "p" means a.m. and p.m. velocities are the same for that day.
Avg. Max. Velocity: Flood 3.3 Kts., Ebb 4.2 Kts.
Max. Flood 2 hrs. 45 min. after Flood Starts, ±15 min.
Max. Ebb 3 hrs. 25 min. after Ebb Starts, ±15 min.
Use THE RACE tables with current charts pp. 92-97

See pp. 22-29 for Current Change at other points.

2014 CURRENT TABLE
THE RACE, LONG ISLAND SOUND

41°13.69'N, 72°03.75'W 0.2 nm E.N.E. of Valiant Rock

Daylight Saving Time

SEPTEMBER

Day of Month	Day of Week	Northwest Flood Starts a.m.	p.m.	Kts.	Southeast Ebb Starts a.m.	p.m.	Kts.
1	M	10 59	11 44	a3.2	4 33	4 56	p4.2
2	T	11 56	...	3.2	5 28	5 52	p4.1
3	W	12 43	12 57	p3.2	6 28	6 54	p4.1
4	T	1 45	2 01	p3.3	7 32	7 57	p4.3
5	F	2 44	3 02	p3.5	8 36	9 00	p4.6
6	S	3 41	4 02	p3.8	9 37	10 00	p4.9
7	S	4 35	4 58	p4.1	10 33	10 57	p5.2
8	M	5 28	5 53	4.3	11 27	11 50	5.4
9	T	6 18	6 47	a4.5	...	12 18	5.6
10	W	7 08	7 40	a4.6	12 43	1 09	p5.6
11	T	7 59	8 33	a4.5	1 35	1 59	p5.5
12	F	8 50	9 27	a4.2	2 27	2 50	p5.2
13	S	9 43	10 24	a3.8	3 20	3 43	p4.8
14	S	10 40	11 23	a3.4	4 15	4 37	p4.3
15	M	11 40	...	3.0	5 13	5 34	p3.9
16	T	12 24	12 42	a2.8	6 12	6 33	p3.7
17	W	1 24	1 43	2.6	7 11	7 32	p3.5
18	T	2 21	2 40	2.6	8 08	8 28	p3.5
19	F	3 13	3 32	2.7	9 00	9 19	p3.6
20	S	3 58	4 18	2.8	9 46	10 05	3.7
21	S	4 40	5 00	3.0	10 27	10 46	a3.9
22	M	5 18	5 40	a3.2	11 04	11 25	a4.1
23	T	5 54	6 17	3.3	11 40	...	4.3
24	W	6 29	6 54	a3.5	12 02	12 16	p4.5
25	T	7 04	7 31	a3.6	12 39	12 52	p4.7
26	F	7 40	8 09	a3.7	1 16	1 29	p4.7
27	S	8 18	8 50	a3.7	1 56	2 10	p4.7
28	S	8 59	9 35	a3.6	2 38	2 53	p4.6
29	M	9 46	10 24	a3.5	3 24	3 41	p4.5
30	T	10 39	11 20	a3.3	4 14	4 34	p4.3

Daylight Saving Time

OCTOBER

Day of Month	Day of Week	Northwest Flood Starts a.m.	p.m.	Kts.	Southeast Ebb Starts a.m.	p.m.	Kts.
1	W	11 38	...	3.2	5 10	5 32	p4.2
2	T	12 20	12 42	p3.1	6 11	6 35	p4.1
3	F	1 22	1 47	p3.2	7 15	7 40	p4.2
4	S	2 23	2 51	p3.4	8 18	8 45	p4.5
5	S	3 20	3 49	p3.7	9 18	9 46	a4.8
6	M	4 14	4 46	4.0	10 15	10 42	a5.2
7	T	5 07	5 40	a4.3	11 08	11 36	a5.5
8	W	5 58	6 32	a4.4	11 58	...	5.6
9	T	6 48	7 23	a4.5	12 28	12 47	p5.6
10	F	7 38	8 14	a4.3	1 18	1 36	p5.4
11	S	8 28	9 05	a4.0	2 08	2 25	p5.1
12	S	9 20	9 57	a3.6	2 59	3 15	p4.7
13	M	10 14	10 51	a3.2	3 50	4 06	p4.2
14	T	11 11	11 46	a2.8	4 43	4 58	p3.8
15	W	...	12 09	2.5	5 36	5 53	3.4
16	T	12 42	1 08	a2.6	6 30	6 48	3.2
17	F	1 36	2 04	a2.5	7 23	7 43	3.2
18	S	2 26	2 55	a2.6	8 13	8 36	a3.4
19	S	3 12	3 41	a2.7	9 00	9 24	a3.6
20	M	3 55	4 24	a2.9	9 42	10 08	a3.9
21	T	4 35	5 05	a3.1	10 23	10 50	a4.2
22	W	5 14	5 44	3.3	11 02	11 30	a4.4
23	T	5 52	6 23	3.5	11 40	...	4.7
24	F	6 30	7 03	a3.7	12 10	12 20	p4.9
25	S	7 10	7 44	a3.8	12 51	1 01	p4.9
26	S	7 52	8 27	a3.8	1 34	1 45	p4.9
27	M	8 38	9 13	a3.7	2 19	2 31	p4.9
28	T	9 28	10 04	a3.6	3 07	3 21	p4.7
29	W	10 23	11 00	a3.4	4 00	4 16	p4.5
30	T	11 25	11 59	3.2	4 56	5 16	p4.3
31	F	...	12 30	3.1	5 57	6 19	4.1

The Kts. (knots) columns show the **maximum** predicted velocities of the stronger one of the Flood Currents and the stronger one of the Ebb Currents for each day.
The letter "a" means the velocity shown should occur **after** the **a.m.** Current Change. The letter "p" means the velocity shown should occur **after** the **p.m.** Current Change (even if next morning). No "a" or "p" means a.m. and p.m. velocities are the same for that day.
Avg. Max. Velocity: Flood 3.3 Kts., Ebb 4.2 Kts.
Max. Flood 2 hrs. 45 min. after Flood Starts, ±15 min.
Max. Ebb 3 hrs. 25 min. after Ebb Starts, ±15 min.
Use THE RACE tables with current charts pp. 92-97

See pp. 22-29 for Current Change at other points.

2014 CURRENT TABLE
THE RACE, LONG ISLAND SOUND

41°13.69'N, 72°03.75'W 0.2 nm E.N.E. of Valiant Rock

***Standard Time starts Nov. 2 at 2 a.m.** | **Standard Time**

NOVEMBER

DAY OF MONTH	DAY OF WEEK	CURRENT TURNS TO NORTHWEST Flood Starts a.m.	p.m.	Kts.	CURRENT TURNS TO SOUTHEAST Ebb Starts a.m.	p.m.	Kts.
1	S	1 01	**1 35**	a3.3	6 59	**7 25**	a4.3
2	S	1 01	***1 38**	a3.4	*7 01	***7 30**	a4.5
3	M	1 59	**2 38**	a3.7	8 01	**8 32**	a4.8
4	T	2 56	**3 35**	a3.9	8 57	**9 29**	a5.1
5	W	3 48	**4 28**	a4.1	9 50	**10 23**	a5.4
6	T	4 40	**5 19**	a4.2	10 40	**11 13**	a5.4
7	F	5 30	**6 08**	a4.2	11 28	**...**	5.4
8	S	6 19	**6 55**	a4.0	12 02	**12 15**	p5.2
9	S	7 08	**7 42**	a3.8	12 49	**1 01**	p4.9
10	M	7 57	**8 30**	a3.4	1 36	**1 47**	p4.5
11	T	8 47	**9 17**	3.1	2 23	**2 33**	p4.1
12	W	9 39	**10 06**	p2.8	3 10	**3 20**	3.7
13	T	10 33	**10 55**	p2.6	3 57	**4 09**	3.4
14	F	11 27	**11 45**	p2.6	4 45	**4 59**	a3.3
15	S	...	**12 21**	2.2	5 33	**5 51**	a3.3
16	S	12 34	**1 11**	a2.6	6 21	**6 44**	a3.4
17	M	1 21	**1 59**	a2.6	7 09	**7 36**	a3.6
18	T	2 06	**2 45**	a2.8	7 55	**8 25**	a3.9
19	W	2 50	**3 28**	a3.0	8 39	**9 12**	a4.2
20	T	3 33	**4 11**	a3.3	9 23	**9 57**	a4.5
21	F	4 15	**4 53**	3.5	10 07	**10 42**	a4.8
22	S	4 59	**5 36**	3.7	10 51	**11 27**	a5.0
23	S	5 43	**6 20**	a3.9	11 37	**...**	5.1
24	M	6 30	**7 06**	a4.0	12 13	**12 24**	p5.2
25	T	7 19	**7 54**	a3.9	1 01	**1 13**	p5.1
26	W	8 12	**8 46**	a3.8	1 52	**2 05**	p4.9
27	T	9 09	**9 41**	3.6	2 45	**3 01**	p4.7
28	F	10 11	**10 39**	p3.5	3 42	**4 01**	a4.5
29	S	11 17	**11 40**	p3.5	4 41	**5 04**	a4.4
30	S	...	**12 22**	3.1	5 42	**6 10**	a4.4

DECEMBER

DAY OF MONTH	DAY OF WEEK	CURRENT TURNS TO NORTHWEST Flood Starts a.m.	p.m.	Kts.	CURRENT TURNS TO SOUTHEAST Ebb Starts a.m.	p.m.	Kts.
1	M	12 41	**1 26**	a3.5	6 43	**7 16**	a4.6
2	T	1 40	**2 27**	a3.6	7 43	**8 18**	a4.8
3	W	2 37	**3 23**	a3.8	8 40	**9 16**	a5.0
4	T	3 33	**4 17**	a3.9	9 33	**10 09**	a5.1
5	F	4 24	**5 05**	a3.9	10 23	**10 58**	a5.2
6	S	5 13	**5 52**	a3.9	11 10	**11 45**	a5.1
7	S	6 01	**6 36**	a3.8	11 54	**...**	4.9
8	M	6 48	**7 19**	3.6	12 29	**12 37**	p4.7
9	T	7 33	**8 01**	p3.4	1 12	**1 19**	p4.4
10	W	8 19	**8 43**	p3.2	1 54	**2 01**	p4.1
11	T	9 05	**9 25**	p3.0	2 35	**2 42**	a3.8
12	F	9 53	**10 08**	p2.8	3 17	**3 25**	a3.6
13	S	10 43	**10 53**	p2.7	3 59	**4 11**	a3.5
14	S	11 33	**11 40**	p2.7	4 43	**5 00**	a3.5
15	M	...	**12 24**	2.3	5 30	**5 52**	a3.5
16	T	12 28	**1 14**	a2.7	6 18	**6 47**	a3.6
17	W	1 17	**2 03**	a2.8	7 08	**7 41**	a3.9
18	T	2 05	**2 50**	a3.0	7 58	**8 34**	a4.2
19	F	2 54	**3 37**	a3.3	8 48	**9 25**	a4.5
20	S	3 42	**4 24**	a3.6	9 37	**10 15**	a4.8
21	S	4 30	**5 11**	3.8	10 26	**11 04**	a5.1
22	M	5 20	**5 58**	4.0	11 15	**11 53**	a5.3
23	T	6 10	**6 45**	a4.2	...	**12 05**	5.4
24	W	7 02	**7 35**	4.1	12 43	**12 56**	p5.3
25	T	7 56	**8 26**	p4.1	1 34	**1 49**	p5.1
26	F	8 54	**9 20**	p3.9	2 28	**2 45**	a4.9
27	S	9 56	**10 18**	p3.7	3 24	**3 45**	a4.8
28	S	11 01	**11 19**	p3.6	4 22	**4 48**	a4.6
29	M	...	**12 07**	3.1	5 22	**5 53**	a4.5
30	T	12 21	**1 12**	a3.5	6 24	**7 00**	a4.5
31	W	1 22	**2 13**	a3.5	7 25	**8 03**	a4.6

The Kts. (knots) columns show the **maximum** predicted velocities of the stronger one of the Flood Currents and the stronger one of the Ebb Currents for each day.
The letter "a" means the velocity shown should occur **after** the **a.m.** Current Change. The letter "p" means the velocity shown should occur **after** the **p.m.** Current Change (even if next morning). No "a" or "p" means a.m. and p.m. velocities are the same for that day.
Avg. Max. Velocity: Flood 3.3 Kts., Ebb 4.2 Kts.
Max. Flood 2 hrs. 45 min. after Flood Starts, ±15 min.
Max. Ebb 3 hrs. 25 min. after Ebb Starts, ±15 min.
Use THE RACE tables with current charts pp. 92-97

See pp. 22-29 for Current Change at other points.

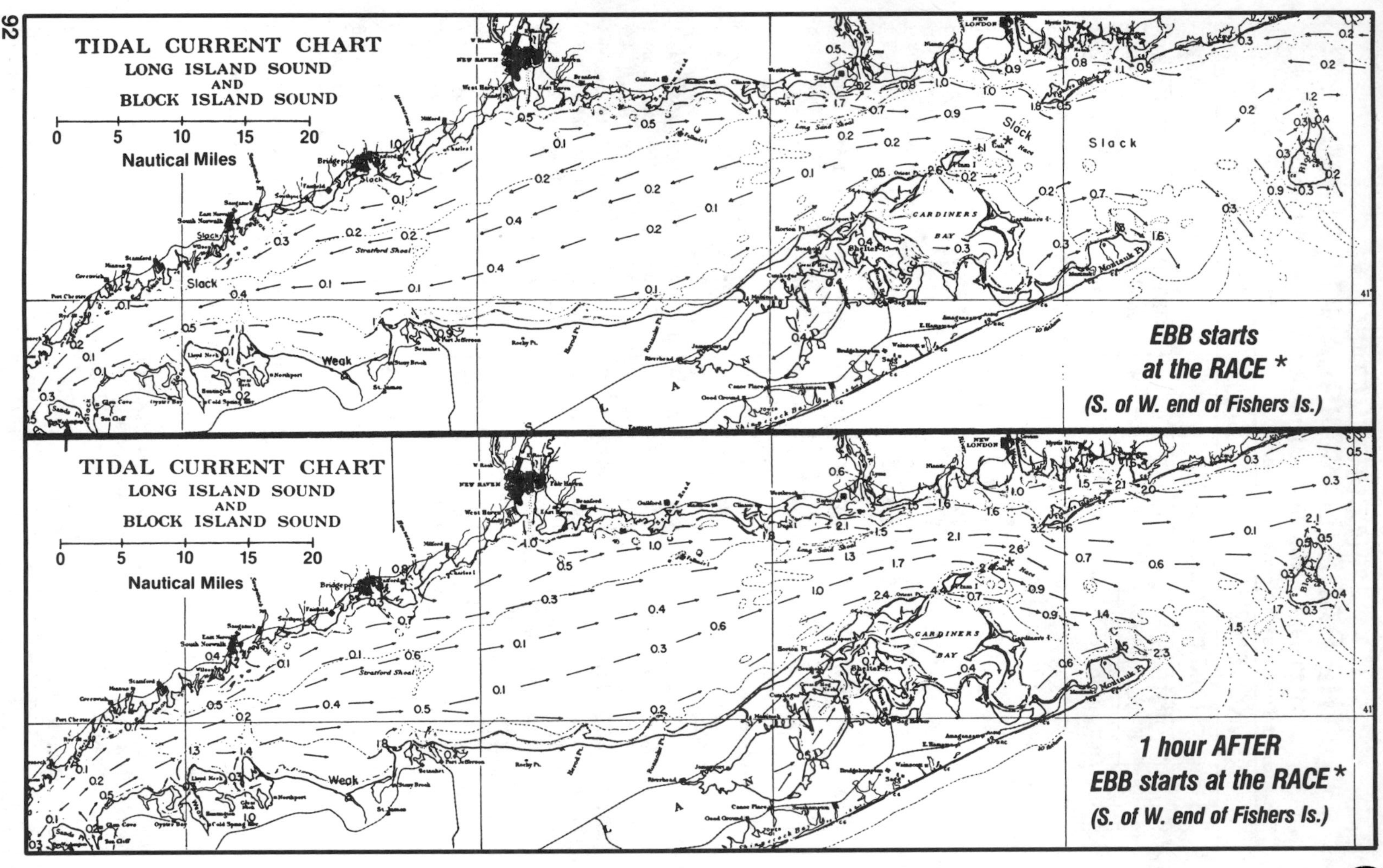
TIDAL CURRENT CHART
LONG ISLAND SOUND
AND
BLOCK ISLAND SOUND
0 5 10 15 20
Nautical Miles
EBB starts
at the RACE *
(S. of W. end of Fishers Is.)
TIDAL CURRENT CHART
LONG ISLAND SOUND
AND
BLOCK ISLAND SOUND
0 5 10 15 20
Nautical Miles
1 hour AFTER
EBB starts at the RACE *
(S. of W. end of Fishers Is.)

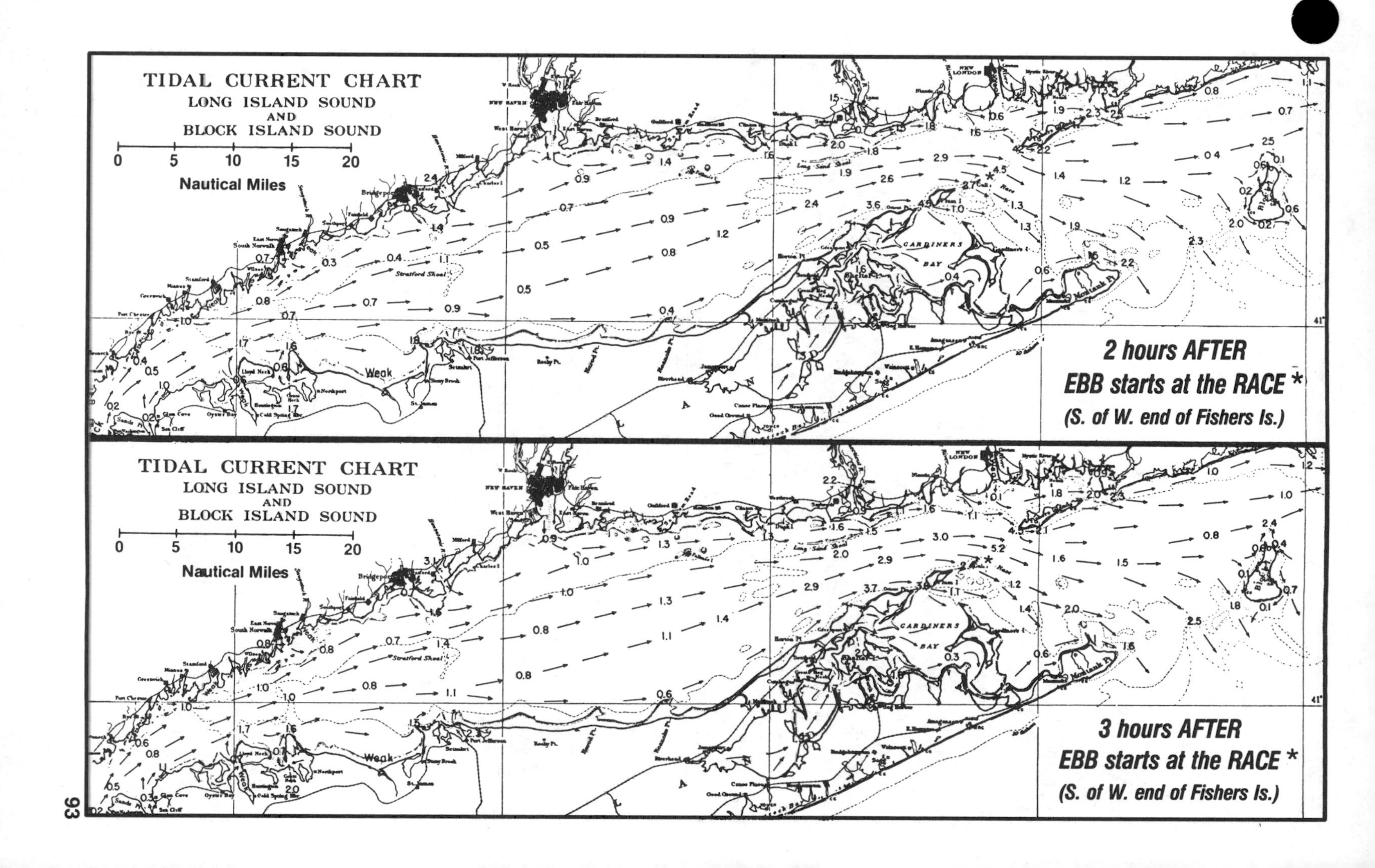
TIDAL CURRENT CHART
LONG ISLAND SOUND
AND
BLOCK ISLAND SOUND
0 5 10 15 20
Nautical Miles
CARDINERS BAY
Stratford Shoal
Long Sand Shoal
Weak
2 hours AFTER
EBB starts at the RACE *
(S. of W. end of Fishers Is.)
TIDAL CURRENT CHART
LONG ISLAND SOUND
AND
BLOCK ISLAND SOUND
0 5 10 15 20
Nautical Miles
CARDINERS BAY
Stratford Shoal
Long Sand Shoal
Weak
3 hours AFTER
EBB starts at the RACE *
(S. of W. end of Fishers Is.)

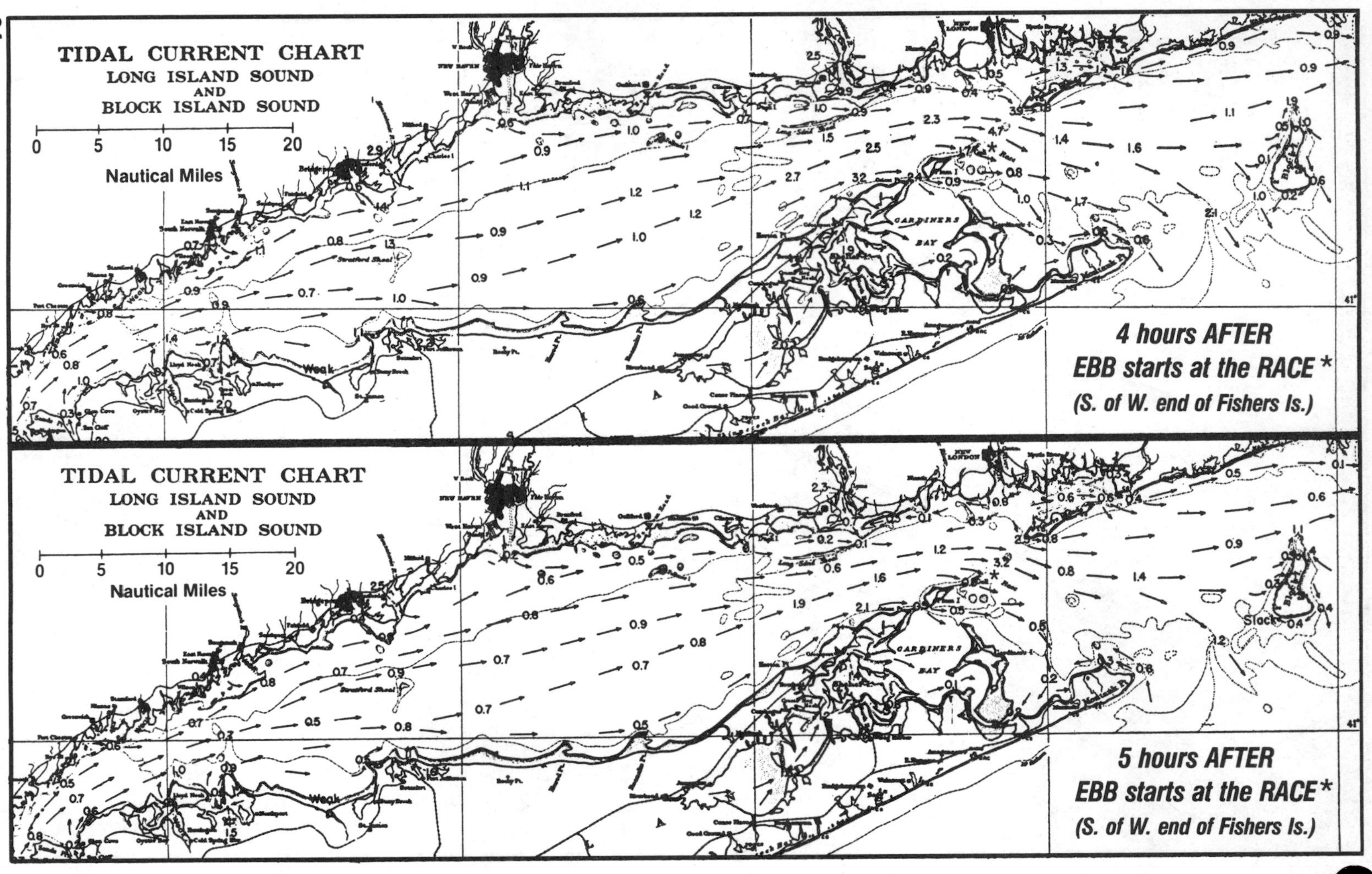
TIDAL CURRENT CHART
LONG ISLAND SOUND
AND
BLOCK ISLAND SOUND
0 5 10 15 20
Nautical Miles
GARDINERS BAY
Stratford Shoal
Weak
4 hours AFTER
EBB starts at the RACE*
(S. of W. end of Fishers Is.)
TIDAL CURRENT CHART
LONG ISLAND SOUND
AND
BLOCK ISLAND SOUND
0 5 10 15 20
Nautical Miles
GARDINERS BAY
Stratford Shoal
Weak
Slack
5 hours AFTER
EBB starts at the RACE*
(S. of W. end of Fishers Is.)

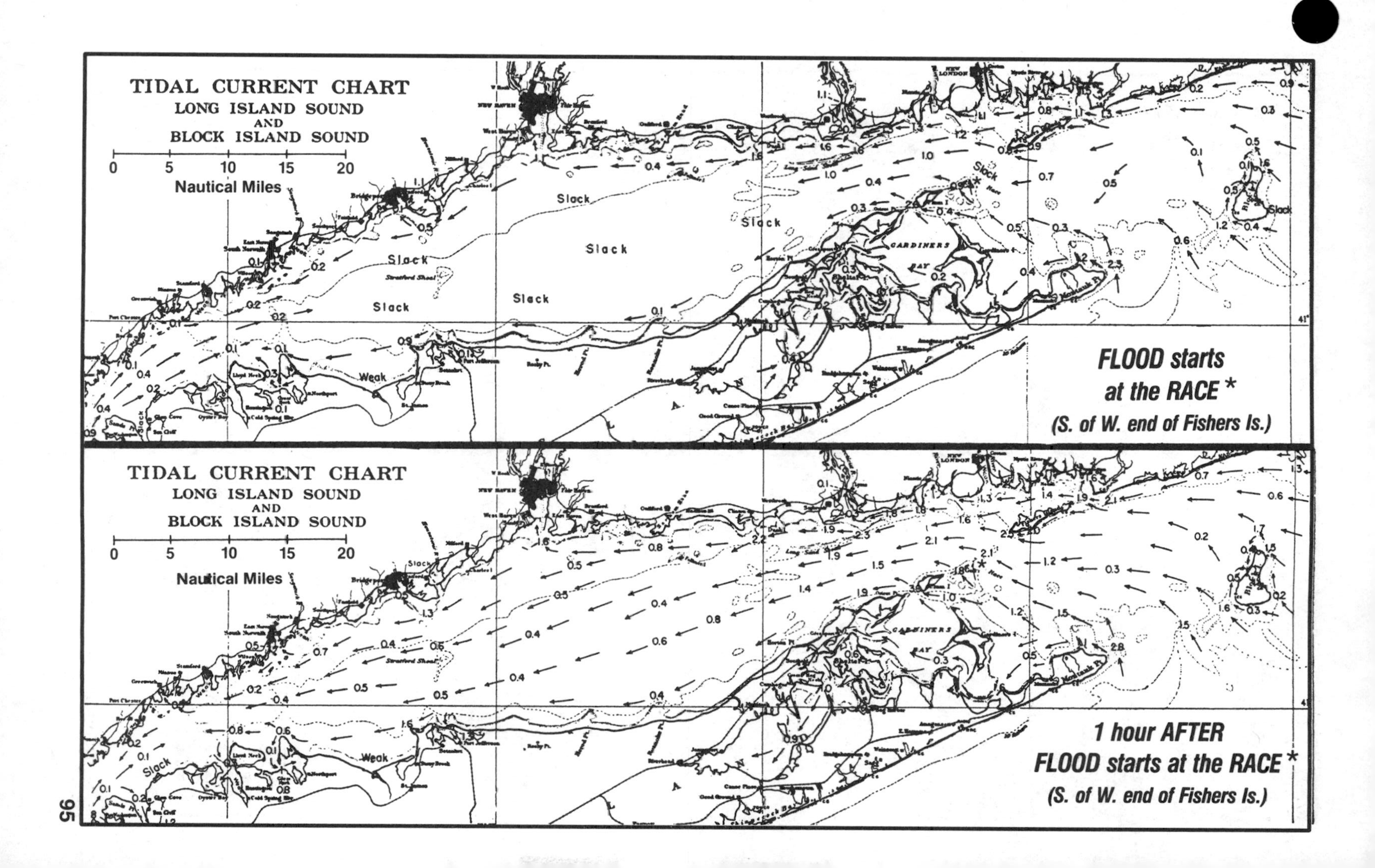
TIDAL CURRENT CHART
LONG ISLAND SOUND
AND
BLOCK ISLAND SOUND
0 5 10 15 20
Nautical Miles
FLOOD starts
at the RACE *
(S. of W. end of Fishers Is.)
TIDAL CURRENT CHART
LONG ISLAND SOUND
AND
BLOCK ISLAND SOUND
0 5 10 15 20
Nautical Miles
1 hour AFTER
FLOOD starts at the RACE *
(S. of W. end of Fishers Is.)

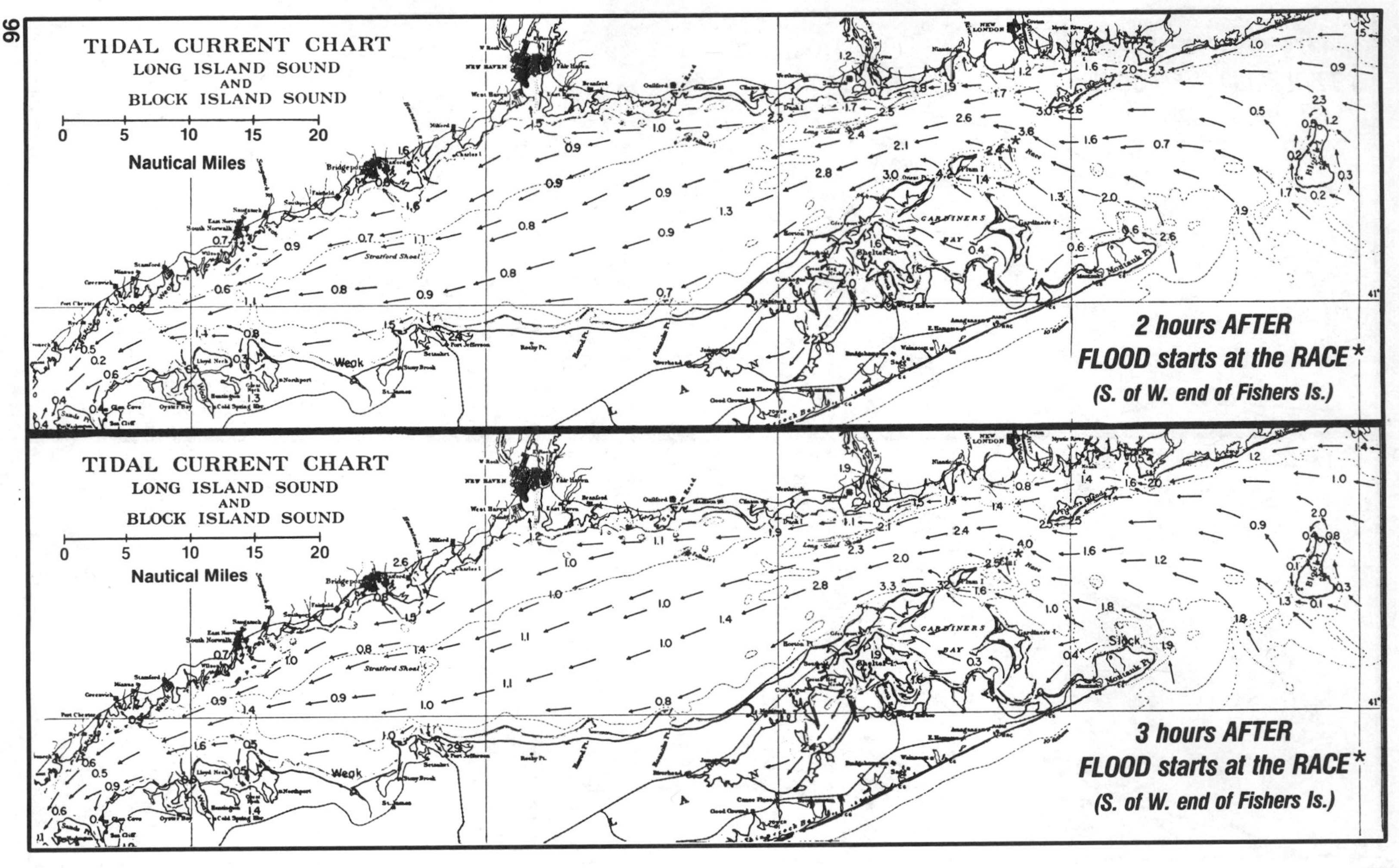

TIDAL CURRENT CHART
LONG ISLAND SOUND
AND
BLOCK ISLAND SOUND
0 5 10 15 20
Nautical Miles
NEW LONDON
NEW HAVEN
Bridgeport
GARDINERS BAY
Stratford Shoal
Weak
2 hours AFTER
FLOOD starts at the RACE*
(S. of W. end of Fishers Is.)
TIDAL CURRENT CHART
LONG ISLAND SOUND
AND
BLOCK ISLAND SOUND
0 5 10 15 20
Nautical Miles
NEW LONDON
NEW HAVEN
Bridgeport
GARDINERS BAY
Stratford Shoal
Weak
Slack
3 hours AFTER
FLOOD starts at the RACE*
(S. of W. end of Fishers Is.)

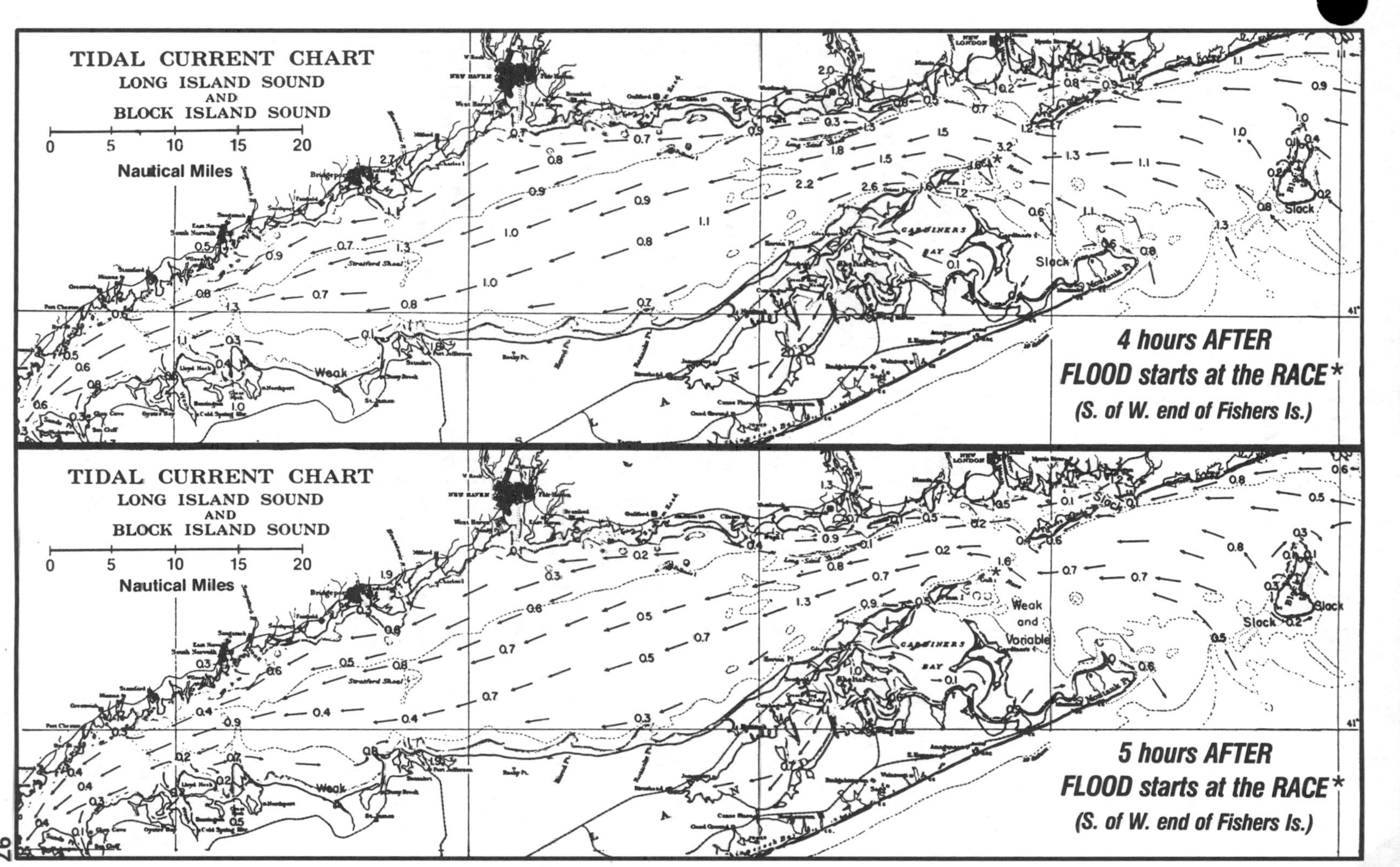
TIDAL CURRENT CHART
LONG ISLAND SOUND
AND
BLOCK ISLAND SOUND
0 5 10 15 20
Nautical Miles
4 hours AFTER
FLOOD starts at the RACE*
(S. of W. end of Fishers Is.)
TIDAL CURRENT CHART
LONG ISLAND SOUND
AND
BLOCK ISLAND SOUND
0 5 10 15 20
Nautical Miles
5 hours AFTER
FLOOD starts at the RACE*
(S. of W. end of Fishers Is.)

2014 HIGH & LOW WATER BRIDGEPORT, CT

41°10.4'N, 73°10.9'W

Standard Time

DAY OF MONTH	DAY OF WEEK	JANUARY HIGH a.m.	Ht.	p.m.	Ht.	LOW a.m.	p.m.
1	W	10 38	8.2	11 11	7.4	4 28	5 07
2	T	11 30	8.3	...	...	5 21	5 58
3	F	12 03	7.5	12 23	8.2	6 15	6 49
4	S	12 56	7.6	1 17	7.9	7 10	7 41
5	S	1 48	7.6	2 11	7.5	8 06	8 33
6	M	2 43	7.4	3 08	7.1	9 04	9 28
7	T	3 39	7.2	4 07	6.7	10 05	10 25
8	W	4 38	7.1	5 08	6.3	11 07	11 23
9	T	5 37	6.9	6 10	6.1	...	12 09
10	F	6 36	6.8	7 09	6.1	12 21	1 08
11	S	7 31	6.8	8 04	6.1	1 18	2 03
12	S	8 23	6.8	8 54	6.2	2 11	2 53
13	M	9 10	6.9	9 40	6.3	2 59	3 38
14	T	9 54	6.9	10 22	6.5	3 44	4 18
15	W	10 34	6.9	11 01	6.6	4 25	4 56
16	T	11 13	6.9	11 39	6.6	5 04	5 33
17	F	11 50	6.9	...	...	5 43	6 09
18	S	12 16	6.7	12 27	6.8	6 21	6 44
19	S	12 53	6.7	1 04	6.7	6 59	7 21
20	M	1 30	6.7	1 43	6.5	7 40	7 59
21	T	2 09	6.6	2 26	6.3	8 23	8 41
22	W	2 51	6.6	3 13	6.2	9 11	9 27
23	T	3 39	6.6	4 06	6.0	10 04	10 19
24	F	4 32	6.6	5 05	5.9	11 03	11 17
25	S	5 30	6.8	6 08	6.0	...	12 05
26	S	6 32	7.0	7 10	6.1	12 18	1 07
27	M	7 33	7.3	8 10	6.5	1 20	2 07
28	T	8 32	7.6	9 07	6.9	2 20	3 04
29	W	9 29	7.9	10 01	7.3	3 18	3 57
30	T	10 23	8.1	10 53	7.6	4 13	4 49
31	F	11 15	8.2	11 43	7.8	5 06	5 38

Standard Time

DAY OF MONTH	DAY OF WEEK	FEBRUARY HIGH a.m.	Ht.	p.m.	Ht.	LOW a.m.	p.m.
1	S	...	...	12 06	8.1	5 59	6 27
2	S	12 33	7.9	12 57	7.8	6 52	7 16
3	M	1 24	7.8	1 49	7.5	7 45	8 06
4	T	2 16	7.6	2 43	7.0	8 40	8 58
5	W	3 08	7.3	3 38	6.6	9 35	9 51
6	T	4 03	7.0	4 36	6.2	10 34	10 48
7	F	5 02	6.7	5 37	5.9	11 35	11 48
8	S	6 01	6.5	6 37	5.9	...	12 35
9	S	7 00	6.4	7 34	5.9	12 47	1 32
10	M	7 55	6.4	8 26	6.1	1 43	2 24
11	T	8 45	6.5	9 13	6.3	2 34	3 09
12	W	9 30	6.7	9 56	6.5	3 20	3 51
13	T	10 12	6.8	10 35	6.7	4 02	4 29
14	F	10 50	6.9	11 12	6.8	4 42	5 05
15	S	11 27	6.9	11 48	6.9	5 20	5 41
16	S	...	...	12 03	6.9	5 57	6 16
17	M	12 23	7.0	12 40	6.8	6 35	6 52
18	T	12 59	7.0	1 18	6.7	7 14	7 29
19	W	1 37	7.0	1 59	6.6	7 56	8 10
20	T	2 18	7.0	2 46	6.4	8 42	8 56
21	F	3 05	6.9	3 39	6.2	9 35	9 49
22	S	4 00	6.8	4 38	6.1	10 35	10 50
23	S	5 02	6.8	5 43	6.1	11 39	11 55
24	M	6 08	6.9	6 48	6.3	...	12 44
25	T	7 13	7.2	7 50	6.6	1 01	1 46
26	W	8 15	7.5	8 48	7.1	2 04	2 44
27	T	9 13	7.8	9 42	7.5	3 04	3 38
28	F	10 07	8.0	10 33	7.9	3 59	4 28

Dates when Ht. of **Low** Water is below Mean Lower Low with Ht. of lowest given for each period and Date of lowest in ( ):

1st - 7th: -1.4' (2nd)
14th - 18th: -0.2'
27th - 31st: -1.4' (31st)

1st - 5th: -1.4' (1st)
14th - 17th: -0.2'
25th - 28th: -0.9' (27th - 28th)

Average Rise and Fall 6.8 ft.

When a high tide exceeds avg. ht., the *following* low tide will be lower than avg.

2014 HIGH & LOW WATER BRIDGEPORT, CT

41°10.4'N, 73°10.9'W

***Daylight Time starts March 9 at 2 a.m.** **Daylight Saving Time**

DAY OF MONTH	DAY OF WEEK	MARCH HIGH a.m.	Ht.	p.m.	Ht.	LOW a.m.	p.m.	DAY OF MONTH	DAY OF WEEK	APRIL HIGH a.m.	Ht.	p.m.	Ht.	LOW a.m.	p.m.
1	S	10 58	8.0	11 22	8.1	4 52	5 16	1	T	12 46	8.2	1 15	7.5	7 11	7 24
2	S	11 48	8.0	...	...	5 42	6 03	2	W	1 31	8.0	2 02	7.3	7 57	8 09
3	M	12 10	8.1	12 36	7.7	6 32	6 50	3	T	2 16	7.7	2 49	7.0	8 44	8 55
4	T	12 58	8.0	1 26	7.4	7 22	7 38	4	F	3 04	7.3	3 38	6.6	9 32	9 44
5	W	1 45	7.7	2 15	7.0	8 11	8 25	5	S	3 52	6.9	4 29	6.3	10 21	10 35
6	T	2 35	7.3	3 07	6.6	9 03	9 17	6	S	4 45	6.5	5 23	6.2	11 14	11 32
7	F	3 27	6.9	4 02	6.2	9 58	10 12	7	M	5 42	6.2	6 20	6.1	...	12 10
8	S	4 23	6.5	5 01	6.0	10 56	11 11	8	T	6 41	6.1	7 17	6.2	12 31	1 06
9	S	*6 23	6.3	*7 00	5.9	...	*12 55	9	W	7 39	6.1	8 11	6.3	1 30	1 59
10	M	7 23	6.1	7 58	6.0	1 11	1 52	10	T	8 34	6.2	9 00	6.6	2 25	2 49
11	T	8 21	6.2	8 52	6.2	2 09	2 45	11	F	9 23	6.4	9 46	6.9	3 15	3 35
12	W	9 14	6.3	9 40	6.4	3 02	3 33	12	S	10 09	6.6	10 28	7.2	4 01	4 17
13	T	10 01	6.5	10 24	6.7	3 51	4 16	13	S	10 51	6.8	11 07	7.4	4 45	4 58
14	F	10 43	6.7	11 04	6.9	4 35	4 56	14	M	11 32	6.9	11 46	7.6	5 26	5 38
15	S	11 23	6.8	11 42	7.1	5 15	5 34	15	T	...	...	12 12	7.1	6 07	6 18
16	S	...	...	12 01	6.9	5 54	6 10	16	W	12 24	7.7	12 53	7.1	6 48	6 58
17	M	12 18	7.3	12 38	7.0	6 33	6 47	17	T	1 04	7.8	1 36	7.1	7 31	7 42
18	T	12 53	7.4	1 16	7.0	7 11	7 24	18	F	1 47	7.8	2 22	7.0	8 17	8 28
19	W	1 30	7.4	1 56	6.9	7 51	8 03	19	S	2 34	7.7	3 12	6.9	9 06	9 20
20	T	2 09	7.4	2 39	6.8	8 34	8 46	20	S	3 27	7.5	4 07	6.8	10 00	10 18
21	F	2 53	7.4	3 26	6.6	9 22	9 35	21	M	4 25	7.3	5 06	6.8	10 59	11 22
22	S	3 42	7.2	4 20	6.5	10 15	10 30	22	T	5 28	7.2	6 09	6.8	...	12 01
23	S	4 39	7.1	5 20	6.4	11 15	11 33	23	W	6 35	7.1	7 12	7.0	12 28	1 03
24	M	5 43	7.0	6 25	6.4	...	12 19	24	T	7 40	7.1	8 13	7.3	1 35	2 04
25	T	6 50	7.0	7 30	6.6	12 40	1 24	25	F	8 42	7.2	9 10	7.7	2 37	3 00
26	W	7 57	7.1	8 32	7.0	1 47	2 26	26	S	9 39	7.3	10 02	8.0	3 35	3 53
27	T	8 59	7.3	9 30	7.4	2 51	3 23	27	S	10 32	7.4	10 51	8.1	4 29	4 43
28	F	9 57	7.6	10 23	7.8	3 50	4 16	28	M	11 22	7.4	11 38	8.2	5 19	5 30
29	S	10 50	7.7	11 13	8.1	4 45	5 06	29	T	...	...	12 09	7.4	6 06	6 15
30	S	11 41	7.8	...	...	5 36	5 53	30	W	12 22	8.1	12 54	7.3	6 50	6 59
31	M	12 01	8.2	12 28	7.7	6 24	6 39								

Dates when Ht. of **Low** Water is below Mean Lower Low with Ht. of lowest given for each period and Date of lowest in ( ):

March	April
1st - 5th: -1.2' (1st - 3rd)	1st - 3rd: -0.9' (1st)
17th - 21st: -0.4' (19th)	15th - 20th: -0.5' (16th - 18th)
27th - 31st: -1.1' (31st)	26th - 30th: -0.7' (28th - 29th)

Average Rise and Fall 6.8 ft.

When a high tide exceeds avg. ht., the *following* low tide will be lower than avg.

2014 HIGH & LOW WATER
BRIDGEPORT, CT
41°10.4'N, 73°10.9'W

DAY OF MONTH	DAY OF WEEK	MAY (Daylight Saving Time)						DAY OF MONTH	DAY OF WEEK	JUNE (Daylight Saving Time)					
		HIGH				LOW				HIGH				LOW	
		a.m.	Ht.	**p.m.**	Ht.	a.m.	**p.m.**			a.m.	Ht.	**p.m.**	Ht.	a.m.	**p.m.**
1	**T**	1 05	7.9	**1 38**	7.1	7 34	**7 42**	1	**S**	2 06	7.2	**2 40**	6.8	8 32	**8 43**
2	**F**	1 49	7.6	**2 23**	6.9	8 17	**8 26**	2	**M**	2 49	7.0	**3 24**	6.7	9 13	**9 29**
3	**S**	2 33	7.3	**3 08**	6.7	9 01	**9 12**	3	**T**	3 34	6.7	**4 09**	6.7	9 56	**10 17**
4	**S**	3 20	6.9	**3 57**	6.5	9 47	**10 02**	4	**W**	4 23	6.5	**4 58**	6.6	10 42	**11 10**
5	**M**	4 08	6.6	**4 46**	6.4	10 34	**10 54**	5	**T**	5 12	6.3	**5 46**	6.7	11 29	**...**
6	**T**	5 01	6.3	**5 38**	6.4	11 24	**11 49**	6	**F**	6 05	6.1	**6 37**	6.8	12 03	**12 20**
7	**W**	5 56	6.1	**6 32**	6.4	...	**12 16**	7	**S**	7 00	6.1	**7 28**	6.9	12 58	**1 11**
8	**T**	6 53	6.1	**7 25**	6.6	12 46	**1 09**	8	**S**	7 55	6.2	**8 18**	7.1	1 52	**2 03**
9	**F**	7 48	6.1	**8 15**	6.8	1 42	**2 00**	9	**M**	8 49	6.4	**9 07**	7.4	2 45	**2 54**
10	**S**	8 40	6.3	**9 02**	7.0	2 35	**2 49**	10	**T**	9 40	6.6	**9 56**	7.7	3 36	**3 44**
11	**S**	9 29	6.5	**9 47**	7.3	3 24	**3 35**	11	**W**	10 31	6.8	**10 44**	8.0	4 26	**4 34**
12	**M**	10 16	6.7	**10 31**	7.6	4 11	**4 20**	12	**T**	11 20	7.1	**11 33**	8.2	5 15	**5 24**
13	**T**	11 01	6.9	**11 13**	7.8	4 56	**5 05**	13	**F**	...	...	**12 09**	7.3	6 04	**6 14**
14	**W**	11 46	7.1	**11 57**	8.0	5 41	**5 49**	14	**S**	12 23	8.3	**12 59**	7.5	6 54	**7 06**
15	**T**	...	...	**12 31**	7.2	6 26	**6 35**	15	**S**	1 14	8.3	**1 50**	7.6	7 44	**8 00**
16	**F**	12 42	8.1	**1 18**	7.3	7 12	**7 23**	16	**M**	2 06	8.2	**2 42**	7.7	8 35	**8 55**
17	**S**	1 29	8.1	**2 07**	7.3	8 01	**8 14**	17	**T**	3 01	8.0	**3 37**	7.7	9 28	**9 54**
18	**S**	2 20	8.0	**2 58**	7.3	8 52	**9 08**	18	**W**	3 58	7.7	**4 34**	7.7	10 23	**10 55**
19	**M**	3 14	7.8	**3 54**	7.2	9 46	**10 07**	19	**T**	4 58	7.3	**5 32**	7.6	11 19	**11 58**
20	**T**	4 13	7.5	**4 52**	7.2	10 43	**11 10**	20	**F**	5 59	7.0	**6 31**	7.6	...	**12 17**
21	**W**	5 15	7.3	**5 53**	7.3	11 42	**...**	21	**S**	7 01	6.8	**7 30**	7.6	1 01	**1 15**
22	**T**	6 18	7.1	**6 53**	7.4	12 15	**12 41**	22	**S**	8 02	6.7	**8 26**	7.6	2 02	**2 12**
23	**F**	7 22	7.0	**7 53**	7.6	1 20	**1 40**	23	**M**	9 00	6.7	**9 20**	7.6	3 00	**3 07**
24	**S**	8 23	6.9	**8 49**	7.8	2 21	**2 37**	24	**T**	9 53	6.7	**10 09**	7.6	3 53	**3 58**
25	**S**	9 20	7.0	**9 41**	7.9	3 19	**3 30**	25	**W**	10 43	6.8	**10 56**	7.6	4 41	**4 46**
26	**M**	10 13	7.0	**10 30**	7.9	4 12	**4 20**	26	**T**	11 28	6.8	**11 39**	7.5	5 26	**5 31**
27	**T**	11 03	7.1	**11 16**	7.9	5 01	**5 07**	27	**F**	...	...	**12 11**	6.9	6 08	**6 13**
28	**W**	11 49	7.1	**...**	...	5 46	**5 52**	28	**S**	12 21	7.4	**12 52**	6.9	6 47	**6 54**
29	**T**	12 01	7.8	**12 33**	7.0	6 30	**6 35**	29	**S**	1 01	7.3	**1 32**	6.9	7 25	**7 34**
30	**F**	12 42	7.7	**1 15**	7.0	7 11	**7 17**	30	**M**	1 41	7.2	**2 11**	6.9	8 02	**8 15**
31	**S**	1 24	7.5	**1 58**	6.9	7 51	**8 00**								

Dates when Ht. of **Low** Water is below Mean Lower Low with Ht. of lowest given for each period and Date of lowest in ( ):

1st: -0.4'
14th - 20th: -0.7' (16th)
26th - 29th: -0.4' (27th - 28th)

12th - 19th: -0.8' (14th - 16th)

Average Rise and Fall 6.8 ft.

When a high tide exceeds avg. ht., the *following* low tide will be lower than avg.

2014 HIGH & LOW WATER
BRIDGEPORT, CT
41°10.4'N, 73°10.9'W

DAY OF MONTH	DAY OF WEEK	JULY (Daylight Saving Time)						DAY OF MONTH	DAY OF WEEK	AUGUST (Daylight Saving Time)					
		HIGH				LOW				HIGH				LOW	
		a.m.	Ht.	p.m.	Ht.	a.m.	p.m.			a.m.	Ht.	p.m.	Ht.	a.m.	p.m.
1	T	2 21	7.0	2 52	6.9	8 40	8 58	1	F	3 12	6.7	3 37	7.1	9 25	9 55
2	W	3 02	6.8	3 33	6.9	9 20	9 42	2	S	3 57	6.5	4 21	7.0	10 08	10 44
3	T	3 45	6.6	4 16	6.9	10 02	10 30	3	S	4 46	6.4	5 10	7.1	10 56	11 39
4	F	4 33	6.4	5 03	6.9	10 47	11 22	4	M	5 41	6.3	6 05	7.1	11 51	...
5	S	5 22	6.3	5 51	6.9	11 35	...	5	T	6 39	6.3	7 02	7.3	12 37	12 48
6	S	6 17	6.2	6 42	7.0	12 15	12 27	6	W	7 40	6.4	8 02	7.5	1 38	1 48
7	M	7 13	6.2	7 36	7.2	1 12	1 21	7	T	8 40	6.6	9 01	7.8	2 37	2 49
8	T	8 11	6.3	8 31	7.5	2 09	2 17	8	F	9 37	7.0	9 58	8.1	3 34	3 47
9	W	9 07	6.5	9 25	7.8	3 04	3 13	9	S	10 32	7.4	10 53	8.4	4 28	4 43
10	T	10 02	6.8	10 19	8.1	3 59	4 08	10	S	11 25	7.8	11 47	8.5	5 20	5 38
11	F	10 55	7.2	11 12	8.3	4 51	5 02	11	M	...	...	12 16	8.1	6 11	6 32
12	S	11 47	7.5	...	...	5 43	5 56	12	T	12 39	8.5	1 07	8.3	7 00	7 26
13	S	12 05	8.5	12 38	7.8	6 33	6 49	13	W	1 31	8.3	1 58	8.4	7 50	8 19
14	M	12 57	8.5	1 30	8.0	7 24	7 44	14	T	2 23	8.0	2 50	8.2	8 40	9 15
15	T	1 50	8.3	2 22	8.0	8 14	8 39	15	F	3 17	7.6	3 44	8.0	9 32	10 11
16	W	2 44	8.0	3 16	8.0	9 06	9 36	16	S	4 13	7.2	4 39	7.7	10 27	11 10
17	T	3 39	7.7	4 11	7.9	9 59	10 36	17	S	5 11	6.8	5 37	7.4	11 24	...
18	F	4 37	7.3	5 08	7.8	10 54	11 36	18	M	6 12	6.6	6 37	7.2	12 11	12 23
19	S	5 37	6.9	6 06	7.6	11 51	...	19	T	7 12	6.4	7 36	7.0	1 11	1 23
20	S	6 38	6.7	7 05	7.5	12 38	12 50	20	W	8 11	6.4	8 33	7.0	2 09	2 20
21	M	7 39	6.5	8 03	7.4	1 39	1 48	21	T	9 05	6.5	9 24	7.0	3 03	3 13
22	T	8 37	6.5	8 58	7.3	2 37	2 44	22	F	9 53	6.7	10 11	7.1	3 51	4 01
23	W	9 31	6.6	9 48	7.3	3 30	3 37	23	S	10 37	6.9	10 54	7.2	4 34	4 45
24	T	10 20	6.7	10 35	7.3	4 19	4 25	24	S	11 18	7.0	11 34	7.2	5 13	5 26
25	F	11 05	6.8	11 18	7.3	5 02	5 09	25	M	11 56	7.2	...	...	5 50	6 05
26	S	11 46	6.9	11 58	7.3	5 43	5 50	26	T	12 11	7.2	12 33	7.3	6 25	6 43
27	S	...	...	12 26	7.0	6 20	6 30	27	W	12 48	7.2	1 09	7.3	7 00	7 20
28	M	12 37	7.3	1 04	7.0	6 56	7 09	28	T	1 25	7.1	1 45	7.3	7 36	7 59
29	T	1 15	7.2	1 41	7.1	7 32	7 48	29	F	2 03	7.0	2 22	7.3	8 13	8 40
30	W	1 53	7.0	2 18	7.1	8 08	8 28	30	S	2 43	6.8	3 01	7.3	8 52	9 24
31	T	2 32	6.9	2 57	7.1	8 45	9 10	31	S	3 27	6.7	3 46	7.2	9 36	10 14

Dates when Ht. of **Low** Water is below Mean Lower Low with Ht. of lowest given for each period and Date of lowest in ( ):

11th - 17th: -1.0' (14th)

9th - 15th: -1.0' (12th)

Average Rise and Fall 6.8 ft.

When a high tide exceeds avg. ht., the *following* low tide will be lower than avg.

2014 HIGH & LOW WATER BRIDGEPORT, CT

41°10.4'N, 73°10.9'W

Daylight Saving Time | **Daylight Saving Time**

DAY OF MONTH	DAY OF WEEK	SEPTEMBER						DAY OF MONTH	DAY OF WEEK	OCTOBER					
		HIGH				LOW				HIGH				LOW	
		a.m.	Ht.	**p.m.**	Ht.	a.m.	**p.m.**			a.m.	Ht.	**p.m.**	Ht.	a.m.	**p.m.**
1	M	4 17	6.5	**4 37**	7.2	10 26	**11 09**	1	W	4 52	6.5	**5 13**	7.2	11 03	**11 49**
2	T	5 12	6.4	**5 34**	7.2	11 22	**...**	2	T	5 54	6.6	**6 18**	7.2	...	**12 08**
3	W	6 13	6.4	**6 37**	7.3	12 10	**12 24**	3	F	6 57	6.8	**7 23**	7.4	12 52	**1 14**
4	T	7 17	6.6	**7 41**	7.5	1 14	**1 29**	4	S	8 00	7.1	**8 26**	7.6	1 54	**2 18**
5	F	8 18	6.9	**8 42**	7.7	2 14	**2 31**	5	S	8 57	7.6	**9 24**	7.8	2 50	**3 17**
6	S	9 16	7.3	**9 40**	8.0	3 12	**3 31**	6	M	9 51	8.0	**10 18**	8.0	3 44	**4 12**
7	S	10 11	7.8	**10 36**	8.3	4 06	**4 27**	7	T	10 42	8.4	**11 10**	8.0	4 35	**5 05**
8	M	11 03	8.2	**11 28**	8.4	4 57	**5 21**	8	W	11 31	8.5	**...**	...	5 23	**5 56**
9	T	11 54	8.4	**...**	...	5 47	**6 14**	9	T	12 01	8.0	**12 19**	8.6	6 11	**6 46**
10	W	12 19	8.3	**12 43**	8.6	6 36	**7 06**	10	F	12 49	7.8	**1 07**	8.4	6 58	**7 34**
11	T	1 10	8.1	**1 33**	8.5	7 24	**7 57**	11	S	1 38	7.6	**1 55**	8.1	7 46	**8 24**
12	F	2 01	7.8	**2 23**	8.3	8 13	**8 50**	12	S	2 27	7.2	**2 44**	7.7	8 35	**9 14**
13	S	2 53	7.5	**3 14**	7.9	9 04	**9 44**	13	M	3 18	6.9	**3 36**	7.3	9 26	**10 07**
14	S	3 46	7.1	**4 08**	7.5	9 57	**10 40**	14	T	4 12	6.6	**4 31**	6.9	10 21	**11 02**
15	M	4 43	6.7	**5 06**	7.1	10 54	**11 39**	15	W	5 08	6.4	**5 29**	6.6	11 18	**11 58**
16	T	5 42	6.5	**6 05**	6.9	11 53	**...**	16	T	6 05	6.3	**6 28**	6.4	...	**12 18**
17	W	6 42	6.4	**7 05**	6.7	12 38	**12 53**	17	F	7 02	6.4	**7 25**	6.4	12 53	**1 16**
18	T	7 39	6.4	**8 02**	6.7	1 35	**1 51**	18	S	7 55	6.6	**8 18**	6.5	1 46	**2 10**
19	F	8 33	6.6	**8 55**	6.8	2 28	**2 44**	19	S	8 44	6.8	**9 07**	6.6	2 34	**2 59**
20	S	9 21	6.8	**9 42**	6.9	3 16	**3 33**	20	M	9 29	7.1	**9 52**	6.7	3 18	**3 45**
21	S	10 05	7.0	**10 25**	7.0	3 59	**4 17**	21	T	10 11	7.3	**10 34**	6.9	4 00	**4 27**
22	M	10 46	7.2	**11 05**	7.1	4 38	**4 58**	22	W	10 50	7.5	**11 14**	7.0	4 40	**5 08**
23	T	11 24	7.4	**11 43**	7.1	5 16	**5 37**	23	T	11 28	7.6	**11 53**	7.0	5 19	**5 47**
24	W	...	...	**12 01**	7.5	5 52	**6 15**	24	F	...	...	**12 05**	7.7	5 57	**6 27**
25	T	12 21	7.1	**12 36**	7.5	6 28	**6 53**	25	S	12 33	7.0	**12 44**	7.7	6 37	**7 09**
26	F	12 58	7.1	**1 12**	7.5	7 05	**7 32**	26	S	1 14	7.0	**1 24**	7.7	7 18	**7 52**
27	S	1 37	7.0	**1 50**	7.5	7 43	**8 14**	27	M	1 57	6.9	**2 09**	7.6	8 03	**8 40**
28	S	2 18	6.9	**2 32**	7.4	8 25	**8 59**	28	T	2 45	6.8	**2 59**	7.5	8 52	**9 32**
29	M	3 04	6.7	**3 19**	7.3	9 11	**9 50**	29	W	3 38	6.7	**3 55**	7.3	9 48	**10 29**
30	T	3 55	6.6	**4 12**	7.3	10 04	**10 47**	30	T	4 36	6.7	**4 57**	7.2	10 49	**11 30**
								31	F	5 37	6.8	**6 02**	7.1	11 55	**...**

Dates when Ht. of **Low** Water is below Mean Lower Low with Ht. of lowest given for each period and Date of lowest in ( ):

7th - 12th: -0.8' (9th - 10th)

5th - 11th: -0.9' (8th)
25th: -0.2'

Average Rise and Fall 6.8 ft.

When a high tide exceeds avg. ht., the *following* low tide will be lower than avg.

2014 HIGH & LOW WATER
BRIDGEPORT, CT

41°10.4'N, 73°10.9'W

*Standard Time starts Nov. 2 at 2 a.m. Standard Time

DAY OF MONTH	DAY OF WEEK	NOVEMBER						DAY OF MONTH	DAY OF WEEK	DECEMBER					
		HIGH				LOW				HIGH				LOW	
		a.m.	Ht.	p.m.	Ht.	a.m.	p.m.			a.m.	Ht.	p.m.	Ht.	a.m.	p.m.
1	S	6 40	7.0	7 07	7.1	12 31	1 01	1	M	6 21	7.4	6 51	6.8	12 09	12 49
2	S	*6 40	7.3	*7 09	7.2	1 31	*1 04	2	T	7 19	7.6	7 50	6.9	1 06	1 48
3	M	7 38	7.7	8 07	7.4	1 28	2 03	3	W	8 13	7.8	8 45	7.0	2 02	2 44
4	T	8 33	8.0	9 02	7.5	2 23	2 59	4	T	9 05	7.9	9 37	7.0	2 55	3 36
5	W	9 22	8.3	9 53	7.6	3 13	3 50	5	F	9 53	7.9	10 24	7.0	3 43	4 23
6	T	10 11	8.3	10 41	7.5	4 02	4 39	6	S	10 39	7.8	11 10	7.0	4 30	5 08
7	F	10 57	8.3	11 29	7.4	4 49	5 26	7	S	11 23	7.7	11 54	6.9	5 15	5 51
8	S	11 43	8.1	...	...	5 35	6 12	8	M	...	...	12 06	7.4	5 59	6 33
9	S	12 15	7.2	12 29	7.8	6 21	6 58	9	T	12 37	6.8	12 49	7.1	6 42	7 15
10	M	1 02	7.0	1 15	7.4	7 07	7 44	10	W	1 21	6.6	1 33	6.8	7 26	7 57
11	T	1 50	6.7	2 04	7.0	7 56	8 32	11	T	2 05	6.5	2 19	6.5	8 12	8 40
12	W	2 39	6.5	2 54	6.7	8 46	9 21	12	F	2 51	6.4	3 07	6.2	9 01	9 26
13	T	3 31	6.4	3 48	6.4	9 40	10 13	13	S	3 40	6.3	3 58	6.0	9 53	10 15
14	F	4 24	6.3	4 44	6.2	10 36	11 05	14	S	4 30	6.3	4 52	5.8	10 47	11 05
15	S	5 18	6.4	5 40	6.1	11 33	11 57	15	M	5 22	6.4	5 47	5.8	11 42	11 57
16	S	6 11	6.5	6 35	6.1	...	12 29	16	T	6 13	6.5	6 41	5.8	...	12 37
17	M	7 01	6.7	7 26	6.2	12 47	1 20	17	W	7 04	6.7	7 34	6.0	12 49	1 29
18	T	7 48	6.9	8 14	6.4	1 35	2 09	18	T	7 53	6.9	8 24	6.2	1 39	2 19
19	W	8 32	7.1	9 00	6.5	2 20	2 54	19	F	8 40	7.2	9 13	6.4	2 28	3 08
20	T	9 15	7.4	9 43	6.7	3 04	3 38	20	S	9 26	7.4	9 59	6.7	3 16	3 55
21	F	9 56	7.5	10 26	6.8	3 46	4 21	21	S	10 13	7.7	10 46	6.9	4 03	4 41
22	S	10 37	7.7	11 09	6.9	4 29	5 04	22	M	10 59	7.9	11 33	7.1	4 51	5 28
23	S	11 19	7.8	11 53	7.0	5 12	5 48	23	T	11 47	7.9	...	...	5 40	6 16
24	M	...	...	12 04	7.8	5 57	6 34	24	W	12 21	7.2	12 37	7.9	6 30	7 05
25	T	12 39	7.0	12 52	7.7	6 45	7 23	25	T	1 11	7.3	1 30	7.7	7 23	7 56
26	W	1 28	7.0	1 44	7.6	7 38	8 15	26	F	2 03	7.3	2 25	7.4	8 20	8 50
27	T	2 22	6.9	2 41	7.3	8 34	9 11	27	S	2 59	7.3	3 24	7.1	9 20	9 46
28	F	3 19	6.9	3 41	7.1	9 36	10 09	28	S	3 57	7.2	4 25	6.7	10 23	10 45
29	S	4 19	7.0	4 45	6.9	10 41	11 09	29	M	4 58	7.2	5 29	6.5	11 27	11 45
30	S	5 20	7.2	5 49	6.8	11 46	...	30	T	5 59	7.2	6 32	6.4	...	12 31
								31	W	6 58	7.3	7 33	6.4	12 45	1 32

Dates when Ht. of **Low** Water is below Mean Lower Low with Ht. of lowest given for each period and Date of lowest in ( ):

3rd - 9th: -0.8' (5th - 6th)
21st - 27th: -0.6' (23rd - 24th)

2nd - 8th: -0.7' (4th - 6th)
19th - 28th: -1.0' (23rd - 24th)
31st: -0.2'

Average Rise and Fall 6.8 ft.

When a high tide exceeds avg. ht., the *following* low tide will be lower than avg.

2014 HIGH & LOW WATER
KINGS POINT, NY
40°48.6'N, 73°45.9'W

DAY OF MONTH	DAY OF WEEK	JANUARY (Standard Time)						DAY OF MONTH	DAY OF WEEK	FEBRUARY (Standard Time)					
		HIGH				LOW				HIGH				LOW	
		a.m.	Ht.	**p.m.**	Ht.	a.m.	**p.m.**			a.m.	Ht.	**p.m.**	Ht.	a.m.	**p.m.**
1	**W**	10 27	9.0	**11 06**	8.0	4 36	**5 33**	1	**S**	...	...	**12 11**	8.8	6 29	**6 59**
2	**T**	11 21	9.0	**...**	...	5 34	**6 24**	2	**S**	12 43	8.5	**1 05**	8.5	7 24	**7 47**
3	**F**	12 01	8.2	**12 17**	8.9	6 31	**7 15**	3	**M**	1 36	8.4	**2 01**	8.1	8 20	**8 39**
4	**S**	12 58	8.2	**1 16**	8.5	7 33	**8 10**	4	**T**	2 32	8.2	**3 01**	7.6	9 22	**9 37**
5	**S**	1 56	8.1	**2 16**	8.1	8 37	**9 08**	5	**W**	3 30	7.8	**4 03**	7.1	10 24	**10 38**
6	**M**	2 59	8.0	**3 23**	7.6	9 46	**10 10**	6	**T**	4 33	7.5	**5 10**	6.7	11 27	**11 42**
7	**T**	4 04	7.8	**4 33**	7.2	10 53	**11 13**	7	**F**	5 40	7.2	**6 16**	6.5	...	**12 27**
8	**W**	5 10	7.6	**5 42**	6.9	11 57	**...**	8	**S**	6 45	7.1	**7 18**	6.5	12 43	**1 25**
9	**T**	6 15	7.6	**6 47**	6.8	12 14	**12 57**	9	**S**	7 45	7.1	**8 14**	6.7	1 41	**2 20**
10	**F**	7 16	7.5	**7 47**	6.8	1 13	**1 54**	10	**M**	8 38	7.2	**9 04**	6.9	2 35	**3 10**
11	**S**	8 12	7.6	**8 41**	6.9	2 09	**2 48**	11	**T**	9 26	7.3	**9 49**	7.1	3 24	**3 57**
12	**S**	9 02	7.6	**9 29**	7.0	3 01	**3 38**	12	**W**	10 08	7.4	**10 30**	7.2	4 09	**4 39**
13	**M**	9 48	7.7	**10 14**	7.1	3 49	**4 24**	13	**T**	10 46	7.5	**11 06**	7.3	4 49	**5 17**
14	**T**	10 30	7.7	**10 55**	7.2	4 34	**5 06**	14	**F**	11 18	7.4	**11 36**	7.3	5 24	**5 48**
15	**W**	11 08	7.6	**11 32**	7.2	5 13	**5 45**	15	**S**	11 38	7.4	**11 51**	7.4	5 47	**6 05**
16	**T**	11 39	7.5	**11 59**	7.1	5 47	**6 18**	16	**S**	11 50	7.4	**...**	...	5 57	**6 12**
17	**F**	11 57	7.4	**...**	...	6 05	**6 37**	17	**M**	12 06	7.5	**12 18**	7.5	6 23	**6 39**
18	**S**	12 24	7.1	**12 12**	7.3	6 14	**6 40**	18	**T**	12 36	7.6	**12 54**	7.5	6 58	**7 15**
19	**S**	12 39	7.1	**12 42**	7.3	6 43	**7 06**	19	**W**	1 14	7.7	**1 36**	7.4	7 39	**7 56**
20	**M**	1 08	7.2	**1 19**	7.2	7 21	**7 42**	20	**T**	1 57	7.8	**2 23**	7.2	8 24	**8 42**
21	**T**	1 45	7.3	**2 02**	7.1	8 03	**8 24**	21	**F**	2 44	7.8	**3 13**	7.0	9 14	**9 33**
22	**W**	2 27	7.4	**2 48**	7.0	8 50	**9 10**	22	**S**	3 37	7.7	**4 09**	6.8	10 12	**10 30**
23	**T**	3 14	7.4	**3 39**	6.8	9 41	**10 01**	23	**S**	4 35	7.5	**5 13**	6.7	11 18	**11 36**
24	**F**	4 05	7.4	**4 35**	6.7	10 39	**10 57**	24	**M**	5 41	7.5	**6 28**	6.8	...	**12 50**
25	**S**	5 02	7.4	**5 37**	6.6	11 43	**11 58**	25	**T**	6 59	7.6	**7 54**	7.1	12 57	**2 22**
26	**S**	6 05	7.6	**6 47**	6.7	...	**1 00**	26	**W**	8 20	8.0	**9 02**	7.7	2 34	**3 24**
27	**M**	7 14	7.8	**8 01**	7.0	1 06	**2 31**	27	**T**	9 24	8.4	**9 58**	8.2	3 41	**4 17**
28	**T**	8 23	8.2	**9 08**	7.4	2 22	**3 37**	28	**F**	10 20	8.7	**10 48**	8.6	4 37	**5 07**
29	**W**	9 26	8.5	**10 05**	7.9	3 37	**4 32**								
30	**T**	10 23	8.8	**10 59**	8.2	4 40	**5 22**								
31	**F**	11 17	8.9	**11 51**	8.4	5 35	**6 11**								

Dates when Ht. of **Low** Water is below Mean Lower Low with Ht. of lowest given for each period and Date of lowest in ( ):

January:
1st - 19th: -1.5' (2nd)
27th - 31st: -1.6' (31st)

February:
1st - 5th: -1.6' (1st)
11th - 20th: -0.3' (12th - 14th, 17th - 19th)
25th - 28th: -1.4' (28th)

Average Rise and Fall 7.1 ft.

When a high tide exceeds avg. ht., the *following* low tide will be lower than avg.

2014 HIGH & LOW WATER KINGS POINT, NY

40°48.6'N, 73°45.9'W

***Daylight Time starts March 9 at 2 a.m.** **Daylight Saving Time**

DAY OF MONTH	DAY OF WEEK	MARCH						DAY OF MONTH	DAY OF WEEK	APRIL					
		HIGH				LOW				HIGH				LOW	
		a.m.	Ht.	p.m.	Ht.	a.m.	p.m.			a.m.	Ht.	p.m.	Ht.	a.m.	p.m.
1	S	11 11	8.8	11 36	8.8	5 29	5 53	1	T	1 02	8.9	1 34	8.2	7 52	8 00
2	S	...	...	12 01	8.7	6 19	6 38	2	W	1 45	8.6	2 21	7.9	8 39	8 40
3	M	12 23	8.8	12 51	8.4	7 09	7 23	3	T	2 28	8.2	3 09	7.5	9 26	9 18
4	T	1 12	8.6	1 42	8.0	8 01	8 10	4	F	3 12	7.8	4 00	7.2	10 18	9 56
5	W	1 59	8.2	2 34	7.6	8 54	8 59	5	S	3 59	7.3	4 55	6.8	11 13	11 03
6	T	2 51	7.8	3 31	7.1	9 52	9 58	6	S	4 58	6.9	5 57	6.6	...	12 11
7	F	3 49	7.3	4 34	6.7	10 52	11 04	7	M	6 16	6.6	7 01	6.5	12 23	1 09
8	S	4 57	6.9	5 40	6.5	11 53	...	8	T	7 27	6.5	8 00	6.7	1 27	2 04
9	S	*7 07	6.7	*7 43	6.5	12 08	*1 51	9	W	8 27	6.6	8 54	6.9	2 25	2 55
10	M	8 11	6.7	8 41	6.6	2 08	2 46	10	T	9 19	6.8	9 40	7.2	3 17	3 42
11	T	9 07	6.8	9 33	6.8	3 04	3 37	11	F	10 04	7.0	10 20	7.4	4 04	4 24
12	W	9 57	7.0	10 18	7.1	3 54	4 23	12	S	10 43	7.2	10 52	7.7	4 46	5 00
13	T	10 40	7.2	10 59	7.3	4 40	5 06	13	S	11 14	7.3	11 12	7.9	5 22	5 26
14	F	11 19	7.4	11 34	7.5	5 21	5 43	14	M	11 35	7.5	11 31	8.2	5 51	5 43
15	S	11 50	7.4	11 59	7.6	5 56	6 12	15	T	11 58	7.7	...	...	6 14	6 12
16	S	...	...	12 10	7.5	6 22	6 27	16	W	12 03	8.4	12 32	7.8	6 44	6 48
17	M	12 11	7.8	12 26	7.6	6 38	6 42	17	T	12 42	8.6	1 13	7.8	7 21	7 30
18	T	12 33	8.0	12 55	7.6	7 03	7 13	18	F	1 25	8.6	1 58	7.8	8 04	8 15
19	W	1 07	8.1	1 32	7.7	7 38	7 50	19	S	2 13	8.5	2 48	7.7	8 51	9 05
20	T	1 47	8.2	2 15	7.6	8 18	8 33	20	S	3 05	8.3	3 42	7.5	9 45	10 02
21	F	2 32	8.2	3 02	7.5	9 04	9 20	21	M	4 01	8.0	4 43	7.4	10 48	11 12
22	S	3 21	8.1	3 54	7.3	9 55	10 13	22	T	5 06	7.7	5 58	7.3	...	12 20
23	S	4 15	7.9	4 52	7.1	10 54	11 14	23	W	6 26	7.5	7 30	7.5	1 01	1 45
24	M	5 16	7.6	5 59	7.0	...	12 07	24	T	8 02	7.5	8 42	7.9	2 21	2 50
25	T	6 27	7.5	7 25	7.1	12 30	1 56	25	F	9 12	7.8	9 40	8.4	3 24	3 47
26	W	7 59	7.6	8 52	7.5	2 21	3 08	26	S	10 09	8.0	10 31	8.7	4 21	4 39
27	T	9 20	7.9	9 54	8.0	3 35	4 07	27	S	11 00	8.2	11 18	8.9	5 14	5 29
28	F	10 20	8.2	10 46	8.5	4 34	4 59	28	M	11 48	8.2	...	...	6 03	6 15
29	S	11 12	8.5	11 34	8.8	5 28	5 48	29	T	12 01	8.9	12 33	8.2	6 50	6 58
30	S	...	...	12 01	8.5	6 18	6 34	30	W	12 43	8.7	1 18	8.0	7 35	7 39
31	M	12 19	8.9	12 48	8.5	7 06	7 18								

Dates when Ht. of **Low** Water is below Mean Lower Low with Ht. of lowest given for each period and Date of lowest in ( ):

1st - 6th: -1.5' (2nd)
16th - 21t: -0.5' (19th - 20th)
27th - 31st: -1.4' (30th - 31st)

1st - 3rd: -1.2' (1st)
14th - 19th: -0.6' (17th - 18th)
25th - 30th: -1.2' (28th)

Average Rise and Fall 7.1 ft.

When a high tide exceeds avg. ht., the *following* low tide will be lower than avg.

2014 HIGH & LOW WATER
KINGS POINT, NY
40°48.6'N, 73°45.9'W

DAY OF MONTH	DAY OF WEEK	MAY HIGH a.m.	Ht.	**p.m.**	Ht.	LOW a.m.	**p.m.**	DAY OF MONTH	DAY OF WEEK	JUNE HIGH a.m.	Ht.	**p.m.**	Ht.	LOW a.m.	**p.m.**
		Daylight Saving Time								Daylight Saving Time					
1	T	1 22	8.4	2 01	7.8	8 18	8 15	1	S	2 07	7.6	2 57	7.3	9 05	8 30
2	F	2 00	8.1	2 44	7.5	9 00	8 39	2	M	2 37	7.4	3 30	7.2	9 10	9 07
3	S	2 36	7.7	3 28	7.2	9 41	9 02	3	T	3 13	7.2	4 01	7.1	9 37	9 51
4	S	3 14	7.3	4 15	7.0	10 23	9 42	4	W	3 57	6.9	4 40	7.0	10 19	10 43
5	M	3 55	7.0	5 04	6.8	11 04	10 30	5	T	4 43	6.7	5 23	7.0	11 04	11 37
6	T	4 44	6.7	6 02	6.7	11 59	11 32	6	F	5 35	6.6	6 12	7.1	11 55	...
7	W	5 49	6.5	7 04	6.8	...	12 58	7	S	6 33	6.5	7 05	7.3	12 36	12 49
8	T	7 23	6.4	7 59	6.9	1 27	1 51	8	S	7 36	6.6	7 58	7.5	1 41	1 44
9	F	8 25	6.5	8 47	7.1	2 26	2 38	9	M	8 38	6.8	8 49	7.9	2 49	2 39
10	S	9 15	6.7	9 24	7.4	3 17	3 17	10	T	9 31	7.1	9 38	8.3	3 49	3 34
11	S	9 55	6.9	9 52	7.8	4 02	3 49	11	W	10 19	7.4	10 26	8.6	4 41	4 27
12	M	10 26	7.2	10 20	8.1	4 41	4 23	12	T	11 06	7.7	11 14	8.9	5 29	5 20
13	T	10 56	7.5	10 55	8.5	5 16	5 02	13	F	11 53	8.0	...	...	6 16	6 12
14	W	11 31	7.7	11 36	8.7	5 50	5 43	14	S	12 04	9.1	12 43	8.1	7 03	7 05
15	T	...	...	12 11	7.9	6 28	6 27	15	S	12 56	9.0	1 36	8.2	7 53	8 01
16	F	12 20	8.9	12 57	8.0	7 10	7 14	16	M	1 50	8.9	2 33	8.3	8 46	9 03
17	S	1 08	8.9	1 46	8.0	7 56	8 04	17	T	2 48	8.6	3 34	8.2	9 44	10 18
18	S	1 59	8.7	2 39	7.9	8 46	8 59	18	W	3 52	8.2	4 41	8.2	10 50	11 35
19	M	2 54	8.5	3 37	7.8	9 44	10 05	19	T	5 05	7.8	5 52	8.2	11 59	...
20	T	3 54	8.1	4 44	7.8	10 56	11 37	20	F	6 23	7.5	7 02	8.2	12 45	1 04
21	W	5 04	7.8	6 03	7.8	...	12 17	21	S	7 35	7.4	8 06	8.2	1 49	2 05
22	T	6 32	7.5	7 21	8.0	1 00	1 26	22	S	8 39	7.4	9 04	8.3	2 49	3 03
23	F	7 53	7.5	8 26	8.2	2 08	2 28	23	M	9 36	7.5	9 57	8.4	3 45	3 58
24	S	8 58	7.6	9 23	8.5	3 08	3 25	24	T	10 28	7.6	10 45	8.4	4 38	4 50
25	S	9 54	7.8	10 14	8.7	4 05	4 19	25	W	11 16	7.7	11 30	8.3	5 27	5 37
26	M	10 45	7.9	11 02	8.7	4 57	5 09	26	T	...	...	12 01	7.7	6 12	6 22
27	T	11 33	7.9	11 45	8.6	5 46	5 56	27	F	12 11	8.2	12 42	7.6	6 55	7 01
28	W	...	...	12 18	7.9	6 32	6 40	28	S	12 49	8.0	1 21	7.6	7 34	7 35
29	T	12 26	8.4	1 01	7.8	7 16	7 20	29	S	1 20	7.8	1 56	7.5	8 08	7 51
30	F	1 04	8.2	1 42	7.6	7 57	7 55	30	M	1 42	7.6	2 24	7.4	8 27	8 04
31	S	1 38	7.9	2 21	7.4	8 35	8 12								

Dates when Ht. of **Low** Water is below Mean Lower Low with Ht. of lowest given for each period and Date of lowest in ( ):

1st - 2nd: -0.6' (1st)
14th - 19th: -0.7' (16th - 17th)
24th - 30th: -0.9' (27th)

12th - 18th: -0.9' (15th)
22nd - 28th: -0.6' (25th)

Average Rise and Fall 7.1 ft.

When a high tide exceeds avg. ht., the *following* low tide will be lower than avg.

2014 HIGH & LOW WATER KINGS POINT, NY

40°48.6'N, 73°45.9'W

		Daylight Saving Time								**Daylight Saving Time**					
DAY OF MONTH	DAY OF WEEK	**JULY**						DAY OF MONTH	DAY OF WEEK	**AUGUST**					
		HIGH				LOW				HIGH				LOW	
		a.m.	Ht.	**p.m.**	Ht.	a.m.	**p.m.**			a.m.	Ht.	**p.m.**	Ht.	a.m.	**p.m.**
1	**T**	2 06	7.5	**2 44**	7.4	8 30	**8 38**	1	**F**	2 47	7.4	**3 11**	7.8	9 06	**9 33**
2	**W**	2 39	7.3	**3 13**	7.4	9 00	**9 19**	2	**S**	3 31	7.3	**3 54**	7.8	9 49	**10 21**
3	**T**	3 19	7.2	**3 50**	7.4	9 39	**10 05**	3	**S**	4 18	7.1	**4 42**	7.8	10 37	**11 14**
4	**F**	4 05	7.0	**4 34**	7.4	10 25	**10 56**	4	**M**	5 12	7.0	**5 36**	7.8	11 31	**...**
5	**S**	4 52	6.9	**5 20**	7.5	11 12	**11 50**	5	**T**	6 08	6.9	**6 33**	7.9	12 12	**12 27**
6	**S**	5 45	6.8	**6 12**	7.6	...	**12 04**	6	**W**	7 12	7.0	**7 37**	8.0	1 18	**1 30**
7	**M**	6 43	6.7	**7 08**	7.7	12 48	**1 00**	7	**T**	8 22	7.2	**8 43**	8.3	2 35	**2 38**
8	**T**	7 46	6.9	**8 07**	8.0	1 52	**1 58**	8	**F**	9 30	7.6	**9 48**	8.7	3 55	**3 51**
9	**W**	8 50	7.1	**9 06**	8.4	3 03	**3 00**	9	**S**	10 30	8.1	**10 47**	9.0	4 56	**5 02**
10	**T**	9 50	7.5	**10 03**	8.7	4 13	**4 03**	10	**S**	11 25	8.5	**11 43**	9.2	5 49	**6 03**
11	**F**	10 45	7.8	**10 58**	9.0	5 12	**5 05**	11	**M**	...	...	**12 17**	8.8	6 39	**6 59**
12	**S**	11 38	8.2	**11 52**	9.1	6 04	**6 05**	12	**T**	12 38	9.1	**1 10**	9.0	7 27	**7 54**
13	**S**	...	...	**12 31**	8.4	6 54	**7 02**	13	**W**	1 33	9.0	**2 04**	9.0	8 16	**8 52**
14	**M**	12 46	9.1	**1 26**	8.6	7 44	**8 01**	14	**T**	2 30	8.6	**2 59**	8.9	9 07	**9 52**
15	**T**	1 43	9.0	**2 22**	8.7	8 35	**9 04**	15	**F**	3 30	8.2	**3 58**	8.6	10 04	**10 56**
16	**W**	2 42	8.6	**3 22**	8.6	9 30	**10 10**	16	**S**	4 34	7.8	**5 02**	8.3	11 07	**11 59**
17	**T**	3 45	8.2	**4 24**	8.5	10 30	**11 18**	17	**S**	5 42	7.4	**6 09**	7.9	...	**12 13**
18	**F**	4 54	7.8	**5 30**	8.3	11 35	**...**	18	**M**	6 49	7.2	**7 16**	7.7	1 01	**1 16**
19	**S**	6 05	7.5	**6 37**	8.2	12 24	**12 39**	19	**T**	7 53	7.1	**8 19**	7.7	2 00	**2 16**
20	**S**	7 14	7.3	**7 43**	8.1	1 27	**1 41**	20	**W**	8 51	7.2	**9 15**	7.7	2 56	**3 13**
21	**M**	8 18	7.2	**8 43**	8.0	2 26	**2 41**	21	**T**	9 43	7.4	**10 05**	7.8	3 48	**4 04**
22	**T**	9 15	7.3	**9 38**	8.1	3 22	**3 37**	22	**F**	10 30	7.6	**10 50**	7.9	4 37	**4 52**
23	**W**	10 07	7.4	**10 27**	8.1	4 15	**4 29**	23	**S**	11 13	7.7	**11 30**	7.9	5 21	**5 35**
24	**T**	10 55	7.6	**11 12**	8.1	5 04	**5 16**	24	**S**	11 52	7.8	**...**	...	6 01	**6 14**
25	**F**	11 39	7.6	**11 53**	8.0	5 49	**6 00**	25	**M**	12 06	7.8	**12 26**	7.9	6 36	**6 46**
26	**S**	...	...	**12 19**	7.7	6 30	**6 40**	26	**T**	12 35	7.7	**12 50**	7.8	7 02	**7 04**
27	**S**	12 30	7.9	**12 56**	7.7	7 07	**7 13**	27	**W**	12 49	7.7	**1 01**	7.9	7 05	**7 14**
28	**M**	1 00	7.7	**1 26**	7.6	7 37	**7 29**	28	**T**	1 08	7.6	**1 23**	8.0	7 23	**7 44**
29	**T**	1 17	7.6	**1 44**	7.6	7 46	**7 39**	29	**F**	1 39	7.6	**1 57**	8.0	7 56	**8 22**
30	**W**	1 36	7.5	**2 02**	7.6	7 56	**8 10**	30	**S**	2 19	7.6	**2 37**	8.1	8 36	**9 04**
31	**T**	2 08	7.5	**2 33**	7.7	8 27	**8 49**	31	**S**	3 02	7.5	**3 22**	8.1	9 20	**9 52**

Dates when Ht. of **Low** Water is below Mean Lower Low with Ht. of lowest given for each period and Date of lowest in ( ):

July:
11th - 17th: -1.0' (13th - 15th)
23rd - 26th: -0.3' (24th)

August:
9th - 15th: -1.1' (12th)

Average Rise and Fall 7.1 ft.

When a high tide exceeds avg. ht., the *following* low tide will be lower than avg.

2014 HIGH & LOW WATER KINGS POINT, NY

40°48.6'N, 73°45.9'W

Daylight Saving Time

DAY OF MONTH	DAY OF WEEK	SEPTEMBER HIGH a.m.	Ht.	p.m.	Ht.	LOW a.m.	p.m.
1	M	3 51	7.3	**4 12**	8.0	10 08	**10 45**
2	T	4 44	7.2	**5 06**	7.9	11 03	**11 46**
3	W	5 42	7.1	**6 07**	7.9	...	**12 03**
4	T	6 50	7.1	**7 17**	8.0	12 57	**1 13**
5	F	8 05	7.4	**8 31**	8.2	2 25	**2 34**
6	S	9 19	7.9	**9 41**	8.6	3 42	**3 58**
7	S	10 19	8.4	**10 41**	8.9	4 40	**5 01**
8	M	11 11	8.9	**11 35**	9.0	5 31	**5 57**
9	T	...	...	**12 01**	9.2	6 19	**6 50**
10	W	12 27	9.0	**12 50**	9.3	7 06	**7 42**
11	T	1 19	8.8	**1 40**	9.2	7 53	**8 35**
12	F	2 13	8.5	**2 32**	8.9	8 41	**9 30**
13	S	3 10	8.1	**3 27**	8.5	9 35	**10 30**
14	S	4 10	7.6	**4 28**	8.0	10 37	**11 31**
15	M	5 14	7.3	**5 36**	7.6	11 44	**...**
16	T	6 20	7.1	**6 45**	7.4	12 32	**12 48**
17	W	7 23	7.0	**7 49**	7.3	1 30	**1 48**
18	T	8 21	7.2	**8 46**	7.4	2 25	**2 44**
19	F	9 13	7.4	**9 37**	7.5	3 17	**3 36**
20	S	10 00	7.6	**10 22**	7.6	4 04	**4 23**
21	S	10 42	7.8	**11 02**	7.7	4 48	**5 06**
22	M	11 20	8.0	**11 38**	7.7	5 26	**5 44**
23	T	11 50	8.0	**...**	...	5 59	**6 16**
24	W	12 05	7.7	**12 07**	8.1	6 19	**6 35**
25	T	12 18	7.7	**12 21**	8.2	6 26	**6 50**
26	F	12 39	7.7	**12 49**	8.3	6 53	**7 20**
27	S	1 13	7.7	**1 26**	8.3	7 29	**7 58**
28	S	1 53	7.6	**2 09**	8.3	8 09	**8 41**
29	M	2 39	7.6	**2 56**	8.3	8 55	**9 30**
30	T	3 28	7.4	**3 48**	8.1	9 46	**10 25**

Daylight Saving Time

DAY OF MONTH	DAY OF WEEK	OCTOBER HIGH a.m.	Ht.	p.m.	Ht.	LOW a.m.	p.m.
1	W	4 23	7.3	**4 45**	8.0	10 43	**11 28**
2	T	5 25	7.2	**5 49**	7.8	11 49	**...**
3	F	6 36	7.3	**7 04**	7.8	12 48	**1 13**
4	S	8 02	7.7	**8 29**	8.0	2 22	**2 52**
5	S	9 12	8.2	**9 38**	8.4	3 27	**3 57**
6	M	10 08	8.7	**10 34**	8.6	4 22	**4 54**
7	T	10 58	9.1	**11 26**	8.7	5 13	**5 47**
8	W	11 45	9.3	**...**	...	6 01	**6 37**
9	T	12 15	8.7	**12 31**	9.3	6 47	**7 26**
10	F	1 05	8.5	**1 17**	9.1	7 31	**8 16**
11	S	1 55	8.2	**2 04**	8.7	8 17	**9 07**
12	S	2 47	7.8	**2 54**	8.2	9 04	**10 01**
13	M	3 43	7.5	**3 50**	7.8	10 01	**10 59**
14	T	4 43	7.2	**4 55**	7.3	11 08	**11 57**
15	W	5 45	7.0	**6 05**	7.0	...	**12 13**
16	T	6 47	6.9	**7 11**	6.9	12 54	**1 14**
17	F	7 45	7.0	**8 10**	6.9	1 48	**2 10**
18	S	8 38	7.2	**9 02**	7.1	2 39	**3 02**
19	S	9 25	7.5	**9 48**	7.2	3 25	**3 49**
20	M	10 07	7.7	**10 29**	7.3	4 08	**4 33**
21	T	10 42	7.9	**11 04**	7.4	4 46	**5 11**
22	W	11 08	8.0	**11 30**	7.5	5 16	**5 44**
23	T	11 23	8.2	**11 46**	7.5	5 30	**6 06**
24	F	11 46	8.3	**...**	...	5 52	**6 27**
25	S	12 13	7.6	**12 20**	8.5	6 26	**7 00**
26	S	12 50	7.7	**1 01**	8.5	7 05	**7 39**
27	M	1 32	7.7	**1 46**	8.5	7 48	**8 24**
28	T	2 20	7.6	**2 36**	8.4	8 36	**9 14**
29	W	3 12	7.5	**3 30**	8.1	9 30	**10 11**
30	T	4 09	7.4	**4 30**	7.9	10 32	**11 20**
31	F	5 14	7.4	**5 38**	7.6	11 51	**...**

Dates when Ht. of **Low** Water is below Mean Lower Low with Ht. of lowest given for each period and Date of lowest in ( ):

7th - 12th: -1.1' (9th - 10th)

5th - 11th: -1.2' (8th)
24th - 26th: -0.2'

Average Rise and Fall 7.1 ft.

When a high tide exceeds avg. ht., the *following* low tide will be lower than avg.

2014 HIGH & LOW WATER KINGS POINT, NY

40°48.6'N, 73°45.9'W

***Standard Time starts Nov. 2 at 2 a.m.** **Standard Time**

DAY OF MONTH	DAY OF WEEK	NOVEMBER						DAY OF MONTH	DAY OF WEEK	DECEMBER					
		HIGH				LOW				HIGH				LOW	
		a.m.	Ht.	**p.m.**	Ht.	a.m.	**p.m.**			a.m.	Ht.	**p.m.**	Ht.	a.m.	**p.m.**
1	**S**	6 34	7.5	**7 03**	7.6	12 53	**1 37**	1	**M**	6 47	7.9	**7 20**	7.4	12 51	**1 36**
2	**S**	*6 59	7.9	***7 28**	7.7	*1 09	***1 48**	2	**T**	7 50	8.2	**8 22**	7.5	1 52	**2 35**
3	**M**	8 03	8.3	**8 32**	8.0	2 10	**2 48**	3	**W**	8 45	8.5	**9 16**	7.7	2 48	**3 29**
4	**T**	8 58	8.8	**9 27**	8.2	3 06	**3 44**	4	**T**	9 36	8.7	**10 07**	7.8	3 42	**4 21**
5	**W**	9 46	9.0	**10 16**	8.3	3 56	**4 35**	5	**F**	10 21	8.7	**10 52**	7.8	4 30	**5 09**
6	**T**	10 32	9.1	**11 04**	8.3	4 44	**5 24**	6	**S**	11 05	8.5	**11 37**	7.7	5 16	**5 55**
7	**F**	11 15	9.0	**11 51**	8.1	5 30	**6 11**	7	**S**	11 46	8.3	**...**	...	6 00	**6 38**
8	**S**	11 59	8.7	**...**	...	6 14	**6 57**	8	**M**	12 21	7.5	**12 25**	8.0	6 40	**7 20**
9	**S**	12 37	7.9	**12 41**	8.4	6 56	**7 44**	9	**T**	1 03	7.3	**1 00**	7.6	7 14	**7 59**
10	**M**	1 25	7.6	**1 25**	7.9	7 37	**8 31**	10	**W**	1 44	7.1	**1 33**	7.3	7 31	**8 32**
11	**T**	2 14	7.3	**2 10**	7.5	8 15	**9 20**	11	**T**	2 23	7.0	**2 08**	7.0	7 57	**8 39**
12	**W**	3 06	7.0	**3 00**	7.1	9 00	**10 13**	12	**F**	3 01	6.8	**2 48**	6.7	8 38	**9 08**
13	**T**	4 01	6.8	**4 02**	6.8	10 20	**11 06**	13	**S**	3 40	6.7	**3 34**	6.5	9 27	**9 52**
14	**F**	5 00	6.8	**5 14**	6.5	11 27	**11 59**	14	**S**	4 23	6.7	**4 26**	6.3	10 23	**10 42**
15	**S**	5 58	6.8	**6 19**	6.5	...	**12 26**	15	**M**	5 13	6.7	**5 28**	6.2	11 30	**11 35**
16	**S**	6 52	7.0	**7 16**	6.5	12 49	**1 19**	16	**T**	6 08	6.8	**6 46**	6.2	...	**1 01**
17	**M**	7 41	7.2	**8 06**	6.7	1 35	**2 09**	17	**W**	7 00	7.0	**7 47**	6.4	12 31	**2 00**
18	**T**	8 22	7.4	**8 49**	6.8	2 17	**2 54**	18	**T**	7 46	7.3	**8 30**	6.6	1 26	**2 49**
19	**W**	8 56	7.6	**9 24**	7.0	2 52	**3 34**	19	**F**	8 28	7.7	**9 08**	6.9	2 19	**3 33**
20	**T**	9 18	7.9	**9 50**	7.2	3 16	**4 09**	20	**S**	9 10	8.1	**9 48**	7.2	3 09	**4 13**
21	**F**	9 43	8.2	**10 16**	7.4	3 45	**4 38**	21	**S**	9 54	8.4	**10 29**	7.5	3 58	**4 54**
22	**S**	10 17	8.4	**10 50**	7.5	4 22	**5 08**	22	**M**	10 40	8.6	**11 15**	7.7	4 46	**5 36**
23	**S**	10 57	8.6	**11 30**	7.6	5 03	**5 45**	23	**T**	11 28	8.7	**...**	...	5 35	**6 20**
24	**M**	11 42	8.6	**...**	...	5 46	**6 27**	24	**W**	12 03	7.8	**12 19**	8.6	6 25	**7 07**
25	**T**	12 16	7.7	**12 30**	8.6	6 33	**7 14**	25	**T**	12 54	7.9	**1 12**	8.4	7 20	**7 58**
26	**W**	1 06	7.7	**1 22**	8.4	7 24	**8 05**	26	**F**	1 50	7.9	**2 10**	8.0	8 21	**8 57**
27	**T**	2 00	7.6	**2 18**	8.1	8 22	**9 04**	27	**S**	2 50	7.8	**3 13**	7.6	9 41	**10 08**
28	**F**	2 59	7.6	**3 20**	7.7	9 32	**10 19**	28	**S**	3 58	7.8	**4 29**	7.2	11 05	**11 23**
29	**S**	4 08	7.5	**4 33**	7.4	11 13	**11 42**	29	**M**	5 16	7.7	**5 53**	7.0	...	**12 15**
30	**S**	5 31	7.7	**6 04**	7.3	...	**12 31**	30	**T**	6 30	7.8	**7 06**	7.0	12 31	**1 19**
								31	**W**	7 35	7.9	**8 08**	7.1	1 33	**2 18**

Dates when Ht. of **Low** Water is below Mean Lower Low with Ht. of lowest given for each period and Date of lowest in ( ):

November	December
3rd - 9th: -1.2' (6th)	1st - 8th: -1.2' (5th)
21st - 27th: -0.6' (23rd - 24th)	20th - 31st: -1.0' (23rd - 24th)

Average Rise and Fall 7.1 ft.

When a high tide exceeds avg. ht., the *following* low tide will be lower than avg.

2014 CURRENT TABLE
HELL GATE, NY (EAST RIVER)

40°46.7'N, 73°56.3'W Off Mill Rock

Standard Time — Standard Time

JANUARY

DAY OF MONTH	DAY OF WEEK	CURRENT TURNS TO NORTHEAST Flood Starts			SOUTHWEST Ebb Starts		
		a.m.	**p.m.**	Kts.	a.m.	**p.m.**	Kts.
1	W	3 20	**3 56**	3.9	9 23	**9 51**	a5.3
2	T	4 13	**4 49**	a4.0	10 16	**10 44**	a5.4
3	F	5 07	**5 41**	a4.0	11 10	**11 37**	a5.3
4	S	6 03	**6 36**	a3.9	...	**12 03**	5.2
5	S	6 58	**7 30**	a3.7	12 30	**12 58**	a5.1
6	M	7 57	**8 27**	a3.5	1 25	**1 54**	a4.9
7	T	8 58	**9 27**	3.3	2 22	**2 52**	a4.7
8	W	10 00	**10 27**	a3.2	3 20	**3 51**	a4.5
9	T	11 02	**11 25**	3.1	4 19	**4 50**	a4.4
10	F	...	**12 01**	3.0	5 17	**5 46**	a4.4
11	S	12 21	**12 55**	3.1	6 12	**6 39**	a4.4
12	S	1 12	**1 43**	a3.2	7 03	**7 28**	a4.5
13	M	1 58	**2 27**	a3.3	7 49	**8 12**	a4.6
14	T	2 40	**3 08**	a3.4	8 33	**8 54**	a4.7
15	W	3 20	**3 47**	3.4	9 14	**9 34**	a4.8
16	T	3 59	**4 25**	a3.5	9 54	**10 13**	4.8
17	F	4 36	**5 02**	a3.5	10 32	**10 51**	a4.9
18	S	5 13	**5 38**	a3.5	11 11	**11 29**	a4.9
19	S	5 50	**6 15**	a3.5	11 49	**...**	4.8
20	M	6 28	**6 52**	a3.4	12 07	**12 28**	4.8
21	T	7 08	**7 31**	a3.3	12 46	**1 09**	4.7
22	W	7 53	**8 15**	a3.2	1 28	**1 54**	a4.7
23	T	8 43	**9 05**	3.1	2 15	**2 44**	a4.6
24	F	9 41	**10 03**	p3.1	3 08	**3 40**	a4.6
25	S	10 44	**11 05**	p3.2	4 08	**4 42**	a4.6
26	S	11 49	**...**	3.1	5 11	**5 45**	4.6
27	M	12 09	**12 52**	3.3	6 15	**6 47**	a4.8
28	T	1 11	**1 51**	3.6	7 16	**7 47**	4.9
29	W	2 10	**2 46**	3.8	8 15	**8 43**	5.1
30	T	3 05	**3 40**	a4.0	9 10	**9 37**	a5.3
31	F	3 59	**4 31**	a4.1	10 04	**10 29**	5.3

FEBRUARY

DAY OF MONTH	DAY OF WEEK	CURRENT TURNS TO NORTHEAST Flood Starts			SOUTHWEST Ebb Starts		
		a.m.	**p.m.**	Kts.	a.m.	**p.m.**	Kts.
1	S	4 52	**5 22**	a4.1	10 57	**11 20**	5.3
2	S	5 45	**6 13**	a4.0	11 48	**...**	5.2
3	M	6 38	**7 05**	a3.9	12 12	**12 40**	a5.1
4	T	7 33	**7 59**	a3.6	1 04	**1 32**	a5.0
5	W	8 28	**8 54**	a3.4	1 57	**2 26**	a4.7
6	T	9 27	**9 51**	3.1	2 52	**3 22**	a4.5
7	F	10 27	**10 50**	3.0	3 49	**4 19**	a4.3
8	S	11 26	**11 47**	p3.0	4 46	**5 16**	a4.1
9	S	...	**12 21**	2.9	5 43	**6 10**	4.1
10	M	12 40	**1 11**	3.0	6 35	**7 00**	4.2
11	T	1 27	**1 56**	3.1	7 24	**7 46**	4.3
12	W	2 11	**2 38**	3.3	8 08	**8 29**	4.5
13	T	2 51	**3 17**	3.4	8 50	**9 09**	4.6
14	F	3 29	**3 54**	3.5	9 30	**9 47**	p4.8
15	S	4 06	**4 29**	a3.6	10 09	**10 25**	p4.9
16	S	4 43	**5 04**	a3.7	10 47	**11 02**	4.9
17	M	5 19	**5 39**	a3.7	11 24	**11 39**	4.9
18	T	5 56	**6 15**	a3.6	...	**12 03**	4.9
19	W	6 35	**6 53**	a3.5	12 18	**12 43**	a4.9
20	T	7 18	**7 36**	a3.4	1 00	**1 27**	a4.8
21	F	8 08	**8 27**	3.3	1 47	**2 17**	a4.7
22	S	9 06	**9 28**	3.2	2 41	**3 15**	a4.6
23	S	10 13	**10 36**	p3.2	3 44	**4 20**	a4.5
24	M	11 22	**11 46**	p3.4	4 51	**5 27**	4.5
25	T	...	**12 29**	3.4	5 59	**6 32**	4.6
26	W	12 52	**1 30**	3.6	7 03	**7 32**	4.8
27	T	1 53	**2 27**	3.8	8 03	**8 29**	5.0
28	F	2 50	**3 20**	4.0	8 58	**9 22**	p5.2

The Kts. (knots) columns show the **maximum** predicted velocities of the stronger one of the Flood Currents and the stronger one of the Ebb Currents for each day.
The letter "a" means the velocity shown should occur **after** the **a.m.** Current Change. The letter "p" means the velocity shown should occur **after** the **p.m.** Current Change (even if next morning). No "a" or "p" means a.m. and p.m. velocities are the same for that day.
Avg. Max. Velocity: Flood 3.4 Kts., Ebb 4.6 Kts.
Max. Flood 3 hrs. after Flood Starts, ±10 min.
Max. Ebb 3 hrs. after Ebb Starts, ±10 min.

At **City Island** the Current turns 2 hours before Hell Gate. At **Throg's Neck** the Current turns 1 hour before Hell Gate. At **Whitestone Pt.** the Current turns 35 min. before Hell Gate. At **College Pt.** the Current turns 20 min. before Hell Gate.

2014 CURRENT TABLE
HELL GATE, NY (EAST RIVER)
40°46.7'N, 73°56.3'W Off Mill Rock

***Daylight Time starts March 9 at 2 a.m.** **Daylight Saving Time**

MARCH

DAY OF MONTH	DAY OF WEEK	CURRENT TURNS TO NORTHEAST Flood Starts			CURRENT TURNS TO SOUTHWEST Ebb Starts		
		a.m.	**p.m.**	Kts.	a.m.	**p.m.**	Kts.
1	**S**	3 43	**4 10**	4.1	9 51	**10 13**	p5.3
2	**S**	4 34	**4 59**	4.1	10 41	**11 02**	5.2
3	**M**	5 24	**5 47**	a4.1	11 30	**11 51**	5.1
4	**T**	6 15	**6 37**	a3.9	...	**12 19**	4.9
5	**W**	7 04	**7 26**	a3.7	12 40	**1 09**	a4.9
6	**T**	7 55	**8 17**	a3.4	1 30	**1 59**	a4.7
7	**F**	8 49	**9 12**	3.1	2 22	**2 52**	a4.4
8	**S**	9 46	**10 09**	2.9	3 17	**3 47**	a4.1
9	**S**	*11 44	**...**	2.8	*5 13	***5 43**	a4.0
10	**M**	12 07	**12 40**	a2.9	6 10	**6 38**	3.9
11	**T**	1 02	**1 32**	2.9	7 04	**7 29**	4.0
12	**W**	1 51	**2 19**	3.1	7 54	**8 16**	p4.2
13	**T**	2 36	**3 02**	p3.3	8 40	**9 00**	p4.4
14	**F**	3 18	**3 41**	3.4	9 23	**9 40**	p4.6
15	**S**	3 57	**4 19**	3.6	10 03	**10 19**	p4.8
16	**S**	4 35	**4 55**	a3.7	10 42	**10 57**	p4.9
17	**M**	5 12	**5 30**	a3.8	11 21	**11 35**	p5.0
18	**T**	5 49	**6 06**	a3.8	11 59	**...**	4.9
19	**W**	6 27	**6 43**	3.7	12 13	**12 39**	a5.0
20	**T**	7 08	**7 23**	a3.7	12 53	**1 21**	a5.0
21	**F**	7 53	**8 09**	3.5	1 37	**2 07**	a4.9
22	**S**	8 45	**9 03**	3.4	2 27	**2 59**	a4.8
23	**S**	9 44	**10 07**	3.3	3 23	**3 58**	a4.6
24	**M**	10 52	**11 18**	p3.3	4 28	**5 04**	a4.5
25	**T**	...	**12 02**	3.2	5 37	**6 12**	a4.5
26	**W**	12 30	**1 09**	3.4	6 45	**7 17**	p4.6
27	**T**	1 37	**2 11**	3.6	7 49	**8 17**	p4.8
28	**F**	2 38	**3 07**	3.8	8 48	**9 12**	p5.0
29	**S**	3 33	**3 59**	4.0	9 42	**10 04**	p5.1
30	**S**	4 25	**4 48**	4.1	10 33	**10 53**	p5.2
31	**M**	5 14	**5 35**	a4.1	11 21	**11 41**	p5.2

APRIL

DAY OF MONTH	DAY OF WEEK	CURRENT TURNS TO NORTHEAST Flood Starts			CURRENT TURNS TO SOUTHWEST Ebb Starts		
		a.m.	**p.m.**	Kts.	a.m.	**p.m.**	Kts.
1	**T**	6 02	**6 21**	a4.0	...	**12 09**	5.0
2	**W**	6 49	**7 07**	a3.9	12 27	**12 55**	a5.0
3	**T**	7 36	**7 54**	a3.6	1 14	**1 41**	a4.9
4	**F**	8 25	**8 43**	a3.4	2 01	**2 29**	a4.6
5	**S**	9 13	**9 34**	3.1	2 50	**3 19**	a4.4
6	**S**	10 06	**10 28**	2.9	3 41	**4 11**	a4.1
7	**M**	11 01	**11 25**	2.8	4 35	**5 05**	a4.0
8	**T**	11 57	**...**	2.8	5 31	**5 59**	a3.9
9	**W**	12 20	**12 49**	2.9	6 25	**6 51**	p4.0
10	**T**	1 12	**1 38**	3.0	7 16	**7 39**	p4.2
11	**F**	1 59	**2 22**	3.2	8 04	**8 24**	p4.4
12	**S**	2 43	**3 04**	3.4	8 49	**9 06**	p4.6
13	**S**	3 24	**3 43**	p3.6	9 31	**9 46**	p4.8
14	**M**	4 04	**4 21**	3.7	10 12	**10 26**	p5.0
15	**T**	4 44	**4 59**	3.8	10 52	**11 06**	p5.1
16	**W**	5 24	**5 38**	3.8	11 33	**11 48**	p5.1
17	**T**	6 06	**6 19**	3.8	...	**12 16**	4.9
18	**F**	6 50	**7 04**	3.7	12 31	**1 01**	a5.1
19	**S**	7 39	**7 54**	3.6	1 19	**1 50**	a5.0
20	**S**	8 33	**8 52**	a3.5	2 11	**2 44**	a4.8
21	**M**	9 33	**9 58**	3.3	3 10	**3 45**	a4.7
22	**T**	10 40	**11 09**	3.3	4 14	**4 50**	a4.5
23	**W**	11 48	**...**	3.3	5 22	**5 56**	a4.5
24	**T**	12 19	**12 53**	3.4	6 29	**6 59**	p4.6
25	**F**	1 24	**1 53**	p3.6	7 31	**7 57**	p4.8
26	**S**	2 23	**2 47**	p3.8	8 29	**8 52**	p4.9
27	**S**	3 17	**3 38**	p3.9	9 22	**9 42**	p5.0
28	**M**	4 07	**4 26**	3.9	10 11	**10 30**	p5.1
29	**T**	4 55	**5 12**	3.9	10 58	**11 16**	p5.1
30	**W**	5 41	**5 57**	3.8	11 43	**...**	4.9

The Kts. (knots) columns show the **maximum** predicted velocities of the stronger one of the Flood Currents and the stronger one of the Ebb Currents for each day.
The letter "a" means the velocity shown should occur **after** the **a.m.** Current Change. The letter "p" means the velocity shown should occur **after** the **p.m.** Current Change (even if next morning). No "a" or "p" means a.m. and p.m. velocities are the same for that day.
Avg. Max. Velocity: Flood 3.4 Kts., Ebb 4.6 Kts.
Max. Flood 3 hrs. after Flood Starts, ±10 min.
Max. Ebb 3 hrs. after Ebb Starts, ±10 min.

See pp. 22-29 for Current Change at other points.

2014 CURRENT TABLE
HELL GATE, NY (EAST RIVER)
40°46.7'N, 73°56.3'W Off Mill Rock

MAY (Daylight Saving Time)								JUNE (Daylight Saving Time)							
DAY OF MONTH	DAY OF WEEK	CURRENT TURNS TO NORTHEAST Flood Starts			CURRENT TURNS TO SOUTHWEST Ebb Starts			DAY OF MONTH	DAY OF WEEK	CURRENT TURNS TO NORTHEAST Flood Starts			CURRENT TURNS TO SOUTHWEST Ebb Starts		
		a.m.	**p.m.**	Kts.	a.m.	**p.m.**	Kts.			a.m.	**p.m.**	Kts.	a.m.	**p.m.**	Kts.
1	T	6 25	**6 41**	3.7	12 01	**12 27**	a5.0	1	S	7 27	**7 41**	3.3	12 58	**1 22**	a4.7
2	F	7 10	**7 25**	3.5	12 45	**1 11**	a4.8	2	M	8 10	**8 25**	a3.2	1 41	**2 06**	a4.6
3	S	7 55	**8 11**	3.3	1 29	**1 56**	a4.6	3	T	8 54	**9 11**	3.0	2 25	**2 50**	a4.4
4	S	8 42	**9 00**	3.1	2 15	**2 42**	a4.4	4	W	9 41	**10 01**	2.9	3 11	**3 37**	a4.3
5	M	9 30	**9 50**	a3.0	3 03	**3 31**	a4.3	5	T	10 29	**10 52**	2.9	4 00	**4 26**	4.2
6	T	10 21	**10 43**	2.8	3 53	**4 22**	a4.1	6	F	11 18	**11 44**	2.9	4 51	**5 16**	4.2
7	W	11 13	**11 37**	2.8	4 45	**5 14**	4.0	7	S	...	**12 08**	3.0	5 43	**6 07**	p4.3
8	T	...	**12 05**	2.9	5 39	**6 05**	p4.1	8	S	12 37	**12 57**	p3.1	6 35	**6 58**	p4.5
9	F	12 30	**12 55**	p3.0	6 31	**6 54**	p4.2	9	M	1 28	**1 46**	p3.3	7 26	**7 48**	p4.7
10	S	1 19	**1 41**	3.1	7 21	**7 41**	p4.4	10	T	2 18	**2 33**	p3.5	8 17	**8 37**	p4.9
11	S	2 06	**2 25**	3.3	8 08	**8 27**	p4.7	11	W	3 07	**3 21**	p3.7	9 06	**9 26**	p5.1
12	M	2 51	**3 07**	p3.5	8 54	**9 11**	p4.9	12	T	3 56	**4 09**	p3.8	9 55	**10 16**	p5.2
13	T	3 35	**3 49**	p3.7	9 38	**9 54**	p5.0	13	F	4 45	**4 59**	p3.9	10 44	**11 06**	p5.2
14	W	4 19	**4 32**	p3.8	10 22	**10 39**	p5.1	14	S	5 34	**5 50**	p3.9	11 35	**11 57**	p5.2
15	T	5 03	**5 16**	3.8	11 07	**11 25**	p5.2	15	S	6 26	**6 43**	p3.9	...	**12 26**	5.1
16	F	5 49	**6 03**	3.8	11 53	**...**	5.0	16	M	7 19	**7 39**	p3.8	12 51	**1 19**	a5.2
17	S	6 38	**6 53**	p3.8	12 13	**12 42**	a5.2	17	T	8 14	**8 38**	3.6	1 46	**2 15**	a5.0
18	S	7 29	**7 47**	3.6	1 03	**1 34**	a5.1	18	W	9 12	**9 41**	3.5	2 44	**3 14**	a4.9
19	M	8 25	**8 47**	3.5	1 58	**2 30**	a4.9	19	T	10 13	**10 45**	a3.4	3 44	**4 14**	a4.7
20	T	9 25	**9 52**	3.4	2 57	**3 30**	a4.8	20	F	11 15	**11 50**	3.3	4 46	**5 15**	p4.6
21	W	10 29	**11 00**	3.3	4 00	**4 33**	a4.6	21	S	...	**12 15**	3.3	5 47	**6 15**	p4.5
22	T	11 34	**...**	3.3	5 05	**5 36**	4.5	22	S	12 51	**1 13**	p3.4	6 47	**7 13**	p4.6
23	F	12 07	**12 36**	p3.4	6 09	**6 38**	p4.6	23	M	1 49	**2 07**	p3.4	7 43	**8 06**	p4.7
24	S	1 10	**1 34**	p3.5	7 10	**7 35**	p4.7	24	T	2 41	**2 57**	p3.5	8 34	**8 56**	p4.7
25	S	2 08	**2 28**	p3.6	8 06	**8 29**	p4.8	25	W	3 29	**3 43**	p3.6	9 22	**9 42**	p4.8
26	M	3 01	**3 18**	p3.7	8 58	**9 19**	p4.9	26	T	4 14	**4 27**	p3.6	10 06	**10 26**	p4.8
27	T	3 50	**4 05**	3.7	9 46	**10 06**	p4.9	27	F	4 56	**5 09**	p3.6	10 49	**11 08**	p4.8
28	W	4 36	**4 50**	3.7	10 32	**10 50**	p4.9	28	S	5 37	**5 49**	3.5	11 30	**11 49**	p4.8
29	T	5 20	**5 33**	p3.7	11 15	**11 33**	p4.9	29	S	6 17	**6 29**	p3.5	...	**12 11**	4.6
30	F	6 02	**6 16**	3.6	11 58	**...**	4.7	30	M	6 56	**7 09**	3.4	12 30	**12 51**	a4.7
31	S	6 44	**6 58**	3.4	12 16	**12 40**	a4.8								

The Kts. (knots) columns show the **maximum** predicted velocities of the stronger one of the Flood Currents and the stronger one of the Ebb Currents for each day.
The letter "a" means the velocity shown should occur **after** the **a.m.** Current Change. The letter "p" means the velocity shown should occur **after** the **p.m.** Current Change (even if next morning). No "a" or "p" means a.m. and p.m. velocities are the same for that day.
Avg. Max. Velocity: Flood 3.4 Kts., Ebb 4.6 Kts.
Max. Flood 3 hrs. after Flood Starts, ±10 min.
Max. Ebb 3 hrs. after Ebb Starts, ±10 min.

At **City Island** the Current turns 2 hours before Hell Gate. At **Throg's Neck** the Current turns 1 hour before Hell Gate. At **Whitestone Pt.** the Current turns 35 min. before Hell Gate. At **College Pt.** the Current turns 20 min. before Hell Gate.

2014 CURRENT TABLE
HELL GATE, NY (EAST RIVER)
40°46.7'N, 73°56.3'W Off Mill Rock

Daylight Saving Time

JULY

DAY OF MONTH	DAY OF WEEK	CURRENT TURNS TO NORTHEAST Flood Starts a.m.	p.m.	Kts.	CURRENT TURNS TO SOUTHWEST Ebb Starts a.m.	p.m.	Kts.
1	T	7 35	**7 49**	3.3	1 10	**1 31**	a4.7
2	W	8 16	**8 31**	3.2	1 51	**2 13**	a4.6
3	T	8 57	**9 16**	3.1	2 34	**2 56**	a4.5
4	F	9 42	**10 05**	3.0	3 19	**3 42**	4.4
5	S	10 28	**10 56**	3.0	4 08	**4 31**	p4.4
6	S	11 18	**11 52**	3.0	4 59	**5 24**	p4.4
7	M	...	**12 12**	3.1	5 54	**6 19**	p4.5
8	T	12 48	**1 06**	p3.3	6 50	**7 15**	p4.7
9	W	1 45	**2 01**	p3.5	7 46	**8 10**	p4.8
10	T	2 39	**2 55**	p3.7	8 40	**9 05**	p5.0
11	F	3 33	**3 48**	p3.9	9 34	**9 59**	p5.1
12	S	4 25	**4 42**	p4.0	10 26	**10 52**	p5.2
13	S	5 17	**5 35**	p4.0	11 19	**11 45**	p5.2
14	M	6 09	**6 29**	p4.0	...	**12 11**	5.2
15	T	7 02	**7 24**	3.9	12 38	**1 05**	a5.2
16	W	7 56	**8 22**	a3.8	1 33	**1 59**	a5.1
17	T	8 51	**9 21**	3.6	2 29	**2 56**	a4.9
18	F	9 49	**10 23**	a3.5	3 26	**3 54**	a4.7
19	S	10 49	**11 25**	3.3	4 26	**4 54**	p4.5
20	S	11 49	...	3.3	5 26	**5 54**	p4.4
21	M	12 26	**12 48**	p3.3	6 25	**6 51**	p4.4
22	T	1 24	**1 42**	p3.3	7 21	**7 45**	p4.4
23	W	2 16	**2 32**	p3.4	8 12	**8 35**	p4.5
24	T	3 04	**3 18**	p3.5	8 59	**9 21**	p4.5
25	F	3 47	**4 00**	p3.5	9 43	**10 04**	p4.6
26	S	4 28	**4 40**	p3.6	10 24	**10 45**	p4.7
27	S	5 07	**5 19**	p3.6	11 04	**11 24**	p4.7
28	M	5 44	**5 56**	p3.6	11 43	...	4.7
29	T	6 21	**6 34**	p3.6	12 03	**12 21**	4.7
30	W	6 57	**7 11**	3.5	12 42	**1 00**	4.7
31	T	7 34	**7 50**	3.4	1 22	**1 39**	a4.7

Daylight Saving Time

AUGUST

DAY OF MONTH	DAY OF WEEK	CURRENT TURNS TO NORTHEAST Flood Starts a.m.	p.m.	Kts.	CURRENT TURNS TO SOUTHWEST Ebb Starts a.m.	p.m.	Kts.
1	F	8 11	**8 31**	3.3	2 02	**2 20**	4.6
2	S	8 52	**9 17**	3.2	2 45	**3 04**	4.5
3	S	9 37	**10 09**	a3.2	3 32	**3 54**	4.4
4	M	10 31	**11 09**	a3.2	4 24	**4 49**	p4.4
5	T	11 28	...	3.2	5 22	**5 49**	p4.5
6	W	12 11	**12 31**	p3.3	6 22	**6 51**	p4.6
7	T	1 13	**1 33**	p3.5	7 23	**7 51**	p4.7
8	F	2 13	**2 32**	p3.8	8 21	**8 49**	p4.9
9	S	3 10	**3 29**	p4.0	9 17	**9 45**	p5.1
10	S	4 04	**4 24**	p4.1	10 11	**10 39**	p5.2
11	M	4 56	**5 18**	p4.2	11 04	**11 32**	5.2
12	T	5 48	**6 11**	p4.2	11 56	...	5.2
13	W	6 39	**7 04**	4.0	12 25	**12 48**	a5.2
14	T	7 31	**7 59**	3.9	1 17	**1 41**	5.0
15	F	8 24	**8 55**	a3.7	2 11	**2 35**	4.8
16	S	9 20	**9 54**	a3.5	3 07	**3 32**	4.5
17	S	10 18	**10 54**	a3.3	4 04	**4 31**	4.3
18	M	11 18	**11 55**	a3.2	5 03	**5 31**	4.1
19	T	...	**12 17**	3.2	6 02	**6 29**	p4.1
20	W	12 53	**1 13**	p3.2	6 58	**7 24**	p4.1
21	T	1 45	**2 03**	p3.3	7 49	**8 14**	p4.2
22	F	2 33	**2 49**	p3.4	8 36	**8 59**	p4.3
23	S	3 16	**3 30**	p3.5	9 19	**9 41**	p4.5
24	S	3 55	**4 09**	p3.6	9 59	**10 21**	p4.6
25	M	4 33	**4 47**	p3.7	10 38	**10 59**	p4.7
26	T	5 09	**5 23**	p3.7	11 15	**11 37**	4.7
27	W	5 44	**5 59**	3.7	11 52	...	4.8
28	T	6 18	**6 35**	p3.7	12 15	**12 30**	p4.8
29	F	6 53	**7 12**	3.6	12 53	**1 08**	4.7
30	S	7 29	**7 52**	3.5	1 33	**1 48**	p4.7
31	S	8 09	**8 38**	3.4	2 15	**2 33**	p4.6

The Kts. (knots) columns show the **maximum** predicted velocities of the stronger one of the Flood Currents and the stronger one of the Ebb Currents for each day.
The letter "a" means the velocity shown should occur **after** the **a.m.** Current Change. The letter "p" means the velocity shown should occur **after** the **p.m.** Current Change (even if next morning). No "a" or "p" means a.m. and p.m. velocities are the same for that day.
Avg. Max. Velocity: Flood 3.4 Kts., Ebb 4.6 Kts.
Max. Flood 3 hrs. after Flood Starts, ±10 min.
Max. Ebb 3 hrs. after Ebb Starts, ±10 min.

See pp. 22-29 for Current Change at other points.

2014 CURRENT TABLE

HELL GATE, NY (EAST RIVER)

40°46.7'N, 73°56.3'W Off Mill Rock

Daylight Saving Time

SEPTEMBER

DAY OF MONTH	DAY OF WEEK	CURRENT TURNS TO NORTHEAST Flood Starts a.m.	p.m.	Kts.	CURRENT TURNS TO SOUTHWEST Ebb Starts a.m.	p.m.	Kts.
1	**M**	8 55	**9 31**	a3.3	3 03	**3 24**	p4.5
2	**T**	9 50	**10 33**	a3.3	3 57	**4 22**	p4.4
3	**W**	10 55	**11 40**	a3.3	4 57	**5 27**	p4.4
4	**T**	...	**12 05**	3.4	6 01	**6 32**	p4.5
5	**F**	12 47	**1 12**	p3.6	7 05	**7 36**	p4.6
6	**S**	1 50	**2 14**	p3.8	8 05	**8 35**	p4.8
7	**S**	2 48	**3 12**	p4.0	9 01	**9 31**	p5.0
8	**M**	3 43	**4 07**	p4.2	9 55	**10 25**	5.1
9	**T**	4 35	**5 00**	4.2	10 47	**11 16**	a5.2
10	**W**	5 25	**5 51**	4.2	11 38	**...**	5.2
11	**T**	6 15	**6 43**	4.1	12 07	**12 28**	5.1
12	**F**	7 05	**7 34**	a4.0	12 58	**1 19**	4.9
13	**S**	7 56	**8 27**	a3.8	1 49	**2 11**	4.7
14	**S**	8 49	**9 23**	a3.5	2 43	**3 06**	4.4
15	**M**	9 46	**10 21**	a3.3	3 38	**4 03**	4.1
16	**T**	10 44	**11 20**	a3.1	4 35	**5 02**	3.9
17	**W**	11 44	**...**	3.1	5 33	**6 00**	p3.9
18	**T**	12 18	**12 40**	p3.1	6 29	**6 55**	p3.9
19	**F**	1 10	**1 31**	p3.2	7 21	**7 45**	p4.0
20	**S**	1 58	**2 17**	p3.4	8 07	**8 30**	p4.2
21	**S**	2 41	**2 59**	p3.5	8 50	**9 12**	p4.4
22	**M**	3 21	**3 38**	p3.6	9 29	**9 52**	4.5
23	**T**	3 58	**4 15**	p3.7	10 08	**10 31**	p4.7
24	**W**	4 34	**4 52**	p3.8	10 45	**11 08**	a4.8
25	**T**	5 09	**5 28**	p3.8	11 22	**11 46**	4.8
26	**F**	5 44	**6 05**	p3.8	11 59	**...**	4.9
27	**S**	6 19	**6 43**	3.7	12 25	**12 38**	p4.8
28	**S**	6 57	**7 25**	3.6	1 05	**1 20**	p4.8
29	**M**	7 39	**8 13**	a3.5	1 48	**2 06**	p4.7
30	**T**	8 28	**9 08**	a3.4	2 37	**2 59**	p4.5

Daylight Saving Time

OCTOBER

DAY OF MONTH	DAY OF WEEK	CURRENT TURNS TO NORTHEAST Flood Starts a.m.	p.m.	Kts.	CURRENT TURNS TO SOUTHWEST Ebb Starts a.m.	p.m.	Kts.
1	**W**	9 28	**10 11**	a3.3	3 33	**4 00**	p4.4
2	**T**	10 36	**11 20**	3.3	4 36	**5 07**	p4.4
3	**F**	11 49	**...**	3.4	5 42	**6 15**	p4.4
4	**S**	12 30	**12 59**	p3.6	6 46	**7 19**	p4.6
5	**S**	1 32	**2 01**	p3.8	7 47	**8 19**	p4.8
6	**M**	2 30	**2 58**	p4.0	8 43	**9 14**	4.9
7	**T**	3 24	**3 52**	p4.1	9 36	**10 06**	a5.1
8	**W**	4 15	**4 43**	p4.2	10 27	**10 57**	a5.2
9	**T**	5 04	**5 33**	a4.2	11 16	**11 46**	a5.2
10	**F**	5 52	**6 22**	a4.1	...	**12 05**	5.1
11	**S**	6 41	**7 11**	a3.9	12 34	**12 54**	4.9
12	**S**	7 30	**8 01**	a3.7	1 23	**1 43**	p4.7
13	**M**	8 20	**8 53**	a3.4	2 13	**2 34**	4.4
14	**T**	9 14	**9 48**	a3.2	3 05	**3 28**	4.1
15	**W**	10 10	**10 44**	a3.0	3 59	**4 24**	3.9
16	**T**	11 08	**11 40**	2.9	4 55	**5 20**	3.8
17	**F**	...	**12 04**	3.0	5 49	**6 15**	p3.9
18	**S**	12 33	**12 56**	p3.1	6 41	**7 06**	p4.0
19	**S**	1 22	**1 44**	p3.2	7 28	**7 53**	p4.2
20	**M**	2 06	**2 27**	p3.4	8 12	**8 36**	p4.4
21	**T**	2 47	**3 08**	p3.5	8 53	**9 17**	4.5
22	**W**	3 26	**3 47**	p3.7	9 33	**9 57**	4.7
23	**T**	4 03	**4 25**	3.7	10 11	**10 36**	a4.9
24	**F**	4 39	**5 04**	p3.8	10 50	**11 15**	a5.0
25	**S**	5 16	**5 43**	3.8	11 29	**11 56**	a5.0
26	**S**	5 55	**6 25**	a3.8	...	**12 10**	5.0
27	**M**	6 37	**7 10**	a3.7	12 38	**12 55**	p4.9
28	**T**	7 23	**8 00**	a3.6	1 24	**1 44**	p4.8
29	**W**	8 17	**8 57**	a3.5	2 15	**2 39**	p4.7
30	**T**	9 19	**10 01**	3.3	3 12	**3 40**	4.5
31	**F**	10 29	**11 08**	3.3	4 15	**4 47**	4.4

The Kts. (knots) columns show the **maximum** predicted velocities of the stronger one of the Flood Currents and the stronger one of the Ebb Currents for each day.
The letter "a" means the velocity shown should occur **after** the **a.m.** Current Change. The letter "p" means the velocity shown should occur **after** the **p.m.** Current Change (even if next morning). No "a" or "p" means a.m. and p.m. velocities are the same for that day.
Avg. Max. Velocity: Flood 3.4 Kts., Ebb 4.6 Kts.
Max. Flood 3 hrs. after Flood Starts, ±10 min.
Max. Ebb 3 hrs. after Ebb Starts, ±10 min.

At **City Island** the Current turns 2 hours before Hell Gate. At **Throg's Neck** the Current turns 1 hour before Hell Gate. At **Whitestone Pt.** the Current turns 35 min. before Hell Gate. At **College Pt.** the Current turns 20 min. before Hell Gate.

2014 CURRENT TABLE

HELL GATE, NY (EAST RIVER)

40°46.7'N, 73°56.3'W Off Mill Rock

***Standard Time starts Nov. 2 at 2 a.m.** — **Standard Time**

NOVEMBER

DAY OF MONTH	DAY OF WEEK	CURRENT TURNS TO NORTHEAST Flood Starts a.m.	**p.m.**	Kts.	CURRENT TURNS TO SOUTHWEST Ebb Starts a.m.	**p.m.**	Kts.
1	**S**	11 40	**...**	3.3	5 21	**5 54**	p4.5
2	**S**	12 15	***-A-**	3.4	*5 25	***5 58**	p4.6
3	**M**	12 17	**12 50**	p3.7	6 26	**6 58**	4.7
4	**T**	1 15	**1 47**	3.8	7 22	**7 53**	4.9
5	**W**	2 08	**2 39**	3.9	8 15	**8 44**	a5.1
6	**T**	2 58	**3 29**	4.0	9 05	**9 33**	a5.2
7	**F**	3 46	**4 17**	a4.0	9 53	**10 21**	a5.1
8	**S**	4 33	**5 04**	a3.9	10 39	**11 07**	a5.1
9	**S**	5 20	**5 51**	a3.8	11 26	**11 53**	a4.9
10	**M**	6 06	**6 37**	a3.6	...	**12 12**	4.7
11	**T**	6 54	**7 26**	a3.4	12 39	**12 59**	4.5
12	**W**	7 44	**8 16**	a3.1	1 27	**1 48**	4.3
13	**T**	8 36	**9 08**	a3.0	2 16	**2 39**	4.1
14	**F**	9 30	**10 01**	2.8	3 08	**3 32**	4.0
15	**S**	10 25	**10 53**	2.8	4 00	**4 25**	4.0
16	**S**	11 18	**11 43**	2.9	4 51	**5 17**	p4.1
17	**M**	...	**12 08**	3.0	5 41	**6 06**	4.2
18	**T**	12 29	**12 54**	p3.2	6 27	**6 53**	4.4
19	**W**	1 12	**1 38**	3.3	7 12	**7 38**	a4.6
20	**T**	1 54	**2 20**	p3.5	7 55	**8 21**	a4.8
21	**F**	2 34	**3 02**	3.6	8 37	**9 03**	a4.9
22	**S**	3 14	**3 44**	3.7	9 19	**9 45**	a5.1
23	**S**	3 56	**4 27**	a3.8	10 02	**10 29**	a5.1
24	**M**	4 39	**5 13**	a3.8	10 47	**11 14**	a5.1
25	**T**	5 25	**6 01**	a3.7	11 35	**...**	5.1
26	**W**	6 16	**6 53**	a3.6	12 03	**12 26**	p5.0
27	**T**	7 12	**7 50**	a3.5	12 55	**1 21**	4.8
28	**F**	8 15	**8 51**	a3.4	1 52	**2 22**	4.7
29	**S**	9 22	**9 56**	3.3	2 54	**3 26**	4.6
30	**S**	10 31	**11 01**	3.3	3 58	**4 31**	a4.6

DECEMBER

DAY OF MONTH	DAY OF WEEK	CURRENT TURNS TO NORTHEAST Flood Starts a.m.	**p.m.**	Kts.	CURRENT TURNS TO SOUTHWEST Ebb Starts a.m.	**p.m.**	Kts.
1	**M**	11 37	**...**	3.3	5 02	**5 35**	4.6
2	**T**	12 03	**12 38**	p3.5	6 02	**6 35**	4.7
3	**W**	1 00	**1 35**	3.6	6 59	**7 30**	a4.9
4	**T**	1 54	**2 28**	3.7	7 52	**8 21**	a5.0
5	**F**	2 43	**3 16**	a3.8	8 42	**9 09**	a5.0
6	**S**	3 30	**4 02**	a3.8	9 29	**9 55**	a5.0
7	**S**	4 16	**4 46**	a3.7	10 14	**10 39**	a5.0
8	**M**	5 00	**5 30**	a3.7	10 58	**11 22**	a4.9
9	**T**	5 44	**6 13**	a3.5	11 41	**...**	4.8
10	**W**	6 28	**6 57**	a3.4	12 05	**12 24**	p4.7
11	**T**	7 13	**7 42**	a3.2	12 48	**1 08**	4.5
12	**F**	8 00	**8 28**	a3.0	1 33	**1 54**	4.4
13	**S**	8 49	**9 17**	a2.9	2 19	**2 42**	4.3
14	**S**	9 40	**10 07**	2.8	3 07	**3 32**	4.2
15	**M**	10 33	**10 57**	2.8	3 58	**4 24**	4.2
16	**T**	11 26	**11 47**	p2.9	4 49	**5 16**	a4.3
17	**W**	...	**12 17**	3.0	5 39	**6 07**	4.4
18	**T**	12 34	**1 06**	3.1	6 29	**6 57**	a4.6
19	**F**	1 21	**1 53**	3.3	7 18	**7 45**	a4.8
20	**S**	2 06	**2 40**	3.5	8 05	**8 33**	a5.0
21	**S**	2 52	**3 26**	a3.7	8 53	**9 20**	a5.1
22	**M**	3 38	**4 12**	a3.8	9 41	**10 07**	a5.2
23	**T**	4 26	**5 00**	a3.8	10 29	**10 56**	a5.3
24	**W**	5 16	**5 50**	a3.8	11 19	**11 46**	a5.2
25	**T**	6 08	**6 42**	a3.8	...	**12 11**	5.1
26	**F**	7 04	**7 38**	a3.6	12 39	**1 06**	5.0
27	**S**	8 05	**8 37**	a3.5	1 34	**2 04**	a4.9
28	**S**	9 09	**9 39**	3.3	2 34	**3 06**	a4.8
29	**M**	10 15	**10 43**	3.2	3 36	**4 09**	a4.6
30	**T**	11 21	**11 45**	p3.3	4 39	**5 12**	a4.6
31	**W**	...	**12 23**	3.3	5 40	**6 12**	a4.6

A also at 11:48 a.m. 3.5

The Kts. (knots) columns show the **maximum** predicted velocities of the stronger one of the Flood Currents and the stronger one of the Ebb Currents for each day.

The letter "a" means the velocity shown should occur **after** the **a.m.** Current Change. The letter "p" means the velocity shown should occur **after** the **p.m.** Current Change (even if next morning). No "a" or "p" means a.m. and p.m. velocities are the same for that day.

Avg. Max. Velocity: Flood 3.4 Kts., Ebb 4.6 Kts.

Max. Flood 3 hrs. after Flood Starts, ±10 min.

Max. Ebb 3 hrs. after Ebb Starts, ±10 min.

See pp. 22-29 for Current Change at other points.

2014 CURRENT TABLE

THE NARROWS, NY HARBOR

40°36.56'N, 74°02.77'W Mid-Channel

JANUARY

Standard Time

DAY OF MONTH	DAY OF WEEK	CURRENT TURNS TO					
		NORTH Flood Starts			SOUTH Ebb Starts		
		a.m.	**p.m.**	Kts.	a.m.	**p.m.**	Kts.
1	**W**	3 15	**4 05**	a2.4	9 20	**9 46**	a2.6
2	**T**	4 06	**4 53**	a2.4	10 10	**10 37**	a2.5
3	**F**	4 59	**5 44**	a2.2	11 02	**11 33**	a2.4
4	**S**	5 58	**6 41**	a2.0	11 56	**...**	2.3
5	**S**	7 01	**7 41**	a1.8	12 31	**12 53**	p2.1
6	**M**	8 09	**8 43**	a1.6	1 31	**1 50**	p1.9
7	**T**	9 19	**9 44**	1.5	2 32	**2 48**	p1.8
8	**W**	10 27	**10 44**	p1.5	3 35	**3 49**	p1.7
9	**T**	11 31	**11 41**	p1.4	4 43	**4 54**	a1.7
10	**F**	...	**12 33**	1.3	5 50	**5 59**	a1.7
11	**S**	12 35	**1 33**	a1.4	6 50	**6 58**	a1.8
12	**S**	1 26	**2 27**	a1.4	7 40	**7 50**	a1.8
13	**M**	2 12	**3 13**	a1.4	8 21	**8 34**	a1.8
14	**T**	2 52	**3 51**	a1.4	8 55	**9 10**	a1.8
15	**W**	3 26	**4 22**	a1.3	9 25	**9 43**	a1.7
16	**T**	3 56	**4 49**	a1.3	9 54	**10 14**	a1.7
17	**F**	4 26	**5 15**	a1.4	10 24	**10 47**	a1.7
18	**S**	4 59	**5 44**	a1.4	10 58	**11 25**	a1.8
19	**S**	5 38	**6 18**	a1.5	11 37	**...**	1.8
20	**M**	6 23	**6 58**	1.5	12 07	**12 20**	p1.9
21	**T**	7 14	**7 44**	p1.6	12 53	**1 07**	p2.0
22	**W**	8 10	**8 34**	p1.8	1 41	**1 55**	p2.0
23	**T**	9 09	**9 26**	p1.9	2 31	**2 46**	1.9
24	**F**	10 10	**10 20**	p2.0	3 23	**3 41**	1.9
25	**S**	11 10	**11 16**	p2.0	4 20	**4 42**	a2.0
26	**S**	...	**12 09**	1.6	5 20	**5 45**	a2.1
27	**M**	12 12	**1 06**	a2.1	6 19	**6 45**	a2.3
28	**T**	1 08	**2 01**	a2.2	7 16	**7 42**	a2.4
29	**W**	2 03	**2 52**	a2.3	8 09	**8 35**	a2.5
30	**T**	2 58	**3 41**	a2.4	9 01	**9 26**	a2.6
31	**F**	3 51	**4 30**	a2.4	9 51	**10 18**	a2.6

FEBRUARY

Standard Time

DAY OF MONTH	DAY OF WEEK	CURRENT TURNS TO					
		NORTH Flood Starts			SOUTH Ebb Starts		
		a.m.	**p.m.**	Kts.	a.m.	**p.m.**	Kts.
1	**S**	4 45	**5 19**	a2.3	10 43	**11 12**	a2.5
2	**S**	5 41	**6 13**	a2.1	11 36	**...**	2.4
3	**M**	6 42	**7 11**	a1.9	12 09	**12 32**	2.2
4	**T**	7 48	**8 13**	1.7	1 08	**1 29**	2.0
5	**W**	8 55	**9 14**	1.6	2 08	**2 26**	1.8
6	**T**	10 03	**10 16**	1.5	3 09	**3 26**	1.7
7	**F**	11 09	**11 15**	1.4	4 14	**4 31**	a1.7
8	**S**	...	**12 12**	1.3	5 21	**5 39**	a1.7
9	**S**	12 11	**1 12**	a1.4	6 24	**6 42**	a1.8
10	**M**	1 04	**2 06**	a1.4	7 17	**7 35**	a1.8
11	**T**	1 52	**2 51**	a1.4	8 00	**8 17**	a1.8
12	**W**	2 34	**3 26**	a1.4	8 35	**8 51**	a1.8
13	**T**	3 09	**3 54**	a1.4	9 04	**9 21**	a1.7
14	**F**	3 39	**4 17**	a1.4	9 31	**9 49**	a1.7
15	**S**	4 08	**4 40**	a1.4	9 59	**10 19**	a1.8
16	**S**	4 39	**5 06**	1.5	10 30	**10 53**	1.8
17	**M**	5 14	**5 37**	1.6	11 06	**11 33**	1.9
18	**T**	5 55	**6 16**	p1.7	11 47	**...**	2.0
19	**W**	6 42	**7 00**	p1.8	12 17	**12 33**	p2.0
20	**T**	7 36	**7 50**	p1.8	1 04	**1 21**	2.0
21	**F**	8 35	**8 45**	p1.9	1 54	**2 13**	1.9
22	**S**	9 37	**9 44**	p1.9	2 46	**3 08**	a1.9
23	**S**	10 41	**10 46**	p1.9	3 43	**4 09**	a1.9
24	**M**	11 42	**11 48**	p1.9	4 46	**5 16**	a1.9
25	**T**	...	**12 40**	1.5	5 50	**6 21**	a2.1
26	**W**	12 48	**1 35**	a2.0	6 51	**7 20**	2.2
27	**T**	1 47	**2 27**	a2.1	7 47	**8 15**	a2.4
28	**F**	2 43	**3 17**	a2.2	8 40	**9 07**	a2.5

The Kts. (knots) columns show the **maximum** predicted velocities of the stronger one of the Flood Currents and the stronger one of the Ebb Currents for each day.
The letter "a" means the velocity shown should occur **after** the **a.m.** Current Change. The letter "p" means the velocity shown should occur **after** the **p.m.** Current Change (even if next morning). No "a" or "p" means a.m. and p.m. velocities are the same for that day.
Avg. Max. Velocity: Flood 1.7 Kts., Ebb 2.0 Kts.
Max. Flood 2 hrs. 25 min. after Flood Starts, ±30 min.
Max. Ebb 3 hrs. 15 min. after Ebb Starts, ±10 min.

At **The Battery, Desbrosses St., & Chelsea Dock** Current turns 1 1/2 hrs. after the Narrows. At **42nd St.** and the **George Washington Bridge**, the Current turns 1 3/4 hrs. after the Narrows. See pp. 22-29 for Current Change at other points.

South Shore /

2014 CURRENT TABLE
THE NARROWS, NY HARBOR

40°36.56'N, 74°02.77'W Mid-Channel

***Daylight Time starts March 9 at 2 a.m.** **Daylight Saving Time**

MARCH

DAY OF MONTH	DAY OF WEEK	CURRENT TURNS TO NORTH Flood Starts			CURRENT TURNS TO SOUTH Ebb Starts		
		a.m.	**p.m.**	Kts.	a.m.	**p.m.**	Kts.
1	S	3 37	**4 04**	a2.3	9 31	**9 57**	2.5
2	S	4 29	**4 52**	a2.2	10 21	**10 49**	a2.5
3	M	5 22	**5 43**	a2.1	11 12	**11 44**	a2.4
4	T	6 20	**6 39**	a1.9	...	**12 07**	2.3
5	W	7 22	**7 37**	1.7	12 40	**1 03**	2.1
6	T	8 28	**8 39**	1.6	1 38	**2 01**	a1.9
7	F	9 36	**9 42**	1.5	2 37	**3 00**	a1.8
8	S	10 42	**10 44**	1.4	3 38	**4 04**	a1.7
9	S	...	***12 43**	1.3	*5 42	***6 11**	a1.7
10	M	12 42	**1 39**	1.3	6 46	**7 14**	a1.8
11	T	1 35	**2 30**	1.3	7 41	**8 06**	a1.8
12	W	2 24	**3 12**	1.4	8 27	**8 48**	a1.8
13	T	3 07	**3 47**	1.4	9 04	**9 23**	a1.8
14	F	3 44	**4 15**	1.4	9 35	**9 53**	1.8
15	S	4 17	**4 40**	1.5	10 04	**10 22**	p1.9
16	S	4 48	**5 05**	p1.6	10 33	**10 53**	1.9
17	M	5 20	**5 32**	p1.7	11 05	**11 26**	2.0
18	T	5 54	**6 04**	p1.8	11 40	**...**	2.0
19	W	6 34	**6 42**	p1.8	12 04	**12 20**	2.0
20	T	7 20	**7 26**	p1.8	12 47	**1 06**	a2.1
21	F	8 12	**8 16**	p1.8	1 33	**1 55**	a2.0
22	S	9 11	**9 14**	p1.7	2 23	**2 48**	a2.0
23	S	10 13	**10 18**	p1.7	3 16	**3 44**	a1.8
24	M	11 17	**11 26**	p1.7	4 14	**4 46**	a1.7
25	T	...	**12 19**	1.3	5 17	**5 54**	a1.7
26	W	12 32	**1 17**	a1.7	6 25	**7 01**	1.8
27	T	1 35	**2 12**	a1.7	7 28	**8 02**	2.0
28	F	2 35	**3 03**	a1.9	8 26	**8 57**	2.2
29	S	3 30	**3 52**	2.0	9 19	**9 47**	p2.4
30	S	4 22	**4 39**	2.1	10 09	**10 37**	2.4
31	M	5 12	**5 25**	2.1	10 57	**11 26**	2.4

APRIL

DAY OF MONTH	DAY OF WEEK	CURRENT TURNS TO NORTH Flood Starts			CURRENT TURNS TO SOUTH Ebb Starts		
		a.m.	**p.m.**	Kts.	a.m.	**p.m.**	Kts.
1	T	6 03	**6 12**	2.0	11 47	**...**	2.3
2	W	6 57	**7 04**	1.8	12 16	**12 40**	a2.3
3	T	7 56	**8 00**	p1.7	1 10	**1 35**	a2.1
4	F	9 01	**9 02**	1.5	2 05	**2 33**	a1.9
5	S	10 05	**10 04**	1.4	3 00	**3 31**	a1.8
6	S	11 07	**11 06**	1.4	3 57	**4 31**	a1.7
7	M	...	**12 05**	1.3	4 56	**5 33**	a1.7
8	T	12 04	**12 56**	1.3	5 56	**6 33**	a1.7
9	W	12 58	**1 42**	1.3	6 52	**7 24**	1.7
10	T	1 46	**2 22**	p1.4	7 40	**8 08**	1.8
11	F	2 31	**2 58**	p1.5	8 22	**8 46**	1.9
12	S	3 12	**3 30**	p1.6	8 58	**9 20**	p2.0
13	S	3 49	**4 01**	p1.8	9 32	**9 53**	p2.1
14	M	4 25	**4 31**	p1.9	10 06	**10 27**	p2.2
15	T	5 01	**5 03**	p2.0	10 41	**11 02**	p2.2
16	W	5 38	**5 37**	p2.0	11 19	**11 41**	p2.2
17	T	6 19	**6 16**	p1.9	...	**12 01**	2.0
18	F	7 04	**7 01**	p1.8	12 23	**12 47**	a2.1
19	S	7 57	**7 54**	p1.7	1 11	**1 38**	a2.1
20	S	8 55	**8 56**	p1.6	2 02	**2 32**	a1.9
21	M	9 57	**10 05**	p1.5	2 56	**3 30**	a1.8
22	T	11 00	**11 17**	p1.5	3 54	**4 32**	a1.6
23	W	...	**12 01**	1.3	4 57	**5 39**	a1.6
24	T	12 25	**12 58**	a1.4	6 04	**6 46**	1.7
25	F	1 27	**1 51**	1.5	7 08	**7 47**	p1.9
26	S	2 25	**2 41**	p1.7	8 06	**8 41**	p2.1
27	S	3 19	**3 30**	p1.9	8 59	**9 31**	p2.2
28	M	4 10	**4 15**	p2.0	9 48	**10 17**	p2.3
29	T	4 58	**5 00**	p2.0	10 35	**11 03**	2.2
30	W	5 46	**5 44**	p1.9	11 23	**11 50**	p2.2

The Kts. (knots) columns show the **maximum** predicted velocities of the stronger one of the Flood Currents and the stronger one of the Ebb Currents for each day.
The letter "a" means the velocity shown should occur **after** the **a.m.** Current Change. The letter "p" means the velocity shown should occur **after** the **p.m.** Current Change (even if next morning). No "a" or "p" means a.m. and p.m. velocities are the same for that day.
Avg. Max. Velocity: Flood 1.7 Kts., Ebb 2.0 Kts.
Max. Flood 2 hrs. 25 min. after Flood Starts, ±30 min.
Max. Ebb 3 hrs. 15 min. after Ebb Starts, ±10 min.

See pp. 22-29 for Current Change at other points.

2014 CURRENT TABLE
THE NARROWS, NY HARBOR
40°36.56'N, 74°02.77'W Mid-Channel

Daylight Saving Time — **Daylight Saving Time**

MAY								JUNE							
DAY OF MONTH	DAY OF WEEK	CURRENT TURNS TO						DAY OF MONTH	DAY OF WEEK	CURRENT TURNS TO					
		NORTH Flood Starts			SOUTH Ebb Starts					NORTH Flood Starts			SOUTH Ebb Starts		
		a.m.	**p.m.**	Kts.	a.m.	**p.m.**	Kts.			a.m.	**p.m.**	Kts.	a.m.	**p.m.**	Kts.
1	T	6 36	**6 31**	p1.7	...	**12 14**	2.0	1	S	7 55	**7 39**	p1.4	12 55	**1 28**	a1.8
2	F	7 31	**7 23**	p1.6	12 38	**1 07**	a2.0	2	M	8 45	**8 35**	p1.3	1 43	**2 19**	a1.8
3	S	8 29	**8 20**	1.4	1 30	**2 02**	a1.9	3	T	9 35	**9 32**	p1.3	2 31	**3 08**	a1.7
4	S	9 29	**9 22**	1.3	2 22	**2 57**	a1.8	4	W	10 23	**10 30**	p1.4	3 19	**3 57**	a1.7
5	M	10 25	**10 22**	1.3	3 14	**3 52**	a1.7	5	T	11 07	**11 24**	1.4	4 08	**4 47**	a1.7
6	T	11 17	**11 19**	1.3	4 07	**4 46**	a1.6	6	F	11 51	**...**	1.5	4 59	**5 39**	a1.8
7	W	...	**12 04**	1.3	5 01	**5 41**	a1.7	7	S	12 16	**12 35**	p1.7	5 52	**6 31**	1.9
8	T	12 13	**12 48**	p1.4	5 55	**6 33**	1.7	8	S	1 07	**1 18**	p1.8	6 45	**7 21**	p2.1
9	F	1 03	**1 28**	p1.5	6 47	**7 21**	p1.9	9	M	1 57	**2 01**	p2.0	7 36	**8 07**	p2.3
10	S	1 50	**2 07**	p1.6	7 34	**8 04**	p2.0	10	T	2 46	**2 45**	p2.2	8 24	**8 52**	p2.4
11	S	2 35	**2 45**	p1.8	8 17	**8 44**	p2.2	11	W	3 33	**3 29**	p2.3	9 11	**9 35**	p2.5
12	M	3 18	**3 22**	p2.0	8 58	**9 23**	p2.3	12	T	4 19	**4 13**	p2.3	9 56	**10 19**	p2.5
13	T	4 00	**3 59**	p2.1	9 38	**10 01**	p2.4	13	F	5 04	**4 58**	p2.3	10 42	**11 04**	p2.4
14	W	4 41	**4 36**	p2.2	10 18	**10 40**	p2.4	14	S	5 49	**5 46**	p2.1	11 29	**11 51**	p2.3
15	T	5 22	**5 16**	p2.2	11 00	**11 21**	p2.3	15	S	6 37	**6 38**	p1.9	...	**12 20**	2.0
16	F	6 05	**5 58**	p2.0	11 44	**...**	2.0	16	M	7 29	**7 38**	p1.7	12 43	**1 15**	a2.2
17	S	6 52	**6 47**	p1.9	12 06	**12 33**	a2.2	17	T	8 27	**8 45**	p1.5	1 37	**2 13**	a2.0
18	S	7 45	**7 44**	p1.7	12 55	**1 27**	a2.1	18	W	9 27	**9 55**	1.4	2 33	**3 12**	a1.8
19	M	8 43	**8 50**	p1.5	1 48	**2 23**	a1.9	19	T	10 28	**11 05**	a1.4	3 29	**4 12**	a1.7
20	T	9 44	**10 02**	p1.4	2 44	**3 21**	a1.7	20	F	11 27	**...**	1.4	4 28	**5 17**	a1.6
21	W	10 46	**11 13**	p1.3	3 41	**4 22**	a1.6	21	S	12 10	**12 24**	p1.4	5 31	**6 24**	1.6
22	T	11 45	**...**	1.3	4 42	**5 28**	1.5	22	S	1 13	**1 18**	p1.4	6 35	**7 27**	p1.7
23	F	12 19	**12 41**	1.3	5 47	**6 35**	1.6	23	M	2 12	**2 10**	p1.5	7 36	**8 22**	p1.8
24	S	1 20	**1 33**	p1.4	6 50	**7 36**	p1.8	24	T	3 08	**2 59**	p1.5	8 31	**9 09**	p1.9
25	S	2 18	**2 24**	p1.6	7 49	**8 30**	p1.9	25	W	3 58	**3 44**	p1.6	9 21	**9 50**	p1.9
26	M	3 12	**3 12**	p1.7	8 42	**9 18**	p2.0	26	T	4 43	**4 25**	p1.5	10 06	**10 28**	p1.9
27	T	4 02	**3 57**	p1.8	9 31	**10 02**	2.0	27	F	5 24	**5 03**	p1.5	10 47	**11 05**	p1.8
28	W	4 49	**4 39**	p1.7	10 18	**10 44**	p2.0	28	S	6 01	**5 40**	p1.4	11 28	**11 42**	p1.8
29	T	5 34	**5 20**	p1.7	11 03	**11 25**	p2.0	29	S	6 39	**6 18**	p1.4	...	**12 09**	1.6
30	F	6 19	**6 03**	p1.6	11 50	**...**	1.8	30	M	7 16	**7 01**	p1.4	12 21	**12 51**	a1.8
31	S	7 05	**6 48**	p1.5	12 09	**12 38**	a1.9								

The Kts. (knots) columns show the **maximum** predicted velocities of the stronger one of the Flood Currents and the stronger one of the Ebb Currents for each day.
The letter "a" means the velocity shown should occur **after** the **a.m.** Current Change. The letter "p" means the velocity shown should occur **after** the **p.m.** Current Change (even if next morning). No "a" or "p" means a.m. and p.m. velocities are the same for that day.
Avg. Max. Velocity: Flood 1.7 Kts., Ebb 2.0 Kts.
Max. Flood 2 hrs. 25 min. after Flood Starts, ±30 min.
Max. Ebb 3 hrs. 15 min. after Ebb Starts, ±10 min.

At **The Battery, Desbrosses St., & Chelsea Dock** Current turns 1 1/2 hrs. after the Narrows. At **42nd St.** and the **George Washington Bridge**, the Current turns 1 3/4 hrs. after the Narrows. See pp. 22-29 for Current Change at other points.

2014 CURRENT TABLE
THE NARROWS, NY HARBOR

40°36.56'N, 74°02.77'W Mid-Channel

Daylight Saving Time — Daylight Saving Time

JULY								AUGUST							
		CURRENT TURNS TO								CURRENT TURNS TO					
DAY OF MONTH	DAY OF WEEK	NORTH Flood Starts			SOUTH Ebb Starts			DAY OF MONTH	DAY OF WEEK	NORTH Flood Starts			SOUTH Ebb Starts		
		a.m.	**p.m.**	Kts.	a.m.	**p.m.**	Kts.			a.m.	**p.m.**	Kts.	a.m.	**p.m.**	Kts.
1	T	7 56	**7 50**	p1.4	1 03	**1 36**	a1.8	1	F	8 26	**8 51**	a1.6	1 50	**2 23**	a1.9
2	W	8 38	**8 43**	p1.4	1 48	**2 22**	a1.8	2	S	9 13	**9 48**	a1.7	2 37	**3 11**	a1.9
3	T	9 22	**9 38**	1.4	2 33	**3 08**	a1.8	3	S	10 03	**10 47**	a1.8	3 26	**4 01**	1.9
4	F	10 09	**10 35**	a1.6	3 20	**3 56**	a1.8	4	M	10 58	**11 48**	a1.9	4 19	**4 56**	1.9
5	S	10 55	**11 30**	a1.7	4 09	**4 47**	1.8	5	T	11 52	**...**	2.0	5 17	**5 54**	p2.0
6	S	11 44	**...**	1.9	5 02	**5 42**	1.9	6	W	12 45	**12 49**	p2.0	6 20	**6 54**	p2.2
7	M	12 26	**12 33**	p2.0	5 59	**6 37**	p2.1	7	T	1 42	**1 45**	p2.1	7 21	**7 51**	p2.4
8	T	1 20	**1 22**	p2.1	6 56	**7 30**	p2.3	8	F	2 36	**2 40**	p2.3	8 19	**8 45**	p2.5
9	W	2 13	**2 13**	p2.2	7 52	**8 21**	p2.4	9	S	3 28	**3 35**	p2.4	9 12	**9 37**	p2.6
10	T	3 05	**3 03**	p2.3	8 44	**9 10**	p2.5	10	S	4 17	**4 28**	p2.4	10 03	**10 27**	p2.6
11	F	3 55	**3 53**	p2.4	9 34	**9 58**	p2.6	11	M	5 04	**5 20**	p2.3	10 53	**11 17**	p2.5
12	S	4 42	**4 43**	p2.4	10 23	**10 46**	p2.5	12	T	5 51	**6 13**	p2.2	11 44	**...**	2.4
13	S	5 29	**5 34**	p2.2	11 13	**11 36**	p2.4	13	W	6 41	**7 11**	p2.0	12 08	**12 38**	a2.4
14	M	6 16	**6 28**	p2.1	...	**12 04**	2.2	14	T	7 35	**8 14**	1.8	1 02	**1 35**	a2.3
15	T	7 08	**7 27**	p1.9	12 28	**12 59**	a2.3	15	F	8 35	**9 21**	a1.7	1 58	**2 34**	a2.1
16	W	8 04	**8 33**	p1.7	1 22	**1 57**	a2.2	16	S	9 37	**10 30**	a1.6	2 54	**3 34**	a1.9
17	T	9 03	**9 41**	1.5	2 18	**2 56**	a2.0	17	S	10 41	**11 38**	a1.5	3 53	**4 37**	1.7
18	F	10 05	**10 50**	a1.5	3 14	**3 56**	a1.8	18	M	11 44	**...**	1.4	4 57	**5 46**	p1.7
19	S	11 06	**11 57**	a1.5	4 12	**5 01**	a1.7	19	T	12 44	**12 45**	p1.4	6 07	**6 55**	p1.7
20	S	...	**12 06**	1.4	5 14	**6 09**	1.6	20	W	1 46	**1 43**	p1.4	7 16	**7 55**	p1.8
21	M	1 01	**1 03**	p1.4	6 21	**7 15**	p1.7	21	T	2 44	**2 36**	p1.4	8 16	**8 45**	p1.8
22	T	2 03	**1 58**	p1.4	7 27	**8 12**	p1.8	22	F	3 34	**3 23**	1.4	9 04	**9 24**	p1.8
23	W	3 01	**2 50**	p1.4	8 25	**9 00**	p1.8	23	S	4 14	**4 02**	1.4	9 42	**9 57**	p1.8
24	T	3 52	**3 35**	p1.4	9 15	**9 40**	p1.8	24	S	4 45	**4 34**	p1.4	10 14	**10 24**	p1.7
25	F	4 35	**4 15**	p1.4	9 57	**10 15**	p1.8	25	M	5 10	**5 03**	p1.4	10 42	**10 51**	p1.7
26	S	5 10	**4 49**	p1.4	10 33	**10 46**	p1.7	26	T	5 32	**5 32**	p1.4	11 10	**11 20**	p1.8
27	S	5 40	**5 21**	p1.4	11 06	**11 17**	p1.7	27	W	5 55	**6 03**	p1.5	11 40	**11 52**	p1.8
28	M	6 08	**5 53**	p1.4	11 39	**11 50**	p1.7	28	T	6 22	**6 40**	1.5	...	**12 16**	1.8
29	T	6 35	**6 29**	p1.4	...	**12 14**	1.6	29	F	6 57	**7 23**	a1.6	12 30	**12 57**	1.9
30	W	7 06	**7 11**	p1.4	12 26	**12 54**	a1.8	30	S	7 38	**8 14**	a1.7	1 13	**1 42**	2.0
31	T	7 43	**7 58**	p1.5	1 07	**1 37**	a1.9	31	S	8 26	**9 10**	a1.8	2 00	**2 31**	p2.0

The Kts. (knots) columns show the **maximum** predicted velocities of the stronger one of the Flood Currents and the stronger one of the Ebb Currents for each day.
The letter "a" means the velocity shown should occur **after** the **a.m.** Current Change. The letter "p" means the velocity shown should occur **after** the **p.m.** Current Change (even if next morning). No "a" or "p" means a.m. and p.m. velocities are the same for that day.
Avg. Max. Velocity: Flood 1.7 Kts., Ebb 2.0 Kts.
Max. Flood 2 hrs. 25 min. after Flood Starts, ±30 min.
Max. Ebb 3 hrs. 15 min. after Ebb Starts, ±10 min.

See pp. 22-29 for Current Change at other points.

2014 CURRENT TABLE
THE NARROWS, NY HARBOR

40°36.56'N, 74°02.77'W Mid-Channel

Daylight Saving Time

SEPTEMBER

Day of Month	Day of Week	Current Turns to North, Flood Starts a.m.	**p.m.**	Kts.	Current Turns to South, Ebb Starts a.m.	**p.m.**	Kts.
1	M	9 19	**10 10**	a1.8	2 50	**3 21**	1.9
2	T	10 17	**11 13**	a1.9	3 44	**4 16**	p1.9
3	W	11 19	...	1.9	4 43	**5 16**	p1.9
4	T	12 15	**12 22**	p1.9	5 48	**6 21**	p2.0
5	F	1 12	**1 22**	p2.0	6 54	**7 23**	p2.2
6	S	2 07	**2 21**	p2.1	7 54	**8 21**	p2.4
7	S	3 00	**3 18**	p2.2	8 49	**9 14**	p2.5
8	M	3 50	**4 12**	p2.3	9 41	**10 05**	p2.6
9	T	4 37	**5 04**	p2.3	10 31	**10 55**	p2.6
10	W	5 24	**5 56**	2.2	11 22	**11 45**	2.5
11	T	6 12	**6 51**	a2.1	...	**12 14**	2.4
12	F	7 05	**7 51**	a2.0	12 38	**1 10**	a2.4
13	S	8 03	**8 57**	a1.8	1 34	**2 08**	a2.2
14	S	9 06	**10 06**	a1.6	2 32	**3 07**	a1.9
15	M	10 12	**11 14**	a1.5	3 32	**4 08**	1.7
16	T	11 18	...	1.4	4 37	**5 14**	p1.7
17	W	12 19	**12 21**	1.4	5 46	**6 22**	p1.8
18	T	1 18	**1 19**	1.3	6 54	**7 24**	p1.8
19	F	2 12	**2 12**	1.4	7 53	**8 15**	p1.8
20	S	2 59	**2 59**	1.4	8 39	**8 56**	p1.8
21	S	3 38	**3 38**	1.4	9 16	**9 28**	p1.8
22	M	4 08	**4 12**	1.4	9 46	**9 56**	p1.8
23	T	4 32	**4 42**	1.4	10 14	**10 23**	1.8
24	W	4 55	**5 10**	1.5	10 41	**10 52**	1.8
25	T	5 19	**5 41**	a1.6	11 11	**11 24**	1.9
26	F	5 47	**6 17**	a1.7	11 45	...	2.0
27	S	6 21	**6 58**	a1.8	12 01	**12 25**	p2.0
28	S	7 01	**7 47**	a1.8	12 43	**1 09**	p2.0
29	M	7 49	**8 42**	a1.8	1 31	**1 58**	p2.0
30	T	8 45	**9 42**	a1.8	2 23	**2 50**	p1.9

Daylight Saving Time

OCTOBER

Day of Month	Day of Week	Current Turns to North, Flood Starts a.m.	**p.m.**	Kts.	Current Turns to South, Ebb Starts a.m.	**p.m.**	Kts.
1	W	9 47	**10 45**	a1.7	3 18	**3 45**	p1.8
2	T	10 54	**11 47**	a1.7	4 17	**4 45**	p1.8
3	F	...	**12 01**	1.7	5 22	**5 51**	p1.9
4	S	12 46	**1 06**	p1.7	6 29	**6 56**	p2.0
5	S	1 40	**2 05**	p1.9	7 31	**7 56**	p2.2
6	M	2 32	**3 02**	p2.0	8 28	**8 51**	p2.4
7	T	3 22	**3 56**	p2.1	9 20	**9 42**	p2.5
8	W	4 10	**4 47**	2.2	10 10	**10 31**	2.5
9	T	4 57	**5 38**	a2.2	10 59	**11 21**	a2.5
10	F	5 44	**6 31**	a2.1	11 50	...	2.4
11	S	6 35	**7 29**	a2.0	12 13	**12 43**	a2.3
12	S	7 31	**8 32**	a1.8	1 09	**1 39**	2.1
13	M	8 34	**9 39**	a1.6	2 08	**2 36**	1.9
14	T	9 40	**10 44**	a1.5	3 09	**3 35**	p1.8
15	W	10 46	**11 45**	1.4	4 10	**4 35**	p1.7
16	T	11 48	...	1.4	5 15	**5 38**	p1.7
17	F	12 40	**12 45**	a1.4	6 18	**6 38**	p1.8
18	S	1 29	**1 37**	a1.4	7 14	**7 30**	p1.8
19	S	2 12	**2 23**	1.4	8 00	**8 13**	p1.8
20	M	2 49	**3 05**	a1.5	8 38	**8 50**	1.8
21	T	3 21	**3 42**	a1.5	9 12	**9 22**	1.9
22	W	3 50	**4 16**	a1.6	9 43	**9 53**	a2.0
23	T	4 18	**4 49**	a1.7	10 13	**10 25**	a2.0
24	F	4 46	**5 23**	a1.9	10 46	**11 00**	a2.1
25	S	5 18	**6 00**	a1.9	11 21	**11 38**	a2.1
26	S	5 54	**6 41**	a1.9	...	**12 01**	2.1
27	M	6 36	**7 29**	a1.9	12 22	**12 45**	p2.1
28	T	7 25	**8 23**	a1.8	1 12	**1 34**	p2.0
29	W	8 23	**9 22**	a1.7	2 05	**2 27**	p1.9
30	T	9 29	**10 23**	a1.6	3 01	**3 23**	p1.8
31	F	10 40	**11 24**	a1.5	3 59	**4 22**	p1.7

The Kts. (knots) columns show the **maximum** predicted velocities of the stronger one of the Flood Currents and the stronger one of the Ebb Currents for each day.
The letter "a" means the velocity shown should occur **after** the **a.m.** Current Change. The letter "p" means the velocity shown should occur **after** the **p.m.** Current Change (even if next morning). No "a" or "p" means a.m. and p.m. velocities are the same for that day.
Avg. Max. Velocity: Flood 1.7 Kts., Ebb 2.0 Kts.
Max. Flood 2 hrs. 25 min. after Flood Starts, ±30 min.
Max. Ebb 3 hrs. 15 min. after Ebb Starts, ±10 min.

At **The Battery, Desbrosses St., & Chelsea Dock** Current turns 1 1/2 hrs. after the Narrows. At **42nd St.** and the **George Washington Bridge**, the Current turns 1 3/4 hrs. after the Narrows. See pp. 22-29 for Current Change at other points.

2014 CURRENT TABLE
THE NARROWS, NY HARBOR

40°36.56'N, 74°02.77'W Mid-Channel

***Standard Time starts Nov. 2 at 2 a.m.** — **Standard Time**

NOVEMBER

Day of Month	Day of Week	CURRENT TURNS TO NORTH Flood Starts a.m.	p.m.	Kts.	CURRENT TURNS TO SOUTH Ebb Starts a.m.	p.m.	Kts.
1	S	11 49	**...**	1.5	5 03	**5 27**	p1.7
2	S	12 21	***-A-**	1.4	*5 09	***5 32**	p1.8
3	M	12 16	**12 52**	p1.6	6 12	**6 33**	p2.0
4	T	1 08	**1 50**	1.7	7 09	**7 28**	p2.2
5	W	1 57	**2 42**	a1.9	8 01	**8 20**	p2.3
6	T	2 46	**3 33**	a2.0	8 51	**9 09**	2.3
7	F	3 33	**4 22**	a2.1	9 38	**9 59**	a2.3
8	S	4 19	**5 13**	a2.0	10 26	**10 50**	a2.3
9	S	5 08	**6 07**	a1.9	11 16	**11 44**	a2.2
10	M	6 01	**7 06**	a1.7	...	**12 09**	2.0
11	T	7 00	**8 07**	a1.5	12 42	**1 03**	p1.9
12	W	8 03	**9 06**	1.4	1 40	**1 58**	p1.8
13	T	9 07	**10 02**	1.4	2 37	**2 53**	p1.7
14	F	10 07	**10 52**	p1.4	3 33	**3 48**	p1.7
15	S	11 02	**11 37**	p1.4	4 29	**4 43**	p1.7
16	S	11 53	**...**	1.3	5 22	**5 35**	p1.8
17	M	12 18	**12 41**	a1.4	6 11	**6 22**	1.8
18	T	12 56	**1 25**	a1.5	6 54	**7 05**	1.9
19	W	1 32	**2 08**	a1.7	7 33	**7 45**	2.0
20	T	2 08	**2 48**	a1.8	8 09	**8 23**	a2.2
21	F	2 43	**3 26**	a2.0	8 46	**9 01**	a2.2
22	S	3 19	**4 05**	a2.1	9 22	**9 40**	a2.3
23	S	3 56	**4 45**	a2.1	10 01	**10 22**	a2.3
24	M	4 36	**5 28**	a2.0	10 42	**11 08**	a2.2
25	T	5 21	**6 15**	a1.9	11 29	**11 59**	a2.1
26	W	6 13	**7 09**	a1.7	...	**12 19**	2.0
27	T	7 14	**8 06**	a1.6	12 53	**1 13**	p1.9
28	F	8 22	**9 05**	a1.5	1 49	**2 08**	p1.7
29	S	9 32	**10 04**	a1.4	2 46	**3 05**	p1.6
30	S	10 40	**11 01**	1.4	3 47	**4 06**	p1.6

DECEMBER

Day of Month	Day of Week	CURRENT TURNS TO NORTH Flood Starts a.m.	p.m.	Kts.	CURRENT TURNS TO SOUTH Ebb Starts a.m.	p.m.	Kts.
1	M	11 43	**11 55**	p1.4	4 53	**5 10**	p1.7
2	T	...	**12 43**	1.3	5 56	**6 12**	p1.8
3	W	12 48	**1 40**	a1.6	6 55	**7 09**	1.9
4	T	1 39	**2 35**	a1.7	7 47	**8 02**	2.0
5	F	2 27	**3 23**	a1.8	8 35	**8 52**	a2.1
6	S	3 14	**4 11**	a1.8	9 21	**9 41**	a2.1
7	S	4 00	**4 57**	a1.8	10 05	**10 30**	a2.1
8	M	4 45	**5 46**	a1.7	10 51	**11 20**	a2.0
9	T	5 33	**6 37**	a1.6	11 39	**...**	2.0
10	W	6 26	**7 29**	a1.5	12 12	**12 29**	p1.9
11	T	7 22	**8 21**	a1.4	1 05	**1 19**	p1.8
12	F	8 21	**9 10**	1.3	1 56	**2 08**	p1.7
13	S	9 18	**9 56**	p1.4	2 45	**2 57**	p1.7
14	S	10 13	**10 40**	p1.5	3 35	**3 46**	p1.7
15	M	11 05	**11 22**	p1.6	4 25	**4 37**	p1.8
16	T	11 55	**...**	1.3	5 16	**5 29**	1.8
17	W	12 04	**12 44**	a1.7	6 05	**6 19**	a2.0
18	T	12 46	**1 31**	a1.8	6 52	**7 07**	a2.1
19	F	1 28	**2 17**	a2.0	7 35	**7 53**	a2.3
20	S	2 11	**3 02**	a2.1	8 18	**8 38**	a2.4
21	S	2 55	**3 45**	a2.2	9 00	**9 22**	a2.4
22	M	3 38	**4 27**	a2.2	9 43	**10 07**	a2.4
23	T	4 24	**5 12**	a2.1	10 27	**10 55**	a2.3
24	W	5 12	**5 59**	a2.0	11 16	**11 46**	a2.2
25	T	6 07	**6 52**	a1.8	...	**12 07**	2.1
26	F	7 08	**7 48**	a1.6	12 41	**1 01**	p1.9
27	S	8 15	**8 46**	a1.5	1 37	**1 55**	p1.8
28	S	9 24	**9 45**	1.4	2 34	**2 51**	p1.6
29	M	10 31	**10 43**	p1.4	3 34	**3 50**	p1.6
30	T	11 35	**11 40**	p1.4	4 39	**4 53**	p1.6
31	W	...	**12 37**	1.2	5 45	**5 57**	1.6

A also at *11:53 a.m. 1.5

The Kts. (knots) columns show the **maximum** predicted velocities of the stronger one of the Flood Currents and the stronger one of the Ebb Currents for each day.

The letter "a" means the velocity shown should occur **after** the **a.m.** Current Change. The letter "p" means the velocity shown should occur **after** the **p.m.** Current Change (even if next morning). No "a" or "p" means a.m. and p.m. velocities are the same for that day.

Avg. Max. Velocity: Flood 1.7 Kts., Ebb 2.0 Kts.

Max. Flood 2 hrs. 25 min. after Flood Starts, ±30 min.

Max. Ebb 3 hrs. 15 min. after Ebb Starts, ±10 min.

See pp. 22-29 for Current Change at other points.

2014 HIGH & LOW WATER THE BATTERY, NY HARBOR

40°42'N, 74°00.9'W

Standard Time

DAY OF MONTH	DAY OF WEEK	JANUARY HIGH a.m.	Ht.	p.m.	Ht.	LOW a.m.	p.m.
1	W	7 33	5.8	8 03	4.9	1 34	2 17
2	T	8 25	5.9	8 58	4.9	2 27	3 07
3	F	9 19	5.8	9 55	4.9	3 19	3 56
4	S	10 17	5.6	10 55	4.9	4 12	4 47
5	S	11 14	5.3	11 53	4.8	5 04	5 37
6	M	...	...	12 11	4.9	6 02	6 31
7	T	12 49	4.7	1 06	4.6	7 05	7 30
8	W	1 44	4.6	2 02	4.3	8 11	8 29
9	T	2 40	4.5	3 01	4.0	9 14	9 26
10	F	3 38	4.5	4 01	3.9	10 12	10 18
11	S	4 36	4.5	5 00	3.9	11 04	11 07
12	S	5 29	4.6	5 53	3.9	11 53	11 53
13	M	6 16	4.7	6 40	4.1	...	12 39
14	T	6 57	4.7	7 23	4.2	12 39	1 23
15	W	7 36	4.8	8 03	4.2	1 22	2 05
16	T	8 12	4.8	8 42	4.2	2 04	2 45
17	F	8 46	4.7	9 21	4.2	2 44	3 22
18	S	9 19	4.6	9 59	4.1	3 22	3 57
19	S	9 50	4.4	10 35	4.0	3 57	4 30
20	M	10 22	4.3	11 10	4.0	4 31	5 00
21	T	10 58	4.1	11 46	4.1	5 05	5 28
22	W	11 41	4.0	...	...	5 46	6 02
23	T	12 27	4.1	12 31	3.9	6 46	6 56
24	F	1 15	4.3	1 27	3.8	8 10	8 18
25	S	2 11	4.4	2 32	3.8	9 20	9 28
26	S	3 16	4.6	3 45	3.9	10 21	10 30
27	M	4 26	4.9	4 59	4.1	11 18	11 27
28	T	5 31	5.2	6 02	4.5	...	12 13
29	W	6 29	5.5	6 58	4.8	12 24	1 06
30	T	7 21	5.8	7 50	5.0	1 19	1 57
31	F	8 12	5.8	8 42	5.2	2 12	2 47

Standard Time

DAY OF MONTH	DAY OF WEEK	FEBRUARY HIGH a.m.	Ht.	p.m.	Ht.	LOW a.m.	p.m.
1	S	9 04	5.7	9 35	5.2	3 04	3 35
2	S	9 57	5.5	10 30	5.1	3 54	4 22
3	M	10 52	5.2	11 25	5.0	4 45	5 09
4	T	11 48	4.8	...	...	5 40	6 00
5	W	12 18	4.8	12 40	4.5	6 37	6 53
6	T	1 11	4.6	1 34	4.1	7 40	7 53
7	F	2 04	4.4	2 31	3.9	8 44	8 52
8	S	3 01	4.2	3 31	3.7	9 43	9 48
9	S	4 01	4.2	4 32	3.7	10 36	10 39
10	M	4 59	4.3	5 28	3.9	11 25	11 28
11	T	5 50	4.4	6 16	4.0	...	12 11
12	W	6 34	4.6	6 59	4.2	12 14	12 55
13	T	7 13	4.7	7 39	4.4	12 58	1 36
14	F	7 49	4.7	8 15	4.4	1 41	2 15
15	S	8 22	4.7	8 50	4.5	2 21	2 52
16	S	8 53	4.7	9 22	4.5	3 00	3 27
17	M	9 22	4.5	9 53	4.5	3 37	3 59
18	T	9 54	4.4	10 25	4.5	4 12	4 28
19	W	10 31	4.3	11 03	4.5	4 47	4 56
20	T	11 16	4.1	11 50	4.5	5 27	5 30
21	F	...	...	12 09	4.0	6 21	6 20
22	S	12 43	4.5	1 08	3.9	7 40	7 44
23	S	1 43	4.6	2 14	3.9	8 55	9 06
24	M	2 51	4.7	3 29	4.0	9 59	10 12
25	T	4 06	4.9	4 44	4.3	10 57	11 12
26	W	5 15	5.1	5 48	4.7	11 52	...
27	T	6 15	5.4	6 44	5.1	12 09	12 45
28	F	7 07	5.7	7 34	5.4	1 04	1 35

Dates when Ht. of **Low** Water is below Mean Lower Low with Ht. of lowest given for each period and Date of lowest in ( ):

1st - 7th: -1.3' (2nd - 3rd)
14th - 18th: -0.3' (15th - 17th)
27th - 31st: -1.4' (31st)

1st - 4th: -1.4' (1st)
13th - 17th: -0.3' (14th - 16th)
25th - 28th: -1.1' (28th)

Average Rise and Fall 4.6 ft.

When a high tide exceeds avg. ht., the *following* low tide will be lower than avg.

2014 HIGH & LOW WATER
THE BATTERY, NY HARBOR
40°42'N, 74°00.9'W

***Daylight Time starts March 9 at 2 a.m.** **Daylight Saving Time**

DAY OF MONTH	DAY OF WEEK	MARCH						DAY OF MONTH	DAY OF WEEK	APRIL					
		HIGH				LOW				HIGH				LOW	
		a.m.	Ht.	**p.m.**	Ht.	a.m.	**p.m.**			a.m.	Ht.	**p.m.**	Ht.	a.m.	**p.m.**
1	S	7 57	5.7	**8 23**	5.5	1 57	**2 24**	1	T	10 15	5.2	**10 33**	5.6	4 18	**4 29**
2	S	8 46	5.6	**9 12**	5.5	2 48	**3 10**	2	W	11 05	5.0	**11 21**	5.3	5 04	**5 12**
3	M	9 37	5.4	**10 02**	5.4	3 37	**3 55**	3	T	11 57	4.7	**...**	...	5 50	**5 54**
4	T	10 30	5.1	**10 54**	5.2	4 26	**4 41**	4	F	12 11	5.0	**12 50**	4.4	6 39	**6 41**
5	W	11 21	4.8	**11 45**	4.9	5 14	**5 26**	5	S	12 59	4.7	**1 40**	4.2	7 30	**7 32**
6	T	...	...	**12 14**	4.4	6 07	**6 16**	6	S	1 49	4.5	**2 32**	4.1	8 29	**8 33**
7	F	12 35	4.6	**1 07**	4.1	7 05	**7 13**	7	M	2 40	4.2	**3 26**	4.0	9 29	**9 37**
8	S	1 27	4.4	**2 01**	3.9	8 08	**8 15**	8	T	3 34	4.1	**4 22**	4.0	10 24	**10 35**
9	S	*3 21	4.2	***3 59**	3.8	*10 09	***10 15**	9	W	4 33	4.1	**5 18**	4.1	11 14	**11 27**
10	M	4 20	4.1	**4 59**	3.8	11 03	**11 09**	10	T	5 32	4.2	**6 09**	4.4	11 59	**...**
11	T	5 21	4.1	**5 56**	4.0	11 52	**11 59**	11	F	6 24	4.3	**6 54**	4.6	12 15	**12 42**
12	W	6 17	4.3	**6 47**	4.2	...	**12 38**	12	S	7 08	4.5	**7 33**	4.9	1 01	**1 23**
13	T	7 04	4.4	**7 30**	4.4	12 46	**1 21**	13	S	7 47	4.6	**8 08**	5.1	1 46	**2 04**
14	F	7 45	4.6	**8 09**	4.6	1 31	**2 02**	14	M	8 22	4.7	**8 40**	5.3	2 30	**2 44**
15	S	8 21	4.7	**8 44**	4.8	2 14	**2 42**	15	T	8 57	4.8	**9 13**	5.4	3 13	**3 23**
16	S	8 54	4.8	**9 16**	4.9	2 57	**3 19**	16	W	9 34	4.8	**9 48**	5.5	3 55	**4 02**
17	M	9 25	4.7	**9 46**	5.0	3 37	**3 55**	17	T	10 15	4.7	**10 29**	5.5	4 38	**4 41**
18	T	9 57	4.7	**10 16**	5.0	4 16	**4 29**	18	F	11 03	4.6	**11 18**	5.4	5 21	**5 22**
19	W	10 32	4.6	**10 52**	5.0	4 54	**5 02**	19	S	...	...	**12 01**	4.5	6 08	**6 09**
20	T	11 13	4.5	**11 35**	5.0	5 33	**5 35**	20	S	12 15	5.3	**1 01**	4.5	7 03	**7 09**
21	F	...	...	**12 04**	4.3	6 17	**6 15**	21	M	1 16	5.1	**2 02**	4.5	8 07	**8 25**
22	S	12 26	4.9	**1 01**	4.2	7 11	**7 09**	22	T	2 19	5.0	**3 05**	4.6	9 14	**9 39**
23	S	1 24	4.9	**2 02**	4.2	8 22	**8 33**	23	W	3 25	4.9	**4 10**	4.7	10 17	**10 44**
24	M	2 27	4.8	**3 09**	4.2	9 34	**9 52**	24	T	4 33	4.9	**5 15**	5.0	11 13	**11 43**
25	T	3 36	4.8	**4 20**	4.3	10 38	**10 58**	25	F	5 39	5.0	**6 15**	5.3	...	**12 06**
26	W	4 49	4.9	**5 31**	4.6	11 36	**11 58**	26	S	6 39	5.1	**7 08**	5.6	12 38	**12 56**
27	T	5 58	5.1	**6 33**	5.0	...	**12 30**	27	S	7 32	5.2	**7 56**	5.8	1 31	**1 45**
28	F	6 58	5.3	**7 27**	5.4	12 54	**1 21**	28	M	8 20	5.2	**8 40**	5.8	2 23	**2 33**
29	S	7 51	5.5	**8 16**	5.6	1 48	**2 11**	29	T	9 07	5.1	**9 23**	5.8	3 11	**3 18**
30	S	8 39	5.5	**9 02**	5.8	2 40	**2 59**	30	W	9 54	5.0	**10 06**	5.6	3 58	**4 02**
31	M	9 27	5.4	**9 47**	5.7	3 30	**3 45**								

Dates when Ht. of **Low** Water is below Mean Lower Low with Ht. of lowest given for each period and Date of lowest in ( ):

1st - 5th: -1.2' (1st - 2nd)
17th - 19th -0.2'
27th - 31st: -0.8' (29th - 31st)

1st - 3rd: -0.8' (1st)
15th - 18th: -0.3' (16th - 17th)
25th - 30th: -0.5' (28th - 30th)

Average Rise and Fall 4.6 ft.

When a high tide exceeds avg. ht., the *following* low tide will be lower than avg.

2014 HIGH & LOW WATER
THE BATTERY, NY HARBOR

40°42'N, 74°00.9'W

Day of Month	Day of Week	MAY HIGH a.m.	Ht.	p.m.	Ht.	LOW a.m.	p.m.	Day of Month	Day of Week	JUNE HIGH a.m.	Ht.	p.m.	Ht.	LOW a.m.	p.m.
		Daylight Saving Time								Daylight Saving Time					
1	T	10 42	4.8	10 50	5.3	4 43	4 44	1	S	11 56	4.4	11 50	4.8	5 43	5 39
2	F	11 32	4.6	11 36	5.1	5 26	5 26	2	M	...	...	12 44	4.3	6 24	6 20
3	S	...	...	12 23	4.4	6 10	6 08	3	T	12 34	4.6	1 30	4.3	7 07	7 07
4	S	12 25	4.8	1 14	4.3	6 58	6 54	4	W	1 18	4.4	2 15	4.3	7 56	8 06
5	M	1 11	4.5	2 02	4.2	7 48	7 49	5	T	1 59	4.3	2 57	4.3	8 46	9 10
6	T	1 59	4.3	2 51	4.1	8 44	8 54	6	F	2 43	4.1	3 41	4.4	9 38	10 10
7	W	2 47	4.2	3 41	4.2	9 39	9 55	7	S	3 31	4.1	4 27	4.6	10 27	11 04
8	T	3 38	4.1	4 32	4.3	10 29	10 50	8	S	4 27	4.1	5 16	4.8	11 14	11 55
9	F	4 34	4.1	5 23	4.5	11 15	11 40	9	M	5 27	4.1	6 05	5.1	...	12 01
10	S	5 30	4.2	6 09	4.7	11 59	...	10	T	6 25	4.3	6 51	5.4	12 44	12 47
11	S	6 21	4.3	6 51	5.0	12 28	12 41	11	W	7 17	4.5	7 37	5.7	1 34	1 36
12	M	7 07	4.5	7 29	5.3	1 15	1 24	12	T	8 06	4.7	8 22	6.0	2 24	2 27
13	T	7 49	4.6	8 06	5.6	2 02	2 08	13	F	8 55	4.9	9 10	6.0	3 14	3 18
14	W	8 30	4.7	8 45	5.8	2 49	2 53	14	S	9 47	5.0	10 02	6.0	4 03	4 09
15	T	9 13	4.8	9 26	5.8	3 35	3 38	15	S	10 44	5.0	10 59	5.9	4 51	5 00
16	F	10 01	4.8	10 14	5.8	4 21	4 24	16	M	11 45	5.1	11 59	5.7	5 40	5 53
17	S	10 55	4.8	11 08	5.7	5 07	5 12	17	T	...	...	12 45	5.1	6 32	6 52
18	S	11 56	4.8	...	...	5 56	6 04	18	W	12 58	5.5	1 43	5.1	7 27	7 56
19	M	12 08	5.5	12 57	4.8	6 50	7 04	19	T	1 56	5.2	2 38	5.2	8 26	9 04
20	T	1 10	5.3	1 57	4.8	7 49	8 13	20	F	2 53	4.9	3 35	5.2	9 26	10 08
21	W	2 10	5.2	2 55	4.9	8 52	9 23	21	S	3 53	4.7	4 33	5.2	10 23	11 07
22	T	3 11	5.0	3 55	5.0	9 52	10 28	22	S	4 55	4.5	5 31	5.3	11 16	...
23	F	4 14	4.8	4 56	5.2	10 49	11 26	23	M	5 56	4.5	6 26	5.3	12 02	12 06
24	S	5 17	4.8	5 55	5.4	11 41	...	24	T	6 52	4.5	7 14	5.4	12 54	12 55
25	S	6 18	4.8	6 48	5.5	12 21	12 31	25	W	7 42	4.6	7 58	5.4	1 43	1 42
26	M	7 12	4.8	7 35	5.7	1 13	1 19	26	T	8 28	4.6	8 39	5.4	2 30	2 28
27	T	8 01	4.8	8 19	5.7	2 04	2 07	27	F	9 12	4.6	9 19	5.3	3 14	3 13
28	W	8 47	4.8	9 00	5.6	2 52	2 53	28	S	9 56	4.6	9 57	5.2	3 57	3 55
29	T	9 33	4.7	9 41	5.5	3 37	3 37	29	S	10 40	4.5	10 37	5.0	4 37	4 35
30	F	10 19	4.6	10 22	5.3	4 20	4 19	30	M	11 26	4.5	11 16	4.8	5 15	5 13
31	S	11 07	4.5	11 05	5.0	5 02	4 59								

Dates when Ht. of **Low** Water is below Mean Lower Low with Ht. of lowest given for each period and Date of lowest in ( ):

May:
1st: -0.3'
14th - 19th: -0.5' (16th - 17th)
27th - 30th: -0.2'

June:
12th - 19th: -0.7' (14th - 16th)

Average Rise and Fall 4.6 ft.

When a high tide exceeds avg. ht., the *following* low tide will be lower than avg.

2014 HIGH & LOW WATER THE BATTERY, NY HARBOR

40°42'N, 74°00.9'W

Day of Month	Day of Week	JULY HIGH a.m.	Ht.	p.m.	Ht.	LOW a.m.	p.m.	Day of Month	Day of Week	AUGUST HIGH a.m.	Ht.	p.m.	Ht.	LOW a.m.	p.m.
		Daylight Saving Time								Daylight Saving Time					
1	T	...	...	**12 10**	4.4	5 52	**5 51**	1	F	...	...	**12 42**	4.6	6 18	**6 39**
2	W	12 01	4.6	**12 53**	4.4	6 28	**6 30**	2	S	12 29	4.4	**1 18**	4.6	6 46	**7 31**
3	T	12 34	4.5	**1 32**	4.4	7 04	**7 17**	3	S	1 14	4.2	**1 59**	4.7	7 25	**8 44**
4	F	1 13	4.3	**2 10**	4.4	7 45	**8 20**	4	M	2 05	4.2	**2 48**	4.8	8 33	**9 55**
5	S	1 54	4.2	**2 47**	4.5	8 35	**9 27**	5	T	3 02	4.1	**3 43**	5.0	9 52	**10 55**
6	S	2 40	4.1	**3 31**	4.7	9 33	**10 27**	6	W	4 09	4.2	**4 48**	5.2	10 57	**11 51**
7	M	3 36	4.1	**4 23**	4.9	10 30	**11 23**	7	T	5 22	4.4	**5 55**	5.5	11 55	**...**
8	T	4 40	4.1	**5 21**	5.2	11 23	**...**	8	F	6 29	4.7	**6 55**	5.8	12 45	**12 52**
9	W	5 48	4.3	**6 19**	5.5	12 16	**12 17**	9	S	7 28	5.0	**7 49**	6.1	1 38	**1 49**
10	T	6 50	4.5	**7 13**	5.8	1 09	**1 11**	10	S	8 21	5.4	**8 41**	6.2	2 30	**2 44**
11	F	7 45	4.8	**8 05**	6.1	2 01	**2 06**	11	M	9 13	5.6	**9 33**	6.2	3 20	**3 37**
12	S	8 38	5.0	**8 56**	6.2	2 52	**3 00**	12	T	10 06	5.7	**10 26**	6.0	4 08	**4 29**
13	S	9 31	5.2	**9 49**	6.1	3 42	**3 53**	13	W	11 02	5.7	**11 22**	5.7	4 56	**5 21**
14	M	10 27	5.3	**10 45**	6.0	4 31	**4 46**	14	T	11 58	5.7	**...**	...	5 44	**6 15**
15	T	11 26	5.4	**11 43**	5.8	5 20	**5 39**	15	F	12 20	5.4	**12 54**	5.5	6 33	**7 12**
16	W	...	...	**12 25**	5.4	6 09	**6 35**	16	S	1 16	5.0	**1 49**	5.3	7 27	**8 15**
17	T	12 41	5.5	**1 21**	5.4	7 01	**7 36**	17	S	2 12	4.7	**2 43**	5.2	8 25	**9 20**
18	F	1 38	5.1	**2 16**	5.3	7 58	**8 41**	18	M	3 09	4.5	**3 38**	5.0	9 27	**10 21**
19	S	2 34	4.8	**3 11**	5.2	8 57	**9 46**	19	T	4 08	4.3	**4 37**	4.9	10 25	**11 16**
20	S	3 31	4.5	**4 07**	5.1	9 55	**10 46**	20	W	5 08	4.2	**5 35**	4.9	11 18	**...**
21	M	4 31	4.4	**5 05**	5.1	10 51	**11 41**	21	T	6 06	4.3	**6 28**	4.9	12 06	**12 07**
22	T	5 33	4.3	**6 02**	5.1	11 42	**...**	22	F	6 57	4.5	**7 14**	5.1	12 52	**12 54**
23	W	6 30	4.3	**6 53**	5.2	12 31	**12 31**	23	S	7 42	4.7	**7 54**	5.1	1 36	**1 39**
24	T	7 21	4.5	**7 38**	5.2	1 19	**1 18**	24	S	8 22	4.8	**8 31**	5.2	2 18	**2 23**
25	F	8 06	4.6	**8 18**	5.3	2 05	**2 04**	25	M	9 01	4.9	**9 06**	5.2	2 58	**3 05**
26	S	8 49	4.6	**8 56**	5.3	2 48	**2 48**	26	T	9 37	4.9	**9 38**	5.1	3 36	**3 45**
27	S	9 30	4.7	**9 33**	5.2	3 29	**3 30**	27	W	10 11	4.9	**10 09**	4.9	4 11	**4 23**
28	M	10 10	4.7	**10 08**	5.1	4 08	**4 10**	28	T	10 44	4.9	**10 39**	4.7	4 44	**5 00**
29	T	10 50	4.6	**10 42**	4.9	4 44	**4 48**	29	F	11 15	4.9	**11 13**	4.6	5 14	**5 36**
30	W	11 30	4.6	**11 16**	4.7	5 18	**5 24**	30	S	11 50	4.9	**11 55**	4.4	5 40	**6 13**
31	T	11 59	4.6	**11 50**	4.5	5 50	**6 00**	31	S	...	...	**12 31**	4.9	6 09	**7 01**

Dates when Ht. of **Low** Water is below Mean Lower Low with Ht. of lowest given for each period and Date of lowest in ( ):

11th - 17th: -0.9' (14th - 15th)

9th - 15th: -1.0' (12th)

Average Rise and Fall 4.6 ft.

When a high tide exceeds avg. ht., the *following* low tide will be lower than avg.

2014 HIGH & LOW WATER
THE BATTERY, NY HARBOR
40°42'N, 74°00.9'W

Daylight SavingTime | Daylight Saving Time

DAY OF MONTH	DAY OF WEEK	SEPTEMBER						DAY OF MONTH	DAY OF WEEK	OCTOBER					
		HIGH				LOW				HIGH				LOW	
		a.m.	Ht.	p.m.	Ht.	a.m.	p.m.			a.m.	Ht.	p.m.	Ht.	a.m.	p.m.
1	M	12 45	4.3	1 20	4.9	6 49	8 10	1	W	1 32	4.3	1 57	5.1	7 44	9 04
2	T	1 42	4.2	2 16	5.0	7 54	9 26	2	T	2 37	4.3	3 02	5.1	9 14	10 09
3	W	2 44	4.2	3 18	5.1	9 28	10 31	3	F	3 45	4.5	4 11	5.2	10 25	11 07
4	T	3 54	4.3	4 28	5.2	10 40	11 30	4	S	4 56	4.8	5 21	5.3	11 28	...
5	F	5 07	4.6	5 36	5.5	11 40	...	5	S	5 59	5.1	6 22	5.6	12 01	12 23
6	S	6 14	4.9	6 38	5.8	12 23	12 37	6	M	6 55	5.5	7 17	5.7	12 51	1 18
7	S	7 12	5.3	7 33	6.0	1 15	1 33	7	T	7 46	5.9	8 07	5.8	1 41	2 11
8	M	8 04	5.7	8 25	6.1	2 06	2 27	8	W	8 33	6.1	8 56	5.7	2 30	3 03
9	T	8 53	5.9	9 15	6.1	2 55	3 20	9	T	9 20	6.1	9 45	5.5	3 18	3 53
10	W	9 43	6.0	10 06	5.9	3 43	4 11	10	F	10 08	6.0	10 37	5.3	4 04	4 42
11	T	10 35	5.9	11 00	5.6	4 30	5 02	11	S	10 58	5.7	11 31	4.9	4 49	5 30
12	F	11 28	5.8	11 56	5.2	5 16	5 53	12	S	11 50	5.4	...	...	5 35	6 20
13	S	...	...	12 23	5.5	6 04	6 47	13	M	12 28	4.7	12 44	5.1	6 22	7 14
14	S	12 53	4.9	1 18	5.3	6 54	7 46	14	T	1 23	4.4	1 38	4.8	7 15	8 13
15	M	1 49	4.6	2 12	5.0	7 51	8 49	15	W	2 18	4.3	2 31	4.6	8 16	9 14
16	T	2 45	4.4	3 06	4.8	8 54	9 51	16	T	3 12	4.2	3 24	4.4	9 20	10 10
17	W	3 42	4.3	4 03	4.7	9 56	10 47	17	F	4 07	4.2	4 20	4.4	10 19	11 00
18	T	4 40	4.2	5 01	4.6	10 51	11 36	18	S	5 02	4.3	5 15	4.4	11 11	11 44
19	F	5 37	4.4	5 56	4.7	11 41	...	19	S	5 54	4.5	6 06	4.5	11 59	...
20	S	6 28	4.6	6 44	4.8	12 21	12 28	20	M	6 39	4.8	6 51	4.6	12 26	12 44
21	S	7 13	4.8	7 25	5.0	1 03	1 13	21	T	7 19	5.0	7 30	4.7	1 06	1 28
22	M	7 53	5.0	8 03	5.0	1 44	1 56	22	W	7 55	5.2	8 06	4.8	1 46	2 11
23	T	8 29	5.1	8 37	5.0	2 23	2 39	23	T	8 27	5.3	8 39	4.8	2 25	2 54
24	W	9 02	5.2	9 08	5.0	3 01	3 20	24	F	8 57	5.4	9 12	4.7	3 03	3 36
25	T	9 32	5.2	9 38	4.9	3 37	3 59	25	S	9 28	5.4	9 48	4.6	3 40	4 17
26	F	10 01	5.2	10 10	4.7	4 11	4 38	26	S	10 04	5.4	10 30	4.5	4 18	4 59
27	S	10 32	5.2	10 46	4.6	4 42	5 16	27	M	10 47	5.3	11 22	4.4	4 56	5 43
28	S	11 10	5.1	11 33	4.4	5 13	5 56	28	T	11 41	5.2	...	...	5 38	6 33
29	M	11 59	5.1	...	...	5 48	6 45	29	W	12 24	4.3	12 42	5.1	6 30	7 34
30	T	12 30	4.3	12 55	5.1	6 32	7 50	30	T	1 28	4.3	1 46	5.0	7 42	8 42
								31	F	2 31	4.4	2 50	5.0	9 01	9 46

Dates when Ht. of **Low** Water is below Mean Lower Low with Ht. of lowest given for each period and Date of lowest in ( ):

7th - 12th: -0.9' (10th) | 6th - 11th: -0.7' (8th - 9th)

Average Rise and Fall 4.6 ft.

When a high tide exceeds avg. ht., the *following* low tide will be lower than avg.

2014 HIGH & LOW WATER
THE BATTERY, NY HARBOR
40°42'N, 74°00.9'W

***Standard Time starts Nov. 2 at 2 a.m.** **Standard Time**

DAY OF MONTH	DAY OF WEEK	NOVEMBER					
		HIGH				LOW	
		a.m.	Ht.	**p.m.**	Ht.	a.m.	**p.m.**
1	S	3 36	4.6	**3 56**	5.0	10 11	**10 44**
2	S	*3 41	4.9	***4 02**	5.0	*10 12	***10 38**
3	M	4 42	5.2	**5 04**	5.1	11 09	**11 28**
4	T	5 39	5.5	**6 01**	5.2	...	**12 04**
5	W	6 28	5.8	**6 51**	5.3	12 18	**12 55**
6	T	7 15	5.9	**7 39**	5.2	1 06	**1 46**
7	F	7 59	5.9	**8 26**	5.1	1 53	**2 34**
8	S	8 44	5.7	**9 15**	4.9	2 40	**3 21**
9	S	9 30	5.5	**10 07**	4.6	3 24	**4 07**
10	M	10 18	5.2	**11 01**	4.4	4 08	**4 53**
11	T	11 09	4.9	**11 55**	4.2	4 52	**5 41**
12	W	...	...	**12 01**	4.6	5 39	**6 33**
13	T	12 47	4.1	**12 51**	4.4	6 34	**7 29**
14	F	1 37	4.0	**1 40**	4.2	7 37	**8 25**
15	S	2 28	4.1	**2 31**	4.1	8 39	**9 17**
16	S	3 20	4.1	**3 25**	4.0	9 35	**10 03**
17	M	4 11	4.3	**4 19**	4.1	10 25	**10 46**
18	T	4 58	4.5	**5 09**	4.2	11 12	**11 27**
19	W	5 41	4.8	**5 54**	4.3	11 58	**...**
20	T	6 19	5.1	**6 34**	4.4	12 08	**12 43**
21	F	6 54	5.3	**7 12**	4.5	12 50	**1 29**
22	S	7 29	5.4	**7 50**	4.5	1 32	**2 13**
23	S	8 05	5.5	**8 31**	4.5	2 15	**2 58**
24	M	8 47	5.5	**9 19**	4.5	2 59	**3 43**
25	T	9 35	5.4	**10 15**	4.4	3 43	**4 29**
26	W	10 32	5.3	**11 17**	4.4	4 31	**5 19**
27	T	11 34	5.1	**...**	...	5 26	**6 15**
28	F	12 20	4.5	**12 35**	5.0	6 32	**7 18**
29	S	1 20	4.5	**1 37**	4.8	7 45	**8 21**
30	S	2 21	4.7	**2 39**	4.7	8 54	**9 20**

DAY OF MONTH	DAY OF WEEK	DECEMBER					
		HIGH				LOW	
		a.m.	Ht.	**p.m.**	Ht.	a.m.	**p.m.**
1	M	3 22	4.8	**3 43**	4.6	9 56	**10 14**
2	T	4 24	5.1	**4 46**	4.6	10 53	**11 06**
3	W	5 20	5.3	**5 44**	4.7	11 47	**11 55**
4	T	6 12	5.4	**6 36**	4.7	...	**12 39**
5	F	6 58	5.5	**7 23**	4.7	12 44	**1 28**
6	S	7 41	5.5	**8 09**	4.7	1 31	**2 16**
7	S	8 23	5.4	**8 55**	4.5	2 17	**3 01**
8	M	9 06	5.2	**9 43**	4.4	3 01	**3 44**
9	T	9 50	4.9	**10 32**	4.2	3 43	**4 26**
10	W	10 36	4.7	**11 22**	4.1	4 24	**5 08**
11	T	11 22	4.4	**...**	...	5 06	**5 52**
12	F	12 11	4.0	**12 08**	4.2	5 51	**6 39**
13	S	12 57	3.9	**12 52**	4.0	6 46	**7 31**
14	S	1 43	3.9	**1 38**	3.8	7 50	**8 25**
15	M	2 29	4.0	**2 26**	3.7	8 53	**9 15**
16	T	3 17	4.1	**3 20**	3.7	9 48	**10 02**
17	W	4 07	4.3	**4 18**	3.7	10 39	**10 47**
18	T	4 56	4.5	**5 13**	3.9	11 27	**11 32**
19	F	5 41	4.8	**6 02**	4.1	...	**12 15**
20	S	6 24	5.1	**6 47**	4.3	12 19	**1 04**
21	S	7 06	5.4	**7 31**	4.4	1 06	**1 51**
22	M	7 49	5.5	**8 17**	4.6	1 55	**2 38**
23	T	8 35	5.6	**9 08**	4.6	2 43	**3 25**
24	W	9 26	5.5	**10 04**	4.6	3 32	**4 12**
25	T	10 23	5.4	**11 04**	4.6	4 22	**5 01**
26	F	11 22	5.1	**...**	...	5 16	**5 53**
27	S	12 04	4.7	**12 21**	4.9	6 18	**6 51**
28	S	1 02	4.7	**1 20**	4.6	7 26	**7 53**
29	M	2 00	4.7	**2 20**	4.4	8 34	**8 54**
30	T	3 00	4.7	**3 23**	4.2	9 38	**9 50**
31	W	4 02	4.8	**4 27**	4.2	10 36	**10 44**

Dates when Ht. of **Low** Water is below Mean Lower Low with Ht. of lowest given for each period and Date of lowest in ( ):

2nd - 9th: -0.6' (5th - 7th)
22nd - 27th: -0.4' (23rd - 25th)
30th: -0.2'

1st - 9th: -0.6' (5th - 6th)
20th - 31st: -0.9' (23rd - 24th)

Average Rise and Fall 4.6 ft.

When a high tide exceeds avg. ht., the *following* low tide will be lower than avg.

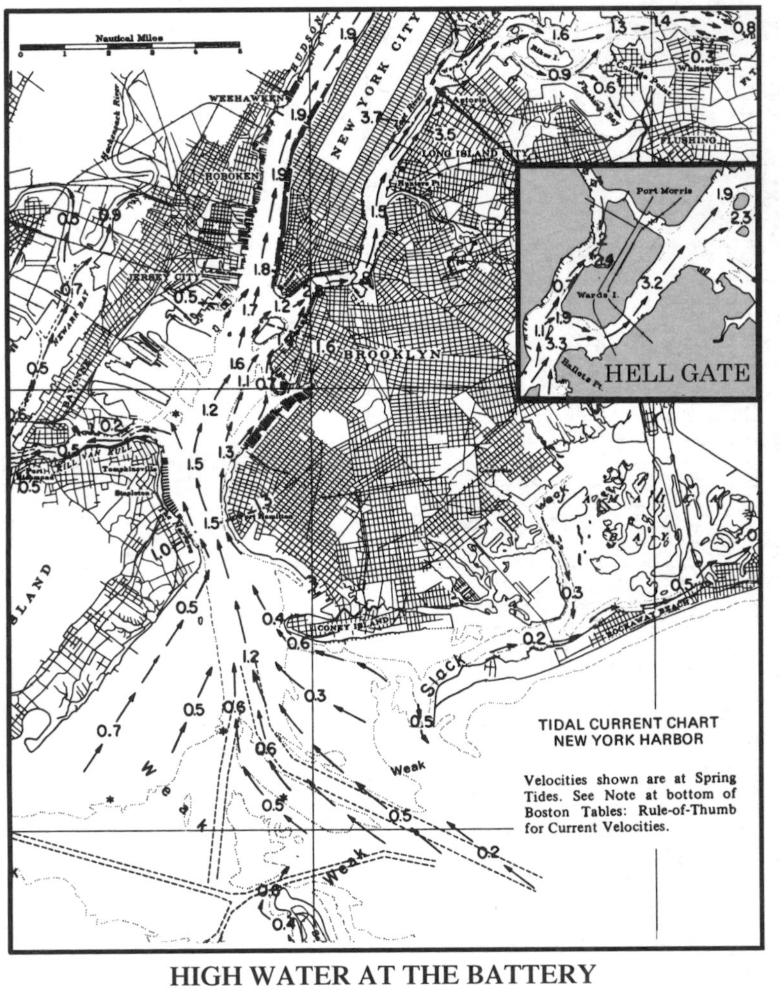

HIGH WATER AT THE BATTERY

NEW YORK BAY CURRENTS

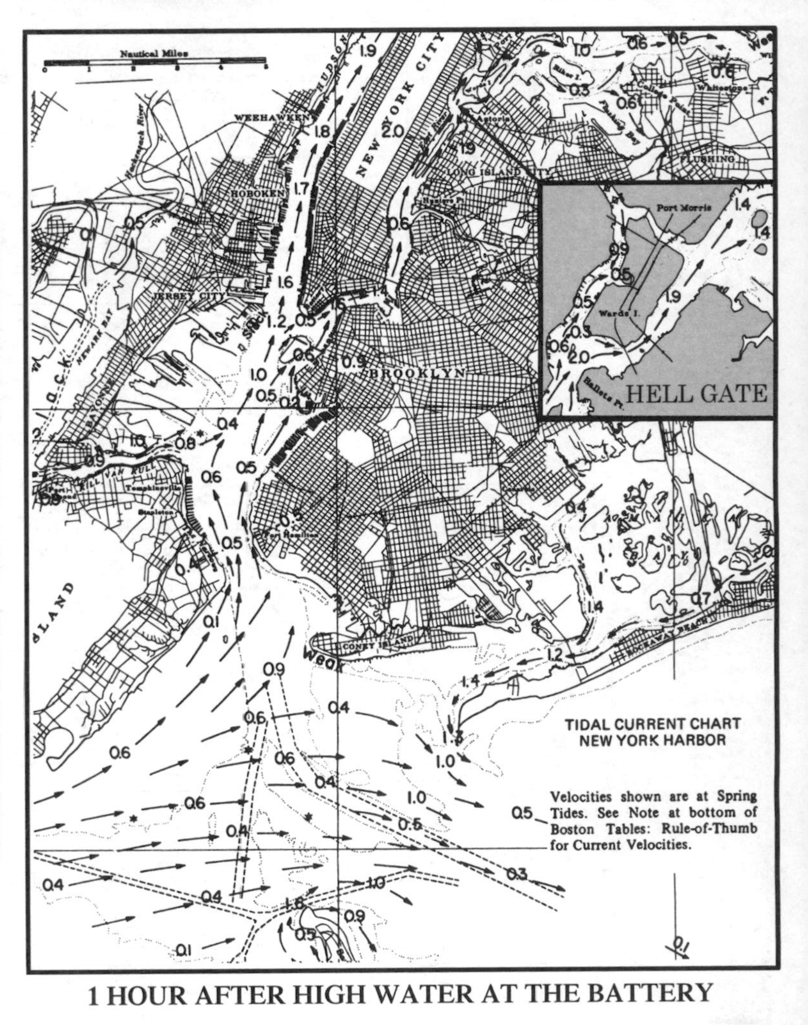

1 HOUR AFTER HIGH WATER AT THE BATTERY

NEW YORK BAY CURRENTS

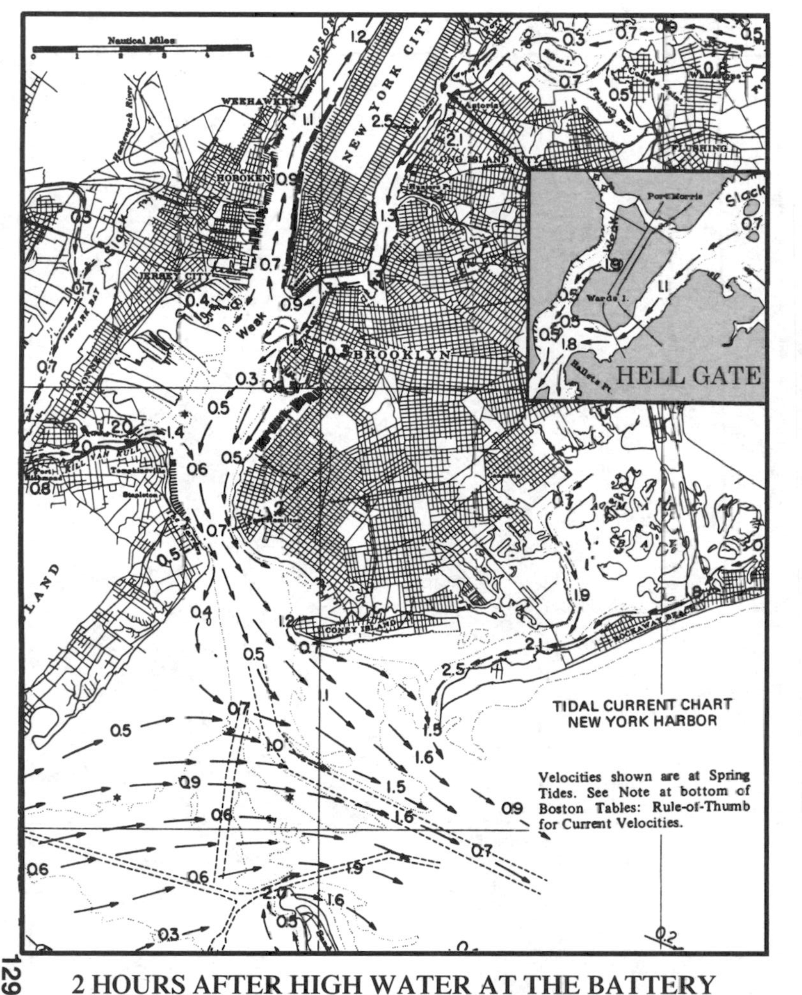

2 HOURS AFTER HIGH WATER AT THE BATTERY

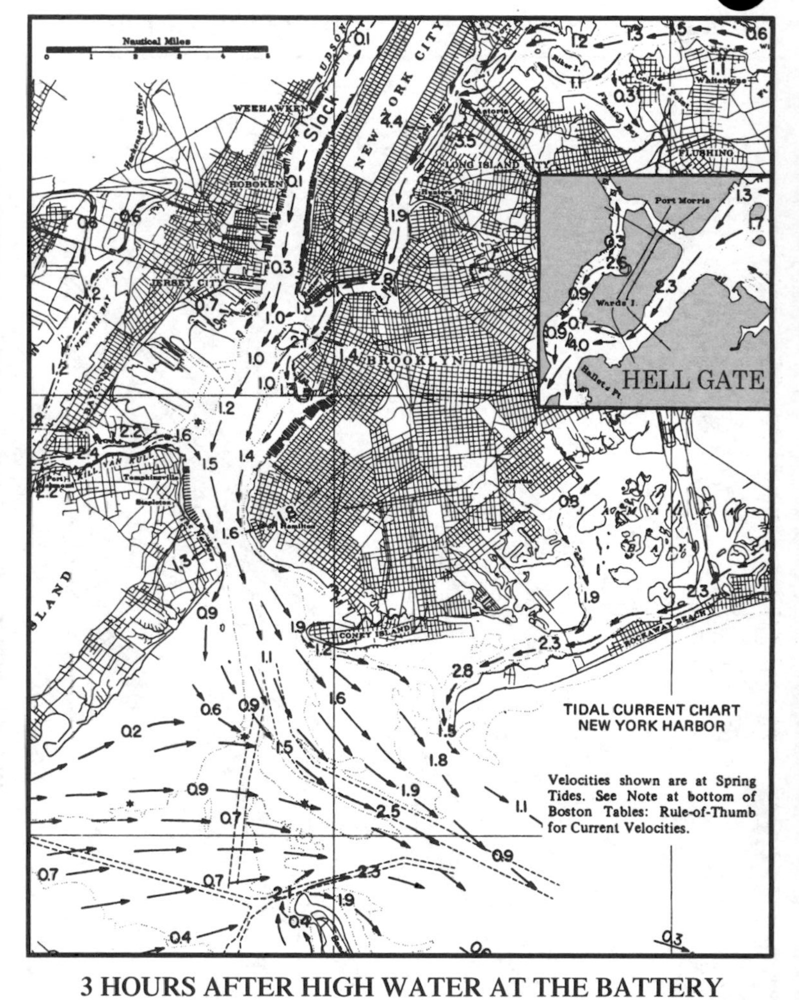

3 HOURS AFTER HIGH WATER AT THE BATTERY

NEW YORK BAY CURRENTS

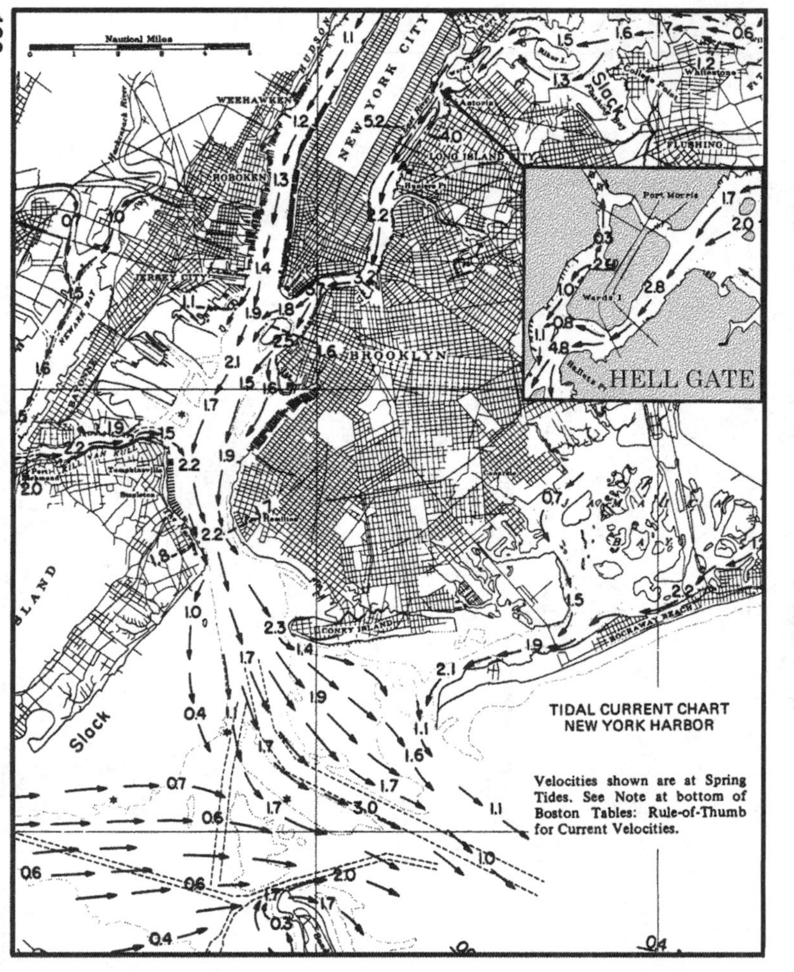

4 HOURS AFTER HIGH WATER AT THE BATTERY

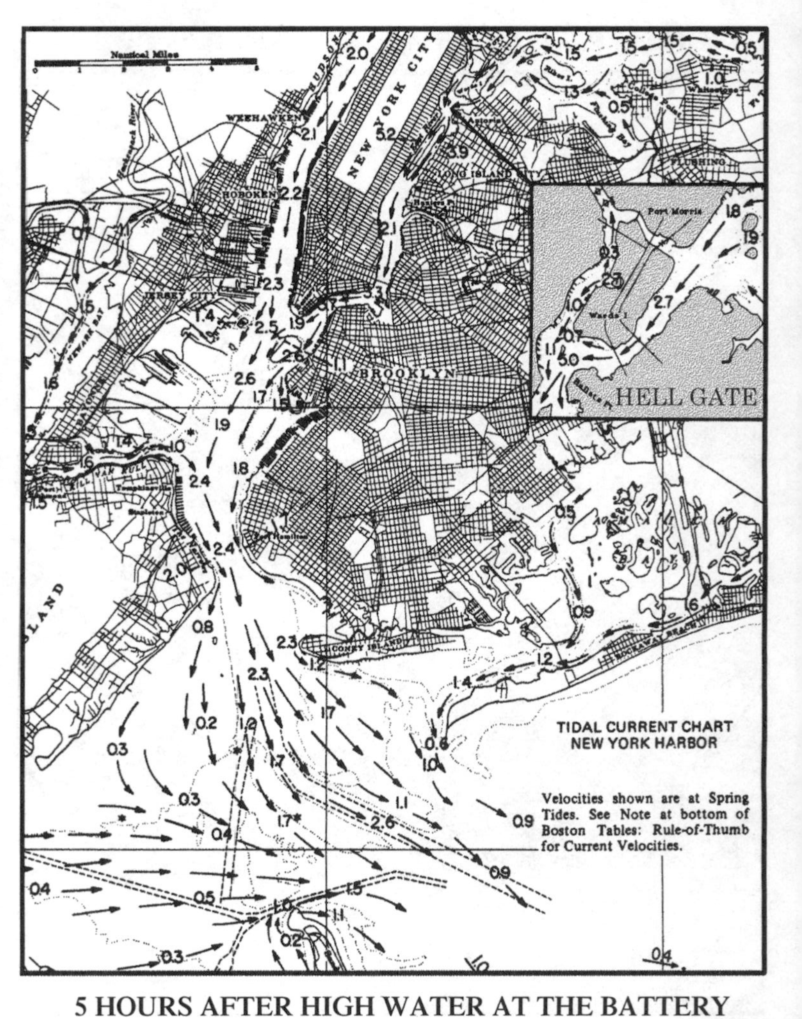

5 HOURS AFTER HIGH WATER AT THE BATTERY

NEW YORK BAY CURRENTS

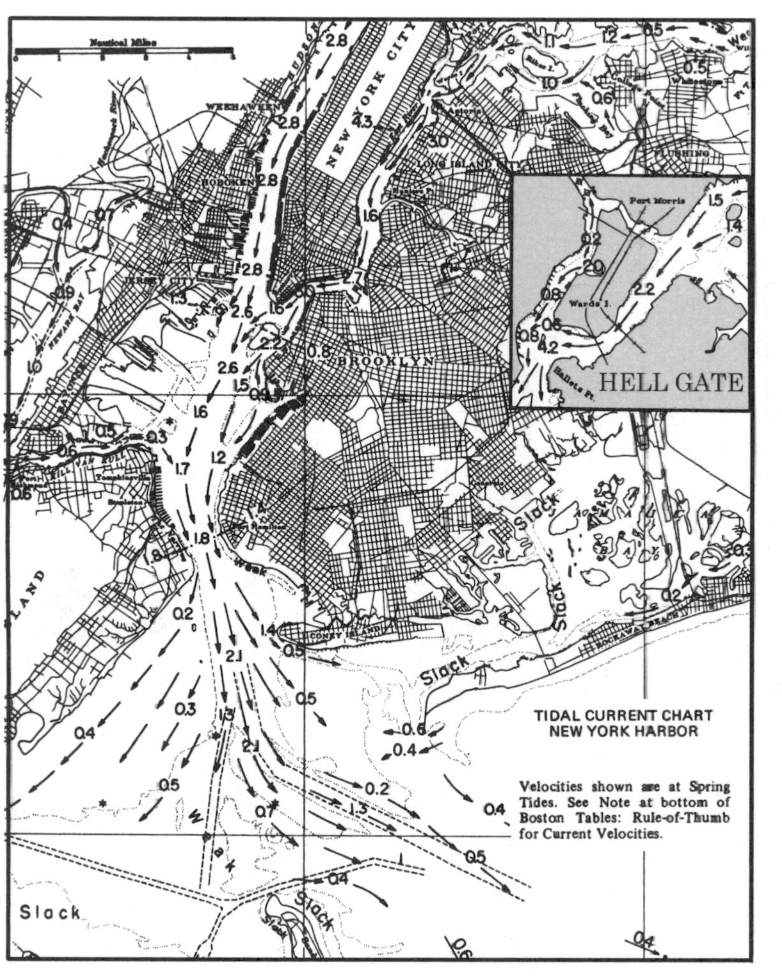

LOW WATER AT THE BATTERY

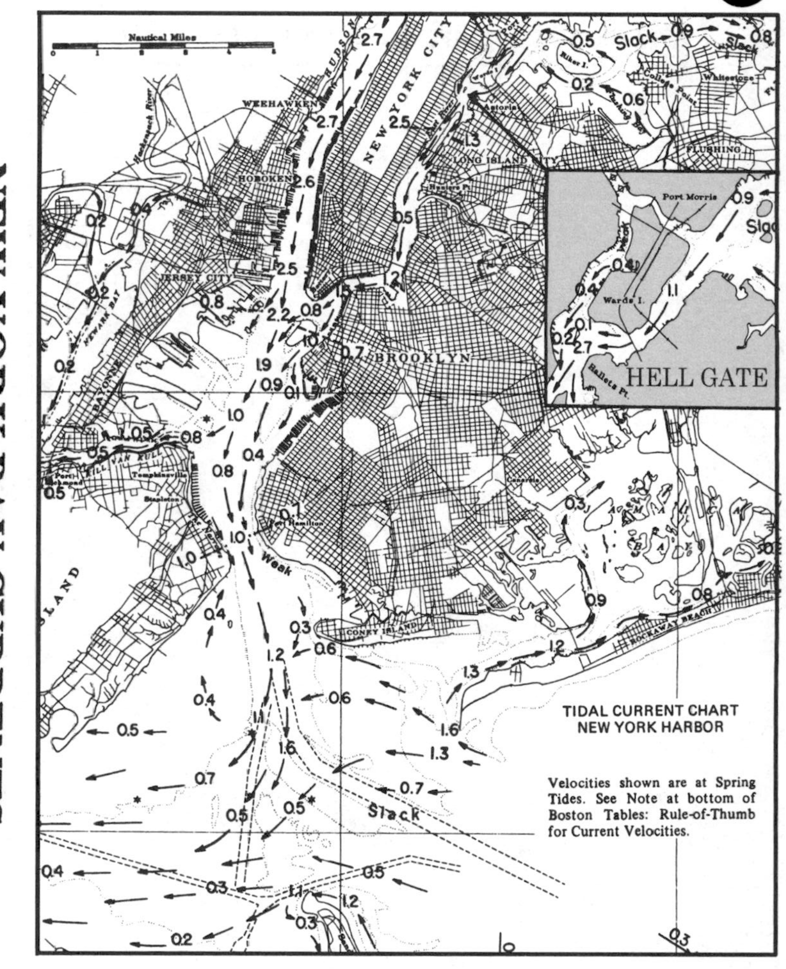

1 HOUR AFTER LOW WATER AT THE BATTERY

NEW YORK BAY CURRENTS

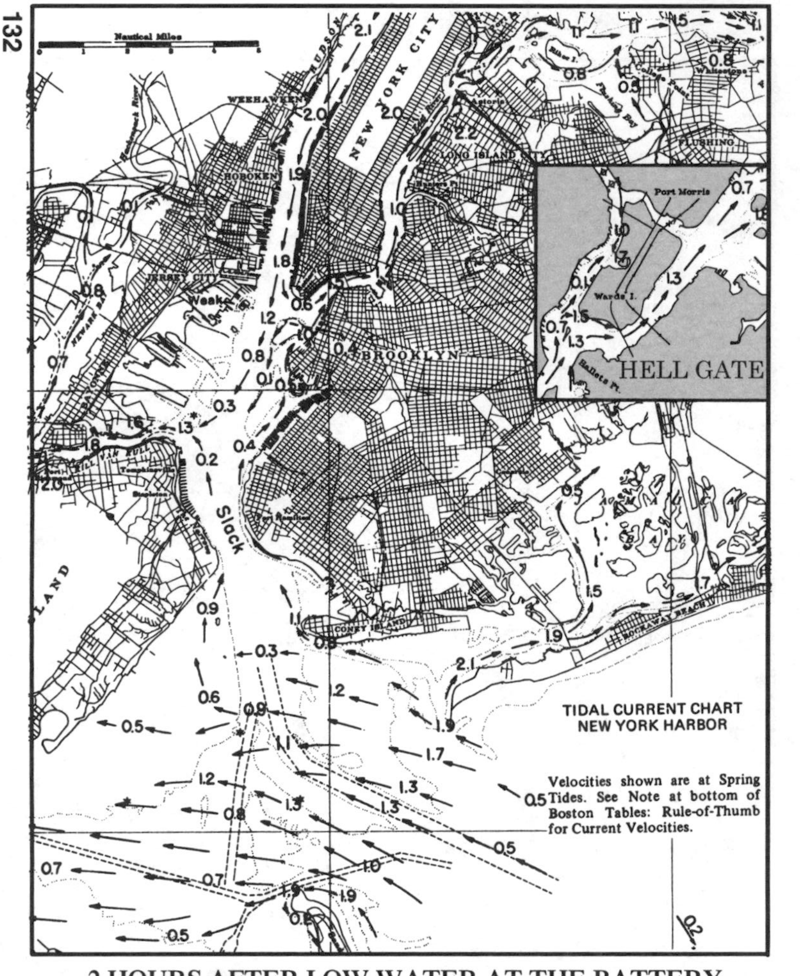

2 HOURS AFTER LOW WATER AT THE BATTERY

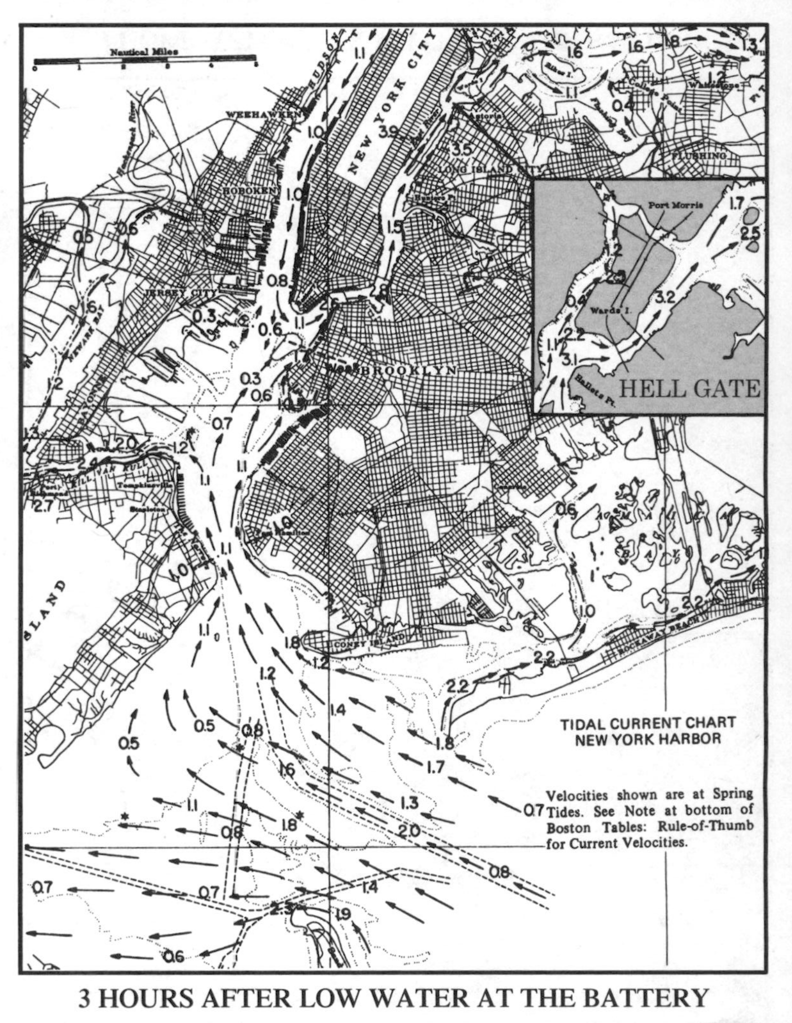

3 HOURS AFTER LOW WATER AT THE BATTERY

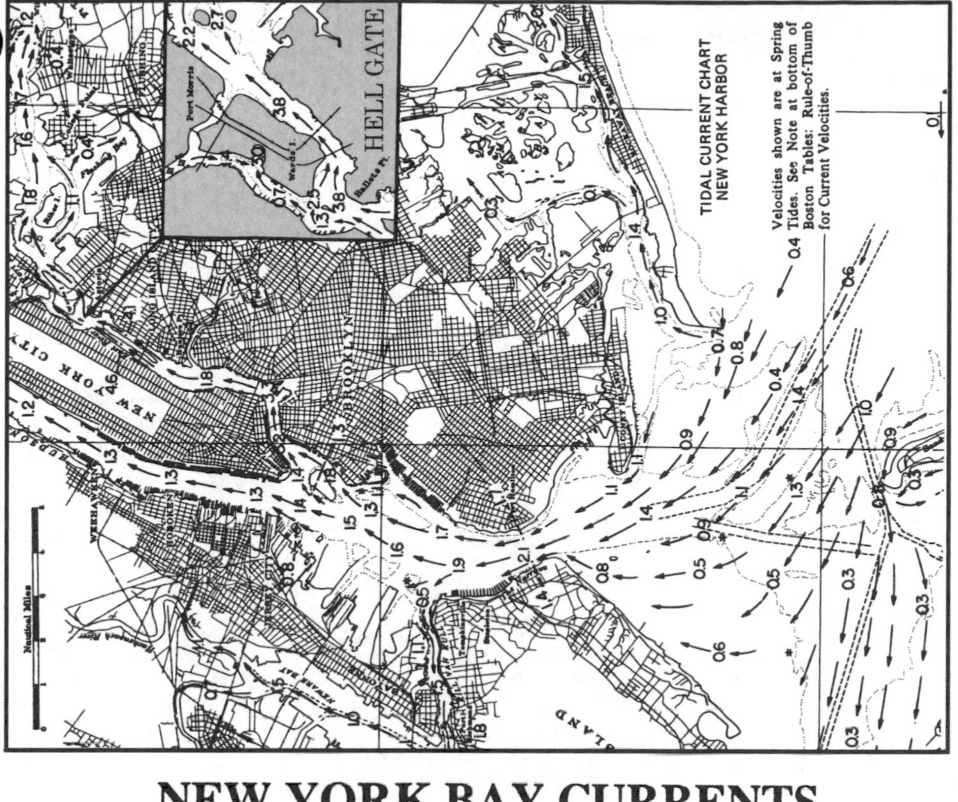

5 HOURS AFTER LOW WATER AT THE BATTERY

NEW YORK BAY CURRENTS

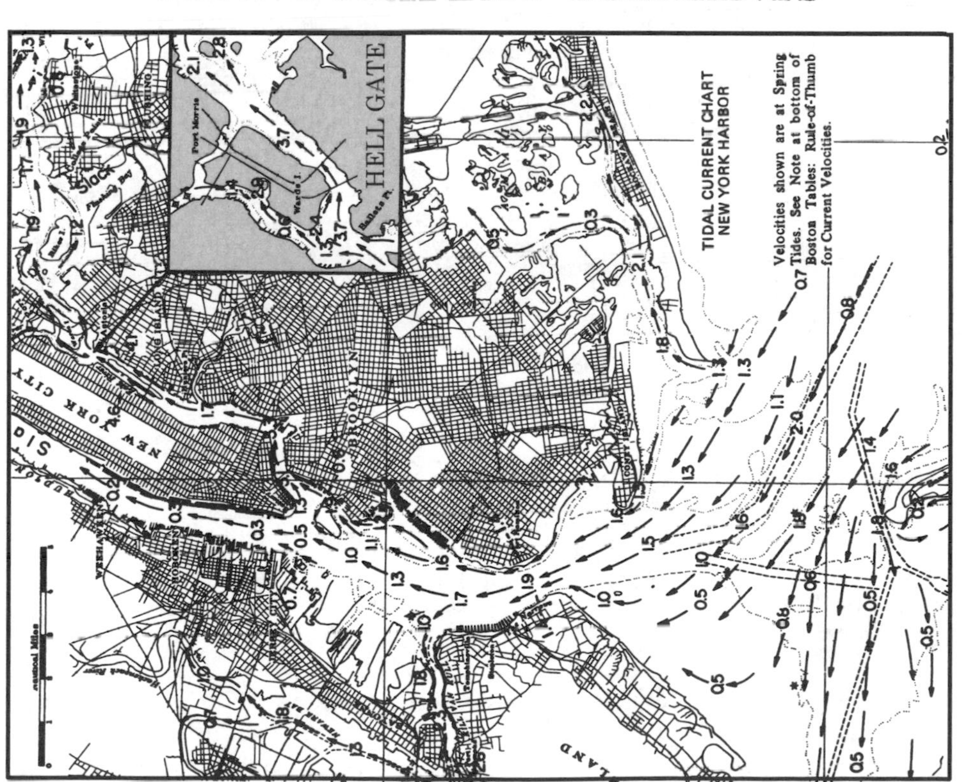

4 HOURS AFTER LOW WATER AT THE BATTERY

2014 HIGH & LOW WATER SANDY HOOK, NJ

40°28'N, 74°00.6'W

Standard Time **Standard Time**

DAY OF MONTH	DAY OF WEEK	JANUARY						DAY OF MONTH	DAY OF WEEK	FEBRUARY					
		HIGH				LOW				HIGH				LOW	
		a.m.	Ht.	**p.m.**	Ht.	a.m.	**p.m.**			a.m.	Ht.	**p.m.**	Ht.	a.m.	**p.m.**
1	**W**	7 12	6.1	**7 39**	5.1	1 03	**1 48**	1	**S**	8 38	6.0	**9 07**	5.5	2 35	**3 06**
2	**T**	8 03	6.2	**8 32**	5.2	1 57	**2 39**	2	**S**	9 29	5.8	**10 00**	5.4	3 26	**3 52**
3	**F**	8 55	6.1	**9 27**	5.1	2 50	**3 28**	3	**M**	10 22	5.4	**10 53**	5.2	4 16	**4 38**
4	**S**	9 50	5.8	**10 25**	5.1	3 43	**4 18**	4	**T**	11 16	5.0	**11 47**	5.0	5 08	**5 26**
5	**S**	10 45	5.5	**11 21**	5.0	4 34	**5 06**	5	**W**	...	...	**12 08**	4.6	6 02	**6 16**
6	**M**	11 40	5.1	**...**	...	5 30	**5 58**	6	**T**	12 38	4.8	**1 01**	4.3	7 04	**7 13**
7	**T**	12 17	4.9	**12 35**	4.8	6 31	**6 54**	7	**F**	1 31	4.6	**1 55**	4.0	8 10	**8 14**
8	**W**	1 11	4.8	**1 30**	4.4	7 38	**7 54**	8	**S**	2 24	4.4	**2 52**	3.8	9 12	**9 12**
9	**T**	2 05	4.7	**2 25**	4.2	8 44	**8 52**	9	**S**	3 21	4.3	**3 52**	3.8	10 07	**10 05**
10	**F**	3 00	4.6	**3 23**	4.0	9 43	**9 45**	10	**M**	4 18	4.4	**4 49**	3.9	10 56	**10 54**
11	**S**	3 57	4.6	**4 21**	4.0	10 36	**10 34**	11	**T**	5 11	4.5	**5 40**	4.1	11 41	**11 40**
12	**S**	4 51	4.7	**5 16**	4.0	11 24	**11 20**	12	**W**	5 58	4.7	**6 25**	4.3	...	**12 24**
13	**M**	5 39	4.8	**6 05**	4.1	...	**12 10**	13	**T**	6 39	4.8	**7 06**	4.5	12 25	**1 05**
14	**T**	6 23	4.9	**6 49**	4.3	12 05	**12 53**	14	**F**	7 17	4.9	**7 43**	4.6	1 08	**1 44**
15	**W**	7 03	5.0	**7 30**	4.3	12 49	**1 34**	15	**S**	7 52	4.9	**8 19**	4.6	1 50	**2 21**
16	**T**	7 40	5.0	**8 10**	4.3	1 31	**2 14**	16	**S**	8 25	4.9	**8 54**	4.6	2 29	**2 56**
17	**F**	8 16	4.9	**8 48**	4.3	2 12	**2 51**	17	**M**	8 58	4.8	**9 28**	4.6	3 07	**3 28**
18	**S**	8 51	4.8	**9 26**	4.2	2 50	**3 25**	18	**T**	9 33	4.6	**10 04**	4.6	3 43	**3 59**
19	**S**	9 24	4.6	**10 03**	4.2	3 27	**3 58**	19	**W**	10 12	4.5	**10 45**	4.6	4 20	**4 30**
20	**M**	9 59	4.5	**10 42**	4.2	4 03	**4 29**	20	**T**	10 58	4.3	**11 33**	4.7	5 00	**5 06**
21	**T**	10 38	4.3	**11 23**	4.2	4 40	**5 00**	21	**F**	11 51	4.2	**...**	...	5 50	**5 54**
22	**W**	11 23	4.2	**...**	...	5 21	**5 37**	22	**S**	12 27	4.7	**12 51**	4.1	6 59	**7 05**
23	**T**	12 08	4.3	**12 14**	4.1	6 15	**6 27**	23	**S**	1 27	4.8	**1 55**	4.1	8 17	**8 27**
24	**F**	12 58	4.4	**1 11**	4.0	7 28	**7 37**	24	**M**	2 32	4.9	**3 05**	4.2	9 26	**9 38**
25	**S**	1 54	4.6	**2 14**	4.0	8 43	**8 51**	25	**T**	3 41	5.1	**4 15**	4.5	10 26	**10 40**
26	**S**	2 57	4.8	**3 23**	4.1	9 48	**9 55**	26	**W**	4 47	5.4	**5 19**	4.9	11 22	**11 38**
27	**M**	4 03	5.1	**4 32**	4.3	10 47	**10 55**	27	**T**	5 46	5.7	**6 15**	5.3	...	**12 15**
28	**T**	5 06	5.5	**5 35**	4.6	11 43	**11 52**	28	**F**	6 40	5.9	**7 07**	5.6	12 34	**1 06**
29	**W**	6 03	5.8	**6 31**	5.0	...	**12 37**								
30	**T**	6 56	6.0	**7 24**	5.3	12 48	**1 29**								
31	**F**	7 47	6.1	**8 15**	5.4	1 43	**2 19**								

Dates when Ht. of **Low** Water is below Mean Lower Low with Ht. of lowest given for each period and Date of lowest in ( ):

1st - 7th: -1.3' (2nd - 3rd)
14th - 18th: -0.3' (16th - 17th)
27th - 31st: -1.4' (31st)

1st - 5th: -1.4' (1st)
14th - 17th: -0.3' (15th - 16th)
25th - 28th: -1.1' (28th)

Average Rise and Fall 4.6 ft.

When a high tide exceeds avg. ht., the *following* low tide will be lower than avg.

South Shore / N.Y. Harbor

2014 HIGH & LOW WATER SANDY HOOK, NJ

40°28'N, 74°00.6'W

***Daylight Time starts March 9 at 2 a.m.** **Daylight Saving Time**

DAY OF MONTH	DAY OF WEEK	MARCH						DAY OF MONTH	DAY OF WEEK	APRIL					
		HIGH				LOW				HIGH				LOW	
		a.m.	Ht.	p.m.	Ht.	a.m.	p.m.			a.m.	Ht.	p.m.	Ht.	a.m.	p.m.
1	S	7 30	6.0	7 56	5.8	1 28	1 55	1	T	9 44	5.5	10 05	5.8	3 49	3 57
2	S	8 18	5.9	8 44	5.8	2 19	2 41	2	W	10 32	5.2	10 51	5.5	4 34	4 39
3	M	9 07	5.7	9 33	5.6	3 08	3 25	3	T	11 22	4.8	11 38	5.2	5 18	5 20
4	T	9 58	5.3	10 23	5.4	3 56	4 09	4	F	...	...	12 14	4.5	6 04	6 03
5	W	10 49	4.9	11 13	5.1	4 42	4 52	5	S	12 27	4.9	1 05	4.3	6 51	6 49
6	T	11 41	4.6	...	...	5 32	5 37	6	S	1 16	4.6	1 56	4.1	7 47	7 47
7	F	12 03	4.8	12 33	4.2	6 27	6 30	7	M	2 06	4.4	2 49	4.0	8 49	8 54
8	S	12 54	4.5	1 26	4.0	7 30	7 32	8	T	2 59	4.3	3 43	4.1	9 49	9 57
9	S	1 46	4.3	*3 21	3.9	*9 34	*9 36	9	W	3 54	4.2	4 38	4.2	10 41	10 52
10	M	3 41	4.2	4 19	3.9	10 32	10 34	10	T	4 50	4.3	5 31	4.4	11 27	11 42
11	T	4 39	4.2	5 16	4.0	11 22	11 26	11	F	5 44	4.4	6 18	4.7	...	12 10
12	W	5 35	4.4	6 09	4.2	...	12 07	12	S	6 31	4.6	7 01	5.0	12 28	12 51
13	T	6 25	4.6	6 55	4.5	12 13	12 50	13	S	7 14	4.8	7 39	5.3	1 14	1 32
14	F	7 09	4.8	7 36	4.8	12 59	1 31	14	M	7 54	4.9	8 16	5.5	1 58	2 12
15	S	7 48	4.9	8 13	4.9	1 43	2 10	15	T	8 32	5.0	8 52	5.6	2 42	2 52
16	S	8 24	5.0	8 48	5.1	2 25	2 48	16	W	9 11	5.0	9 30	5.7	3 25	3 31
17	M	8 59	4.9	9 22	5.2	3 06	3 24	17	T	9 54	4.9	10 13	5.7	4 08	4 11
18	T	9 34	4.9	9 56	5.2	3 46	3 59	18	F	10 41	4.8	11 02	5.6	4 51	4 53
19	W	10 11	4.8	10 34	5.2	4 25	4 33	19	S	11 36	4.7	11 57	5.4	5 37	5 40
20	T	10 54	4.6	11 18	5.1	5 04	5 08	20	S	...	...	12 35	4.6	6 29	6 35
21	F	11 44	4.5	...	...	5 47	5 48	21	M	12 57	5.3	1 36	4.6	7 30	7 45
22	S	12 10	5.1	12 40	4.4	6 38	6 39	22	T	1 58	5.2	2 37	4.7	8 38	9 03
23	S	1 08	5.0	1 41	4.3	7 42	7 51	23	W	3 00	5.2	3 39	4.9	9 44	10 12
24	M	2 10	5.0	2 45	4.3	8 57	9 14	24	T	4 03	5.1	4 42	5.1	10 43	11 14
25	T	3 15	5.0	3 52	4.5	10 06	10 25	25	F	5 06	5.2	5 42	5.4	11 36	...
26	W	4 22	5.1	4 59	4.8	11 06	11 27	26	S	6 06	5.3	6 37	5.7	12 10	12 26
27	T	5 28	5.3	6 02	5.2	...	12 01	27	S	7 00	5.4	7 26	5.9	1 03	1 14
28	F	6 27	5.5	6 57	5.6	12 25	12 52	28	M	7 49	5.4	8 11	6.0	1 54	2 01
29	S	7 21	5.7	7 47	5.9	1 19	1 41	29	T	8 36	5.3	8 54	6.0	2 42	2 46
30	S	8 10	5.8	8 34	6.0	2 11	2 29	30	W	9 22	5.2	9 37	5.8	3 29	3 30
31	M	8 57	5.7	9 19	6.0	3 01	3 14								

Dates when Ht. of **Low** Water is below Mean Lower Low with Ht. of lowest given for each period and Date of lowest in ( ):

1st - 5th: -1.2' (1st - 2nd)
17th - 19th: -0.2'
27th - 31st: -0.9' (30th)

1st - 3rd: -0.7' (1st)
15th - 18th: -0.3' (16th - 17th)
25th - 30th: -0.5' (27th - 29th)

Average Rise and Fall 4.6 ft.

When a high tide exceeds avg. ht., the *following* low tide will be lower than avg.

South Shore / N.Y. Harbor

2014 HIGH & LOW WATER SANDY HOOK, NJ

40°28'N, 74°00.6'W

Daylight Saving Time | **Daylight Saving Time**

DAY OF MONTH	DAY OF WEEK	MAY						DAY OF MONTH	DAY OF WEEK	JUNE					
		HIGH				LOW				HIGH				LOW	
		a.m.	Ht.	**p.m.**	Ht.	a.m.	**p.m.**			a.m.	Ht.	**p.m.**	Ht.	a.m.	**p.m.**
1	**T**	10 08	5.0	**10 20**	5.5	4 12	**4 11**	1	**S**	11 18	4.5	**11 18**	5.0	5 10	**5 05**
2	**F**	10 57	4.7	**11 05**	5.2	4 54	**4 51**	2	**M**	...	...	**12 06**	4.4	5 48	**5 45**
3	**S**	11 46	4.5	**11 51**	4.9	5 36	**5 32**	3	**T**	12 02	4.8	**12 53**	4.3	6 28	**6 29**
4	**S**	...	...	**12 37**	4.3	6 20	**6 16**	4	**W**	12 47	4.6	**1 39**	4.3	7 12	**7 23**
5	**M**	12 39	4.7	**1 26**	4.2	7 06	**7 05**	5	**T**	1 31	4.4	**2 22**	4.4	8 01	**8 26**
6	**T**	1 27	4.5	**2 15**	4.2	8 00	**8 08**	6	**F**	2 17	4.3	**3 08**	4.5	8 56	**9 30**
7	**W**	2 15	4.3	**3 04**	4.2	8 58	**9 14**	7	**S**	3 06	4.2	**3 56**	4.7	9 49	**10 28**
8	**T**	3 05	4.3	**3 54**	4.4	9 52	**10 14**	8	**S**	4 01	4.2	**4 46**	5.0	10 39	**11 21**
9	**F**	3 58	4.3	**4 45**	4.6	10 40	**11 06**	9	**M**	5 00	4.3	**5 38**	5.3	11 27	**...**
10	**S**	4 53	4.3	**5 34**	4.9	11 25	**11 55**	10	**T**	5 57	4.5	**6 28**	5.6	12 12	**12 15**
11	**S**	5 47	4.5	**6 20**	5.2	...	**12 08**	11	**W**	6 51	4.7	**7 16**	5.9	1 03	**1 04**
12	**M**	6 36	4.6	**7 03**	5.5	12 43	**12 52**	12	**T**	7 42	4.9	**8 04**	6.2	1 54	**1 56**
13	**T**	7 22	4.8	**7 45**	5.8	1 30	**1 36**	13	**F**	8 32	5.1	**8 52**	6.3	2 44	**2 48**
14	**W**	8 06	4.9	**8 26**	6.0	2 18	**2 21**	14	**S**	9 23	5.2	**9 43**	6.3	3 34	**3 39**
15	**T**	8 51	5.0	**9 09**	6.0	3 05	**3 08**	15	**S**	10 17	5.2	**10 37**	6.1	4 23	**4 31**
16	**F**	9 38	5.0	**9 57**	6.0	3 51	**3 54**	16	**M**	11 15	5.2	**11 34**	5.9	5 11	**5 23**
17	**S**	10 31	5.0	**10 49**	5.9	4 38	**4 42**	17	**T**	...	...	**12 14**	5.2	6 01	**6 19**
18	**S**	11 28	4.9	**11 47**	5.7	5 26	**5 33**	18	**W**	12 31	5.7	**1 11**	5.3	6 54	**7 22**
19	**M**	...	...	**12 28**	4.9	6 18	**6 30**	19	**T**	1 28	5.4	**2 07**	5.3	7 52	**8 30**
20	**T**	12 46	5.5	**1 27**	5.0	7 15	**7 36**	20	**F**	2 24	5.1	**3 02**	5.3	8 52	**9 38**
21	**W**	1 45	5.4	**2 25**	5.0	8 17	**8 49**	21	**S**	3 21	4.9	**3 58**	5.4	9 50	**10 39**
22	**T**	2 43	5.2	**3 23**	5.2	9 20	**9 57**	22	**S**	4 19	4.7	**4 55**	5.4	10 44	**11 34**
23	**F**	3 43	5.0	**4 22**	5.3	10 18	**10 58**	23	**M**	5 19	4.6	**5 50**	5.5	11 34	**...**
24	**S**	4 43	5.0	**5 20**	5.5	11 11	**11 53**	24	**T**	6 16	4.6	**6 41**	5.5	12 26	**12 22**
25	**S**	5 43	4.9	**6 14**	5.7	...	**12 01**	25	**W**	7 07	4.7	**7 26**	5.6	1 14	**1 09**
26	**M**	6 38	5.0	**7 03**	5.8	12 45	**12 48**	26	**T**	7 54	4.7	**8 08**	5.6	2 01	**1 55**
27	**T**	7 28	5.0	**7 48**	5.9	1 35	**1 34**	27	**F**	8 38	4.7	**8 48**	5.5	2 45	**2 39**
28	**W**	8 15	5.0	**8 31**	5.8	2 22	**2 20**	28	**S**	9 22	4.7	**9 28**	5.4	3 26	**3 21**
29	**T**	9 00	4.9	**9 12**	5.7	3 08	**3 03**	29	**S**	10 05	4.6	**10 06**	5.2	4 05	**4 02**
30	**F**	9 45	4.8	**9 53**	5.5	3 50	**3 45**	30	**M**	10 48	4.6	**10 46**	5.0	4 42	**4 40**
31	**S**	10 31	4.6	**10 35**	5.2	4 31	**4 26**								

Dates when Ht. of **Low** Water is below Mean Lower Low with Ht. of lowest given for each period and Date of lowest in ( ):

1st: -0.3'
14th - 19th: -0.5' (16th - 17th)
25th: -0.2'
27th - 29th: -0.2'

12th - 19th: -0.7' (15th - 16th)

Average Rise and Fall 4.6 ft.

When a high tide exceeds avg. ht., the *following* low tide will be lower than avg.

South Shore / N.Y. Harbor

2014 HIGH & LOW WATER
SANDY HOOK, NJ

40°28'N, 74°00.6'W

Daylight Saving Time **Daylight Saving Time**

DAY OF MONTH	DAY OF WEEK	JULY						DAY OF MONTH	DAY OF WEEK	AUGUST					
		HIGH				LOW				HIGH				LOW	
		a.m.	Ht.	**p.m.**	Ht.	a.m.	**p.m.**			a.m.	Ht.	**p.m.**	Ht.	a.m.	**p.m.**
1	**T**	11 32	4.5	**11 25**	4.8	5 18	**5 18**	1	**F**	...	...	**12 12**	4.7	5 47	**6 09**
2	**W**	...	...	**12 15**	4.4	5 52	**5 58**	2	**S**	12 08	4.5	**12 53**	4.7	6 20	**6 57**
3	**T**	12 06	4.6	**12 57**	4.5	6 27	**6 42**	3	**S**	12 54	4.4	**1 37**	4.8	7 00	**8 00**
4	**F**	12 49	4.5	**1 39**	4.5	7 07	**7 38**	4	**M**	1 47	4.3	**2 29**	5.0	8 01	**9 14**
5	**S**	1 32	4.4	**2 21**	4.6	7 54	**8 43**	5	**T**	2 44	4.3	**3 25**	5.2	9 14	**10 19**
6	**S**	2 21	4.3	**3 08**	4.8	8 52	**9 49**	6	**W**	3 49	4.3	**4 28**	5.4	10 22	**11 18**
7	**M**	3 16	4.2	**4 01**	5.0	9 53	**10 48**	7	**T**	4 58	4.5	**5 32**	5.7	11 23	**...**
8	**T**	4 18	4.3	**4 59**	5.3	10 50	**11 43**	8	**F**	6 03	4.9	**6 32**	6.0	12 14	**12 21**
9	**W**	5 24	4.4	**5 57**	5.7	11 45	**...**	9	**S**	7 02	5.2	**7 27**	6.3	1 08	**1 18**
10	**T**	6 25	4.7	**6 52**	6.0	12 37	**12 40**	10	**S**	7 56	5.6	**8 18**	6.5	2 00	**2 14**
11	**F**	7 21	5.0	**7 45**	6.3	1 30	**1 35**	11	**M**	8 48	5.8	**9 09**	6.5	2 51	**3 08**
12	**S**	8 14	5.3	**8 36**	6.4	2 23	**2 30**	12	**T**	9 39	6.0	**10 01**	6.3	3 40	**4 01**
13	**S**	9 06	5.5	**9 27**	6.4	3 14	**3 24**	13	**W**	10 33	6.0	**10 54**	6.0	4 27	**4 52**
14	**M**	10 00	5.6	**10 20**	6.3	4 03	**4 17**	14	**T**	11 27	5.9	**11 49**	5.6	5 14	**5 44**
15	**T**	10 56	5.6	**11 16**	6.0	4 51	**5 09**	15	**F**	...	...	**12 22**	5.7	6 01	**6 39**
16	**W**	11 53	5.6	**...**	...	5 39	**6 04**	16	**S**	12 44	5.2	**1 16**	5.5	6 51	**7 40**
17	**T**	12 12	5.7	**12 49**	5.5	6 29	**7 02**	17	**S**	1 39	4.9	**2 09**	5.3	7 47	**8 46**
18	**F**	1 07	5.4	**1 44**	5.5	7 23	**8 07**	18	**M**	2 34	4.6	**3 03**	5.1	8 49	**9 51**
19	**S**	2 02	5.0	**2 37**	5.4	8 21	**9 14**	19	**T**	3 31	4.4	**3 58**	5.0	9 50	**10 48**
20	**S**	2 58	4.7	**3 32**	5.3	9 21	**10 17**	20	**W**	4 29	4.3	**4 55**	5.0	10 45	**11 38**
21	**M**	3 55	4.5	**4 28**	5.2	10 18	**11 13**	21	**T**	5 27	4.4	**5 49**	5.0	11 35	**...**
22	**T**	4 55	4.4	**5 24**	5.2	11 10	**11 59**	22	**F**	6 20	4.6	**6 38**	5.2	12 23	**12 22**
23	**W**	5 53	4.4	**6 17**	5.3	11 59	**...**	23	**S**	7 07	4.8	**7 20**	5.3	1 06	**1 07**
24	**T**	6 45	4.5	**7 04**	5.4	12 51	**12 46**	24	**S**	7 49	4.9	**7 59**	5.3	1 47	**1 51**
25	**F**	7 32	4.7	**7 46**	5.4	1 36	**1 31**	25	**M**	8 28	5.0	**8 36**	5.3	2 27	**2 33**
26	**S**	8 15	4.8	**8 25**	5.4	2 18	**2 15**	26	**T**	9 05	5.1	**9 10**	5.2	3 04	**3 13**
27	**S**	8 56	4.8	**9 03**	5.4	2 59	**2 58**	27	**W**	9 40	5.1	**9 44**	5.1	3 39	**3 52**
28	**M**	9 36	4.8	**9 39**	5.2	3 37	**3 38**	28	**T**	10 15	5.0	**10 17**	4.9	4 12	**4 29**
29	**T**	10 15	4.8	**10 14**	5.1	4 12	**4 16**	29	**F**	10 50	5.0	**10 53**	4.7	4 43	**5 06**
30	**W**	10 54	4.7	**10 49**	4.9	4 46	**4 53**	30	**S**	11 28	5.0	**11 36**	4.6	5 13	**5 44**
31	**T**	11 33	4.7	**11 27**	4.7	5 17	**5 30**	31	**S**	...	...	**12 12**	5.0	5 46	**6 29**

Dates when Ht. of **Low** Water is below Mean Lower Low with Ht. of lowest given for each period and Date of lowest in ( ):

11th - 18th: -0.9' (14th - 15th)

9th - 15th: -1.0' (12th)

Average Rise and Fall 4.6 ft.

When a high tide exceeds avg. ht., the *following* low tide will be lower than avg.

2014 HIGH & LOW WATER
SANDY HOOK, NJ

40°28'N, 74°00.6'W

DAY OF MONTH	DAY OF WEEK	SEPTEMBER (Daylight Saving Time)						DAY OF MONTH	DAY OF WEEK	OCTOBER (Daylight Saving Time)					
		HIGH				LOW				HIGH				LOW	
		a.m.	Ht.	p.m.	Ht.	a.m.	p.m.			a.m.	Ht.	p.m.	Ht.	a.m.	p.m.
1	M	12 27	4.4	1 02	5.0	6 26	7 29	1	W	1 11	4.4	1 39	5.2	7 11	8 23
2	T	1 24	4.4	1 59	5.1	7 25	8 44	2	T	2 15	4.5	2 42	5.2	8 34	9 34
3	W	2 25	4.4	3 00	5.2	8 48	9 55	3	F	3 19	4.6	3 47	5.4	9 51	10 35
4	T	3 32	4.5	4 06	5.4	10 05	10 57	4	S	4 26	4.9	4 53	5.5	10 56	11 31
5	F	4 40	4.7	5 11	5.7	11 08	11 52	5	S	5 28	5.3	5 53	5.7	11 53	...
6	S	5 45	5.1	6 13	6.0	...	12 07	6	M	6 26	5.7	6 49	5.9	12 21	12 48
7	S	6 44	5.5	7 08	6.2	12 45	1 03	7	T	7 18	6.1	7 40	6.0	1 11	1 42
8	M	7 37	5.9	8 00	6.4	1 36	1 58	8	W	8 06	6.3	8 29	6.0	2 00	2 34
9	T	8 27	6.2	8 49	6.3	2 26	2 51	9	T	8 53	6.3	9 18	5.8	2 47	3 24
10	W	9 17	6.3	9 39	6.1	3 14	3 43	10	F	9 40	6.2	10 07	5.5	3 33	4 12
11	T	10 07	6.2	10 31	5.8	4 00	4 33	11	S	10 29	5.9	10 59	5.1	4 18	5 00
12	F	10 58	6.0	11 24	5.4	4 46	5 22	12	S	11 19	5.6	11 53	4.8	5 02	5 47
13	S	11 51	5.7	...	...	5 31	6 14	13	M	...	...	12 11	5.2	5 47	6 38
14	S	12 19	5.0	12 45	5.4	6 19	7 10	14	T	12 48	4.5	1 03	4.9	6 36	7 34
15	M	1 15	4.7	1 38	5.1	7 12	8 13	15	W	1 42	4.3	1 55	4.7	7 33	8 36
16	T	2 09	4.5	2 31	4.9	8 13	9 18	16	T	2 35	4.2	2 47	4.6	8 39	9 36
17	W	3 05	4.3	3 25	4.8	9 18	10 16	17	F	3 29	4.3	3 41	4.5	9 42	10 27
18	T	4 01	4.3	4 21	4.7	10 17	11 06	18	S	4 23	4.4	4 34	4.5	10 37	11 12
19	F	4 58	4.4	5 16	4.8	11 09	11 51	19	S	5 15	4.6	5 26	4.6	11 26	11 54
20	S	5 51	4.6	6 06	4.9	11 56	...	20	M	6 02	4.8	6 14	4.7	...	12 12
21	S	6 38	4.9	6 50	5.1	12 32	12 41	21	T	6 45	5.1	6 57	4.8	12 34	12 56
22	M	7 19	5.1	7 30	5.2	1 12	1 24	22	W	7 24	5.3	7 36	4.9	1 13	1 40
23	T	7 57	5.3	8 07	5.2	1 51	2 07	23	T	8 00	5.5	8 13	4.9	1 52	2 23
24	W	8 33	5.4	8 42	5.2	2 29	2 48	24	F	8 35	5.6	8 50	4.9	2 31	3 05
25	T	9 06	5.4	9 15	5.0	3 05	3 28	25	S	9 10	5.6	9 29	4.8	3 10	3 47
26	F	9 39	5.4	9 50	4.9	3 40	4 07	26	S	9 48	5.6	10 12	4.7	3 48	4 29
27	S	10 14	5.3	10 29	4.7	4 13	4 46	27	M	10 32	5.5	11 03	4.5	4 28	5 13
28	S	10 54	5.3	11 15	4.6	4 47	5 26	28	T	11 25	5.4	...	...	5 11	6 01
29	M	11 42	5.2	...	...	5 23	6 12	29	W	12 01	4.5	12 24	5.3	6 01	6 57
30	T	12 10	4.4	12 39	5.2	6 08	7 11	30	T	1 03	4.5	1 26	5.2	7 05	8 03
								31	F	2 05	4.6	2 27	5.2	8 23	9 11

Dates when Ht. of **Low** Water is below Mean Lower Low with Ht. of lowest given for each period and Date of lowest in ( ):

7th - 12th: -0.9' (10th)

6th - 11th: -0.8' (8th)

Average Rise and Fall 4.6 ft.

When a high tide exceeds avg. ht., the *following* low tide will be lower than avg.

2014 HIGH & LOW WATER SANDY HOOK, NJ

40°28'N, 74°00.6'W

***Standard Time starts Nov. 2 at 2 a.m.** **Standard Time**

DAY OF MONTH	DAY OF WEEK	NOVEMBER HIGH a.m.	Ht.	**p.m.**	Ht.	LOW a.m.	**p.m.**
1	**S**	3 07	4.7	**3 29**	5.2	9 37	**10 12**
2	**S**	*3 09	5.0	***3 32**	5.2	*9 42	***10 07**
3	**M**	4 10	5.4	**4 33**	5.3	10 40	**10 58**
4	**T**	5 07	5.7	**5 30**	5.4	11 35	**11 48**
5	**W**	5 58	6.0	**6 21**	5.5	...	**12 26**
6	**T**	6 46	6.1	**7 10**	5.4	12 35	**1 17**
7	**F**	7 31	6.1	**7 57**	5.3	1 22	**2 05**
8	**S**	8 16	6.0	**8 45**	5.0	2 08	**2 52**
9	**S**	9 01	5.7	**9 34**	4.8	2 52	**3 37**
10	**M**	9 48	5.4	**10 26**	4.5	3 35	**4 21**
11	**T**	10 36	5.0	**11 18**	4.3	4 17	**5 05**
12	**W**	11 26	4.7	**...**	...	5 02	**5 53**
13	**T**	12 10	4.2	**12 16**	4.5	5 52	**6 47**
14	**F**	1 01	4.1	**1 05**	4.3	6 52	**7 45**
15	**S**	1 51	4.1	**1 55**	4.2	7 58	**8 39**
16	**S**	2 41	4.2	**2 47**	4.2	8 58	**9 27**
17	**M**	3 31	4.4	**3 40**	4.2	9 51	**10 11**
18	**T**	4 21	4.6	**4 31**	4.3	10 40	**10 53**
19	**W**	5 06	4.9	**5 20**	4.4	11 26	**11 35**
20	**T**	5 49	5.2	**6 04**	4.5	...	**12 12**
21	**F**	6 29	5.5	**6 46**	4.7	12 17	**12 57**
22	**S**	7 08	5.6	**7 28**	4.7	1 00	**1 43**
23	**S**	7 48	5.7	**8 12**	4.7	1 45	**2 28**
24	**M**	8 31	5.7	**8 59**	4.7	2 29	**3 13**
25	**T**	9 19	5.6	**9 53**	4.6	3 15	**3 59**
26	**W**	10 13	5.5	**10 52**	4.6	4 03	**4 48**
27	**T**	11 12	5.3	**11 52**	4.6	4 55	**5 41**
28	**F**	...	...	**12 12**	5.2	5 57	**6 42**
29	**S**	12 51	4.7	**1 11**	5.0	7 09	**7 46**
30	**S**	1 50	4.8	**2 10**	4.9	8 22	**8 47**

DAY OF MONTH	DAY OF WEEK	DECEMBER HIGH a.m.	Ht.	**p.m.**	Ht.	LOW a.m.	**p.m.**
1	**M**	2 50	5.0	**3 11**	4.8	9 27	**9 43**
2	**T**	3 49	5.2	**4 12**	4.8	10 25	**10 34**
3	**W**	4 46	5.4	**5 10**	4.8	11 19	**11 24**
4	**T**	5 40	5.6	**6 04**	4.9	...	**12 11**
5	**F**	6 27	5.7	**6 52**	4.9	12 12	**1 00**
6	**S**	7 11	5.7	**7 38**	4.8	12 59	**1 47**
7	**S**	7 54	5.6	**8 24**	4.7	1 45	**2 31**
8	**M**	8 36	5.4	**9 10**	4.5	2 28	**3 14**
9	**T**	9 19	5.1	**9 57**	4.3	3 10	**3 54**
10	**W**	10 03	4.9	**10 46**	4.2	3 51	**4 34**
11	**T**	10 49	4.6	**11 34**	4.1	4 31	**5 13**
12	**F**	11 34	4.3	**...**	...	5 14	**5 56**
13	**S**	12 21	4.0	**12 20**	4.1	6 04	**6 45**
14	**S**	1 07	4.0	**1 06**	4.0	7 05	**7 40**
15	**M**	1 54	4.1	**1 55**	3.9	8 11	**8 34**
16	**T**	2 42	4.2	**2 48**	3.8	9 11	**9 25**
17	**W**	3 32	4.4	**3 44**	3.9	10 05	**10 12**
18	**T**	4 23	4.7	**4 40**	4.0	10 55	**10 59**
19	**F**	5 13	5.0	**5 33**	4.2	11 44	**11 46**
20	**S**	5 59	5.3	**6 22**	4.4	...	**12 32**
21	**S**	6 44	5.6	**7 09**	4.6	12 35	**1 21**
22	**M**	7 30	5.8	**7 56**	4.8	1 24	**2 09**
23	**T**	8 17	5.8	**8 46**	4.8	2 14	**2 56**
24	**W**	9 07	5.8	**9 39**	4.8	3 03	**3 43**
25	**T**	10 00	5.6	**10 37**	4.8	3 53	**4 31**
26	**F**	10 57	5.4	**11 35**	4.8	4 46	**5 21**
27	**S**	11 55	5.1	**...**	...	5 44	**6 16**
28	**S**	12 33	4.9	**12 52**	4.8	6 51	**7 17**
29	**M**	1 29	4.9	**1 50**	4.6	8 02	**8 19**
30	**T**	2 27	4.9	**2 50**	4.4	9 09	**9 18**
31	**W**	3 26	5.0	**3 51**	4.3	10 08	**10 12**

Dates when Ht. of **Low** Water is below Mean Lower Low with Ht. of lowest given for each period and Date of lowest in ( ):

November:
2nd - 9th: -0.7' (6th)
21st - 27th: -0.5' (24th)
30th: -0.2'

December:
1st - 9th: -0.5' (2nd - 7th)
19th - 31st: -0.9' (23rd - 25th)

Average Rise and Fall 4.6 ft.

When a high tide exceeds avg. ht., the *following* low tide will be lower than avg.

2014 CURRENT TABLE
DELAWARE BAY ENTRANCE

38°46.85'N, 75°02.58'W

Standard Time

JANUARY

DAY OF MONTH	DAY OF WEEK	CURRENT TURNS TO NORTHWEST Flood Starts			CURRENT TURNS TO SOUTHEAST Ebb Starts		
		a.m.	**p.m.**	Kts.	a.m.	**p.m.**	Kts.
1	W	3 00	**3 45**	a2.1	9 12	**9 33**	a2.2
2	T	3 51	**4 35**	a2.1	10 02	**10 26**	a2.2
3	F	4 45	**5 26**	a2.1	10 53	**11 20**	a2.1
4	S	5 41	**6 21**	a2.0	11 46	**...**	2.1
5	S	6 39	**7 15**	a1.9	12 16	**12 42**	p2.0
6	M	7 40	**8 13**	1.8	1 16	**1 40**	p1.9
7	T	8 44	**9 12**	p1.8	2 17	**2 40**	1.8
8	W	9 49	**10 12**	p1.8	3 20	**3 42**	1.8
9	T	10 52	**11 09**	p1.8	4 21	**4 42**	a1.8
10	F	11 51	**...**	1.6	5 20	**5 40**	a1.8
11	S	12 04	**12 46**	a1.8	6 15	**6 35**	a1.8
12	S	12 55	**1 37**	a1.9	7 05	**7 24**	a1.9
13	M	1 42	**2 23**	a1.9	7 51	**8 09**	a1.9
14	T	2 26	**3 05**	a1.8	8 33	**8 50**	a1.9
15	W	3 06	**3 43**	a1.8	9 11	**9 28**	a1.8
16	T	3 43	**4 18**	a1.8	9 45	**10 03**	a1.8
17	F	4 18	**4 51**	a1.8	10 19	**10 38**	a1.8
18	S	4 53	**5 24**	a1.7	10 52	**11 13**	a1.8
19	S	5 29	**5 58**	1.7	11 28	**11 52**	a1.9
20	M	6 09	**6 36**	1.7	...	**12 07**	1.8
21	T	6 54	**7 19**	p1.7	12 35	**12 50**	p1.8
22	W	7 45	**8 07**	p1.7	1 23	**1 39**	p1.8
23	T	8 41	**9 00**	p1.7	2 15	**2 32**	1.7
24	F	9 43	**9 58**	p1.7	3 12	**3 30**	1.7
25	S	10 47	**10 58**	p1.8	4 12	**4 32**	1.7
26	S	11 50	**11 59**	p1.9	5 14	**5 35**	a1.8
27	M	...	**12 51**	1.6	6 14	**6 36**	a1.9
28	T	12 59	**1 47**	a1.9	7 12	**7 35**	a2.0
29	W	1 56	**2 41**	a2.0	8 07	**8 31**	a2.1
30	T	2 51	**3 32**	a2.1	9 00	**9 24**	a2.1
31	F	3 44	**4 21**	a2.1	9 51	**10 16**	a2.2

Standard Time

FEBRUARY

DAY OF MONTH	DAY OF WEEK	CURRENT TURNS TO NORTHWEST Flood Starts			CURRENT TURNS TO SOUTHEAST Ebb Starts		
		a.m.	**p.m.**	Kts.	a.m.	**p.m.**	Kts.
1	S	4 37	**5 10**	a2.1	10 41	**11 08**	a2.2
2	S	5 29	**5 59**	2.0	11 30	**...**	2.1
3	M	6 23	**6 50**	1.9	12 01	**12 21**	2.0
4	T	7 20	**7 43**	p1.9	12 53	**1 13**	1.9
5	W	8 17	**8 37**	p1.8	1 48	**2 08**	1.8
6	T	9 17	**9 34**	p1.7	2 46	**3 06**	a1.8
7	F	10 18	**10 32**	p1.7	3 45	**4 06**	a1.7
8	S	11 18	**11 29**	p1.7	4 43	**5 05**	a1.7
9	S	...	**12 14**	1.5	5 40	**6 02**	a1.7
10	M	12 24	**1 06**	a1.7	6 33	**6 55**	a1.8
11	T	1 14	**1 54**	a1.7	7 21	**7 43**	a1.8
12	W	2 00	**2 36**	a1.8	8 05	**8 26**	a1.8
13	T	2 42	**3 15**	a1.8	8 44	**9 04**	a1.8
14	F	3 21	**3 49**	a1.8	9 20	**9 39**	a1.9
15	S	3 56	**4 21**	a1.8	9 53	**10 13**	a1.9
16	S	4 30	**4 52**	1.8	10 26	**10 47**	a1.9
17	M	5 05	**5 24**	1.8	10 59	**11 23**	1.9
18	T	5 42	**6 00**	p1.8	11 36	**...**	1.9
19	W	6 25	**6 41**	p1.8	12 03	**12 18**	1.9
20	T	7 13	**7 28**	p1.8	12 48	**1 05**	a1.9
21	F	8 08	**8 22**	p1.8	1 40	**1 58**	a1.8
22	S	9 11	**9 24**	p1.7	2 37	**2 58**	a1.8
23	S	10 18	**10 32**	p1.7	3 40	**4 05**	a1.7
24	M	11 27	**11 41**	p1.8	4 47	**5 15**	1.7
25	T	...	**12 32**	1.6	5 54	**6 22**	a1.8
26	W	12 46	**1 32**	a1.9	6 56	**7 24**	1.9
27	T	1 47	**2 26**	a2.0	7 54	**8 21**	2.0
28	F	2 43	**3 16**	2.0	8 48	**9 14**	2.1

The Kts. (knots) columns show the **maximum** predicted velocities of the stronger one of the Flood Currents and the stronger one of the Ebb Currents for each day.
The letter "a" means the velocity shown should occur **after** the **a.m.** Current Change. The letter "p" means the velocity shown should occur **after** the **p.m.** Current Change (even if next morning). No "a" or "p" means a.m. and p.m. velocities are the same for that day.
Avg. Max. Velocity: Flood 1.8 Kts., Ebb 1.9 Kts.
Max. Flood 3 hrs. 5 min. after Flood Starts, ±15 min.
Max. Ebb 3 hrs. 5 min. after Ebb Starts, ±15 min.

See pp. 22-29 for Current Change at other points.

2014 CURRENT TABLE
DELAWARE BAY ENTRANCE

38°46.85'N, 75°02.58'W

***Daylight Time starts March 9 at 2 a.m.** **Daylight Saving Time**

MARCH

DAY OF MONTH	DAY OF WEEK	CURRENT TURNS TO NORTHWEST Flood Starts			CURRENT TURNS TO SOUTHEAST Ebb Starts		
		a.m.	**p.m.**	Kts.	a.m.	**p.m.**	Kts.
1	**S**	3 36	**4 04**	2.1	9 37	**10 03**	a2.2
2	**S**	4 26	**4 50**	2.1	10 25	**10 51**	2.1
3	**M**	5 15	**5 35**	p2.1	11 11	**11 38**	2.1
4	**T**	6 05	**6 22**	p2.0	11 57	**...**	2.0
5	**W**	6 54	**7 09**	p1.9	12 26	**12 45**	a2.0
6	**T**	7 46	**8 00**	p1.8	1 15	**1 35**	a1.8
7	**F**	8 42	**8 55**	p1.7	2 08	**2 29**	a1.7
8	**S**	9 40	**9 53**	p1.6	3 03	**3 28**	a1.7
9	**S**	*11 40	***11 52**	p1.6	*5 01	***5 28**	a1.6
10	**M**	...	**12 37**	1.4	5 59	**6 27**	a1.6
11	**T**	12 50	**1 30**	a1.6	6 54	**7 23**	a1.7
12	**W**	1 43	**2 19**	1.6	7 45	**8 12**	a1.7
13	**T**	2 32	**3 02**	1.7	8 32	**8 56**	a1.8
14	**F**	3 16	**3 41**	1.7	9 13	**9 36**	1.8
15	**S**	3 56	**4 16**	p1.8	9 50	**10 11**	a1.9
16	**S**	4 32	**4 48**	1.8	10 25	**10 45**	1.9
17	**M**	5 07	**5 19**	p1.9	10 58	**11 18**	1.9
18	**T**	5 41	**5 50**	p1.9	11 32	**11 54**	p2.0
19	**W**	6 19	**6 26**	p1.9	...	**12 09**	1.9
20	**T**	7 00	**7 08**	p1.9	12 34	**12 51**	a2.0
21	**F**	7 48	**7 57**	p1.8	1 19	**1 39**	a2.0
22	**S**	8 44	**8 55**	p1.8	2 11	**2 34**	a1.9
23	**S**	9 48	**10 02**	p1.7	3 10	**3 38**	a1.8
24	**M**	10 58	**11 16**	p1.7	4 16	**4 49**	a1.7
25	**T**	...	**12 09**	1.5	5 27	**6 03**	a1.7
26	**W**	12 30	**1 14**	1.7	6 37	**7 12**	a1.8
27	**T**	1 37	**2 14**	1.8	7 41	**8 14**	1.9
28	**F**	2 39	**3 07**	p2.0	8 39	**9 09**	2.0
29	**S**	3 34	**3 56**	p2.1	9 32	**10 00**	2.1
30	**S**	4 25	**4 42**	p2.1	10 21	**10 46**	p2.2
31	**M**	5 12	**5 26**	p2.1	11 06	**11 30**	2.1

APRIL

DAY OF MONTH	DAY OF WEEK	CURRENT TURNS TO NORTHWEST Flood Starts			CURRENT TURNS TO SOUTHEAST Ebb Starts		
		a.m.	**p.m.**	Kts.	a.m.	**p.m.**	Kts.
1	**T**	5 58	**6 08**	p2.1	11 49	**...**	2.0
2	**W**	6 43	**6 50**	p2.0	12 13	**12 32**	a2.1
3	**T**	7 28	**7 34**	p1.8	12 56	**1 15**	a2.0
4	**F**	8 17	**8 23**	p1.7	1 41	**2 02**	a1.8
5	**S**	9 07	**9 14**	p1.6	2 28	**2 54**	a1.7
6	**S**	10 01	**10 12**	p1.5	3 20	**3 50**	a1.7
7	**M**	10 58	**11 13**	p1.5	4 16	**4 50**	a1.6
8	**T**	11 55	**...**	1.4	5 14	**5 50**	a1.6
9	**W**	12 13	**12 49**	1.5	6 11	**6 46**	a1.6
10	**T**	1 09	**1 38**	p1.6	7 04	**7 37**	a1.7
11	**F**	2 00	**2 22**	p1.7	7 53	**8 22**	a1.8
12	**S**	2 46	**3 02**	p1.8	8 37	**9 03**	1.8
13	**S**	3 28	**3 38**	p1.9	9 16	**9 40**	p1.9
14	**M**	4 06	**4 12**	p1.9	9 53	**10 15**	p2.0
15	**T**	4 42	**4 44**	p2.0	10 28	**10 50**	p2.0
16	**W**	5 19	**5 19**	p2.0	11 05	**11 28**	p2.1
17	**T**	5 58	**5 58**	p2.0	11 45	**...**	1.9
18	**F**	6 42	**6 43**	p2.0	12 10	**12 29**	a2.1
19	**S**	7 31	**7 35**	p1.9	12 56	**1 20**	a2.0
20	**S**	8 28	**8 37**	p1.7	1 50	**2 20**	a1.9
21	**M**	9 33	**9 49**	p1.7	2 50	**3 27**	a1.8
22	**T**	10 42	**11 06**	1.6	3 58	**4 40**	a1.8
23	**W**	11 51	**...**	1.6	5 10	**5 53**	a1.8
24	**T**	12 20	**12 55**	p1.8	6 19	**7 00**	1.8
25	**F**	1 27	**1 53**	p1.9	7 23	**8 00**	p2.0
26	**S**	2 27	**2 45**	p2.1	8 21	**8 53**	p2.1
27	**S**	3 21	**3 33**	p2.1	9 13	**9 42**	p2.1
28	**M**	4 10	**4 18**	p2.1	10 01	**10 26**	p2.1
29	**T**	4 56	**5 00**	p2.1	10 45	**11 08**	p2.1
30	**W**	5 40	**5 40**	p2.0	11 26	**11 48**	p2.0

The Kts. (knots) columns show the **maximum** predicted velocities of the stronger one of the Flood Currents and the stronger one of the Ebb Currents for each day.
The letter "a" means the velocity shown should occur **after** the **a.m.** Current Change. The letter "p" means the velocity shown should occur **after** the **p.m.** Current Change (even if next morning). No "a" or "p" means a.m. and p.m. velocities are the same for that day.
Avg. Max. Velocity: Flood 1.8 Kts., Ebb 1.9 Kts.
Max. Flood 3 hrs. 5 min. after Flood Starts, ±15 min.
Max. Ebb 3 hrs. 5 min. after Ebb Starts, ±15 min.

See pp. 22-29 for Current Change at other points.

2014 CURRENT TABLE
DELAWARE BAY ENTRANCE

38°46.85'N, 75°02.58'W

MAY

Daylight Saving Time

DAY OF MONTH	DAY OF WEEK	CURRENT TURNS TO					
		NORTHWEST Flood Starts			SOUTHEAST Ebb Starts		
		a.m.	**p.m.**	Kts.	a.m.	**p.m.**	Kts.
1	**T**	6 22	**6 20**	p1.9	...	**12 06**	1.7
2	**F**	7 03	**7 01**	p1.8	12 27	**12 48**	a1.9
3	**S**	7 46	**7 45**	p1.7	1 07	**1 32**	a1.8
4	**S**	8 33	**8 37**	p1.6	1 50	**2 21**	a1.8
5	**M**	9 22	**9 32**	p1.5	2 38	**3 15**	a1.7
6	**T**	10 15	**10 32**	a1.5	3 31	**4 12**	a1.7
7	**W**	11 10	**11 32**	a1.5	4 27	**5 10**	a1.6
8	**T**	...	**12 02**	1.6	5 23	**6 06**	a1.7
9	**F**	12 30	**12 52**	p1.7	6 18	**6 58**	1.7
10	**S**	1 23	**1 37**	p1.8	7 09	**7 44**	p1.8
11	**S**	2 12	**2 19**	p1.9	7 56	**8 27**	p1.9
12	**M**	2 56	**2 58**	p2.0	8 39	**9 07**	p2.0
13	**T**	3 38	**3 35**	p2.0	9 20	**9 46**	p2.0
14	**W**	4 18	**4 12**	p2.1	9 59	**10 25**	p2.1
15	**T**	4 58	**4 52**	p2.1	10 41	**11 06**	p2.1
16	**F**	5 41	**5 36**	p2.0	11 25	**11 50**	p2.1
17	**S**	6 28	**6 25**	p2.0	...	**12 14**	1.8
18	**S**	7 19	**7 22**	p1.9	12 39	**1 09**	a2.1
19	**M**	8 17	**8 28**	p1.8	1 34	**2 12**	a2.0
20	**T**	9 20	**9 40**	1.7	2 35	**3 20**	a1.9
21	**W**	10 26	**10 55**	a1.7	3 42	**4 31**	a1.8
22	**T**	11 30	**...**	1.8	4 51	**5 40**	a1.8
23	**F**	12 06	**12 32**	p1.9	5 59	**6 44**	1.9
24	**S**	1 12	**1 28**	p2.0	7 02	**7 41**	p2.0
25	**S**	2 11	**2 20**	p2.1	7 59	**8 34**	p2.1
26	**M**	3 04	**3 08**	p2.1	8 51	**9 21**	p2.1
27	**T**	3 53	**3 53**	p2.1	9 39	**10 05**	p2.1
28	**W**	4 39	**4 34**	p2.0	10 22	**10 45**	p2.0
29	**T**	5 21	**5 13**	p1.9	11 03	**11 22**	p2.0
30	**F**	6 00	**5 51**	p1.8	11 42	**11 59**	p1.9
31	**S**	6 39	**6 30**	p1.7	...	**12 22**	1.5

JUNE

Daylight Saving Time

DAY OF MONTH	DAY OF WEEK	CURRENT TURNS TO					
		NORTHWEST Flood Starts			SOUTHEAST Ebb Starts		
		a.m.	**p.m.**	Kts.	a.m.	**p.m.**	Kts.
1	**S**	7 18	**7 12**	p1.6	12 36	**1 04**	a1.8
2	**M**	7 59	**7 59**	p1.6	1 15	**1 50**	a1.8
3	**T**	8 44	**8 52**	1.5	1 59	**2 40**	a1.8
4	**W**	9 33	**9 50**	a1.6	2 48	**3 33**	a1.7
5	**T**	10 22	**10 48**	a1.6	3 40	**4 28**	a1.7
6	**F**	11 13	**11 47**	a1.7	4 34	**5 23**	a1.7
7	**S**	...	**12 03**	1.8	5 29	**6 15**	1.7
8	**S**	12 42	**12 50**	p1.8	6 22	**7 04**	p1.8
9	**M**	1 35	**1 36**	p1.9	7 13	**7 51**	p1.9
10	**T**	2 24	**2 19**	p2.0	8 01	**8 35**	p2.0
11	**W**	3 10	**3 02**	p2.1	8 48	**9 19**	p2.1
12	**T**	3 55	**3 46**	p2.1	9 34	**10 03**	p2.1
13	**F**	4 40	**4 32**	p2.1	10 21	**10 48**	p2.2
14	**S**	5 27	**5 21**	p2.1	11 11	**11 35**	p2.2
15	**S**	6 16	**6 15**	p2.0	...	**12 04**	1.8
16	**M**	7 08	**7 14**	p1.9	12 26	**1 01**	a2.1
17	**T**	8 04	**8 19**	1.8	1 22	**2 03**	a2.1
18	**W**	9 04	**9 28**	a1.8	2 21	**3 09**	a2.0
19	**T**	10 05	**10 39**	a1.8	3 24	**4 15**	a1.9
20	**F**	11 06	**11 47**	a1.9	4 29	**5 20**	a1.9
21	**S**	...	**12 05**	2.0	5 34	**6 22**	1.9
22	**S**	12 51	**1 01**	p2.0	6 36	**7 19**	p1.9
23	**M**	1 50	**1 54**	p2.0	7 34	**8 11**	p2.0
24	**T**	2 44	**2 43**	p2.0	8 27	**8 59**	p2.0
25	**W**	3 34	**3 28**	p2.0	9 15	**9 42**	p2.0
26	**T**	4 19	**4 10**	p1.9	10 00	**10 22**	p2.0
27	**F**	5 00	**4 49**	p1.8	10 41	**10 58**	p1.9
28	**S**	5 38	**5 26**	p1.8	11 19	**11 32**	p1.9
29	**S**	6 14	**6 03**	p1.7	11 57	**...**	1.5
30	**M**	6 49	**6 43**	p1.7	12 07	**12 36**	a1.9

The Kts. (knots) columns show the **maximum** predicted velocities of the stronger one of the Flood Currents and the stronger one of the Ebb Currents for each day.
The letter "a" means the velocity shown should occur **after** the **a.m.** Current Change. The letter "p" means the velocity shown should occur **after** the **p.m.** Current Change (even if next morning). No "a" or "p" means a.m. and p.m. velocities are the same for that day.
Avg. Max. Velocity: Flood 1.8 Kts., Ebb 1.9 Kts.
Max. Flood 3 hrs. 5 min. after Flood Starts, ±15 min.
Max. Ebb 3 hrs. 5 min. after Ebb Starts, ±15 min.

See pp. 22-29 for Current Change at other points.

2014 CURRENT TABLE

DELAWARE BAY ENTRANCE

38°46.85'N, 75°02.58'W

Daylight Saving Time — **Daylight Saving Time**

JULY

DAY OF MONTH	DAY OF WEEK	CURRENT TURNS TO NORTHWEST Flood Starts a.m.	**p.m.**	Kts.	SOUTHEAST Ebb Starts a.m.	**p.m.**	Kts.
1	T	7 26	**7 26**	1.6	12 43	**1 18**	a1.9
2	W	8 06	**8 14**	1.6	1 23	**2 04**	a1.9
3	T	8 49	**9 06**	a1.7	2 07	**2 53**	a1.8
4	F	9 36	**10 04**	a1.7	2 55	**3 45**	a1.8
5	S	10 24	**11 02**	a1.7	3 47	**4 38**	a1.8
6	S	11 14	**...**	1.8	4 41	**5 32**	1.7
7	M	12 01	**12 05**	p1.9	5 36	**6 25**	p1.8
8	T	12 58	**12 56**	p1.9	6 32	**7 17**	p1.9
9	W	1 53	**1 47**	p2.0	7 27	**8 07**	p2.0
10	T	2 45	**2 37**	p2.1	8 21	**8 56**	p2.1
11	F	3 35	**3 27**	p2.1	9 14	**9 45**	p2.1
12	S	4 24	**4 18**	p2.1	10 06	**10 34**	p2.2
13	S	5 13	**5 11**	p2.1	11 00	**11 23**	p2.2
14	M	6 03	**6 06**	p2.0	11 54	**...**	1.9
15	T	6 54	**7 04**	1.9	12 14	**12 51**	a2.2
16	W	7 47	**8 06**	a1.9	1 07	**1 49**	a2.1
17	T	8 42	**9 11**	a1.9	2 03	**2 50**	a2.1
18	F	9 40	**10 17**	a1.9	3 02	**3 53**	a2.0
19	S	10 38	**11 23**	a1.9	4 03	**4 55**	a1.9
20	S	11 36	**...**	1.9	5 06	**5 56**	1.8
21	M	12 27	**12 33**	p1.9	6 07	**6 53**	p1.9
22	T	1 26	**1 27**	p1.9	7 06	**7 46**	p1.9
23	W	2 20	**2 17**	p1.9	8 01	**8 34**	p1.9
24	T	3 10	**3 04**	p1.9	8 51	**9 18**	p1.9
25	F	3 55	**3 47**	p1.9	9 37	**9 57**	p1.9
26	S	4 35	**4 26**	p1.8	10 18	**10 33**	p1.9
27	S	5 12	**5 03**	p1.8	10 56	**11 07**	p1.9
28	M	5 46	**5 39**	p1.7	11 32	**11 39**	p1.9
29	T	6 19	**6 15**	p1.7	...	**12 08**	1.6
30	W	6 52	**6 54**	1.7	12 13	**12 46**	a1.9
31	T	7 27	**7 38**	a1.7	12 50	**1 27**	a1.9

AUGUST

DAY OF MONTH	DAY OF WEEK	CURRENT TURNS TO NORTHWEST Flood Starts a.m.	**p.m.**	Kts.	SOUTHEAST Ebb Starts a.m.	**p.m.**	Kts.
1	F	8 06	**8 26**	a1.7	1 30	**2 12**	a1.9
2	S	8 50	**9 20**	a1.8	2 14	**3 02**	a1.9
3	S	9 38	**10 19**	a1.8	3 04	**3 55**	a1.8
4	M	10 31	**11 23**	a1.8	3 58	**4 52**	1.7
5	T	11 27	**...**	1.8	4 57	**5 50**	1.7
6	W	12 25	**12 25**	p1.9	5 59	**6 48**	p1.8
7	T	1 26	**1 23**	p2.0	7 01	**7 44**	p1.9
8	F	2 23	**2 20**	p2.0	8 01	**8 38**	p2.0
9	S	3 17	**3 15**	p2.1	8 59	**9 30**	p2.1
10	S	4 08	**4 09**	p2.1	9 55	**10 21**	p2.2
11	M	4 57	**5 03**	p2.1	10 48	**11 10**	p2.2
12	T	5 46	**5 56**	p2.1	11 41	**11 59**	p2.2
13	W	6 34	**6 51**	2.0	...	**12 35**	2.0
14	T	7 24	**7 48**	a2.0	12 49	**1 29**	a2.2
15	F	8 16	**8 48**	a2.0	1 41	**2 26**	a2.1
16	S	9 10	**9 50**	a1.9	2 36	**3 24**	a1.9
17	S	10 07	**10 54**	a1.9	3 34	**4 24**	a1.8
18	M	11 05	**11 57**	a1.8	4 35	**5 24**	1.7
19	T	...	**12 03**	1.8	5 37	**6 22**	p1.7
20	W	12 57	**12 59**	p1.8	6 37	**7 16**	p1.8
21	T	1 52	**1 51**	p1.8	7 34	**8 06**	p1.9
22	F	2 41	**2 39**	p1.8	8 25	**8 51**	p1.9
23	S	3 26	**3 23**	p1.8	9 11	**9 31**	p1.9
24	S	4 06	**4 03**	p1.8	9 52	**10 07**	p1.9
25	M	4 42	**4 40**	p1.8	10 29	**10 40**	p1.9
26	T	5 14	**5 14**	p1.8	11 04	**11 11**	p1.9
27	W	5 45	**5 48**	1.7	11 37	**11 43**	p2.0
28	T	6 16	**6 25**	a1.8	...	**12 13**	1.7
29	F	6 49	**7 05**	a1.8	12 17	**12 51**	a2.0
30	S	7 26	**7 50**	a1.8	12 55	**1 34**	a1.9
31	S	8 09	**8 43**	a1.8	1 39	**2 22**	a1.9

The Kts. (knots) columns show the **maximum** predicted velocities of the stronger one of the Flood Currents and the stronger one of the Ebb Currents for each day.
The letter "a" means the velocity shown should occur **after** the **a.m.** Current Change. The letter "p" means the velocity shown should occur **after** the **p.m.** Current Change (even if next morning). No "a" or "p" means a.m. and p.m. velocities are the same for that day.
Avg. Max. Velocity: Flood 1.8 Kts., Ebb 1.9 Kts.
Max. Flood 3 hrs. 5 min. after Flood Starts, ±15 min.
Max. Ebb 3 hrs. 5 min. after Ebb Starts, ±15 min.

See pp. 22-29 for Current Change at other points.

2014 CURRENT TABLE
DELAWARE BAY ENTRANCE

38°46.85'N, 75°02.58'W

Daylight Saving Time

SEPTEMBER

Day of Month	Day of Week	CURRENT TURNS TO NORTHWEST Flood Starts			CURRENT TURNS TO SOUTHEAST Ebb Starts		
		a.m.	**p.m.**	Kts.	a.m.	**p.m.**	Kts.
1	M	8 58	**9 43**	a1.8	2 28	**3 17**	a1.8
2	T	9 54	**10 49**	a1.8	3 24	**4 17**	1.7
3	W	10 57	**11 58**	a1.8	4 27	**5 20**	1.7
4	T	...	**12 04**	1.8	5 35	**6 24**	p1.8
5	F	1 03	**1 08**	p1.9	6 43	**7 26**	p1.9
6	S	2 04	**2 09**	p2.0	7 48	**8 23**	p2.0
7	S	2 59	**3 07**	p2.1	8 48	**9 16**	p2.2
8	M	3 50	**4 01**	p2.1	9 43	**10 07**	p2.2
9	T	4 38	**4 53**	p2.1	10 35	**10 55**	p2.2
10	W	5 25	**5 43**	2.1	11 25	**11 41**	p2.2
11	T	6 11	**6 34**	a2.1	...	**12 14**	2.0
12	F	6 58	**7 27**	a2.0	12 28	**1 05**	a2.1
13	S	7 47	**8 22**	a2.0	1 16	**1 57**	a2.0
14	S	8 38	**9 20**	a1.9	2 07	**2 51**	a1.8
15	M	9 34	**10 21**	a1.8	3 03	**3 49**	a1.7
16	T	10 32	**11 23**	a1.7	4 02	**4 48**	1.6
17	W	11 31	**...**	1.7	5 04	**5 46**	p1.6
18	T	12 23	**12 29**	p1.7	6 06	**6 42**	p1.7
19	F	1 17	**1 23**	p1.7	7 03	**7 33**	p1.8
20	S	2 07	**2 12**	p1.8	7 55	**8 19**	p1.9
21	S	2 51	**2 57**	p1.8	8 41	**9 00**	p1.9
22	M	3 31	**3 38**	p1.8	9 22	**9 37**	p1.9
23	T	4 07	**4 14**	1.8	9 59	**10 10**	p1.9
24	W	4 39	**4 49**	1.8	10 33	**10 42**	p2.0
25	T	5 10	**5 22**	1.8	11 06	**11 14**	p2.0
26	F	5 40	**5 57**	a1.9	11 40	**11 47**	p2.0
27	S	6 12	**6 36**	a1.9	...	**12 18**	1.9
28	S	6 50	**7 21**	a1.9	12 25	**1 01**	a1.9
29	M	7 34	**8 14**	a1.8	1 09	**1 49**	a1.9
30	T	8 26	**9 15**	a1.8	2 00	**2 45**	a1.8

Daylight Saving Time

OCTOBER

Day of Month	Day of Week	CURRENT TURNS TO NORTHWEST Flood Starts			CURRENT TURNS TO SOUTHEAST Ebb Starts		
		a.m.	**p.m.**	Kts.	a.m.	**p.m.**	Kts.
1	W	9 28	**10 24**	a1.7	2 59	**3 48**	1.7
2	T	10 37	**11 36**	a1.7	4 07	**4 56**	p1.7
3	F	11 49	**...**	1.7	5 20	**6 05**	p1.8
4	S	12 44	**12 59**	p1.8	6 31	**7 09**	p1.9
5	S	1 44	**2 01**	p2.0	7 37	**8 08**	p2.0
6	M	2 39	**2 58**	p2.0	8 35	**9 01**	p2.1
7	T	3 30	**3 51**	p2.1	9 29	**9 51**	p2.2
8	W	4 17	**4 40**	2.1	10 19	**10 37**	p2.2
9	T	5 02	**5 29**	a2.1	11 06	**11 22**	2.1
10	F	5 47	**6 16**	a2.1	11 52	**...**	2.0
11	S	6 31	**7 04**	a2.0	12 06	**12 38**	a2.0
12	S	7 17	**7 54**	a1.9	12 51	**1 26**	a1.9
13	M	8 05	**8 48**	a1.8	1 39	**2 16**	1.7
14	T	8 59	**9 45**	a1.7	2 31	**3 10**	1.6
15	W	9 56	**10 44**	a1.6	3 28	**4 07**	p1.6
16	T	10 56	**11 42**	a1.6	4 29	**5 05**	p1.6
17	F	11 55	**...**	1.6	5 30	**6 01**	p1.6
18	S	12 37	**12 50**	p1.6	6 28	**6 53**	p1.7
19	S	1 26	**1 41**	p1.7	7 20	**7 41**	p1.8
20	M	2 12	**2 27**	p1.8	8 06	**8 24**	p1.9
21	T	2 52	**3 09**	1.8	8 48	**9 03**	p1.9
22	W	3 29	**3 47**	1.8	9 26	**9 39**	1.9
23	T	4 02	**4 23**	a1.9	10 01	**10 12**	1.9
24	F	4 34	**4 58**	a1.9	10 36	**10 45**	1.9
25	S	5 06	**5 34**	a1.9	11 11	**11 21**	a2.0
26	S	5 41	**6 14**	a1.9	11 50	**...**	1.9
27	M	6 21	**7 00**	a1.9	12 01	**12 34**	1.9
28	T	7 09	**7 54**	a1.8	12 48	**1 24**	1.8
29	W	8 05	**8 56**	a1.8	1 42	**2 22**	p1.8
30	T	9 11	**10 05**	a1.7	2 44	**3 27**	1.7
31	F	10 25	**11 16**	a1.7	3 55	**4 36**	p1.7

The Kts. (knots) columns show the **maximum** predicted velocities of the stronger one of the Flood Currents and the stronger one of the Ebb Currents for each day.
The letter "a" means the velocity shown should occur **after** the **a.m.** Current Change. The letter "p" means the velocity shown should occur **after** the **p.m.** Current Change (even if next morning). No "a" or "p" means a.m. and p.m. velocities are the same for that day.
Avg. Max. Velocity: Flood 1.8 Kts., Ebb 1.9 Kts.
Max. Flood 3 hrs. 5 min. after Flood Starts, ±15 min.
Max. Ebb 3 hrs. 5 min. after Ebb Starts, ±15 min.

See pp. 22-29 for Current Change at other points.

2014 CURRENT TABLE

DELAWARE BAY ENTRANCE

38°46.85'N, 75°02.58'W

*Standard Time starts Nov. 2 at 2 a.m. — Standard Time

NOVEMBER

DAY OF MONTH	DAY OF WEEK	CURRENT TURNS TO NORTHWEST Flood Starts a.m.	p.m.	Kts.	CURRENT TURNS TO SOUTHEAST Ebb Starts a.m.	p.m.	Kts.
1	S	11 39	...	1.7	5 09	5 46	p1.8
2	S	12 22	*-A-	1.7	*5 20	*5 51	p1.9
3	M	12 23	12 50	p1.9	6 23	6 50	p2.0
4	T	1 19	1 47	2.0	7 21	7 44	p2.1
5	W	2 08	2 38	a2.1	8 13	8 33	2.1
6	T	2 55	3 27	a2.1	9 01	9 19	2.1
7	F	3 40	4 13	a2.1	9 46	10 02	a2.1
8	S	4 23	4 57	a2.0	10 30	10 44	a2.0
9	S	5 05	5 42	a1.9	11 12	11 27	a1.9
10	M	5 48	6 27	a1.8	11 55	...	1.8
11	T	6 33	7 15	a1.7	12 11	12 40	p1.7
12	W	7 23	8 06	a1.6	1 00	1 29	p1.6
13	T	8 18	9 01	a1.5	1 53	2 22	p1.6
14	F	9 16	9 56	a1.5	2 51	3 18	p1.6
15	S	10 15	10 50	1.5	3 49	4 14	p1.6
16	S	11 11	11 40	1.6	4 46	5 08	p1.7
17	M	...	12 04	1.6	5 39	5 58	p1.8
18	T	12 27	12 53	1.7	6 28	6 44	p1.8
19	W	1 10	1 37	a1.8	7 12	7 26	1.8
20	T	1 49	2 18	a1.9	7 52	8 05	1.9
21	F	2 26	2 57	a1.9	8 30	8 43	a2.0
22	S	3 01	3 35	a2.0	9 08	9 20	a2.0
23	S	3 37	4 15	a2.0	9 46	10 00	a2.0
24	M	4 17	4 57	a2.0	10 28	10 44	a2.0
25	T	5 02	5 45	a1.9	11 14	11 34	a2.0
26	W	5 53	6 40	a1.9	...	12 06	1.9
27	T	6 53	7 41	a1.8	12 30	1 04	1.8
28	F	8 00	8 47	a1.7	1 35	2 09	p1.8
29	S	9 13	9 54	1.7	2 44	3 17	p1.8
30	S	10 26	10 59	p1.8	3 56	4 26	p1.8

DECEMBER

DAY OF MONTH	DAY OF WEEK	CURRENT TURNS TO NORTHWEST Flood Starts a.m.	p.m.	Kts.	CURRENT TURNS TO SOUTHEAST Ebb Starts a.m.	p.m.	Kts.
1	M	11 34	...	1.7	5 04	5 31	p1.9
2	T	12 01	12 36	a1.9	6 07	6 30	1.9
3	W	12 55	1 32	a2.0	7 04	7 25	2.0
4	T	1 48	2 25	a2.1	7 55	8 15	a2.1
5	F	2 34	3 12	a2.1	8 43	9 01	a2.1
6	S	3 19	3 57	a2.0	9 27	9 44	a2.0
7	S	4 01	4 39	a2.0	10 08	10 24	a1.9
8	M	4 41	5 19	a1.9	10 47	11 04	a1.9
9	T	5 22	6 00	a1.8	11 26	11 45	a1.8
10	W	6 04	6 42	a1.7	...	12 07	1.7
11	T	6 49	7 27	a1.6	12 30	12 50	p1.7
12	F	7 39	8 15	1.5	1 18	1 38	p1.7
13	S	8 33	9 06	1.5	2 10	2 29	p1.6
14	S	9 30	9 58	p1.6	3 05	3 23	p1.6
15	M	10 27	10 50	p1.6	4 00	4 18	p1.7
16	T	11 22	11 39	p1.7	4 54	5 11	p1.7
17	W	...	12 15	1.6	5 45	6 01	1.7
18	T	12 26	1 03	a1.8	6 33	6 48	1.8
19	F	1 10	1 49	a1.9	7 18	7 33	a1.9
20	S	1 52	2 32	a1.9	8 01	8 16	a2.0
21	S	2 33	3 15	a2.0	8 43	8 59	a2.0
22	M	3 16	3 58	a2.0	9 27	9 44	a2.1
23	T	4 01	4 44	a2.0	10 12	10 32	a2.1
24	W	4 50	5 33	a2.0	11 00	11 24	a2.1
25	T	5 44	6 26	a1.9	11 52	...	2.0
26	F	6 44	7 24	a1.9	12 21	12 49	1.9
27	S	7 49	8 26	1.8	1 23	1 51	p1.9
28	S	8 58	9 30	p1.8	2 30	2 56	1.8
29	M	10 08	10 33	p1.8	3 37	4 02	1.8
30	T	11 15	11 34	p1.9	4 43	5 07	1.8
31	W	...	12 18	1.7	5 46	6 08	a1.9

A also at 11:48 a.m. 1.8

The Kts. (knots) columns show the **maximum** predicted velocities of the stronger one of the Flood Currents and the stronger one of the Ebb Currents for each day.
The letter "a" means the velocity shown should occur **after** the **a.m.** Current Change. The letter "p" means the velocity shown should occur **after** the **p.m.** Current Change (even if next morning). No "a" or "p" means a.m. and p.m. velocities are the same for that day.
Avg. Max. Velocity: Flood 1.8 Kts., Ebb 1.9 Kts.
Max. Flood 3 hrs. 5 min. after Flood Starts, ±15 min.
Max. Ebb 3 hrs. 5 min. after Ebb Starts, ±15 min.

See pp. 22-29 for Current Change at other points.

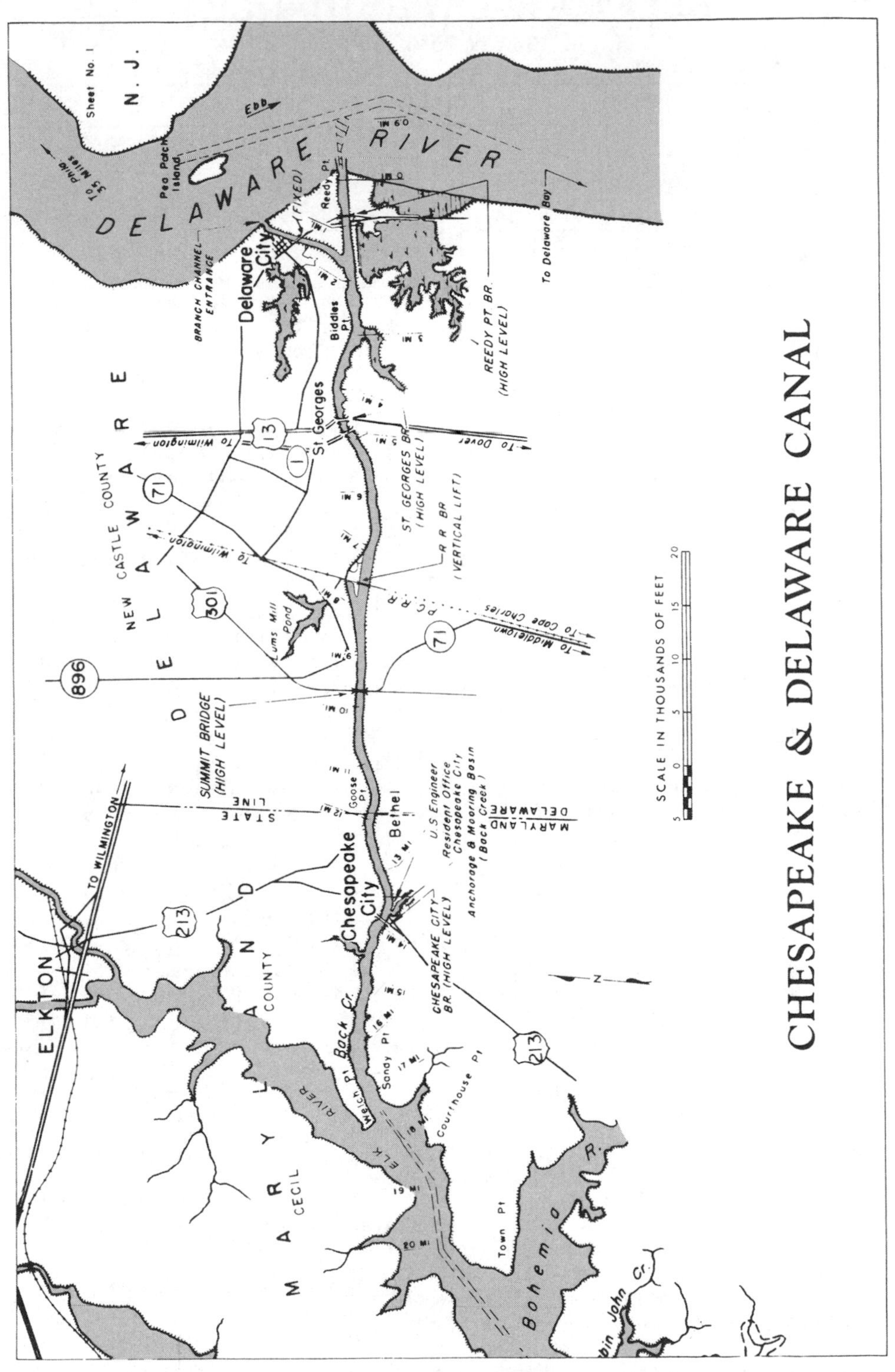

See Chesapeake & Delaware Canal Current Tables, pp. 148-153

CHESAPEAKE & DELAWARE CANAL REGULATIONS

(Traffic Dispatcher is located at Chesapeake City and monitors Channel 13.)

Philadelphia District Engineer issues notices periodically showing available channel depths and navigation conditions.

Projected Channel dimensions are 35 ft. deep and 450 ft. wide. (The branch to Delaware City is 8 ft. deep and 50 ft. wide.) The distance from the Delaware River Ship Channel to the Elk River is 19.1 miles.

1. Traffic controls, located at Reedy Point and Old Town Point Wharf, flash green when Canal is open, flash red when it is closed.
2. Vessel identification and monitoring are performed by TV cameras at Reedy Point and Old Town Point Wharf.
3. The following vessels, tugs and tows are required to have radiotelephones:
   a. Power vessels of 300 gross tons and upward.
   b. Vessels of 100 gross tons and upward carrying 1 or more passengers for hire.
   c. Every towing vessel of 26 feet or over.
4. Vessels listed in 3. will not enter the Canal until radio communication is made with the dispatcher and clearance is received. Ships' captains will tell the dispatcher the estimated time of passing Reedy Point or Town Point. Communication is to be established on Channel 13 (156.65 MHz). Dispatcher also monitors Channel 16 (156.8 MHz) to respond to emergencies.
5. A westbound vessel must be able to pass Reedy Is. or Pea Patch Is. within 120 min. of receiving clearance; an eastbound vessel must be able to pass Arnold Point within 120 min. If passage is not made within these 120 min., a new clearance must be solicited. Vessels must also report to the dispatcher the time of passing the outer end of the jetties at Reedy Point and Old Town Point Wharf.
6. Vessels exceeding 800 feet are required to have operable bow thrusters.
7. Maximum combined extreme breadth of vessels meeting and overtaking each other is 190 feet.
8. Vessels of all types are required to travel at a safe speed to avoid damage by suction or wash to wharves, landings, other boats, etc. Operators of yachts, motorboats, etc. are cautioned that there are many large, deep-draft ocean-going and commercial vessels using the Canal. There is "no anchoring" in the canal at any time. Moor or anchor outside of Reedy Point, near Arnold Point, or in Chesapeake City Basin.
9. Vessels proceeding *with* the current shall have the right-of-way but all small pleasure craft shall relinquish the right-of-way to deeper draft vessels which have a limited maneuvering ability.
10. Vessels under sail will not be permitted in the Canal.
11. Vessels difficult to handle must use the Canal during daylight hours and must have tug assistance. They should clear Reedy Point Bridge (going east) or Chesapeake City Bridge (going west) before dark.

Anchorage and wharfage facilities for small vessels only are at Chesapeake City and permission to use them for more than 24 hours must be obtained from the dispatcher.

The **railroad bridge** has a clearance when closed of 45 ft. at MHW. The bridge monitors Channel 13 and gives 30 minutes notice prior to lowering.

The **five highway bridges** are high level and fixed.

Normal tide range is 5.4 ft. at Delaware R. end of the Canal and 2.6 ft. at Chesapeake City. Local mean low water at Courthouse Pt. is 2.5 ft. and decreases gradually eastward to 0.6 ft. at Delaware R. (See pp. 18 and 19 for times of High Water in this area.)
Note: A violent northeast storm may raise tide 4 to 5 ft. above normal in the Canal; a westerly storm may cause low tide to fall slightly below normal at Chesapeake City and as much as 4.0 ft. below normal at Reedy Point.

2014 CURRENT TABLE
CHESAPEAKE & DELAWARE CANAL
39°31.89'N, 75°49.65'W at Chesapeake City

JANUARY

Standard Time

DAY OF MONTH	DAY OF WEEK	CURRENT TURNS TO EAST — Flood Starts a.m.	p.m.	Kts.	CURRENT TURNS TO WEST — Ebb Starts a.m.	p.m.	Kts.
1	W	3 08	**4 27**	p2.6	8 59	**10 49**	a3.0
2	T	4 06	**5 18**	p2.7	9 57	**11 36**	a3.0
3	F	5 03	**6 09**	p2.7	10 57	**...**	2.9
4	S	6 03	**6 58**	p2.7	12 22	**12 01**	p2.8
5	S	7 00	**7 44**	2.5	1 07	**1 01**	p2.6
6	M	8 00	**8 31**	2.4	1 52	**2 04**	p2.3
7	T	9 02	**9 19**	2.2	2 39	**3 10**	a2.1
8	W	10 07	**10 09**	2.0	3 27	**4 19**	a2.1
9	T	11 12	**11 00**	1.8	4 16	**5 26**	a2.1
10	F	11 59	**11 49**	1.7	5 04	**6 30**	a2.1
11	S	...	**1 05**	1.8	5 50	**7 28**	a2.0
12	S	12 38	**1 53**	p1.8	6 33	**8 20**	a2.0
13	M	1 26	**2 35**	p1.9	7 13	**9 06**	a2.1
14	T	2 13	**3 11**	p1.9	7 52	**9 45**	a2.2
15	W	2 58	**3 45**	p2.0	8 32	**10 22**	a2.2
16	T	3 41	**4 18**	p2.2	9 13	**10 55**	a2.3
17	F	4 24	**4 53**	p2.3	9 56	**11 28**	a2.3
18	S	5 08	**5 29**	p2.4	10 42	**11 59**	a2.4
19	S	5 54	**6 07**	p2.5	11 30	**...**	2.3
20	M	6 40	**6 46**	p2.5	12 31	**12 21**	p2.2
21	T	7 29	**7 26**	p2.4	1 03	**1 15**	a2.1
22	W	8 20	**8 09**	p2.3	1 39	**2 13**	a2.2
23	T	9 17	**8 56**	p2.2	2 19	**3 20**	a2.2
24	F	10 20	**9 50**	p2.1	3 04	**4 35**	a2.3
25	S	11 24	**10 50**	p2.1	3 56	**5 47**	a2.3
26	S	...	**12 27**	1.9	4 52	**6 52**	a2.4
27	M	12 01	**1 27**	2.1	5 52	**7 52**	a2.5
28	T	12 55	**2 25**	p2.3	6 54	**8 45**	a2.7
29	W	1 57	**3 20**	2.4	7 57	**9 34**	a2.8
30	T	2 57	**4 11**	2.5	8 59	**10 19**	a2.9
31	F	3 55	**5 01**	a2.7	9 58	**11 04**	a2.9

FEBRUARY

Standard Time

DAY OF MONTH	DAY OF WEEK	CURRENT TURNS TO EAST — Flood Starts a.m.	p.m.	Kts.	CURRENT TURNS TO WEST — Ebb Starts a.m.	p.m.	Kts.
1	S	4 52	**5 49**	a2.8	10 58	**11 48**	a2.8
2	S	5 49	**6 35**	a2.8	11 57	**...**	2.6
3	M	6 46	**7 20**	a2.6	12 32	**12 57**	p2.4
4	T	7 43	**8 04**	a2.4	1 17	**1 56**	a2.3
5	W	8 40	**8 47**	a2.1	2 01	**2 57**	a2.2
6	T	9 40	**9 34**	a1.9	2 46	**4 00**	a2.1
7	F	10 40	**10 25**	a1.7	3 33	**5 03**	a2.0
8	S	11 36	**11 17**	a1.6	4 21	**6 02**	a1.8
9	S	...	**12 25**	1.6	5 08	**6 55**	a1.8
10	M	12 08	**1 09**	p1.6	5 54	**7 42**	a1.8
11	T	12 57	**1 49**	p1.7	6 38	**8 24**	a1.9
12	W	1 44	**2 27**	p1.9	7 24	**9 00**	a2.0
13	T	2 30	**3 04**	p2.0	8 11	**9 33**	a2.1
14	F	3 14	**3 41**	p2.2	8 59	**10 04**	a2.2
15	S	3 57	**4 19**	p2.3	9 46	**10 35**	a2.2
16	S	4 41	**4 57**	p2.4	10 34	**11 06**	2.2
17	M	5 27	**5 37**	p2.5	11 23	**11 40**	p2.3
18	T	6 13	**6 18**	p2.5	...	**12 15**	2.0
19	W	7 01	**7 00**	p2.4	12 16	**1 08**	a2.4
20	T	7 51	**7 44**	p2.3	12 55	**2 06**	a2.4
21	F	8 46	**8 33**	p2.1	1 38	**3 10**	a2.4
22	S	9 50	**9 31**	p2.1	2 27	**4 20**	a2.3
23	S	10 58	**10 37**	p2.0	3 24	**5 28**	a2.3
24	M	11 59	**11 43**	p2.1	4 31	**6 29**	a2.3
25	T	...	**1 06**	2.1	5 40	**7 24**	a2.4
26	W	12 47	**2 04**	2.3	6 49	**8 14**	a2.5
27	T	1 49	**2 59**	a2.5	7 56	**9 01**	a2.6
28	F	2 48	**3 50**	a2.7	8 59	**9 45**	a2.7

The Kts. (knots) columns show the **maximum** predicted velocities of the stronger one of the Flood Currents and the stronger one of the Ebb Currents for each day.
The letter "a" means the velocity shown should occur **after** the **a.m.** Current Change. The letter "p" means the velocity shown should occur **after** the **p.m.** Current Change (even if next morning). No "a" or "p" means a.m. and p.m. velocities are the same for that day.
Avg. Max. Velocity: Flood 2.0 Kts., Ebb 1.9 Kts.
Max. Flood 3 hrs. 10 min. after Flood Starts ±45 min.
Max. Ebb 2 hrs. 45 min. after Ebb Starts ±45 min.
See pp. 22-29 for Current Change at other points.

Note *from NOS: These predictions should be considered questionable. Caution is advised.*

2014 CURRENT TABLE
CHESAPEAKE & DELAWARE CANAL
39°31.89'N, 75°49.65'W at Chesapeake City

***Daylight Time starts March 9 at 2 a.m.** **Daylight Saving Time**

MARCH

DAY OF MONTH	DAY OF WEEK	CURRENT TURNS TO EAST Flood Starts			CURRENT TURNS TO WEST Ebb Starts		
		a.m.	**p.m.**	Kts.	a.m.	**p.m.**	Kts.
1	**S**	3 45	**4 38**	a2.8	9 59	**10 28**	a2.6
2	**S**	4 41	**5 24**	a2.8	10 56	**11 12**	a2.5
3	**M**	5 36	**6 09**	a2.8	11 54	**11 55**	p2.4
4	**T**	6 31	**6 53**	a2.6	...	**12 50**	2.1
5	**W**	7 22	**7 34**	a2.4	12 38	**1 45**	a2.4
6	**T**	8 13	**8 17**	a2.1	1 20	**2 41**	a2.2
7	**F**	9 04	**9 03**	a1.8	2 01	**3 38**	a2.0
8	**S**	9 55	**9 54**	a1.7	2 43	**4 34**	a1.8
9	**S**	*11 44	***11 49**	a1.5	*4 28	***6 27**	a1.7
10	**M**	...	**12 30**	1.5	5 19	**7 14**	a1.6
11	**T**	12 41	**1 13**	p1.6	6 13	**7 56**	a1.6
12	**W**	1 30	**1 55**	p1.7	7 07	**8 33**	a1.7
13	**T**	2 17	**2 37**	p1.9	8 01	**9 08**	1.8
14	**F**	3 03	**3 19**	p2.1	8 55	**9 40**	p2.0
15	**S**	3 49	**4 01**	p2.2	9 47	**10 12**	p2.2
16	**S**	4 34	**4 43**	p2.3	10 38	**10 44**	p2.4
17	**M**	5 18	**5 24**	2.4	11 27	**11 18**	p2.5
18	**T**	6 04	**6 07**	a2.5	...	**12 18**	1.9
19	**W**	6 50	**6 50**	a2.5	12 01	**1 10**	a2.6
20	**T**	7 38	**7 36**	a2.4	12 33	**2 04**	a2.6
21	**F**	8 28	**8 24**	a2.3	1 17	**3 00**	a2.6
22	**S**	9 22	**9 17**	a2.2	2 05	**4 00**	a2.5
23	**S**	10 25	**10 19**	a2.1	2 59	**5 04**	a2.4
24	**M**	11 33	**11 29**	a2.1	4 03	**6 05**	a2.3
25	**T**	...	**12 39**	2.1	5 18	**7 00**	a2.3
26	**W**	12 36	**1 40**	p2.2	6 35	**7 52**	a2.3
27	**T**	1 40	**2 38**	2.3	7 47	**8 40**	a2.4
28	**F**	2 41	**3 32**	a2.5	8 55	**9 26**	a2.4
29	**S**	3 40	**4 22**	a2.7	9 57	**10 10**	p2.5
30	**S**	4 36	**5 09**	a2.8	10 55	**10 53**	p2.5
31	**M**	5 30	**5 54**	a2.8	11 51	**11 35**	p2.6

APRIL

DAY OF MONTH	DAY OF WEEK	CURRENT TURNS TO EAST Flood Starts			CURRENT TURNS TO WEST Ebb Starts		
		a.m.	**p.m.**	Kts.	a.m.	**p.m.**	Kts.
1	**T**	6 22	**6 38**	a2.7	...	**12 46**	2.0
2	**W**	7 12	**7 22**	a2.5	12 17	**1 40**	a2.5
3	**T**	7 59	**8 05**	a2.3	12 57	**2 31**	a2.4
4	**F**	8 42	**8 49**	a2.1	1 36	**3 21**	a2.2
5	**S**	9 21	**9 34**	a1.9	2 13	**4 10**	a2.0
6	**S**	10 01	**10 24**	a1.8	2 50	**4 58**	a1.8
7	**M**	10 42	**11 18**	a1.7	3 34	**5 44**	a1.7
8	**T**	11 27	**...**	1.7	4 29	**6 26**	a1.6
9	**W**	12 11	**12 13**	p1.8	5 34	**7 04**	1.6
10	**T**	1 02	**12 59**	p1.9	6 39	**7 39**	p1.8
11	**F**	1 50	**1 46**	p2.0	7 41	**8 13**	p2.1
12	**S**	2 38	**2 33**	p2.1	8 40	**8 47**	p2.3
13	**S**	3 26	**3 19**	p2.2	9 36	**9 22**	p2.5
14	**M**	4 13	**4 05**	2.3	10 29	**9 58**	p2.6
15	**T**	4 58	**4 51**	a2.4	11 20	**10 36**	p2.7
16	**W**	5 44	**5 37**	a2.5	-A-	**12 12**	1.7
17	**T**	6 31	**6 25**	a2.6	...	**1 05**	1.6
18	**F**	7 20	**7 15**	a2.6	12 01	**1 58**	a2.8
19	**S**	8 11	**8 08**	a2.5	12 48	**2 51**	a2.8
20	**S**	9 05	**9 06**	a2.4	1 42	**3 45**	a2.7
21	**M**	10 04	**10 10**	a2.3	2 43	**4 41**	a2.5
22	**T**	11 09	**11 19**	a2.2	3 52	**5 36**	a2.4
23	**W**	...	**12 12**	2.2	5 12	**6 29**	a2.2
24	**T**	12 27	**1 12**	p2.2	6 31	**7 18**	2.1
25	**F**	1 31	**2 07**	a2.3	7 43	**8 06**	p2.3
26	**S**	2 32	**3 00**	a2.5	8 50	**8 52**	p2.5
27	**S**	3 31	**3 50**	a2.6	9 51	**9 37**	p2.6
28	**M**	4 26	**4 37**	a2.6	10 48	**10 20**	p2.6
29	**T**	5 17	**5 22**	a2.6	11 42	**11 01**	p2.6
30	**W**	6 05	**6 07**	a2.5	-B-	**12 35**	1.6

A also at 11:16 p.m. 2.8 **B** also at 11:39 p.m. 2.5

The Kts. (knots) columns show the **maximum** predicted velocities of the stronger one of the Flood Currents and the stronger one of the Ebb Currents for each day.
The letter "a" means the velocity shown should occur **after** the **a.m.** Current Change. The letter "p" means the velocity shown should occur **after** the **p.m.** Current Change (even if next morning). No "a" or "p" means a.m. and p.m. velocities are the same for that day.
Avg. Max. Velocity: Flood 2.0 Kts., Ebb 1.9 Kts.
Max. Flood 3 hrs. 10 min. after Flood Starts ±45 min.
Max. Ebb 2 hrs. 45 min. after Ebb Starts ±45 min.
See pp. 22-29 for Current Change at other points.
Note *from NOS: These predictions should be considered questionable. Caution is advised.*

2014 CURRENT TABLE
CHESAPEAKE & DELAWARE CANAL

39°31.89'N, 75°49.65'W at Chesapeake City

MAY — Daylight Saving Time

DAY OF MONTH	DAY OF WEEK	CURRENT TURNS TO EAST Flood Starts a.m.	p.m.	Kts.	CURRENT TURNS TO WEST Ebb Starts a.m.	p.m.	Kts.
1	**T**	6 50	**6 52**	a2.4	...	**1 25**	1.5
2	**F**	7 30	**7 36**	a2.3	12 16	**2 11**	a2.3
3	**S**	8 06	**8 19**	a2.1	12 52	**2 55**	a2.2
4	**S**	8 39	**9 04**	a2.0	1 27	**3 36**	a2.0
5	**M**	9 10	**9 51**	a2.0	2 07	**4 15**	a1.9
6	**T**	9 47	**10 43**	a1.9	2 53	**4 54**	a1.8
7	**W**	10 30	**11 38**	a1.9	3 51	**5 32**	a1.7
8	**T**	11 19	**...**	2.0	5 00	**6 08**	p1.8
9	**F**	12 31	**12 09**	p2.0	6 13	**6 44**	p2.0
10	**S**	1 23	**12 59**	p2.1	7 20	**7 20**	p2.3
11	**S**	2 14	**1 48**	p2.2	8 23	**7 58**	p2.5
12	**M**	3 04	**2 38**	p2.2	9 23	**8 38**	p2.7
13	**T**	3 53	**3 29**	2.3	10 18	**9 20**	p2.8
14	**W**	4 41	**4 20**	a2.4	11 11	**10 03**	p2.9
15	**T**	5 28	**5 11**	a2.5	11 59	**10 49**	p2.9
16	**F**	6 17	**6 03**	a2.6	-A-	**12 54**	1.5
17	**S**	7 07	**6 58**	a2.7	...	**1 45**	1.5
18	**S**	7 58	**7 55**	a2.6	12 32	**2 34**	a2.9
19	**M**	8 51	**8 54**	a2.5	1 31	**3 23**	a2.7
20	**T**	9 46	**9 58**	a2.4	2 36	**4 13**	a2.6
21	**W**	10 44	**11 07**	a2.3	3 48	**5 04**	a2.4
22	**T**	11 43	**...**	2.3	5 06	**5 55**	a2.2
23	**F**	12 15	**12 40**	p2.2	6 23	**6 45**	p2.2
24	**S**	1 20	**1 33**	2.2	7 34	**7 33**	p2.4
25	**S**	2 22	**2 25**	a2.3	8 40	**8 20**	p2.5
26	**M**	3 20	**3 15**	a2.4	9 41	**9 05**	p2.6
27	**T**	4 13	**4 04**	a2.4	10 36	**9 48**	p2.5
28	**W**	5 01	**4 50**	a2.4	11 28	**10 28**	p2.5
29	**T**	5 44	**5 36**	a2.4	-B-	**12 16**	1.4
30	**F**	6 23	**6 20**	a2.3	-C-	**1 01**	1.4
31	**S**	6 57	**7 04**	a2.2	...	**1 42**	1.4

JUNE — Daylight Saving Time

DAY OF MONTH	DAY OF WEEK	CURRENT TURNS TO EAST Flood Starts a.m.	p.m.	Kts.	CURRENT TURNS TO WEST Ebb Starts a.m.	p.m.	Kts.
1	**S**	7 28	**7 47**	a2.2	12 15	**2 19**	a2.2
2	**M**	7 59	**8 30**	a2.2	12 54	**2 53**	a2.1
3	**T**	8 31	**9 15**	a2.2	1 38	**3 26**	a2.0
4	**W**	9 08	**10 06**	a2.2	2 28	**3 59**	a1.9
5	**T**	9 48	**11 01**	a2.2	3 25	**4 34**	p1.9
6	**F**	10 35	**11 59**	a2.2	4 33	**5 12**	p2.0
7	**S**	11 26	**...**	2.1	5 48	**5 52**	p2.2
8	**S**	12 55	**12 18**	p2.1	6 59	**6 34**	p2.3
9	**M**	1 49	**1 10**	p2.2	8 06	**7 17**	p2.5
10	**T**	2 43	**2 03**	p2.2	9 08	**8 03**	p2.7
11	**W**	3 35	**2 59**	2.2	10 05	**8 51**	p2.8
12	**T**	4 25	**3 55**	a2.4	10 57	**9 41**	p2.9
13	**F**	5 14	**4 51**	a2.5	11 47	**10 33**	p3.0
14	**S**	6 04	**5 46**	a2.7	-D-	**12 35**	1.5
15	**S**	6 54	**6 44**	a2.7	...	**1 23**	1.7
16	**M**	7 45	**7 42**	a2.7	12 26	**2 09**	a2.9
17	**T**	8 35	**8 41**	a2.6	1 29	**2 54**	a2.8
18	**W**	9 26	**9 44**	a2.5	2 34	**3 41**	a2.5
19	**T**	10 18	**10 51**	a2.4	3 44	**4 30**	a2.3
20	**F**	11 12	**...**	2.3	4 58	**5 21**	p2.2
21	**S**	12 01	**12 07**	2.1	6 11	**6 12**	p2.3
22	**S**	1 06	**1 00**	2.0	7 21	**7 02**	p2.3
23	**M**	2 07	**1 51**	a2.1	8 25	**7 50**	p2.4
24	**T**	3 03	**2 43**	a2.2	9 25	**8 36**	p2.4
25	**W**	3 53	**3 33**	a2.2	10 17	**9 19**	p2.3
26	**T**	4 37	**4 20**	a2.2	11 04	**9 59**	p2.3
27	**F**	5 15	**5 05**	a2.2	11 46	**10 36**	p2.2
28	**S**	5 49	**5 48**	a2.2	-E-	**12 25**	1.4
29	**S**	6 20	**6 30**	a2.2	...	**1 01**	1.5
30	**M**	6 51	**7 12**	a2.3	12 01	**1 33**	a2.2

A also at 11:38 p.m. 2.9 **B** also at 11:05 p.m. 2.4 **C** also at 11:40 p.m. 2.3
D also at 11:28 p.m. 3.0 **E** also at 11:13 p.m. 2.2

The Kts. (knots) columns show the **maximum** predicted velocities of the stronger one of the Flood Currents and the stronger one of the Ebb Currents for each day.
The letter "a" means the velocity shown should occur **after** the **a.m.** Current Change. The letter "p" means the velocity shown should occur **after** the **p.m.** Current Change (even if next morning). No "a" or "p" means a.m. and p.m. velocities are the same for that day.
Avg. Max. Velocity: Flood 2.0 Kts., Ebb 1.9 Kts.
Max. Flood 3 hrs. 10 min. after Flood Starts ±45 min.
Max. Ebb 2 hrs. 45 min. after Ebb Starts ±45 min.
See pp. 22-29 for Current Change at other points.

 Note *from NOS: These predictions should be considered questionable. Caution is advised.*

2014 CURRENT TABLE
CHESAPEAKE & DELAWARE CANAL

39°31.89'N, 75°49.65'W at Chesapeake City

Daylight Saving Time — **Daylight Saving Time**

JULY

DAY OF MONTH	DAY OF WEEK	CURRENT TURNS TO EAST Flood Starts			CURRENT TURNS TO WEST Ebb Starts		
		a.m.	**p.m.**	Kts.	a.m.	**p.m.**	Kts.
1	T	7 23	**7 55**	a2.3	12 34	**2 04**	a2.2
2	W	7 57	**8 40**	a2.3	1 21	**2 33**	a2.1
3	T	8 34	**9 29**	a2.3	2 12	**3 04**	2.0
4	F	9 15	**10 25**	a2.3	3 08	**3 39**	p2.1
5	S	9 59	**11 24**	a2.2	4 13	**4 19**	p2.2
6	S	10 50	**...**	2.2	5 27	**5 05**	p2.3
7	M	12 25	**-A-**	1.7	6 40	**5 54**	p2.4
8	T	1 23	**12 41**	p2.1	7 47	**6 44**	p2.5
9	W	2 20	**1 39**	p2.1	8 50	**7 37**	p2.6
10	T	3 15	**2 39**	2.2	9 45	**8 33**	p2.8
11	F	4 08	**3 39**	2.3	10 35	**9 30**	p2.9
12	S	5 00	**4 37**	a2.5	11 22	**10 28**	p2.9
13	S	5 50	**5 33**	2.6	11 59	**11 27**	p2.9
14	M	6 39	**6 30**	a2.7	...	**12 52**	1.9
15	T	7 28	**7 28**	a2.7	12 27	**1 36**	a2.9
16	W	8 16	**8 27**	a2.6	1 29	**2 21**	a2.7
17	T	9 03	**9 28**	a2.5	2 33	**3 07**	a2.4
18	F	9 51	**10 33**	a2.3	3 38	**3 55**	p2.2
19	S	10 42	**11 41**	a2.2	4 47	**4 46**	p2.2
20	S	11 35	**...**	2.0	5 57	**5 39**	p2.1
21	M	12 45	**12 29**	a1.9	7 03	**6 31**	p2.1
22	T	1 44	**1 23**	a1.9	8 04	**7 21**	p2.1
23	W	2 38	**2 15**	a1.9	8 59	**8 07**	p2.1
24	T	3 24	**3 05**	a1.9	9 47	**8 52**	p2.1
25	F	4 04	**3 52**	a1.9	10 29	**9 33**	p2.1
26	S	4 38	**4 34**	a2.0	11 06	**10 13**	p2.1
27	S	5 10	**5 15**	a2.1	11 39	**10 54**	p2.2
28	M	5 42	**5 56**	a2.2	11 59	**11 36**	p2.2
29	T	6 15	**6 38**	a2.3	...	**12 39**	1.9
30	W	6 50	**7 22**	a2.4	12 21	**1 09**	a2.2
31	T	7 26	**8 08**	a2.4	1 10	**1 39**	p2.2

AUGUST

DAY OF MONTH	DAY OF WEEK	CURRENT TURNS TO EAST Flood Starts			CURRENT TURNS TO WEST Ebb Starts		
		a.m.	**p.m.**	Kts.	a.m.	**p.m.**	Kts.
1	F	8 05	**8 56**	a2.4	2 01	**2 13**	p2.2
2	S	8 46	**9 49**	a2.3	2 56	**2 51**	p2.3
3	S	9 30	**10 49**	a2.2	3 58	**3 34**	p2.3
4	M	10 23	**11 54**	a2.1	5 09	**4 23**	p2.3
5	T	11 21	**...**	2.0	6 21	**5 20**	p2.3
6	W	12 56	**12 24**	p2.0	7 26	**6 20**	p2.4
7	T	1 56	**1 26**	p2.1	8 25	**7 22**	p2.5
8	F	2 54	**2 28**	p2.3	9 18	**8 25**	p2.7
9	S	3 49	**3 28**	p2.4	10 05	**9 28**	p2.8
10	S	4 41	**4 26**	p2.6	10 50	**10 29**	p2.8
11	M	5 31	**5 22**	p2.7	11 33	**11 28**	p2.8
12	T	6 20	**6 18**	p2.8	...	**12 16**	2.2
13	W	7 07	**7 15**	p2.8	12 28	**1 01**	a2.7
14	T	7 53	**8 13**	p2.6	1 28	**1 46**	a2.5
15	F	8 38	**9 11**	a2.4	2 29	**2 31**	p2.4
16	S	9 24	**10 12**	a2.2	3 30	**3 19**	p2.2
17	S	10 13	**11 14**	a2.0	4 33	**4 09**	p2.1
18	M	11 07	**...**	1.7	5 38	**5 03**	p1.9
19	T	12 15	**12 04**	a1.7	6 39	**5 57**	p1.8
20	W	1 10	**12 58**	a1.7	7 34	**6 49**	p1.7
21	T	1 58	**1 50**	a1.7	8 23	**7 38**	p1.8
22	F	2 40	**2 38**	a1.7	9 06	**8 25**	p1.8
23	S	3 18	**3 23**	a1.8	9 44	**9 10**	p1.9
24	S	3 53	**4 04**	a1.9	10 16	**9 55**	p2.0
25	M	4 27	**4 45**	2.0	10 46	**10 40**	p2.1
26	T	5 02	**5 26**	2.2	11 15	**11 25**	2.1
27	W	5 38	**6 09**	2.3	11 44	**...**	2.2
28	T	6 16	**6 53**	a2.4	12 12	**12 15**	p2.4
29	F	6 56	**7 39**	a2.4	1 02	**12 49**	p2.5
30	S	7 37	**8 27**	a2.4	1 54	**1 27**	p2.5
31	S	8 21	**9 18**	a2.2	2 48	**2 09**	p2.4

A also at 11:45 a.m. 2.1

The Kts. (knots) columns show the **maximum** predicted velocities of the stronger one of the Flood Currents and the stronger one of the Ebb Currents for each day.
The letter "a" means the velocity shown should occur **after** the **a.m.** Current Change. The letter "p" means the velocity shown should occur **after** the **p.m.** Current Change (even if next morning). No "a" or "p" means a.m. and p.m. velocities are the same for that day.
Avg. Max. Velocity: Flood 2.0 Kts., Ebb 1.9 Kts.
Max. Flood 3 hrs. 10 min. after Flood Starts ±45 min.
Max. Ebb 2 hrs. 45 min. after Ebb Starts ±45 min.

See pp. 22-29 for Current Change at other points.

Note *from NOS: These predictions should be considered questionable. Caution is advised.*

2014 CURRENT TABLE
CHESAPEAKE & DELAWARE CANAL

39°31.89'N, 75°49.65'W at Chesapeake City

Daylight Saving Time

SEPTEMBER

DAY OF MONTH	DAY OF WEEK	CURRENT TURNS TO EAST Flood Starts a.m.	p.m.	Kts.	CURRENT TURNS TO WEST Ebb Starts a.m.	p.m.	Kts.
1	M	9 08	**10 16**	a2.1	3 48	**2 56**	p2.4
2	T	10 03	**11 22**	a2.0	4 54	**3 50**	p2.3
3	W	11 07	...	1.9	6 00	**4 54**	p2.3
4	T	12 29	**12 15**	2.0	7 00	**6 04**	p2.3
5	F	1 30	**1 18**	p2.1	7 54	**7 15**	p2.4
6	S	2 30	**2 19**	p2.3	8 44	**8 24**	p2.5
7	S	3 26	**3 19**	p2.6	9 30	**9 29**	p2.6
8	M	4 18	**4 16**	p2.7	10 14	**10 30**	p2.6
9	T	5 07	**5 12**	p2.9	10 57	**11 29**	p2.5
10	W	5 55	**6 07**	p2.9	11 41	...	2.5
11	T	6 41	**7 03**	p2.8	12 27	**12 25**	p2.6
12	F	7 27	**7 58**	p2.6	1 25	**1 10**	p2.5
13	S	8 13	**8 51**	p2.3	2 22	**1 56**	p2.4
14	S	8 59	**9 44**	p2.0	3 19	**2 42**	p2.2
15	M	9 48	**10 38**	p1.8	4 16	**3 30**	p2.0
16	T	10 42	**11 31**	p1.7	5 13	**4 21**	p1.8
17	W	11 39	...	1.3	6 07	**5 16**	p1.6
18	T	12 20	**12 34**	a1.6	6 56	**6 12**	p1.5
19	F	1 03	**1 23**	a1.6	7 39	**7 06**	p1.6
20	S	1 43	**2 09**	a1.7	8 16	**7 58**	1.6
21	S	2 22	**2 53**	a1.8	8 50	**8 49**	1.8
22	M	3 01	**3 36**	1.9	9 21	**9 39**	a2.0
23	T	3 41	**4 18**	2.1	9 51	**10 27**	a2.2
24	W	4 21	**5 01**	p2.3	10 21	**11 15**	a2.4
25	T	5 01	**5 45**	p2.4	10 53	**11 59**	a2.5
26	F	5 43	**6 30**	p2.5	11 28	...	2.6
27	S	6 26	**7 16**	p2.5	12 55	**12 07**	p2.7
28	S	7 12	**8 04**	p2.4	1 48	**12 49**	p2.7
29	M	8 00	**8 54**	p2.3	2 41	**1 36**	p2.6
30	T	8 51	**9 50**	p2.2	3 36	**2 27**	p2.5

Daylight Saving Time

OCTOBER

DAY OF MONTH	DAY OF WEEK	CURRENT TURNS TO EAST Flood Starts a.m.	p.m.	Kts.	CURRENT TURNS TO WEST Ebb Starts a.m.	p.m.	Kts.
1	W	9 50	**10 54**	p2.1	4 35	**3 27**	p2.3
2	T	10 57	...	1.9	5 35	**4 37**	p2.3
3	F	12 01	**12 05**	a2.1	6 30	**5 56**	p2.2
4	S	1 04	**1 10**	p2.2	7 20	**7 11**	p2.2
5	S	2 02	**2 11**	p2.4	8 08	**8 22**	p2.3
6	M	2 58	**3 10**	p2.6	8 54	**9 27**	2.3
7	T	3 50	**4 08**	p2.8	9 39	**10 28**	a2.5
8	W	4 39	**5 03**	p2.8	10 24	**11 25**	a2.6
9	T	5 26	**5 57**	p2.8	11 08	...	2.7
10	F	6 13	**6 50**	p2.7	12 22	**12 01**	p2.7
11	S	7 00	**7 41**	p2.5	1 18	**12 36**	p2.5
12	S	7 48	**8 28**	p2.3	2 11	**1 21**	p2.4
13	M	8 35	**9 11**	p2.0	3 03	**2 04**	p2.1
14	T	9 24	**9 53**	p1.9	3 52	**2 48**	p1.9
15	W	10 15	**10 35**	p1.7	4 41	**3 34**	p1.7
16	T	11 10	**11 17**	p1.7	5 27	**4 29**	p1.6
17	F	11 59	**11 59**	p1.2	6 09	**5 31**	p1.5
18	S	...	**12 52**	1.3	6 47	**6 32**	a1.6
19	S	12 41	**1 38**	a1.8	7 20	**7 31**	a1.8
20	M	1 24	**2 24**	a1.9	7 52	**8 28**	a2.0
21	T	2 08	**3 09**	2.0	8 24	**9 23**	a2.3
22	W	2 53	**3 55**	p2.2	8 58	**10 15**	a2.5
23	T	3 39	**4 40**	p2.4	9 33	**11 06**	a2.6
24	F	4 25	**5 25**	p2.5	10 11	**11 56**	a2.8
25	S	5 12	**6 10**	p2.6	10 51	...	2.8
26	S	6 00	**6 57**	p2.6	12 48	**-A-**	1.5
27	M	6 50	**7 45**	p2.6	1 39	**12 20**	p2.8
28	T	7 43	**8 36**	p2.5	2 29	**1 12**	p2.7
29	W	8 38	**9 30**	p2.4	3 20	**2 09**	p2.5
30	T	9 38	**10 30**	p2.3	4 11	**3 14**	p2.4
31	F	10 45	**11 33**	p2.2	5 04	**4 29**	p2.2

A also at 11:34 a.m. 2.8

The Kts. (knots) columns show the **maximum** predicted velocities of the stronger one of the Flood Currents and the stronger one of the Ebb Currents for each day.
The letter "a" means the velocity shown should occur **after** the **a.m.** Current Change. The letter "p" means the velocity shown should occur **after** the **p.m.** Current Change (even if next morning). No "a" or "p" means a.m. and p.m. velocities are the same for that day.

Avg. Max. Velocity: Flood 2.0 Kts., Ebb 1.9 Kts.

Max. Flood 3 hrs. 10 min. after Flood Starts ±45 min.

Max. Ebb 2 hrs. 45 min. after Ebb Starts ±45 min.

See pp. 22-29 for Current Change at other points.

Note *from NOS: These predictions should be considered questionable. Caution is advised.*

2014 CURRENT TABLE
CHESAPEAKE & DELAWARE CANAL
39°31.89'N, 75°49.65'W at Chesapeake City

***Standard Time starts Nov. 2 at 2 a.m.** — **Standard Time**

NOVEMBER

DAY OF MONTH	DAY OF WEEK	CURRENT TURNS TO EAST Flood Starts			CURRENT TURNS TO WEST Ebb Starts		
		a.m.	**p.m.**	Kts.	a.m.	**p.m.**	Kts.
1	S	11 54	...	2.1	5 56	**5 50**	p2.1
2	S	12 34	***12 01**	2.2	*5 45	***6 06**	p2.1
3	M	12 31	**1 01**	p2.4	6 33	**7 16**	a2.2
4	T	1 27	**2 03**	p2.6	7 20	**8 22**	a2.5
5	W	2 18	**3 00**	p2.7	8 07	**9 22**	a2.7
6	T	3 08	**3 54**	p2.7	8 54	**10 18**	a2.7
7	F	3 57	**4 45**	p2.7	9 38	**11 13**	a2.7
8	S	4 45	**5 34**	p2.6	10 22	**11 59**	a2.6
9	S	5 34	**6 19**	p2.4	11 05	...	2.5
10	M	6 22	**6 59**	p2.2	12 54	-A-	1.5
11	T	7 09	**7 34**	p2.1	1 39	**12 28**	p2.1
12	W	7 55	**8 06**	p2.0	2 21	**1 09**	p1.9
13	T	8 42	**8 39**	p1.9	3 00	**1 53**	p1.7
14	F	9 32	**9 16**	p1.9	3 38	**2 45**	p1.6
15	S	10 24	**9 58**	p1.9	4 15	**3 50**	a1.6
16	S	11 15	**10 45**	p1.9	4 49	**4 59**	a1.8
17	M	11 59	**11 33**	p2.0	5 22	**6 04**	a2.0
18	T	...	**12 55**	1.8	5 57	**7 07**	a2.2
19	W	12 21	**1 44**	a2.1	6 33	**8 07**	a2.4
20	T	1 10	**2 33**	p2.2	7 12	**9 02**	a2.6
21	F	2 01	**3 20**	p2.3	7 54	**9 54**	a2.8
22	S	2 53	**4 07**	p2.5	8 38	**10 44**	a2.9
23	S	3 45	**4 54**	p2.6	9 24	**11 34**	a2.9
24	M	4 37	**5 42**	p2.7	10 12	...	2.9
25	T	5 31	**6 31**	p2.7	12 22	-B-	1.5
26	W	6 27	**7 21**	p2.6	1 09	**12 01**	p2.8
27	T	7 24	**8 12**	p2.5	1 55	**1 01**	p2.6
28	F	8 24	**9 07**	p2.4	2 42	**2 09**	p2.4
29	S	9 30	**10 05**	p2.3	3 30	**3 24**	p2.2
30	S	10 40	**11 03**	p2.3	4 21	**4 44**	2.0

DECEMBER

DAY OF MONTH	DAY OF WEEK	CURRENT TURNS TO EAST Flood Starts			CURRENT TURNS TO WEST Ebb Starts		
		a.m.	**p.m.**	Kts.	a.m.	**p.m.**	Kts.
1	M	11 47	**11 59**	2.2	5 11	**5 58**	a2.2
2	T	...	**12 51**	2.3	6 01	**7 08**	a2.4
3	W	12 53	**1 52**	p2.4	6 51	**8 12**	a2.6
4	T	1 47	**2 50**	p2.5	7 40	**9 11**	a2.7
5	F	2 39	**3 41**	p2.5	8 28	**10 05**	a2.7
6	S	3 29	**4 29**	p2.5	9 13	**10 55**	a2.6
7	S	4 19	**5 12**	p2.4	9 57	**11 42**	a2.5
8	M	5 07	**5 51**	p2.3	10 37	...	2.3
9	T	5 54	**6 24**	p2.2	12 25	-C-	1.5
10	W	6 38	**6 55**	p2.2	1 04	**12 01**	p2.1
11	T	7 21	**7 23**	p2.1	1 39	**12 38**	p2.0
12	F	8 03	**7 54**	p2.1	2 12	**1 23**	p1.8
13	S	8 49	**8 30**	p2.1	2 42	**2 15**	1.7
14	S	9 41	**9 12**	p2.1	3 13	**3 17**	a1.8
15	M	10 36	**9 59**	p2.1	3 48	**4 29**	a2.0
16	T	11 32	**10 51**	p2.1	4 26	**5 40**	a2.1
17	W	-D-	**12 26**	1.8	5 07	**6 46**	a2.3
18	T	...	**1 19**	1.9	5 51	**7 48**	a2.5
19	F	12 36	**2 11**	2.1	6 36	**8 45**	a2.6
20	S	1 32	**3 02**	p2.3	7 25	**9 37**	a2.8
21	S	2 29	**3 50**	p2.4	8 15	**10 25**	a2.9
22	M	3 25	**4 39**	p2.6	9 07	**11 11**	a2.9
23	T	4 20	**5 27**	p2.7	10 01	**11 56**	a2.9
24	W	5 15	**6 16**	p2.7	10 57	...	2.9
25	T	6 11	**7 05**	p2.7	12 40	**12 01**	p2.8
26	F	7 09	**7 54**	p2.6	1 24	**1 00**	p2.6
27	S	8 09	**8 44**	p2.5	2 08	**2 07**	p2.4
28	S	9 14	**9 38**	p2.4	2 55	**3 19**	2.1
29	M	10 24	**10 33**	p2.2	3 46	**4 35**	a2.2
30	T	11 33	**11 29**	2.1	4 39	**5 47**	a2.3
31	W	...	**12 38**	2.1	5 32	**6 55**	a2.4

A also at 11:47 a.m. 2.3 **B** also at 11:03 a.m. 2.9 **C** also at 11:17 a.m. 2.2
D also at 11:43 p.m. 2.1

The Kts. (knots) columns show the **maximum** predicted velocities of the stronger one of the Flood Currents and the stronger one of the Ebb Currents for each day.
The letter "a" means the velocity shown should occur **after** the **a.m.** Current Change. The letter "p" means the velocity shown should occur **after** the **p.m.** Current Change (even if next morning). No "a" or "p" means a.m. and p.m. velocities are the same for that day.
Avg. Max. Velocity: Flood 2.0 Kts., Ebb 1.9 Kts.
Max. Flood 3 hrs. 10 min. after Flood Starts ±45 min.
Max. Ebb 2 hrs. 45 min. after Ebb Starts ±45 min.

See pp. 22-29 for Current Change at other points.

Note *from NOS: These predictions should be considered questionable. Caution is advised.*

Upper Chesapeake Bay Currents

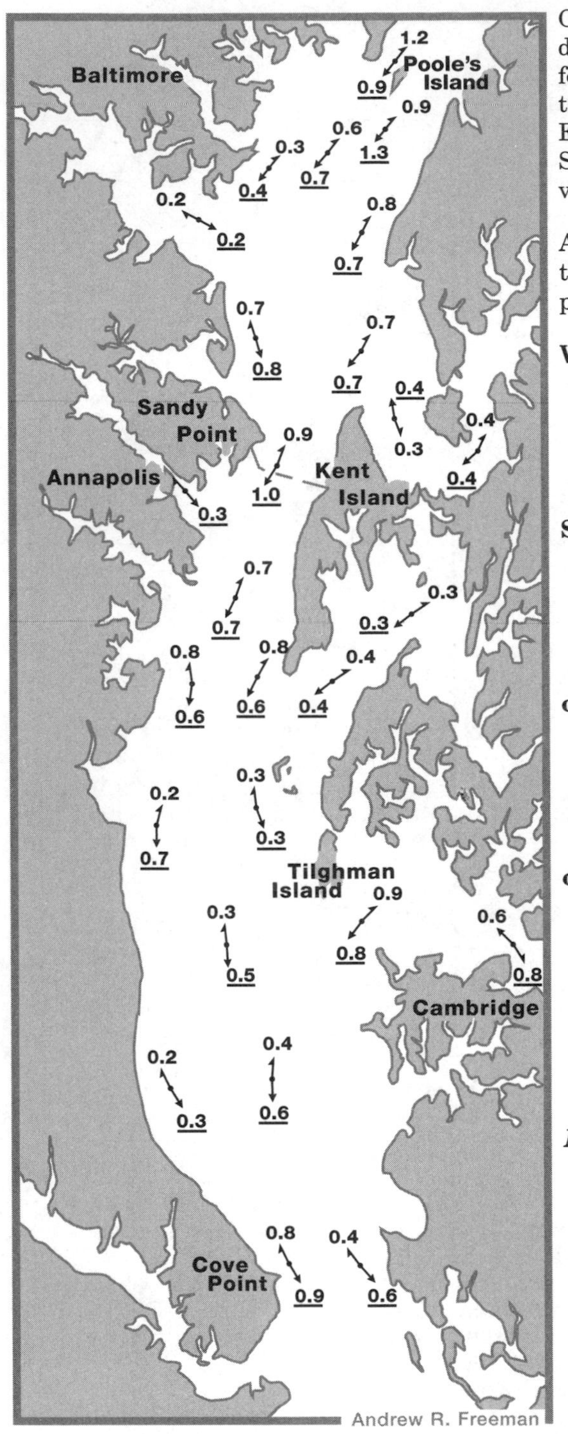

On this Current Diagram, the arrows denote maximum velocities. Refer to the four areas listed below for the specific times. Double-headed arrows are for Ebb and the velocities are underlined. Single-headed arrows are for Flood and velocities and not underlined.

All times below are in hours and relate to the time of **High Water at Baltimore**, pp. 156-159.

West of Pooles Island:
Flood begins 3 1/2 before
Flood max. 1 1/2 before (1.2 kts.)
Ebb begins 2 1/2 after
Ebb max. 4 1/2 after (0.9 kts.)

Sandy Point:
Flood begins 3 1/2 before
Flood max. 1 1/2 before (0.9 kts.)
Ebb begins 1 1/2 after
Ebb max. 4 1/2 after (1.0 kts.)

off Tilghman Island:
Flood begins 5 1/2 before
Flood max. 3 1/2 before (0.3 kts.)
Ebb begins 1/2 after
Ebb max. 3 1/2 after (0.7 kts.)

off Cove Point:
Flood begins 6 1/2 before
Flood max. 4 1/2 before (0.9 kts.)
Ebb begins 1/2 before
Ebb max. 1 1/2 after (0.8 kts.)

Note:
From the beginning of the Flood Current at Cove Point until the Ebb Current begins off Baltimore, a northbound vessel will have over 8 hours of fair current. A vessel bound southward from Sandy Point can expect only 4 hours of fair current.

Relationship of High Water and Ebb Current

Many people wonder why the times of High Water and the start of Ebb Current at the mouths of bays and inlets are not simultaneous. (See p. 10, Why Tides and Currents Often Behave Differently.) The twelve diagrams below show the hourly stages of the Tide in the Ocean and a Bay connected by a narrow Inlet.

Picture the rising Tide, borne by the Flood Current, as a long wave. The wave enters the inlet and the crest reaches its maximum height in or at the inlet. But, the body of water inside the inlet - in the bay - has yet to be filled and the Flood Current continues to pour water through the inlet for a good period after the crest has already passed the inlet. The Ebb Current will not start until the level of the water in the ocean is lower than the water in the bay.

This does not necessarily apply to the mouths of small bays with wide entrances. The narrowness of the inlet and the size of the bay are the controlling factors.

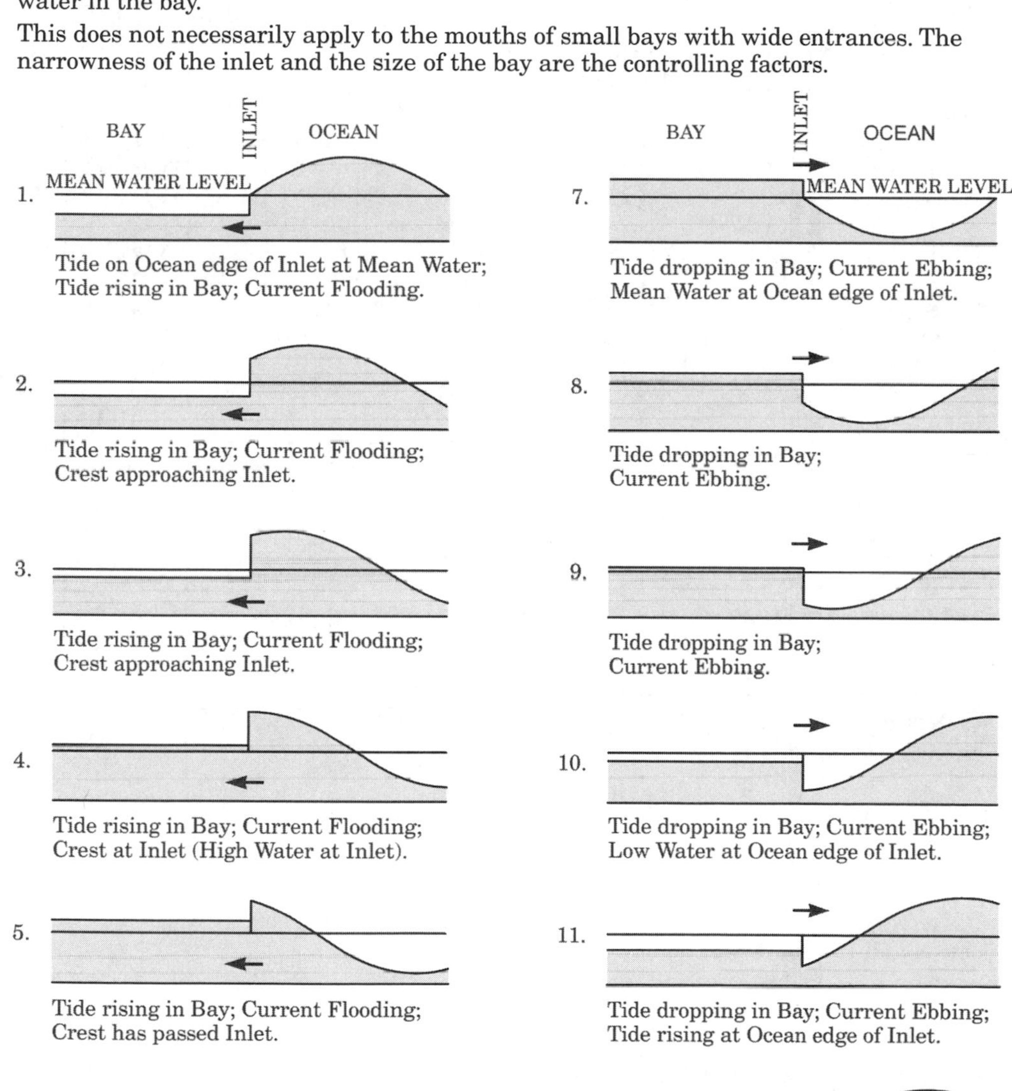

1. Tide on Ocean edge of Inlet at Mean Water; Tide rising in Bay; Current Flooding.

2. Tide rising in Bay; Current Flooding; Crest approaching Inlet.

3. Tide rising in Bay; Current Flooding; Crest approaching Inlet.

4. Tide rising in Bay; Current Flooding; Crest at Inlet (High Water at Inlet).

5. Tide rising in Bay; Current Flooding; Crest has passed Inlet.

6. High Water in Bay; Ebb Current about to start.

7. Tide dropping in Bay; Current Ebbing; Mean Water at Ocean edge of Inlet.

8. Tide dropping in Bay; Current Ebbing.

9. Tide dropping in Bay; Current Ebbing.

10. Tide dropping in Bay; Current Ebbing; Low Water at Ocean edge of Inlet.

11. Tide dropping in Bay; Current Ebbing; Tide rising at Ocean edge of Inlet.

12. Low Water in Bay; Flood Current about to start.

2014 HIGH WATER
BALTIMORE, MD
At Ft. McHenry 39°16'N, 76°34.7'W

Standard Time — Standard Time — *Daylight Time starts March 9 at 2 a.m.

DAY OF MONTH	DAY OF WEEK	JANUARY a.m.	Ht.	**p.m.**	Ht.	DAY OF WEEK	FEBRUARY a.m.	Ht.	**p.m.**	Ht.	DAY OF WEEK	MARCH a.m.	Ht.	**p.m.**	Ht.	DAY OF MONTH
1	**W**	5 43	0.7	**6 34**	1.5	**S**	7 17	1.0	**7 59**	1.2	**S**	6 06	1.2	**6 49**	1.3	1
2	**T**	6 38	0.8	**7 26**	1.4	**S**	8 12	1.0	**8 46**	1.2	**S**	6 59	1.3	**7 36**	1.2	2
3	**F**	7 34	0.8	**8 17**	1.4	**M**	9 07	1.0	**9 34**	1.1	**M**	7 51	1.3	**8 21**	1.1	3
4	**S**	8 31	0.9	**9 09**	1.3	**T**	10 05	1.0	**10 22**	1.0	**T**	8 44	1.3	**9 08**	1.1	4
5	**S**	9 28	0.9	**9 59**	1.2	**W**	11 02	1.0	**11 10**	0.8	**W**	9 35	1.3	**9 53**	1.0	5
6	**M**	10 28	0.9	**10 49**	1.1	**T**	...	...	**12 04**	1.0	**T**	10 28	1.3	**10 42**	0.9	6
7	**T**	11 31	1.0	**11 41**	0.9	**F**	12 02	0.8	**1 07**	1.0	**F**	11 24	1.2	**11 33**	0.9	7
8	**W**	...	...	**12 36**	1.0	**S**	12 56	0.7	**2 12**	1.0	**S**	...	...	**12 23**	1.1	8
9	**T**	12 34	0.8	**1 42**	1.0	**S**	1 53	0.7	**3 14**	1.0	**S**	12 28	0.8	***2 26**	1.1	9
10	**F**	1 28	0.7	**2 46**	1.0	**M**	2 49	0.7	**4 09**	1.0	**M**	2 25	0.8	**3 29**	1.1	10
11	**S**	2 22	0.7	**3 45**	1.1	**T**	3 43	0.7	**4 57**	1.0	**T**	3 22	0.9	**4 27**	1.1	11
12	**S**	3 16	0.6	**4 37**	1.1	**W**	4 33	0.7	**5 37**	1.0	**W**	4 17	0.9	**5 16**	1.1	12
13	**M**	4 07	0.6	**5 22**	1.1	**T**	5 19	0.8	**6 13**	1.0	**T**	5 07	1.0	**5 58**	1.1	13
14	**T**	4 56	0.6	**6 03**	1.1	**F**	6 01	0.8	**6 46**	1.1	**F**	5 53	1.0	**6 35**	1.1	14
15	**W**	5 41	0.6	**6 39**	1.1	**S**	6 41	0.8	**7 17**	1.1	**S**	6 36	1.1	**7 10**	1.1	15
16	**T**	6 24	0.7	**7 13**	1.1	**S**	7 20	0.9	**7 49**	1.0	**S**	7 15	1.2	**7 44**	1.1	16
17	**F**	7 05	0.7	**7 45**	1.1	**M**	7 58	0.9	**8 23**	1.0	**M**	7 53	1.2	**8 19**	1.1	17
18	**S**	7 46	0.7	**8 18**	1.1	**T**	8 38	1.0	**8 59**	1.0	**T**	8 30	1.3	**8 56**	1.1	18
19	**S**	8 26	0.7	**8 51**	1.0	**W**	9 20	1.0	**9 39**	0.9	**W**	9 10	1.4	**9 35**	1.1	19
20	**M**	9 08	0.7	**9 27**	1.0	**T**	10 06	1.1	**10 24**	0.9	**T**	9 52	1.4	**10 19**	1.0	20
21	**T**	9 52	0.8	**10 05**	0.9	**F**	10 57	1.1	**11 14**	0.8	**F**	10 39	1.4	**11 07**	1.0	21
22	**W**	10 39	0.8	**10 48**	0.9	**S**	11 53	1.2	**...**	...	**S**	11 30	1.4	**...**	...	22
23	**T**	11 31	0.9	**11 35**	0.8	**S**	12 09	0.8	**12 55**	1.2	**S**	12 01	1.0	**12 28**	1.4	23
24	**F**	...	...	**12 26**	1.0	**M**	1 10	0.8	**2 01**	1.2	**M**	12 58	1.0	**1 32**	1.4	24
25	**S**	12 29	0.7	**1 26**	1.1	**T**	2 13	0.8	**3 07**	1.2	**T**	2 00	1.0	**2 39**	1.4	25
26	**S**	1 27	0.7	**2 27**	1.1	**W**	3 15	0.9	**4 09**	1.3	**W**	3 03	1.1	**3 45**	1.4	26
27	**M**	2 28	0.6	**3 29**	1.2	**T**	4 14	1.0	**5 07**	1.3	**T**	4 05	1.2	**4 48**	1.3	27
28	**T**	3 29	0.7	**4 28**	1.3	**F**	5 12	1.1	**6 00**	1.3	**F**	5 04	1.4	**5 44**	1.3	28
29	**W**	4 28	0.7	**5 25**	1.3						**S**	5 59	1.5	**6 36**	1.3	29
30	**T**	5 26	0.8	**6 19**	1.3						**S**	6 52	1.6	**7 24**	1.3	30
31	**F**	6 22	0.9	**7 10**	1.3						**M**	7 42	1.6	**8 10**	1.2	31

Dates when Ht. of **Low** Water is below Mean Low with Ht. of lowest given for each period and Date of lowest in ( ):

1st - 31st: -0.6' (30th - 31st)

1st - 16th: -0.5' (1st)
19th - 28th: -0.4' (27th - 28th)

1st - 5th: -0.4' (1st)

Average Rise and Fall 1.1 ft.

When a high tide exceeds avg. ht., the *following* low tide will be lower than avg.

2014 HIGH WATER
BALTIMORE, MD
At Ft. McHenry 39°16'N, 76°34.7'W

DAY OF MONTH	DAY OF WEEK	APRIL Daylight Saving Time				DAY OF WEEK	MAY Daylight Saving Time				DAY OF WEEK	JUNE Daylight Saving Time				DAY OF MONTH
		a.m.	Ht.	**p.m.**	Ht.		a.m.	Ht.	**p.m.**	Ht.		a.m.	Ht.	**p.m.**	Ht.	
1	**T**	8 31	1.6	**8 55**	1.2	**T**	8 53	1.8	**9 15**	1.2	**S**	9 49	1.8	**10 23**	1.3	1
2	**W**	9 18	1.6	**9 40**	1.1	**F**	9 36	1.8	**10 01**	1.2	**M**	10 29	1.7	**11 10**	1.3	2
3	**T**	10 05	1.6	**10 27**	1.1	**S**	10 19	1.7	**10 49**	1.2	**T**	11 09	1.6	**11 59**	1.4	3
4	**F**	10 54	1.5	**11 16**	1.1	**S**	11 05	1.6	**11 40**	1.2	**W**	11 53	1.5	**...**	...	4
5	**S**	11 43	1.4	**...**	...	**M**	11 50	1.5	**...**	...	**T**	12 51	1.4	**12 37**	1.4	5
6	**S**	12 06	1.1	**12 36**	1.3	**T**	12 31	1.3	**12 39**	1.4	**F**	1 44	1.5	**1 25**	1.4	6
7	**M**	1 00	1.1	**1 32**	1.3	**W**	1 26	1.3	**1 30**	1.4	**S**	2 35	1.6	**2 16**	1.3	7
8	**T**	1 57	1.1	**2 31**	1.2	**T**	2 21	1.4	**2 23**	1.3	**S**	3 25	1.7	**3 09**	1.2	8
9	**W**	2 53	1.2	**3 27**	1.2	**F**	3 14	1.4	**3 15**	1.3	**M**	4 13	1.8	**4 04**	1.2	9
10	**T**	3 48	1.2	**4 19**	1.2	**S**	4 04	1.5	**4 05**	1.2	**T**	4 59	1.9	**5 00**	1.2	10
11	**F**	4 38	1.3	**5 05**	1.2	**S**	4 50	1.6	**4 54**	1.2	**W**	5 44	2.0	**5 55**	1.2	11
12	**S**	5 24	1.4	**5 47**	1.2	**M**	5 33	1.7	**5 42**	1.2	**T**	6 31	2.1	**6 49**	1.2	12
13	**S**	6 06	1.5	**6 27**	1.2	**T**	6 15	1.8	**6 29**	1.2	**F**	7 18	2.1	**7 42**	1.2	13
14	**M**	6 46	1.5	**7 07**	1.2	**W**	6 56	1.9	**7 17**	1.2	**S**	8 06	2.1	**8 36**	1.3	14
15	**T**	7 24	1.6	**7 47**	1.2	**T**	7 39	2.0	**8 05**	1.2	**S**	8 56	2.1	**9 31**	1.4	15
16	**W**	8 03	1.7	**8 30**	1.2	**F**	8 24	2.0	**8 56**	1.2	**M**	9 48	2.0	**10 28**	1.5	16
17	**T**	8 45	1.8	**9 15**	1.1	**S**	9 11	2.0	**9 48**	1.3	**T**	10 40	1.9	**11 27**	1.5	17
18	**F**	9 29	1.8	**10 03**	1.1	**S**	10 02	1.9	**10 43**	1.3	**W**	11 34	1.7	**...**	...	18
19	**S**	10 18	1.8	**10 55**	1.2	**M**	10 56	1.8	**11 41**	1.4	**T**	12 28	1.6	**12 29**	1.6	19
20	**S**	11 11	1.7	**11 51**	1.2	**T**	11 53	1.7	**...**	...	**F**	1 31	1.7	**1 26**	1.5	20
21	**M**	...	...	**12 09**	1.6	**W**	12 42	1.5	**12 52**	1.6	**S**	2 34	1.8	**2 23**	1.4	21
22	**T**	12 51	1.3	**1 12**	1.6	**T**	1 44	1.6	**1 53**	1.5	**S**	3 34	1.9	**3 22**	1.3	22
23	**W**	1 53	1.3	**2 17**	1.5	**F**	2 47	1.7	**2 53**	1.4	**M**	4 31	1.9	**4 19**	1.2	23
24	**T**	2 56	1.4	**3 21**	1.4	**S**	3 47	1.8	**3 52**	1.3	**T**	5 23	2.0	**5 15**	1.2	24
25	**F**	3 56	1.6	**4 21**	1.4	**S**	4 44	1.9	**4 47**	1.3	**W**	6 10	2.0	**6 07**	1.2	25
26	**S**	4 54	1.7	**5 17**	1.3	**M**	5 36	1.9	**5 40**	1.2	**T**	6 53	2.0	**6 56**	1.2	26
27	**S**	5 48	1.8	**6 08**	1.3	**T**	6 24	2.0	**6 31**	1.2	**F**	7 33	1.9	**7 43**	1.2	27
28	**M**	6 38	1.8	**6 57**	1.3	**W**	7 09	2.0	**7 19**	1.2	**S**	8 10	1.9	**8 28**	1.3	28
29	**T**	7 25	1.9	**7 44**	1.2	**T**	7 51	2.0	**8 05**	1.2	**S**	8 46	1.8	**9 11**	1.3	29
30	**W**	8 10	1.9	**8 29**	1.2	**F**	8 31	1.9	**8 51**	1.2	**M**	9 21	1.8	**9 55**	1.3	30
31						**S**	9 10	1.8	**9 37**	1.2						31

Dates when Ht. of **Low** Water is below Mean Low with Ht. of lowest given for each period and Date of lowest in ( ):

Average Rise and Fall 1.1 ft.

When a high tide exceeds avg. ht., the *following* low tide will be lower than avg.

2014 HIGH WATER BALTIMORE, MD

At Ft. McHenry 39°16'N, 76°34.7'W

DAY OF MONTH	DAY OF WEEK	JULY Daylight Saving Time				DAY OF WEEK	AUGUST Daylight Saving Time				DAY OF WEEK	SEPTEMBER Daylight Saving Time				DAY OF MONTH
		a.m.	Ht.	**p.m.**	Ht.		a.m.	Ht.	**p.m.**	Ht.		a.m.	Ht.	**p.m.**	Ht.	
1	T	9 56	1.7	**10 40**	1.4	F	10 36	1.6	**11 33**	1.6	M	11 39	1.4	...	...	1
2	W	10 33	1.6	**11 26**	1.4	S	11 17	1.5	...	...	T	12 34	1.9	**12 34**	1.3	2
3	T	11 11	1.6	...	...	S	12 20	1.7	**12 03**	1.4	W	1 30	1.9	**1 36**	1.3	3
4	F	12 14	1.5	**-A-**	...	M	1 11	1.8	**12 57**	1.3	T	2 32	1.9	**2 43**	1.3	4
5	S	1 02	1.6	**12 38**	1.4	T	2 03	1.9	**1 55**	1.3	F	3 32	2.0	**3 47**	1.4	5
6	S	1 52	1.7	**1 29**	1.3	W	2 59	1.9	**2 59**	1.2	S	4 32	2.0	**4 49**	1.5	6
7	M	2 43	1.8	**2 25**	1.2	T	3 57	2.0	**4 03**	1.3	S	5 28	2.0	**5 48**	1.6	7
8	T	3 34	1.9	**3 25**	1.2	F	4 53	2.0	**5 05**	1.3	M	6 20	2.0	**6 45**	1.7	8
9	W	4 26	2.0	**4 26**	1.2	S	5 48	2.1	**6 04**	1.4	T	7 10	1.9	**7 40**	1.8	9
10	T	5 17	2.0	**5 26**	1.2	S	6 41	2.1	**7 01**	1.5	W	7 58	1.8	**8 33**	1.9	10
11	F	6 08	2.1	**6 24**	1.3	M	7 32	2.0	**7 57**	1.6	T	8 44	1.7	**9 27**	2.0	11
12	S	7 00	2.1	**7 21**	1.3	T	8 21	2.0	**8 53**	1.7	F	9 31	1.6	**10 20**	2.0	12
13	S	7 50	2.1	**8 16**	1.4	W	9 09	1.9	**9 49**	1.8	S	10 19	1.5	**11 15**	2.0	13
14	M	8 41	2.0	**9 12**	1.5	T	9 57	1.8	**10 46**	1.9	S	11 10	1.4	...	...	14
15	T	9 31	1.9	**10 10**	1.6	F	10 46	1.6	**11 44**	1.9	M	12 11	1.9	**12 04**	1.3	15
16	W	10 21	1.8	**11 08**	1.7	S	11 37	1.5	...	...	T	1 10	1.8	**1 02**	1.3	16
17	T	11 11	1.7	...	...	S	12 44	1.9	**12 30**	1.4	W	2 11	1.8	**2 04**	1.3	17
18	F	12 09	1.8	**12 03**	1.6	M	1 46	1.9	**1 28**	1.3	T	3 10	1.8	**3 06**	1.3	18
19	S	1 11	1.8	**12 58**	1.4	T	2 48	1.9	**2 29**	1.3	F	4 04	1.7	**4 05**	1.3	19
20	S	2 14	1.9	**1 55**	1.3	W	3 47	1.9	**3 31**	1.2	S	4 51	1.7	**5 00**	1.4	20
21	M	3 15	1.9	**2 54**	1.2	T	4 41	1.9	**4 30**	1.3	S	5 32	1.7	**5 48**	1.5	21
22	T	4 13	1.9	**3 54**	1.2	F	5 28	1.8	**5 24**	1.3	M	6 08	1.7	**6 33**	1.5	22
23	W	5 06	1.9	**4 52**	1.2	S	6 09	1.8	**6 12**	1.4	T	6 42	1.6	**7 13**	1.6	23
24	T	5 53	1.9	**5 45**	1.2	S	6 45	1.8	**6 57**	1.4	W	7 15	1.6	**7 50**	1.7	24
25	F	6 35	1.9	**6 35**	1.3	M	7 19	1.8	**7 39**	1.5	T	7 49	1.6	**8 27**	1.7	25
26	S	7 12	1.9	**7 20**	1.3	T	7 50	1.7	**8 19**	1.5	F	8 24	1.5	**9 03**	1.8	26
27	S	7 47	1.8	**8 04**	1.4	W	8 22	1.7	**8 57**	1.6	S	9 02	1.5	**9 42**	1.8	27
28	M	8 20	1.8	**8 46**	1.4	T	8 54	1.7	**9 35**	1.7	S	9 44	1.4	**10 25**	1.9	28
29	T	8 53	1.8	**9 27**	1.4	F	9 29	1.6	**10 14**	1.7	M	10 30	1.3	**11 13**	1.9	29
30	W	9 25	1.7	**10 08**	1.5	S	10 07	1.5	**10 56**	1.8	T	11 23	1.3	...	...	30
31	T	9 59	1.6	**10 50**	1.6	S	10 50	1.4	**11 42**	1.8						31

A also at 11:53 a.m. 1.5

Dates when Ht. of **Low** Water is below Mean Low with Ht. of lowest given for each period and Date of lowest in ( ):

Average Rise and Fall 1.1 ft.

When a high tide exceeds avg. ht., the *following* low tide will be lower than avg.

2014 HIGH WATER
BALTIMORE, MD

At Ft. McHenry 39°16'N, 76°34.7'W

Daylight Saving Time | ***Standard Time starts Nov. 2 at 2 a.m.** | **Standard Time**

DAY OF MONTH	DAY OF WEEK	OCTOBER				DAY OF WEEK	NOVEMBER				DAY OF WEEK	DECEMBER				DAY OF MONTH
		a.m.	Ht.	**p.m.**	Ht.		a.m.	Ht.	**p.m.**	Ht.		a.m.	Ht.	**p.m.**	Ht.	
1	W	12 06	1.9	**12 21**	1.3	S	1 45	1.6	**2 20**	1.3	M	1 19	1.2	**2 14**	1.3	1
2	T	1 05	1.9	**1 24**	1.3	S	*1 45	1.6	***2 25**	1.4	T	2 16	1.1	**3 16**	1.4	2
3	F	2 07	1.9	**2 30**	1.4	M	2 43	1.5	**3 27**	1.5	W	3 12	1.0	**4 14**	1.5	3
4	S	3 10	1.8	**3 36**	1.5	T	3 39	1.4	**4 25**	1.7	T	4 06	1.0	**5 08**	1.5	4
5	S	4 09	1.8	**4 37**	1.6	W	4 30	1.4	**5 18**	1.8	F	4 56	0.9	**5 56**	1.5	5
6	M	5 04	1.8	**5 35**	1.7	T	5 19	1.3	**6 09**	1.8	S	5 45	0.9	**6 42**	1.5	6
7	T	5 56	1.7	**6 31**	1.8	F	6 07	1.2	**6 57**	1.8	S	6 33	0.8	**7 25**	1.5	7
8	W	6 45	1.7	**7 23**	1.9	S	6 55	1.2	**7 44**	1.8	M	7 19	0.8	**8 07**	1.4	8
9	T	7 32	1.6	**8 14**	2.0	S	7 41	1.1	**8 29**	1.7	T	8 05	0.8	**8 48**	1.4	9
10	F	8 19	1.5	**9 04**	2.0	M	8 29	1.1	**9 15**	1.7	W	8 51	0.8	**9 29**	1.3	10
11	S	9 06	1.4	**9 54**	1.9	T	9 18	1.0	**10 01**	1.6	T	9 39	0.8	**10 09**	1.2	11
12	S	9 54	1.3	**10 44**	1.9	W	10 09	1.0	**10 48**	1.5	F	10 30	0.8	**10 50**	1.1	12
13	M	10 44	1.3	**11 36**	1.8	T	11 04	1.0	**11 36**	1.4	S	11 23	0.8	**11 32**	1.1	13
14	T	11 37	1.2	...	...	F	...	...	**12 02**	1.0	S	...	...	**12 19**	0.9	14
15	W	12 30	1.7	**12 35**	1.2	S	12 25	1.3	**1 02**	1.1	M	12 17	1.0	**1 16**	0.9	15
16	T	1 25	1.6	**1 35**	1.2	S	1 13	1.2	**2 01**	1.1	T	1 03	0.9	**2 11**	1.0	16
17	F	2 20	1.6	**2 37**	1.2	M	2 00	1.2	**2 56**	1.2	W	1 52	0.8	**3 03**	1.1	17
18	S	3 12	1.5	**3 37**	1.3	T	2 46	1.1	**3 46**	1.3	T	2 43	0.8	**3 50**	1.2	18
19	S	3 59	1.5	**4 31**	1.4	W	3 31	1.1	**4 30**	1.4	F	3 35	0.7	**4 36**	1.3	19
20	M	4 42	1.5	**5 20**	1.4	T	4 16	1.1	**5 11**	1.5	S	4 26	0.7	**5 21**	1.3	20
21	T	5 22	1.4	**6 04**	1.5	F	5 00	1.0	**5 50**	1.5	S	5 17	0.7	**6 07**	1.4	21
22	W	6 00	1.4	**6 43**	1.6	S	5 45	1.0	**6 30**	1.6	M	6 08	0.7	**6 53**	1.4	22
23	T	6 38	1.4	**7 21**	1.7	S	6 31	1.0	**7 12**	1.6	T	6 59	0.8	**7 41**	1.4	23
24	F	7 16	1.3	**7 57**	1.7	M	7 19	0.9	**7 57**	1.7	W	7 51	0.8	**8 30**	1.4	24
25	S	7 57	1.3	**8 35**	1.8	T	8 08	0.9	**8 45**	1.6	T	8 46	0.9	**9 20**	1.3	25
26	S	8 39	1.2	**9 17**	1.8	W	9 01	1.0	**9 35**	1.6	F	9 43	0.9	**10 11**	1.2	26
27	M	9 25	1.2	**10 02**	1.8	T	9 58	1.0	**10 29**	1.5	S	10 44	0.9	**11 04**	1.1	27
28	T	10 15	1.2	**10 51**	1.8	F	10 59	1.0	**11 25**	1.4	S	11 47	1.0	**11 58**	1.0	28
29	W	11 10	1.2	**11 46**	1.8	S	...	...	**12 03**	1.1	M	...	...	**12 53**	1.1	29
30	T	...	...	**12 10**	1.2	S	12 22	1.3	**1 09**	1.2	T	12 54	0.9	**2 00**	1.1	30
31	F	12 44	1.7	**1 14**	1.2						W	1 51	0.8	**3 03**	1.2	31

Dates when Ht. of **Low** Water is below Mean Low with Ht. of lowest given for each period and Date of lowest in ( ):

2nd - 9th: -0.3' (5th - 6th)
17th - 31st: -0.4' (22nd - 24th, 30th - 31st)

Average Rise and Fall 1.1 ft.

When a high tide exceeds avg. ht., the *following* low tide will be lower than avg.

2014 HIGH WATER
MIAMI HARBOR ENTRANCE, FL
25°45.8'N, 80°07.8'W

Standard Time (January) · Standard Time (February) · *Daylight Time starts Mar. 9 at 2 a.m. (March)

DAY OF MONTH	DAY OF WEEK	JANUARY a.m.	Ht.	**p.m.**	Ht.	DAY OF WEEK	FEBRUARY a.m.	Ht.	**p.m.**	Ht.	DAY OF WEEK	MARCH a.m.	Ht.	**p.m.**	Ht.	DAY OF MONTH
1	W	8 17	2.9	**8 29**	2.7	S	9 36	2.8	**9 58**	2.7	S	8 26	2.8	**8 51**	2.8	1
2	T	9 07	2.9	**9 22**	2.8	S	10 23	2.7	**10 49**	2.6	S	9 13	2.8	**9 39**	2.8	2
3	F	9 56	2.9	**10 15**	2.8	M	11 10	2.6	**11 40**	2.5	M	9 58	2.7	**10 27**	2.7	3
4	S	10 47	2.8	**11 10**	2.7	T	11 59	2.4	**...**	...	T	10 44	2.6	**11 15**	2.6	4
5	S	11 36	2.7	**...**	...	W	12 33	2.3	**12 48**	2.2	W	11 28	2.4	**...**	...	5
6	M	12 05	2.5	**12 28**	2.6	T	1 29	2.1	**1 41**	2.0	T	12 03	2.4	**12 14**	2.2	6
7	T	1 02	2.4	**1 22**	2.4	F	2 28	2.0	**2 38**	1.9	F	12 53	2.2	**1 03**	2.0	7
8	W	2 02	2.3	**2 18**	2.2	S	3 30	1.9	**3 39**	1.8	S	1 47	2.0	**1 57**	1.9	8
9	T	3 05	2.2	**3 17**	2.1	S	4 32	1.9	**4 39**	1.8	S	*3 46	1.9	***3 57**	1.8	9
10	F	4 08	2.1	**4 16**	2.0	M	5 28	1.9	**5 34**	1.9	M	4 47	1.9	**5 00**	1.8	10
11	S	5 06	2.1	**5 12**	2.0	T	6 16	2.0	**6 23**	1.9	T	5 46	1.9	**5 59**	1.9	11
12	S	5 59	2.2	**6 03**	2.0	W	6 58	2.1	**7 07**	2.0	W	6 37	2.0	**6 51**	2.0	12
13	M	6 45	2.2	**6 48**	2.1	T	7 37	2.2	**7 48**	2.1	T	7 22	2.1	**7 38**	2.1	13
14	T	7 26	2.2	**7 30**	2.1	F	8 14	2.2	**8 27**	2.2	F	8 04	2.2	**8 21**	2.2	14
15	W	8 04	2.3	**8 10**	2.1	S	8 51	2.3	**9 06**	2.2	S	8 42	2.3	**9 02**	2.3	15
16	T	8 41	2.3	**8 49**	2.2	S	9 26	2.3	**9 44**	2.2	S	9 20	2.4	**9 41**	2.4	16
17	F	9 17	2.3	**9 28**	2.2	M	10 01	2.3	**10 22**	2.2	M	9 57	2.4	**10 21**	2.5	17
18	S	9 53	2.3	**10 06**	2.1	T	10 37	2.2	**11 03**	2.2	T	10 34	2.4	**11 01**	2.5	18
19	S	10 29	2.2	**10 45**	2.1	W	11 14	2.1	**11 46**	2.1	W	11 12	2.3	**11 43**	2.4	19
20	M	11 05	2.2	**11 26**	2.0	T	11 54	2.1	**...**	...	T	11 51	2.3	**...**	...	20
21	T	11 42	2.1	**...**	...	F	12 35	2.1	**12 41**	2.0	F	12 27	2.4	**12 35**	2.2	21
22	W	12 10	2.0	**12 22**	2.0	S	1 31	2.0	**1 37**	2.0	S	1 17	2.3	**1 25**	2.2	22
23	T	1 00	2.0	**1 08**	2.0	S	2 35	2.0	**2 44**	2.0	S	2 13	2.2	**2 24**	2.1	23
24	F	1 57	2.0	**2 03**	1.9	M	3 44	2.1	**3 57**	2.1	M	3 16	2.2	**3 33**	2.1	24
25	S	3 01	2.0	**3 07**	2.0	T	4 51	2.2	**5 07**	2.2	T	4 24	2.2	**4 45**	2.2	25
26	S	4 09	2.1	**4 15**	2.0	W	5 52	2.4	**6 10**	2.4	W	5 30	2.3	**5 54**	2.4	26
27	M	5 13	2.2	**5 22**	2.2	T	6 47	2.6	**7 07**	2.6	T	6 30	2.5	**6 57**	2.5	27
28	T	6 13	2.4	**6 24**	2.4	F	7 38	2.7	**8 00**	2.7	F	7 25	2.6	**7 53**	2.7	28
29	W	7 07	2.6	**7 21**	2.5						S	8 16	2.7	**8 44**	2.8	29
30	T	7 59	2.7	**8 15**	2.7						S	9 03	2.8	**9 33**	2.9	30
31	F	8 48	2.8	**9 07**	2.7						M	9 49	2.8	**10 19**	2.9	31

Dates when Ht. of **Low** Water is below Mean Low with Ht. of lowest given for each period and Date of lowest in ( ):

January:
1st - 8th: -0.8' (2nd)
12th: -0.2'
14th - 18th: -0.3' (16th)
24th - 31st: -0.9' (31st)

February:
1st - 6th: -0.8' (1st)
12th - 28th: -0.7' (28th)

March:
1st - 6th: -0.7' (1st - 3rd)
16th - 22nd: -0.3' (18th - 20th)
26th - 31st: -0.6' (30th - 31st)

Average Rise and Fall 2.5 ft.

When a high tide exceeds avg. ht., the *following* low tide will be lower than avg.

2014 HIGH WATER
MIAMI HARBOR ENTRANCE, FL
25°45.8'N, 80°07.8'W

		Daylight Saving Time					Daylight Saving Time					Daylight Saving Time				
DAY OF MONTH	DAY OF WEEK	APRIL				DAY OF WEEK	MAY				DAY OF WEEK	JUNE				DAY OF MONTH
		a.m.	Ht.	**p.m.**	Ht.		a.m.	Ht.	**p.m.**	Ht.		a.m.	Ht.	**p.m.**	Ht.	
1	T	10 33	2.7	**11 04**	2.8	T	10 49	2.5	**11 22**	2.6	S	11 45	2.2	...	...	1
2	W	11 16	2.6	**11 48**	2.6	F	11 30	2.4	...	...	M	12 16	2.3	**12 27**	2.1	2
3	T	11 58	2.4	...	...	S	12 04	2.5	**12 12**	2.3	T	12 56	2.2	**1 12**	2.0	3
4	F	12 33	2.5	**12 43**	2.3	S	12 47	2.3	**12 57**	2.1	W	1 39	2.1	**2 01**	2.0	4
5	S	1 18	2.3	**1 28**	2.1	M	1 29	2.2	**1 43**	2.0	T	2 22	2.1	**2 52**	2.0	5
6	S	2 06	2.1	**2 18**	2.0	T	2 16	2.1	**2 36**	2.0	F	3 11	2.0	**3 48**	2.0	6
7	M	2 59	2.0	**3 15**	1.9	W	3 07	2.1	**3 33**	2.0	S	4 02	2.0	**4 46**	2.1	7
8	T	3 56	2.0	**4 16**	1.9	T	4 00	2.0	**4 32**	2.0	S	4 57	2.1	**5 44**	2.2	8
9	W	4 54	2.0	**5 17**	2.0	F	4 55	2.1	**5 31**	2.1	M	5 52	2.1	**6 40**	2.3	9
10	T	5 49	2.0	**6 14**	2.1	S	5 48	2.1	**6 25**	2.2	T	6 46	2.2	**7 33**	2.5	10
11	F	6 38	2.1	**7 04**	2.2	S	6 38	2.2	**7 15**	2.4	W	7 39	2.3	**8 24**	2.6	11
12	S	7 23	2.2	**7 50**	2.4	M	7 25	2.3	**8 03**	2.5	T	8 30	2.4	**9 13**	2.7	12
13	S	8 06	2.3	**8 33**	2.5	T	8 12	2.4	**8 49**	2.6	F	9 22	2.5	**10 03**	2.8	13
14	M	8 47	2.4	**9 16**	2.6	W	8 57	2.5	**9 35**	2.7	S	10 13	2.6	**10 52**	2.8	14
15	T	9 27	2.5	**9 58**	2.7	T	9 43	2.5	**10 22**	2.8	S	11 06	2.6	**11 42**	2.8	15
16	W	10 08	2.5	**10 41**	2.7	F	10 31	2.6	**11 09**	2.8	M	...	...	**12 01**	2.6	16
17	T	10 50	2.5	**11 26**	2.7	S	11 20	2.6	**11 59**	2.7	T	12 32	2.7	**12 56**	2.6	17
18	F	11 34	2.5	...	...	S	...	...	**12 12**	2.5	W	1 25	2.7	**1 55**	2.5	18
19	S	12 13	2.6	**12 22**	2.4	M	12 50	2.7	**1 08**	2.5	T	2 19	2.5	**2 56**	2.4	19
20	S	1 04	2.5	**1 17**	2.4	T	1 44	2.6	**2 09**	2.4	F	3 16	2.4	**3 59**	2.4	20
21	M	1 59	2.5	**2 18**	2.3	W	2 41	2.5	**3 13**	2.4	S	4 15	2.3	**5 02**	2.4	21
22	T	3 00	2.4	**3 24**	2.3	T	3 41	2.5	**4 19**	2.4	S	5 14	2.3	**6 02**	2.4	22
23	W	4 03	2.4	**4 34**	2.4	F	4 41	2.4	**5 23**	2.5	M	6 11	2.3	**6 58**	2.4	23
24	T	5 07	2.4	**5 40**	2.5	S	5 41	2.4	**6 23**	2.5	T	7 04	2.3	**7 48**	2.4	24
25	F	6 06	2.5	**6 41**	2.6	S	6 36	2.4	**7 18**	2.6	W	7 53	2.3	**8 33**	2.4	25
26	S	7 01	2.6	**7 36**	2.7	M	7 28	2.5	**8 08**	2.6	T	8 38	2.3	**9 15**	2.5	26
27	S	7 52	2.6	**8 27**	2.8	T	8 16	2.5	**8 54**	2.6	F	9 20	2.3	**9 54**	2.4	27
28	M	8 40	2.7	**9 14**	2.8	W	9 01	2.4	**9 37**	2.6	S	10 00	2.3	**10 32**	2.4	28
29	T	9 24	2.6	**9 58**	2.8	T	9 43	2.4	**10 18**	2.6	S	10 40	2.2	**11 09**	2.4	29
30	W	10 07	2.6	**10 41**	2.7	F	10 24	2.3	**10 57**	2.5	M	11 20	2.2	**11 46**	2.3	30
31						S	11 04	2.3	**11 36**	2.4						31

Dates when Ht. of **Low** Water is below Mean Low with Ht. of lowest given for each period and Date of lowest in ( ):

1st - 3rd: -0.5' (1st)
14th - 19th: -0.3' (15th - 18th)
25th - 30th: -0.5' (28th)

1st: -0.3'
13th -19th: -0.5' (16th)
24th - 30th: -0.4' (27th)

10th - 28th: -0.6' (13th - 14th)

Average Rise and Fall 2.5 ft.

When a high tide exceeds avg. ht., the *following* low tide will be lower than avg.

2014 HIGH WATER
MIAMI HARBOR ENTRANCE, FL
25°45.8'N, 80°07.8'W

DAY OF MONTH	DAY OF WEEK	JULY (Daylight Saving Time) a.m.	Ht.	p.m.	Ht.	DAY OF WEEK	AUGUST (Daylight Saving Time) a.m.	Ht.	p.m.	Ht.	DAY OF WEEK	SEPTEMBER (Daylight Saving Time) a.m.	Ht.	p.m.	Ht.	DAY OF MONTH
1	T	...	...	**12 01**	2.1	F	12 28	2.3	**12 55**	2.2	M	1 20	2.5	**2 06**	2.5	1
2	W	12 23	2.3	**12 42**	2.1	S	1 07	2.3	**1 41**	2.2	T	2 12	2.4	**3 06**	2.5	2
3	T	1 02	2.2	**1 26**	2.0	S	1 49	2.2	**2 33**	2.2	W	3 14	2.4	**4 11**	2.6	3
4	F	1 44	2.1	**2 15**	2.0	M	2 40	2.2	**3 33**	2.2	T	4 24	2.5	**5 17**	2.7	4
5	S	2 27	2.1	**3 08**	2.0	T	3 36	2.2	**4 35**	2.3	F	5 32	2.7	**6 18**	2.8	5
6	S	3 16	2.0	**4 06**	2.1	W	4 41	2.3	**5 40**	2.4	S	6 36	2.9	**7 14**	3.0	6
7	M	4 11	2.1	**5 07**	2.2	T	5 48	2.4	**6 40**	2.6	S	7 35	3.1	**8 07**	3.2	7
8	T	5 11	2.1	**6 07**	2.3	F	6 51	2.6	**7 37**	2.8	M	8 30	3.2	**8 56**	3.3	8
9	W	6 12	2.2	**7 05**	2.4	S	7 50	2.7	**8 29**	2.9	T	9 22	3.3	**9 45**	3.3	9
10	T	7 11	2.4	**8 00**	2.6	S	8 46	2.9	**9 20**	3.0	W	10 13	3.4	**10 32**	3.3	10
11	F	8 08	2.5	**8 52**	2.8	M	9 40	3.0	**10 09**	3.1	T	11 03	3.3	**11 20**	3.2	11
12	S	9 03	2.6	**9 42**	2.9	T	10 33	3.1	**10 57**	3.1	F	11 53	3.2	...	...	12
13	S	9 57	2.7	**10 32**	2.9	W	11 25	3.1	**11 46**	3.0	S	12 08	3.0	**12 44**	3.0	13
14	M	10 50	2.8	**11 21**	2.9	T	...	...	**12 17**	3.0	S	12 57	2.9	**1 36**	2.8	14
15	T	11 44	2.8	...	...	F	12 35	2.9	**1 11**	2.8	M	1 49	2.7	**2 32**	2.7	15
16	W	12 11	2.8	**12 39**	2.7	S	1 27	2.7	**2 06**	2.7	T	2 45	2.6	**3 31**	2.6	16
17	T	1 02	2.7	**1 35**	2.6	S	2 20	2.6	**3 05**	2.5	W	3 45	2.5	**4 31**	2.5	17
18	F	1 54	2.6	**2 33**	2.5	M	3 17	2.4	**4 07**	2.4	T	4 46	2.5	**5 28**	2.5	18
19	S	2 49	2.4	**3 34**	2.4	T	4 18	2.3	**5 08**	2.4	F	5 44	2.5	**6 19**	2.6	19
20	S	3 47	2.3	**4 37**	2.3	W	5 18	2.3	**6 06**	2.4	S	6 35	2.6	**7 03**	2.7	20
21	M	4 47	2.2	**5 38**	2.3	T	6 14	2.3	**6 56**	2.4	S	7 21	2.7	**7 44**	2.8	21
22	T	5 45	2.2	**6 34**	2.3	F	7 05	2.4	**7 40**	2.5	M	8 03	2.8	**8 23**	2.8	22
23	W	6 40	2.2	**7 25**	2.3	S	7 50	2.5	**8 19**	2.6	T	8 43	2.9	**9 00**	2.9	23
24	T	7 30	2.2	**8 09**	2.4	S	8 31	2.5	**8 57**	2.6	W	9 22	3.0	**9 37**	2.9	24
25	F	8 15	2.3	**8 50**	2.4	M	9 11	2.6	**9 33**	2.7	T	10 02	3.0	**10 14**	2.9	25
26	S	8 56	2.3	**9 27**	2.5	T	9 49	2.7	**10 09**	2.7	F	10 41	3.0	**10 51**	2.8	26
27	S	9 36	2.3	**10 04**	2.5	W	10 28	2.7	**10 45**	2.7	S	11 22	2.9	**11 30**	2.8	27
28	M	10 15	2.4	**10 40**	2.5	T	11 06	2.7	**11 20**	2.6	S	...	...	**12 05**	2.9	28
29	T	10 54	2.3	**11 16**	2.4	F	11 46	2.6	**11 57**	2.6	M	12 12	2.7	**12 53**	2.8	29
30	W	11 33	2.3	**11 52**	2.4	S	...	...	**12 28**	2.6	T	1 00	2.7	**1 46**	2.8	30
31	T	...	...	**12 13**	2.3	S	12 36	2.5	**1 14**	2.5						31

Dates when Ht. of **Low** Water is below Mean Low with Ht. of lowest given for each period and Date of lowest in ( ):

9th - 19th: -0.6' (11th - 14th) 8th - 15th: -0.5' (10th - 11th) 9th - 11th: -0.2'

Average Rise and Fall 2.5 ft.

When a high tide exceeds avg. ht., the *following* low tide will be lower than avg.

2014 HIGH WATER
MIAMI HARBOR ENTRANCE, FL
25°45.8'N, 80°07.8'W

Daylight Saving Time — ***Standard Time starts Nov. 2 at 2 a.m.** — **Standard Time**

DAY OF MONTH	DAY OF WEEK	OCTOBER				DAY OF WEEK	NOVEMBER				DAY OF WEEK	DECEMBER				DAY OF MONTH
		a.m.	Ht.	**p.m.**	Ht.		a.m.	Ht.	**p.m.**	Ht.		a.m.	Ht.	**p.m.**	Ht.	
1	W	1 57	2.7	**2 46**	2.8	S	4 00	2.8	**4 32**	2.9	M	3 51	2.7	**4 09**	2.7	1
2	T	3 02	2.7	**3 51**	2.8	S	*4 07	2.9	***4 33**	3.0	T	4 53	2.8	**5 07**	2.7	2
3	F	4 12	2.7	**4 55**	2.9	M	5 09	3.1	**5 29**	3.0	W	5 50	2.9	**6 01**	2.7	3
4	S	5 21	2.9	**5 57**	3.0	T	6 07	3.2	**6 22**	3.1	T	6 44	2.9	**6 52**	2.8	4
5	S	6 23	3.1	**6 52**	3.1	W	6 58	3.3	**7 11**	3.2	F	7 31	3.0	**7 39**	2.7	5
6	M	7 21	3.3	**7 44**	3.3	T	7 47	3.4	**7 58**	3.2	S	8 16	2.9	**8 23**	2.7	6
7	T	8 14	3.4	**8 33**	3.3	F	8 34	3.3	**8 44**	3.1	S	8 59	2.9	**9 06**	2.6	7
8	W	9 05	3.5	**9 21**	3.4	S	9 19	3.3	**9 28**	3.0	M	9 40	2.8	**9 48**	2.6	8
9	T	9 53	3.5	**10 07**	3.3	S	10 03	3.1	**10 12**	2.9	T	10 20	2.7	**10 29**	2.4	9
10	F	10 41	3.4	**10 53**	3.2	M	10 47	3.0	**10 57**	2.7	W	11 00	2.6	**11 12**	2.3	10
11	S	11 28	3.3	**11 39**	3.1	T	11 31	2.8	**11 43**	2.6	T	11 40	2.4	**11 56**	2.2	11
12	S	...	...	**12 15**	3.1	W	...	...	**12 17**	2.7	F	...	...	**12 21**	2.3	12
13	M	12 27	2.9	**1 04**	2.9	T	12 32	2.5	**1 04**	2.5	S	12 44	2.1	**1 06**	2.2	13
14	T	1 16	2.7	**1 55**	2.8	F	1 25	2.4	**1 55**	2.5	S	1 36	2.1	**1 54**	2.2	14
15	W	2 09	2.6	**2 49**	2.6	S	2 22	2.3	**2 47**	2.4	M	2 32	2.1	**2 45**	2.1	15
16	T	3 07	2.5	**3 45**	2.6	S	3 21	2.4	**3 41**	2.4	T	3 30	2.1	**3 40**	2.1	16
17	F	4 07	2.5	**4 41**	2.6	M	4 17	2.4	**4 32**	2.5	W	4 28	2.2	**4 35**	2.2	17
18	S	5 06	2.5	**5 33**	2.6	T	5 10	2.6	**5 21**	2.5	T	5 23	2.3	**5 28**	2.3	18
19	S	6 00	2.6	**6 20**	2.7	W	5 58	2.7	**6 07**	2.6	F	6 15	2.5	**6 19**	2.4	19
20	M	6 48	2.7	**7 04**	2.8	T	6 44	2.8	**6 51**	2.7	S	7 04	2.6	**7 09**	2.5	20
21	T	7 32	2.9	**7 45**	2.8	F	7 29	2.9	**7 35**	2.7	S	7 51	2.7	**7 58**	2.6	21
22	W	8 14	3.0	**8 26**	2.9	S	8 13	3.0	**8 20**	2.8	M	8 38	2.8	**8 48**	2.6	22
23	T	8 55	3.1	**9 05**	2.9	S	8 58	3.0	**9 05**	2.8	T	9 25	2.9	**9 38**	2.7	23
24	F	9 37	3.1	**9 45**	2.9	M	9 43	3.0	**9 52**	2.8	W	10 13	2.8	**10 30**	2.7	24
25	S	10 18	3.1	**10 26**	2.9	T	10 31	3.0	**10 42**	2.8	T	11 02	2.8	**11 24**	2.6	25
26	S	11 02	3.1	**11 08**	2.9	W	11 20	2.9	**11 37**	2.7	F	11 52	2.7	...	...	26
27	M	11 47	3.0	**11 55**	2.8	T	...	...	**12 12**	2.8	S	12 21	2.5	**12 46**	2.6	27
28	T	...	...	**12 36**	3.0	F	12 36	2.7	**1 08**	2.8	S	1 21	2.5	**1 42**	2.5	28
29	W	12 47	2.8	**1 30**	2.9	S	1 39	2.6	**2 07**	2.7	M	2 25	2.4	**2 42**	2.4	29
30	T	1 47	2.7	**2 28**	2.9	S	2 45	2.6	**3 08**	2.7	T	3 30	2.4	**3 44**	2.3	30
31	F	2 52	2.7	**3 30**	2.9						W	4 34	2.4	**4 45**	2.3	31

Dates when Ht. of **Low** Water is below Mean Low with Ht. of lowest given for each period and Date of lowest in ( ):

3rd: -0.2'
5th - 7th: -0.2'
20th - 31st: -0.5' (23rd - 24th)

Average Rise and Fall 2.5 ft.

When a high tide exceeds avg. ht., the *following* low tide will be lower than avg.

Anglers United by One Fish

by Lou Tabory

For years striped bass have captivated anglers. Here are four departed friends with a common bond, the pursuit of striped bass.

John Merwin was the least likely angler to be obsessed with striper fishing. John, a serious trout angler, moved to Vermont in his early fishing years. A well-known angling writer, John became fishing editor of Field and Stream. I met him at fishing shows and we became good friends. He worked with different outdoor magazines and helped me tremendously to improve my writing.

John took little interest in saltwater fishing until two events changed him. I goaded him into fishing the surf and we hit a good morning. And while fishing Nauset Beach with Barb and me one evening, his wife Martha took a striper close to 50 pounds. After this John took to surf casting with a vengeance, and in his remaining years he fished the beaches of Cape Cod or Rhode Island each spring and fall. He even started making his own wooden plugs called Old John. Although a diehard fly rod angler, John mostly fished spinning tackle along the beaches. He marveled at the way a striper would crash a surface swimmer. One of his homemade creations looked like a Danny Plug. I have one of these treasures hanging in our living room--a fond memory of a great friend.

I got to know Tim Colemen as a fledgling writer just getting into the business. Tim, the editor of The New England Fishermen, helped me publish my first articles about local saltwater fly fishing. He was a gifted writer and I always admired his work. But I admired even more his skills at catching big stripers from the surf on plugs. I heard stories about fishing on Block Island and knew that conditions were not often suited for fly tackle, but looking back it would have been fun spending time with such great anglers.

The fishing these surf anglers experienced was exceptional for 50-pound plus stripers. Tim's best fish was 67 pounds. Most serious beach anglers hope to someday reach the 50-pound benchmark, the Holy Grail of striper-surf-anglers. Tim had many big fish from the surf, and one night took two 50-pound fish. Yet, he used a simple approach and was very low-key about the trophy stripers he took. He was usually laid back, even in Tim's picture with the big fish there is no fanfare--just an angler after a good night's fishing. Tim was always that easygoing, fun guy, full of life. Pat Abate, one of Tim's fishing partners, told me about a glass jar Tim called "the jar of broken hearts" for all the bent hooks that anglers brought back to the cabin after a night's fishing. Tim passed on while fishing the Rhode Island beaches, still looking for that next 50-pound striper.

Jack Frech, a legend at Montauk known as "The Professor," was a hardnosed angler and one of the first to use a wetsuit, swimming out to offshore rocks, sometimes fishing all night. One night fishing the rocks he tried to grab a large striper. The hooks of the big swimming plug drove into his sleeve. The fish pulled him off the rock but luckily the hooks never penetrated beyond the wetsuit. Jack said that if the barbed hooks had driven into his flesh he might have been in real trouble. After that night Jack always used barbless hooks.

Jack's goal was to land a 50-pound plus striper from the surf. He never did. He landed two 49 1/2-pound fish. Most anglers would have called them 50-pounders, but not Jack.

Jack was an intense angler, fishing every cast as if it was his last. He even counted the number of reel turns to determine how far off the beach fish were holding. He built his own plugs using special heavy hooks. Jack used surf rods to 13 feet long and developed a unique casting style to prevent other anglers from crowding him at the Point. He would leave about 8 feet of line off the rod tip and then use a sweeping, swirling rod motion, rotating the rod tip 360 degrees around his body. This cleared a 20 to 25-foot area in a location where most anglers fished shoulder to shoulder.

Although in his later years Jack mostly fished Nantucket, he was still held in high regard by the hardcore Montauk anglers--he was that good. Barb and I fished Nantucket with him each fall. After Jack died the Island was never the same for us.

Bob Luce fished the rips from Cape Cod's Monomoy Island south to Great Point, Nantucket and then east of the Island. These vast rips are the last sanctuary for stripers. This water gets less pressure than other locations and with good reason. The ever-changing shoals produce strong rips and stand-up waves. There is pea soup fog for days. He spent his life fishing these rips and knew them like his backyard. Few anglers have the knowledge and experience of this extensive fishing location that he did.

Bob usually trolled with wire line and bucktail jigs, but he also did something interesting, trolling with fly tackle and lead core line. This was great sport for kids and anglers new to fly fishing. Through the years he introduced many anglers to saltwater fly fishing. When fishing with a husband and wife Bob positioned the boat as he trolled along the rip's edge so the wife's line was always in a prime location. He did the same with father and son teams. Bob was so good he knew where the fish were holding on each rip. When fly casting he would position the boat in just the right location. He learned quickly how I wanted to fish and always gave me the best angle to present the fly. Each year my wife Barb and I would fish as Bob's guests. Once underway his favorite game was to open my sloppy fly box and criticize its unstructured organization. We saw fishing that we probably will never experience again, but it is his dry sense of humor and wit that we miss most.

All four of these wonderful friendships would not have been possible without a common bond--fishing for striped bass.

Lou Tabory has been an outdoor writer for over 40 years, with articles in all the significant fishing publications. He has authored five books on saltwater flyfishing, including Inshore Fly Fishing. He has fished northeast waters for almost 60 years and is considered one of the early pioneers of Northeast fly fishing. – Eds.

CHARACTERISTICS OF LIGHT SIGNALS

(see footnote on next page for abbreviations used.)

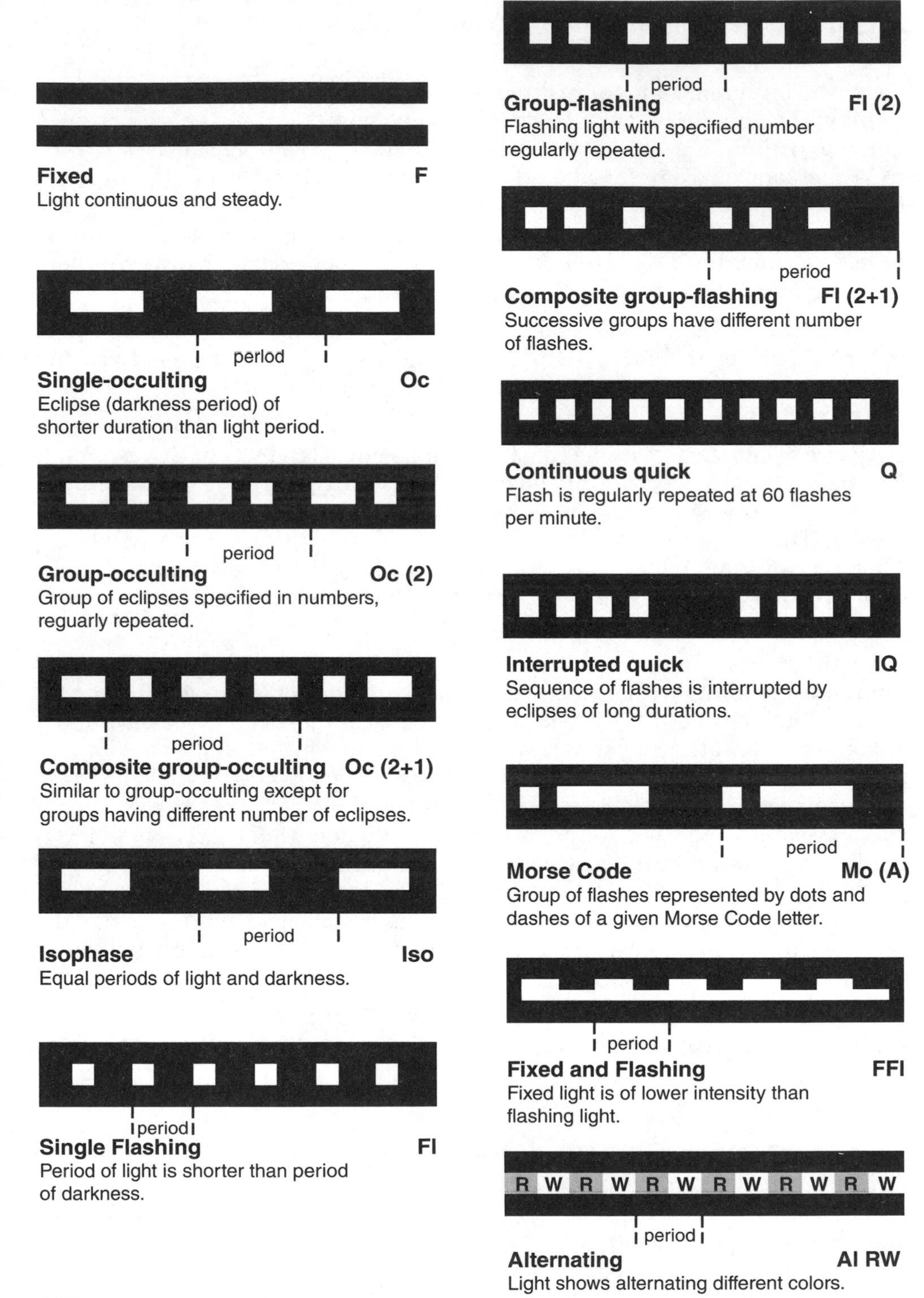

LIGHTS, FOG SIGNALS and OFFSHORE BUOYS

NOVA SCOTIA, EAST COAST

North Canso Lt., W. side of N. entr. to Strait of Canso – Fl. W. ev. 3 s., Obscured S. of 120°, Ht. 36.7 m. (120′), Rge. 10 mi., (45-41-29.8N/61-29-18.1W)

Cranberry Is. Lt., off Cape Canso, S. part of Is. – Fl. W. ev. 15 s., 2 Horns 2 bl. ev. 60 s., Horns point 066° and 141°, Ht. 16.9 m. (56′), Rge. 21 mi., Racon (B), (45-19-29.6N/60-55-38.2W)

White Head Is. Lt., SW side of Is. – Fl. W. ev. 5 s., Horn 1 bl. ev. 30 s., Horn points 190°, Ht. 18.2 m. (60′), Rge. 12 mi., (45-11-49.1N/61-08-10.8W)

Country Is. Lt., S. side of Is. – Fl. W. ev. 20 s., Ht. 16.5 m. (54′), Rge. 10 mi., (45-05-59.8N/61-32-31.9W)

Liscomb Is. Lt., near Cranberry Pt. – Fl. W. ev. 10 s., Horn 1 bl. ev. 30 s., Ht. 21.9 m. (72′), Rge. 14 mi., (44-59-15.8N/61-57-58.4W)

Beaver Is. Lt., E. end of Is. – Fl. W. ev. 7 s., Horn 1 bl. ev. 60 s., Horn points 144°, Ht. 19.9 m. (66′), Rge. 14 mi., (44-49-29.2N/62-20-16W)

Ship Harbour Lt., on Wolfes Pt. – LFl. G. ev. 6 s., Ht. 18.2 m. (60′), Rge. 4 mi., (44-44-55.4N/62-45-23.6W)

Owls Head Lt., at end of head – Fl. W. ev. 4 s., Ht. 25.8 m. (84′), Rge. 6 mi., (44-43-14.6N/62-47-59.5W)

Egg Is. Lt., center of Is. – LFl. W. ev. 6 s., Ht. 23.8 m. (78′), Rge. 14 mi., (44-39-52.7N/62-51-48.4W)

Jeddore Rock Lt., summit of rock – LFl. W. ev. 12 s., Ht. 29.5 m. (97′), Rge. 8 mi., (44-39-47.1N/63-00-37.3)

Bear Cove Lt. & Bell By. "H6," NE of cove, Q. R., Racon (N), Red, (44-32-36.3N/63-31-19.6W)

Sambro Harbor Lt. & Wh. By. "HS," S. of SW breaker, Halifax Hbr. app. – Mo(A)W ev. 6 s., RWS, (44-24-30N/63-33-36.5W)

Chebucto Head Lt., on summit, Halifax Hbr. app. – Fl. W. ev. 20 s., Horn 2 bl. ev. 60 s., Horn points 113°, Ht. 47.8 m. (157′), Rge. 10 mi., Racon (Z), (44-30-26.6N/63-31-21.8W)

Halifax Alpha Lt. & Wh. By. "HA," Halifax app. – Mo(A)W ev. 6 s., RWS, (44-21-45N/63-24-15W)

Sambro Is. Lt., center of Is. – Fl. W. ev. 6 s., Ht. 42.7 m. (145′), Rge. 23 mi., (44-26-12N/63-33-48W)

Ketch Harbour Lt. By. "HE 19," Ketch Harbour entr. – Fl. G. ev 4 s., Green (44-28-19.6N/63-32-16W)

Betty Is. Lt., on Brig Pt. – Fl. W. ev. 15 s., Horn 1 bl. ev. 60 s., Ht. 19.2 m. (63′), Rge. 13 mi., (44-26-19.7N/63-46-00.4W)

Pearl Is. Lt., off St. Margaret's & Mahone Bays – Fl. W. ev. 10 s., Ht. 19.0 m. (63′), Rge. 8 mi., (44-22-57.2N/64-02-54W)

East Ironbound Is. Lt., center of Is. – Iso. W. ev. 6 s., Ht. 44.5 m. (147′), Rge. 13 mi., (44-26-22.4N/64-04-59.7W)

***Abbreviations*: Alt.**, Alternating; **App.**, Approach; **By.**, Buoy; **Ch.**, Channel; **Entr.**, Entrance; **ev.**, every; **F.**, Fixed; **fl.**, flash; **Fl.**, Flashing; **Fl(2)**, Group Flashing; **LFl**, 2 s. flash.; **G.**, Green; **Hbr.**, Harbor or Harbour, **Ht.**, height; **Is.**, Island; **Iso.**, Isophase (Equal interval); **Iso. W.**, Isophase White (Red sector(s) of Lights warn of dangerous angle of approach. Bearings and ranges are <u>from</u> the observer <u>to</u> the aid.); **Jct.**, Junction; **Keyed**, Fog signal is radio activated. During times of reduced visibility, within ½ mile of the fog signal, turn VHF marine radio to channel 83A and 81A as alternate. Key microphone 5–10 times consecutively to activate fog signal for 45 minutes. **Lt.**, Light; **Ltd.**, Lighted.; **mi.**, miles; **Mo(A)**, Morse Code "A," **Mo(U)**, Morse Code "U"; **Oc.**, Occulting; **Pt.**, Point; **Q.**, Quick (Flashing); **RaRef.**, Radar Reflector; **R.**, Red; **rge.**, range; **RWS**, R.&W. Stripes; **RWSRST**, RWS with R. Spherical Topmarks; **s.**, seconds; **Wh.**, Whistle; **W.**, White; **Y.**, Yellow

Notices To Mariners: Keep informed of important changes. Visit **www.navcen.uscg.gov/lnm/** to receive Local Notices to Mariners via email. When reporting discrepancies in navigational aids, contact nearest C.G. unit and give official name of the aid.

Table for Converting Seconds to Decimals of a Minute, p. 262, for standard GPS input of Lat/Lon.

Fish Sense: Strive for Mastery, but Save the Mystery

by Zach Harvey

Long before I recognized a distinction between speed through water and speed over ground, before hatch-matching, controlled-depth trolling, drift conditions, tide lags, rigging needles, or modified Albrights, before there were two moons in a month, there was mystery in my watery fore. There was luck. And always, there was the promise of a good fish at any second.

Through the lens of my near-20-year fishing and fish-writing career, my early fishing looks better and better with each passing season. That is, I'm not altogether convinced that the considerable sum of what I've learned about fish, fishing, and fisheries through the first leg of adulthood has done my love of the rod and reel a lick of good. I look at my three-and-a-half-year-old daughter, Kaya, and I feel a profound sadness I can't quite articulate—a quiet dismay that things have become so dreadfully logical, that I've become so sober, so rational about an activity that once wound me up to the point of total insomnia.

As much as years on deck have honed skills I'll soon use to jump-start my little person's appreciation for fishing, they have also beaten a measure of the mystery, the awe, the thrill of limitless possibility, right out of me. I'm ashamed to admit it, but the process that steered me toward a more rational, more intelligent approach to fishing—that steered me away from foolish notions like luck or karma—has, over a long fetch, filled me with a tired arrogance about the whole undertaking.

It was in the pages of this very book that I found more than a few of my earliest revelations about fishing and sea conditions, the interplay between weather and seas, winds and tides, and the most basic aspects of predicting the all-important "high-percentage" time windows that so dramatically improve catch rates. I have always counted Eldridge among the most important tools for fishing, as the hackneyed little phrase goes, "smarter, not harder…" Trouble is, the longer I fish smart, the more I mutter the other sentiment: "Ignorance is bliss."

At the risk of sounding like some finger-wagging octogenarian lamenting our culture's technophiliac leanings, I fear that my generation will be the last to learn the fishery the way salts have for time out of mind, via apprenticeship. What scares me is just how quickly hard-won pieces of nautical wisdom—real, live skills, some of them bordering on art-forms—are fading out of the fleet as technology advances and such bits of fishing tradition become redundant. (Consider the net impact of GPS/chartplotter technology on the old art of trolling or running tight channels by shore-ranges—or the autopilot's impact on the skill of steering a compass course.) Given the rate at which accuracy of marine weather forecasts has improved, I wonder whether future generations will have the foggiest notion about reading the sky, the clouds, the water, etc. when the VHF fizzles out.

Continued p. 170

East Point Island Lt., Mahone Bay – F.G., Ht. 9.6 m. (31′), Rge. 7 mi., (44-20-59.2N/64-12-15W)

Cross Is. Lt., E. Pt. of Is. – Fl. W. ev. 10 s., Ht. 24.9 m. (82′), Rge. 10 mi., (44-18-43.7N/64-10-06.4W)

West Ironbound Is. Lt., Entr. to La Have R. – Fl. W. ev. 12 s., Ht. 24.3 m. (80′), Rge. 8 mi., (44-13-43.7N/64-16-28W)

Mosher Is. Lt., W. side Entr. to La Have R. – F.W., Horn 1 bl. ev. 20 s., Ht. 23.3 m. (77′), Rge. 13 mi., (44-14-14.6N/64-18-59.1W)

Cherry Cove Lt., betw. Little Hbr. & Back Cove – Iso. G. ev. 4 s., Horn 1 bl. ev. 30 s., Horn points 055°46′, Ht. 6.7 m. (22′), Rge. 8 mi., (44-09-29.8N/64-28-53.2W)

Medway Head Lt., W. side entr. to Pt. Medway – Fl. W. ev. 12 s., Ht. 24.2 m. (80′), Rge. 11 mi., (44-06-10.6N/64-32-23.3W)

Western Head Lt., W. side entr. to Liverpool Bay – Fl. W. ev. 15 s., Horn 1 bl. ev. 60 s., Horn points 104°, Ht. 16.8 m. (55′), Rge. 15 mi., (43-59-20.8N/64-39-44.5W)

Lockeport Lt., on Gull Rock, entr. to hbr. – LFl. W. ev. 15 s., Horn 1 bl. ev. 30 s., Ht. 16.7 m. (56′), Rge. 12 mi., (43-39-18.3N/65-05-55.9W)

Cape Roseway Lt., near SE Pt. of McNutt Is. – Fl. W. ev. 10 s., Ht. 33.1 m. (109′), Rge. 10 mi., (43-37-21.4N/65-15-50W)

Cape Negro Is. Lt., on SE end of Is. – Fl(2) W. ev. 15 s., Horn 1 bl. ev. 60 s., Ht. 28.3 m. (92′), Rge. 10 mi., (43-30-26.2N/65-20-44.2W)

The Salvages Lt., SE end of Is. – LFl. W. ev. 12 s., Horn 3 bl. ev. 60 s., Ht. 15.6 m. (51′), Rge. 10 mi., (43-28-08.1N/65-22-44W)

Baccaro Point Lt., E. side entr. to Barrington Bay – Mo(D)W 10 ev. s., Horn 1 bl. ev. 20 s., Horn points 200°, Ht. 15.0 m. (49′), Rge. 15 mi., (43-26-59N/65-28-15W)

Cape Sable Lt., on cape – Fl. W. ev. 5 s., Horn 1 bl. ev. 60 s., Horn points 150°, Ht. 29.7 m. (97′), Rge. 18 mi., Racon (C), (43-23-24N/65-37-16.9W)

West Head Lt., Cape Sable Is. – F.R., Horn 2 bl. ev. 60 s., Horn points 254°, Ht. 15.6 m. (51′), Rge. 7 mi., (43-27-23.8N/65-39-16.9W)

Outer Island Lt., on S. Pt. of Outer Is. – Fl. W. ev. 10 s., Ht. 13.7 m. (46′), Rge. 10 mi., (43-27-23.2N/65-44-36.2W)

Seal Is. Lt., S. Pt. of Is. – Fl. W. ev. 10 s., Horn 3 bl. ev. 60 s., Horn points 183°, Ht. 33.4 m. (110′), Rge. 19 mi., (43-23-40N/66-00-51W)

NOVA SCOTIA, WEST COAST

Peases Is. Lt., S. Pt. of one of the Tusket Is. – Fl. W. ev. 6 s., Horn 2 bl. ev. 60 s., Ht. 16 m. (53′), Rge. 9 mi., (43-37-42.6N/66-01-34.9W)

Cape Forchu Lt., E. Cape S. Pt. Yarmouth Sd. – LFl. W. ev. 12 s., Ht. 34.5 m. (113′), Rge. 12 mi., Racon (B), (43-47-38.8N/66-09-19.3W)

Lurcher Shoal Bifurcation Light By. "NM," W. of SW shoal – Fl.(2+1) R. ev. 6 s., Racon (K), R.G.R. marked "NM," (43-48-57.2N/66-29-58W)

Cape St. Marys Lt., E. side of Bay – Fl. W. ev. 5 s., Horn 1 bl. ev. 60 s., Horn points 251° 30′, Ht. 31.8 m (105′), Rge. 13 mi., (44-05-09.2N/66-12-39.6W)

Brier Is. Lt., on W. side of Is. R. & W. Tower – Fl(3) W. ev. 18 s., 2 Horns 2 bl. ev. 60 s., Horns point 270° and 315°, Ht. 22.2 m. (72′), Rge. 14 mi., (44-14-55N/66-23-32W)

Boars Head Lt., W. side of N. entr. to Petit Passage – Fl. W. ev. 5 s., Horn 3 bl. ev. 60 s., Horn points 315°, Ht. 28.0 m. (91′), Rge. 16 mi., (44-24-14.5N/66-12-55W)

Prim Pt. Lt., Digby Gut, W. Pt. of entr. to Annapolis Basin – Iso. W. ev. 6 s., Horn 1 bl. ev. 30 s., Horn points 318°, Ht. 24.8 m. (82′), Rge. 12 mi., (44-41-28N/65-47-10.8W)

Ile Haute Lt., on highest Pt. – Fl. W. ev. 4 s., Rge. 7 mi., Ht. 112 m. (367′), (45-15-03.3N/65-00-19.8W)

NEW BRUNSWICK COAST

Cape Enrage Lt., at pitch of cape – Fl. G. ev. 6 s., Horn 3 bl. ev. 60 s., Horn points 220°, Ht. 40.7 m. (134′), Rge. 10 mi., (45-35-38.1N/64-46-47.7W)

For abbreviations see footnote p. 167

Continued from p. 168

There's also, among the most seasoned fishermen, an age-old tradition of simplicity—what I've jokingly called "the fine art of common sense." It has always struck me that those with the most time under their belts tend to be those least willing to lay out ornate theories on fish behavior. They won't hesitate, however, to comment on the human end of the fishing equation. One such nugget of hard-won wisdom—drilled into my concrete skull by three captains widely regarded as "old-school"—gets at the very core of my present crisis: "You won't catch fish if you don't have a line in the water."

I've become much more adept at identifying prime windows for good fishing, as well as flagging total future misfires, in areas I frequent—thanks, often, to data in these very pages. Unfortunately, this is where I feel the aforementioned arrogance. Fishing smarter has become a reality. But there's a fine line between fishing smart and smarting your way right off the ocean—that is, getting so confident in your hunches that you stop feeling a need to prove their merit out on the grounds. That's when it hits you: When did talking yourself out of going become the prime directive?

The trouble with pet theories—fishing predictions in general—is that, for as long as men have baited hooks and dunked them in the briny, fish have made asses of them with alarming frequency. Mother Nature has a cruel sense of humor—especially, I've noticed, when someone has ventured out on a theoretical limb whose breaking load he's eclipsed by 100 pounds. That's really the bottom line here, the lynch pin of this whole entry: In my earliest years at the water's edge, luck was as real a variable as high tide or foot-long sea worms. And where there's luck, there's the very real possibility something amazing will happen at any second, no matter what the apparent odds. That's when every molecule of my being needed to go.

Ironically, the whole idea that it really could happen anywhere, at any time, to anyone—what a more logical angler might dismiss as pure voodoo—isn't a matter of superstition. It's fishing reality. When I start to get smug about what I think I know, I'm making it a point, lately, to run through all the 50-pound stripers caught at dead-low tide at high noon at the swimming beach by the guy with the wrong everything when every piece of the timing puzzle screams "Wrong!" It happens all the time. That's why we keep fishing.

The take-home is just this: By all means use this book to confirm or deny a hundred hair-brained notions about the fishing you do. Do, by all means, learn to incorporate tide and current information into your pre-trip planning. Do, by all means, fish smarter. Just be sure that, in your efforts to fish peak times, you don't get too rigid about when it's "worth going." The simplest angling truth still applies: You'll catch fish because your line's in the water—even on many occasions when the voice of smarter suggests otherwise.

Zach Harvey is a freelance writer, editor, photographer, illustrator, and deckhand, and Fishing Editor for Soundings Magazine. He lives in Wakefield, RI, with his wife, Sarah, and daughter, Kaya.

Quaco Lt., tower on head – Fl. W. ev. 10 s., Horn 1 bl. ev. 30 s., Horn points 130°, Ht. 26.0 m. (86′), Rge. 21 mi., (45-19-25.3N/65-32-08.8W)

Cape Spencer Lt., pitch of cape – Fl. W. ev. 11 s., Horn 3 bl. ev. 60 s., Horn points 165°, Ht. 61.6 m. (203′), Rge. 14 mi., (45-11-42.5N/65-54-35.5W)

Partridge Is. Lt., highest pt. of Is., Saint John Harbour – Fl. W. ev. 7.5 s., Ht. 35.3 m. (116′), Rge. 19 mi., (45-14-21N/66-03-13.8W)

Musquash Head Lt., E. side entr. to Musquash Hbr. – Fl. W. ev. 3 s., Horn 1 bl. ev. 60 s., Horn points 180°, Ht. 35.1 m. (116′), Rge. 20 mi., (45-08-37.1N/66-14-14.2W)

Pt. Lepreau Lt., on point – Fl. W. ev. 5 s., Horn 3 bl. ev. 60 s., Horn points 190°, Ht. 25.5 m. (84′), Rge. 14 mi., (45-03-31.7N/66-27-31.3W)

Pea Pt. Lt., E. side entr. to Letang Hbr. – F.W. visible 251° thru N & E to 161°, Horn 2 bl. ev. 60 s., Horn points 180°, Ht. 17.2 m. (56′), Rge. 12 mi., (45-02-20.4N/66-48-28.2W)

Head Harbour Lt., outer rock of E. Quoddy Head – F.R., Horn 1 bl. ev. 60 s., Horn points 116°, Ht. 17.6 m. (58′), Rge. 13 mi., (44-57-28.6N/66-54-00.2W)

Swallowtail Lt., NE Pt. of Grand Manan – Oc. W. ev. 6 s., Horn 1 bl. ev. 20 s., Horn points 100°, Ht. 37.1 m. (122′), Rge. 12 mi., (44-45-51.1N/66-43-57.5W)

Great Duck Is. Lt., S. end of Is. – Fl. W. ev. 10 s., Horn 1 bl. ev. 60 s., Horn points 120°, Ht. 15.3 m. (50′), Rge. 18 mi., (44-41-03.5N/66-41-34.3W)

Southwest Head Lt., S. end of Grand Manan – Fl. W. ev. 10 s., Horn 1 bl. ev. 60 s., Horn points 240°, Ht. 47.5 m. (156′), Rge. 16 mi., (44-36-02.9N/66-54-19.8W)

Gannet Rock Lt., S. of Grand Manan – Fl. W. ev. 5 s., Horn 3 bl. ev. 60 s., Horn omni-directional, Ht. 28.2 m. (93′), Rge. 19 mi., Racon (G), (44-30-37.1N/66-46-52.9W)

Machias Seal Is. Lt., On Is. summit – Fl. W. ev. 3 s., Horn 2 bl. ev. 60 s. Horn Points 065°, Ht. 25 m. (83′), Rge. 17 mi., (44-30-07N/67-06-04W)

MAINE

West Quoddy Head Lt., Entr. Quoddy Roads – Fl(2) W. ev. 15 s., Horn 2 bl. ev. 30 s., Ht. 83′, Rge. 18 mi., (44-48-54N/66-57-02W)

Libby Island Lt., Entr. Machias Bay – Fl(2) W. ev. 20 s., Horn 1 bl. ev. 15 s., Ht. 91′, Rge. 18 mi., (44-34-06N/67-22-03W)

Moose Peak Lt., E. end Mistake Is. – Fl. W. ev. 30 s., Horn 2 bl. ev. 30 s., Ht. 72′, Rge. 20 mi., (44-28-28N/67-31-55W)

Petit Manan Lt., E. Pt. of Is. – Fl. W. ev. 10 s., Horn 1 bl. ev. 30 s., Ht. 123′, Rge. 19 mi., (44-22-03N/67-51-52W)

Prospect Harbor Point Lt. – Fl. R. ev. 6 s., (2 W. sect.), Ht. 42′, Rge. R. 7 mi., W. 9 mi., ltd. 24 hrs., (44-24-12N/68-00-47W)

Mount Desert Lt., 20 mi. S. of island – Fl. W. ev. 15 s., Horn 2 bl. ev. 30 s., Ht. 75′, Rge. 20 mi., (43-58-07N/68-07-42W)

Great Duck Island Lt., S. end of island – Fl. R. ev. 5 s., Horn 1 bl. ev. 15 s., Ht. 67′, Rge. 19 mi., (44-08-31N/68-14-45W)

Frenchman Bay Ltd. By. "FB," Fl. (2+1) R. ev. 6 s., Rge. 4 mi., R&G Bands, Racon (B), (44-19-21N/68-07-24W)

Egg Rock Lt., Frenchman Bay – Fl. R. ev. 5 s., Horn 2 bl. ev. 30 s., Ht. 64′, Rge. 18 mi., (44-21-14N/68-08-18W)

Baker Island Lt., SW Entr. Somes Sound – Fl. W. ev. 10 s., Ht. 105′, Rge. 10 mi., (44-14-28N/68-11-56W)

Bass Harbor Head Lt., SW Pt. Mt. Desert Is. – Oc. R. ev. 4 s., Ht. 56′, Rge. 13 mi., ltd. 24 hrs., (44-13-19N/68-20-14W)

Blue Hill Bay Lt. #3, on Green Is. – Fl. G. ev. 4 s., Ht. 25′, Rge. 5 mi., SG on tower, (44-14-55N/68-29-52W)

Burnt Coat Harbor Lt. – Oc. W. ev. 4 s., Ht. 75′, Rge. 9 mi., (44-08-03N/68-26-50W)

For abbreviations see footnote p. 167

SEEK LANDFALL
For The Gear You Need To Arrive Alive
MAPTECH
OceanGrafix
CHARTS ON DEMAND
Gill
RESPECT THE ELEMENTS
MUSTANG SURVIVAL
MUSTO
ACR
HARKEN
INNOVATIVE SAILING SOLUTIONS
C-MAP
by JEPPESEN
GARMIN
HENRI LLOYD
SLAM
Landfall
WHERE SAFE VOYAGES BEGIN
Stamford, CT | www.landfallnav.com | 800.941.2219
Photo credit: Gary Felton www.yacht-photography.net

Halibut Rocks Lt., Jericho Bay – Fl. W. ev. 6 s., Horn 1 bl. ev. 10 s., Ht. 25′, Rge. 6 mi., NR on tower, (44-08-03N/68-31-32W)

Eggemoggin Ltd. Bell By. "EG" – Mo(A)W, Rge. 5 mi., RWSRST, (44-19-13N/68-44-34W)

Eggemoggin Reach Bell By. "ER" – RWSRST, (44-18-00N/68-46-29W)

Crotch Island Lt. #21, Deer Is. Thorofare – Fl. G. ev. 4 s., Ht. 20′, Rge. 5 mi., SG on tower, (44-08-46N/68-40-39W)

Saddleback Ledge Lt., Isle au Haut Bay – Fl. W. ev. 6 s., Horn 1 bl. ev. 10 s., Ht. 52′, Rge. 9 mi., (44-00-52N/68-43-35W)

Isle Au Haut Lt., Isle au Haut Bay – Fl. R. ev. 4 s., W. Sect. 034°-060°, Ht. 48′, Rge. R. 6 mi., W. 8 mi., (44-03-53N/68-39-05W)

Deer Island Thorofare Lt., W. end of thorofare – Fl. W. ev. 6 s., Horn 1 bl. ev. 15 s., Ht. 52′, Rge. 8 mi., (44-08-04N/68-42-12W)

Goose Rocks Lt., E. Entr. Fox Is. Thorofare – Fl. R. ev. 6 s., W. Sect. 301°-304°, Horn 1 bl. ev. 10 s., Ht. 51′, Rge. R. 11 mi., W. 12 mi., (44-08-08N/68-49-50W)

Eagle Island Lt., E. Penobscot Bay – Fl. W. ev. 4 s., Ht. 106′, Rge. 9 mi., (44-13-04N/68-46-04W)

Green Ledge Lt. #4, E. Penobscot Bay – Fl. R. ev. 6 s., Ht. 31′, Rge. 5 mi., TR on tower, (44-17-25N/68-49-42W)

Heron Neck Lt., E. Entr. Hurricane Sound – F.R., W. Sect. 030°-063°, Horn 1 bl. ev. 30 s., Ht. 92′, Rge. R. 7 mi., W. 9 mi., (44-01-30N/68-51-44W)

Matinicus Rock Lt., Penobscot Bay App. – Fl. W. ev. 10 s., Horn 1 bl. ev. 15 s., Ht. 90′, Rge. 20 mi., (43-47-01N/68-51-18W)

Grindel Pt. Lt., West Penobscot Bay – Fl. W. ev. 4 s., Ht. 39′, Rge. 4 mi., (44-16-53N/68-56-35W)

Two-Bush Island Lt., Two-Bush Ch. – Fl. W. ev. 5 s., R. Sect. 061°-247°, Horn 1 bl. ev. 15 s., Ht. 65′, Rge. W. 21 mi., R. 15 mi., (43-57-51N/69-04-26W)

Two Bush Island Ltd. Wh. By. "TBI" – Mo(A)W, Rge. 6 mi., RWS, (43-58-17N/69-00-16W)

Whitehead Lt., W. side of S. entr. Muscle Ridge Ch. – Oc.G. ev. 4 s., Horn 2 bl. ev. 30 s., Ht. 75′, Rge. 6 mi., (43-58-43N/69-07-27W)

Owl's Head Lt., S. side Rockland Entr. – F.W., Horn 2 bl. ev. 20 s., Ht. 100′, Rge. 16 mi., Obscured from 324°-354° by Monroe Island, ltd. 24 hrs., (44-05-32N/69-02-38W)

Rockland Harbor Breakwater Lt., S. end of breakwater – Fl. W. ev. 5 s., Horn 1 bl. ev. 15 s., Ht. 39′, Rge. 17 mi., (44-06-15N/69-04-39W)

Lowell Rock Lt. #2, Rockport Entr. – Fl. R. ev. 6 s., Ht. 25′, Rge. 5 mi., TR on spindle, (44-09-46N/69-03-37W)

Browns Head Lt., W. Entr. Fox Is. Thorofare – F.W., 2 R. Sect. 001°-050° and 061°-091°, Horn 1 bl. ev. 10 s., Ht. 39′, Rge. R. 11 mi., F.W. 14 mi., ltd. 24 hrs., (44-06-42N/68-54-34W)

Curtis Island Lt., S. side Camden Entr. – Oc.G. ev. 4 s., Ht. 52′, Rge. 6 mi., (44-12-05N/69-02-56W)

Northeast Point Lt. #2, Camden Entr. – Fl. R. ev. 4 s., Ht. 20′, Rge. 5 mi., TR on white tower, (44-12-31N/69-02-47W)

Dice Head Lt., N. side Entr. to Castine – Fl. W. ev. 6 s., Ht. 134′, Rge. 11 mi., White tower, (44-22-58N/68-49-08W)

Fort Point Lt., W. side Entr. to Penobscot R. – F.W., Horn 1 bl. ev. 10 s., Ht. 88′, Rge. 15 mi., ltd. 24 hrs., (44-28-02N/68-48-42W)

Marshall Point Lt., E. side of Pt. Clyde Hbr. S. Entr. – F.W., Horn 1 bl. ev. 10 s., Ht. 30′, Rge. 13 mi., ltd. 24 hrs., (43-55-03N/69-15-41W)

Marshall Point Ltd. By. "MP" – Mo(A)W, Rge. 6 mi., RWSRST, (43-55-18N/69-10-52W)

For abbreviations see footnote p. 167

Monhegan Island Lt., Penobscot Bay – Fl. W. ev. 15 s., Ht. 178′, Rge. 20 mi., (43-45-53N/69-18-57W)

Franklin Is. Lt., Muscongus Bay – Fl. W. ev. 6 s., Ht. 57′, Rge. 8 mi., (43-53-31N/69-22-29W)

Pemaquid Pt. Lt., W. side Muscongus Bay Entr. – Fl. W. ev. 6 s., Ht. 79′, Rge. 14 mi., (43-50-12N/69-30-21W)

Ram Is. Lt., Fisherman Is. Passage S. side – Iso. R. ev. 6 s., 2 W. Sect. 258°-261° and 030°-046°, Covers fairways, Keyed (VHF 83A) Horn 1 bl. ev. 30 s., Ht. 36′, Rge. W. 11 mi., R. 9 mi., W. 9 mi. (43-48-14N/69-35-57W)

Burnt Is. Lt., Boothbay Hbr. W. side Entr. – Fl. R. ev. 6 s., 2 W. Sect. 307°-316° and 355°-008°, Covers fairways. Horn 1 bl. ev. 10 s., Ht. 61′, Rge. W. 8 mi., R. 6 mi., (43-49-31N/69-38-25W)

The Cuckolds Lt., Boothbay – Fl(2) W. ev. 6 s., Keyed (VHF 83A) Horn 1 bl. ev. 15 s., Ht. 59′, Rge. 12 mi., (43-46-46N/69-39-00W)

Seguin Lt., 2 mi. S. of Kennebec R. mouth – F.W., Horn 2 bl. ev. 20 s., Ht. 180′, Rge. 18 mi., (43-42-27N/69-45-29W)

Hendricks Head Lt., Sheepscot R. mouth E. side – F.W., R. Sect. 180°-000°, Ht. 43′, Rge. R. 7 mi., F.W. 9 mi., (43-49-21N/69-41-23W)

Pond Is. Lt., Kennebec R. mouth W. side – Iso. W. ev. 6 s., Horn 2 bl. ev. 30 s., Ht. 52′, Rge. 9 mi., (43-44-24N/69-46-13W)

Perkins Is. Lt., Kennebec R. – Fl. R. ev. 2.5 s., 2 W. Sect. 018° – 038°, 172° – 188°, Covers fairways, Ht. 41′, Rge. R. 5 mi., W. 6 mi., (43-47-12N/69-47-07W)

Squirrel Pt. Lt., Kennebec R. – Iso. R. ev. 6 s., W. Sect. 321° - 324°, Covers fairway, Ht. 25′, Rge. R. 7 mi., W. 9 mi., (43-48-59N/69-48-09W)

Fuller Rock Lt., off Cape Small – Fl. W. ev. 4 s., Ht. 39′, Rge. 6 mi., NR on tower, (43-41-45N/69-50-01W)

White Bull Ltd. Gong By. "WB" – Mo(A)W, Rge. 6 mi., RWS, (43-42-49N/69-55-13W)

Whaleboat Island Lt., Broad Sd., Casco Bay – Fl. W. ev. 6 s., Ht. 47′, Rge. 6 mi., NR on tower, (43-44-31N/70-03-40W)

Cow Island Ledge Lt., Portland to Merepoint – Fl. W. ev. 6 s., Ht. 23′, Rge. 8 mi., RaRef., NR on spindle, (43-42-11N/70-11-19W)

Halfway Rock Lt., midway betw. Cape Small Pt. and Cape Eliz. – Fl. R. ev. 5 s., Horn 2 bl. ev. 30 s., Ht. 76′, Rge. 19 mi., (43-39-21N/70-02-12W)

Portland Ltd. Wh. By. "P", Portland Hbr. App. – Mo(A)W, Rge. 6 mi., Racon (M), RWSRST, (43-31-36N/70-05-28W)

Ram Island Ledge Lt., N. side of Portland Hbr. Entr. – Fl. (2) W. ev. 6 s., Horn 1 bl. ev. 10 s., Ht. 77′, Rge. 8 mi., (43-37-53N/70-11-15W)

Cape Elizabeth Lt., S. of Portland Hbr. Entr. – Fl(4) W. ev. 15 s., Horn 2 bl. ev. 60 s., Ht. 129′, Rge. 15 mi., ltd. 24 hrs., (43-33-58N/70-12-00W)

Portland Head Lt., SW side Portland Hbr. Entr. – Fl. W. ev. 4 s., Horn 1 bl. ev. 15 s., Ht. 101′, Rge. 24 mi., ltd. 24 hrs., (43-37-23N/70-12-28W)

Spring Pt. Ledge Lt., Portland main ch. W. side – Fl. W. ev. 6 s., 2 R. Sect., 2 W. Sectors 331°-337° Covers fairway entrance, and 074°-288°, Horn 1 bl. ev. 10 s., Ht. 54′, Rge. R. 10 mi., W. 12 mi., ltd. 24 hrs., (43-39-08N/70-13-26W)

Wood Island Lt., S. Entr. Wood Is. Hbr. N. side – Alt. W. and G. ev. 10 s. (Night), Keyed (VHF 83A) Horn 2 bl. ev. 30 s., Ht. 71′, Rge. W. 13 mi., G. 13 mi., (43-27-25N/70-19-45W)

Goat Is. Lt., Cape Porpoise Hbr. Entr. – Fl. W. ev. 6 s., Horn 1 bl. ev. 15 s., Ht. 38′, Rge. 12 mi., (43-21-28N/70-25-30W)

Cape Neddick Lt., On N. side of Nubble – Iso. R. ev. 6 s., Horn 1 bl. ev. 10 s., Ht. 88′, Rge. 13 mi., ltd. 24 hrs., (43-09-55N/70-35-28W)

Jaffrey Point Lt. #4 – Fl. R. ev. 4 s., Ht. 22′, rge. 5 mi., TR on tower, (43-03-18N/70-42-49W)

For abbreviations see footnote p. 167

Boon Is. Lt., 6.5 mi. off coast – Fl. W. ev. 5 s., Horn 1 bl. ev. 10 s., Ht. 137′, Rge. 19 mi., (43-07-17N/70-28-35W)

York Harbor Ltd. Bell By. "YH" – Mo(A)W, Rge. 5 mi., RWSRST, (43-07-45N/70-37-01W)

NEW HAMPSHIRE

Whaleback Lt., Portsmouth Entr. NE side –Fl(2) W. ev. 10 s., Keyed (VHF 83A) Horn 2 bl. ev. 30 s., Ht. 59′, Rge. 11 mi., (43-03-32N/70-41-47W)

Portsmouth Harbor Lt. (New Castle), on Fort Point – F. G., Horn 1 bl. ev. 10 s., Ht. 52′, Rge. 12 mi., (43-04-16N/70-42-31W)

Rye Harbor Entr. Ltd. Wh. By. "RH" – Mo(A)W, Rge. 6 mi., RWSRST, (42-59-38N/70-43-45W)

Isles Of Shoals Lt., 5.5 mi. off coast – Fl. W. ev. 15 s., Horn 1 bl. ev. 30 s., Ht. 82′, Rge. 14 mi., (42-58-02N/70-37-24W)

MASSACHUSETTS

Newburyport Harbor Lt., N. end of Plum Is. – Oc.(2) G. ev. 15 s., Obscured from 165°-192° and 313°-344°, Ht. 50′, Rge. 10 mi., (42-48-55N/70-49-08W)

Merrimack River Entr. Ltd. Wh. By. "MR"– Mo(A)W, Rge. 4 mi., RWSRST, (42-48-34N/70-47-03W)

Ipswich Lt., Ipswich Entr. S. side – Oc.W. ev. 4 s., Ht. 30′, Rge. 5 mi., NR on tower, (42-41-07N/70-45-58W)

Rockport Breakwater Lt. #6, W. side Entr. Rockport inner hbr. – Fl. R. ev. 4 s., Ht. 32′, Rge. 5 mi., TR on tower, (42-39-39N/70-36-43W)

Annisquam Harbor Lt., E. side Entr. – Fl. W. ev. 7.5 s., R. Sector 180°-217°, Horn 2 bl. ev. 60 s., Ht. 45′, Rge. R. 11 mi., W. 14 mi., (42-39-43N/70-40-53W)

Straitsmouth Lt., Rockport Entr. S. side – Fl. G. ev. 6 s., Keyed (VHF 81A) Horn 1 bl. ev. 15 s., Ht. 46′, Rge. 6 mi., (42-39-44N/70-35-17W)

Cape Ann Lt., E. side Thacher Is. – Fl. R. ev. 5 s., Horn 2 bl. ev. 60 s., Ht. 166′, Rge. 17 mi., (42-38-12N/70-34-30W)

Eastern Point Ltd. Wh. By. #2 – Fl. R. ev. 4 s., Rge. 3 mi., Red, (42-34-14N/70-39-50W)

Eastern Point Lt., Gloucester Entr. E. side – Fl. W. ev. 5 s., Ht. 57′, Rge. 20 mi., (42-34-49N/70-39-52W)

Gloucester Breakwater Lt., W. end – Oc.R. ev. 4 s., Horn 1 bl. ev. 10 s., Ht. 45′, Rge. 6 mi., (42-34-57N/70-40-20W)

Bakers Island Lt., Salem Ch. – Alt. Fl. W. and R. ev. 20 s., Horn 1 bl. ev. 30 s., Ht. 111′, Rge. W. 16 mi., R. 14 mi., (42-32-11N/70-47-09W)

Hospital Point Range Front Lt., Beverly Cove W. side – F.W., Ht. 69′, (42-32-47N/70-51-21W)

The Graves Ltd. Wh. By. #5 – Fl. G. ev. 4 s., Rge. 4 mi., Green, (42-22-33N/70-51-28W)

Marblehead Lt., N. point Marblehead Neck – F.G., Ht. 130′, Rge. 7 mi., (42-30-19N/70-50-01W)

The Graves Lt., Boston Hbr. S. Ch. Entr. – Fl(2) W. ev. 12 s., Horn 2 bl. ev. 20 s., Ht. 98′, Rge. 15 mi., (42-21-54N/70-52-09W)

Boston App. Ltd. By. "BG"– Mo(A)W, Rge. 6 mi., RWSRST, (42-23-27N/70-51-29W)

Deer Island Lt., President Roads, Boston Hbr. – Alt. W. and R. ev. 10 s., R. Sect. 198°-222°, Obscured 112°-186°, Horn 1 bl. ev. 10 s., Ht. 53′, Rge. 11 mi., (42-20-23N/70-57-16W)

Deer Island Danger Lt., On south end of spit – F.R., R. Sect. 198°-222°, Ht. 15′, Rge. 6 mi., (42-20-23N/70-57-16W)

Long Island Head Lt., President Roads, Boston Hbr. – Fl. W. ev. 2.5 s., Ht. 120′, Rge. 6 mi., (42-19-49N/70-57-28W)

Boston Ltd. Wh. By. "B", Boston Hbr. Entr. – Mo(A)W, Rge. 6 mi., Racon (B), RWSRST, (42-22-42N/70-46-58W)

For abbreviations see footnote p. 167

Boston App. Ltd. By. "BF" (NOAA-44013) –Fl(4) Y. ev. 20 sec, Rge. 7 mi., Yellow, (42-20-44N/70-39-04W)

Boston North Ch.Entr. Ltd. Wh. By. "NC" – Mo(A)W, Rge. 6 mi., RWSRST, Racon (N), (42-22-32N/70-54-18W)

Minots Ledge Lt., Boston Hbr. Entr. S. side – Fl(1+4+3) W. ev. 45 s., Horn 1 bl. ev. 10 s., Ht. 85′, Rge. 10 mi., (42-16-11N/70-45-33W)

Boston Lt., SE side Little Brewster Is. – Fl. W. ev. 10 s., Horn 1 bl. ev. 30 s., Ht. 102′, Rge. 27 mi., (42-19-41N/70-53-24W)

Scituate App. Ltd. Gong By. "SA"– Mo(A)W, Rge. 4 mi., RWSRST, (42-12-08N/70-41-49W)

Plymouth Lt. (Gurnet), N. side Entr. to hbr. – Fl(3) W. ev. 30 s., R. Sect. 323°-352°, Horn 2 bl. ev. 15 s., Ht. 102′, Rge. R. 15 mi., W. 17 mi., (42-00-13N/70-36-02W)

Race Point Lt., NW Point of Cape Cod – Fl. W. ev. 10 s., Rge. 16 mi., Obscured 220°-292°, Ht. 41′, (42-03-44N/70-14-35W)

Wood End Lt., Entr. to Provincetown – Fl. R. ev. 10 s., Horn 1 bl. ev. 30 s., Ht. 45′, Rge. 13 mi., (42-01-17N/70-11-37W)

Long Point Lt., Provincetown Entr. SW side – Oc.G. ev. 4 s., Horn 1 bl. ev. 15 s., Ht. 36′, Rge. 8 mi., (42-01-59N/70-10-07W)

Mary Ann Rocks Ltd. Wh. By. #12 – Fl. R. ev. 2.5 s., Rge. 4 mi., Red, (41-55-07N/70-30-22W)

Cape Cod Canal App. Ltd. Bell By. "CC" – Mo(A)W, Rge. 4 mi., RWSRST, (41-48-53N/70-27-39W)

Cape Cod Canal Breakwater Lt. #6, E. Entr. – Fl. R. ev. 5 s., Keyed (VHF 83A) Horn 1 bl. ev. 15 s., Ht. 43′, Rge. 9 mi., (41-46-47N/70-29-23W)

Highland Lt., NE side of Cape Cod – Fl. W. ev. 5 s., Ht. 170′, Rge. 18 mi., ltd. 24 hrs., (42-02-22N/70-03-39W)

Nauset Beach Lt., E. side of Cape Cod – Alt. W. R. ev. 10 s., (Tides divide and run in opposite directions abreast of light), Ht. 120′, Rge. W. 24 mi., R. 20 mi., (41-51-36N/69-57-12W)

Chatham Beach Ltd. Wh. By. "C" – Mo(A)W, Rge. 4 mi., RWSRST, (41-39-12N/69-55-30W)

Chatham Lt., W. side of hbr. – Fl(2)W. ev. 10 s., Ht. 80′, Rge. 24 mi., ltd. 24 hrs., (41-40-17N/69-57-01W)

Chatham Inlet Bar Guide Lt., Fl. Y. ev. 2.5 s., Ht. 62′, Rge. 11 mi., (41-40-18N/69-57-00W)

Hyannis Harbor App. Ltd. Bell By. "HH" – Mo(A)W, Rge. 6 mi., RWSRST, (41-35-57N/70-17-22W)

Pollock Rip Ch. Ltd. By. #8 – Fl. R. ev. 6 s., Rge. 3 mi., Red, (41-32-43N/69-58-56W)

Cape Wind Meteorological Lt. Tower "MT"– Fl.Y. ev. 6 s., (41-28-20N/70-18-53W)

Nantucket Lt., (Great Point), Nantucket, N. end of Is., – Fl. W. ev. 5 s., R. sect. 084°-106° (Covers Cross Rip & Tuckernuck Shoals), Ht. 71′, Rge. W. 14 mi., R. 12 mi., (41-23-25N/70-02-54W)

Sankaty Head Lt., E. end of Is. – Fl. W. ev. 7.5 s., Ht. 158′, Rge. 24 mi., (41-17-04N/69-57-58W)

Nantucket East Breakwater Lt. #3, Outer Entr. to hbr. – Fl. G. ev. 4 s., Horn 1 bl. ev. 10 s., Ht. 30′, Rge. 12 mi., (41-18-37N/70-06-00W)

Brant Point Lt., Hbr. Entr. W. side – Oc.R. ev. 4 s., Horn 1 bl. ev. 10 s., Ht. 26′, Rge. 10 mi., (41-17-24N/70-05-25W)

Cape Poge Lt., NE point of Chappaquiddick Is. – Fl. W. ev. 6 s., Ht. 65′, Rge. 9 mi., (41-25-10N/70-27-08W)

Muskeget Ch. Ltd. Wh. By. "MC" – Mo(A)W, Rge. 4 mi., RWSRST, (41-15-00N/70-26-10W)

Edgartown Harbor Lt., Inner end of hbr. W. side – Fl. R. ev. 6 s., Ht. 45′, Rge. 5 mi., (41-23-27N/70-30-11W)

For abbreviations see footnote p. 167

East Chop Lt., E. side Vineyard Haven Hbr. Entr. – Iso. G. ev. 6 s., Ht. 79′, Rge. 9 mi., (41-28-13N/70-34-03W)

West Chop Lt., W. side Vineyard Haven Hbr. Entr. – Oc.W. ev. 4 s., R. Sect. 281°-331° (covers Squash Meadow and Norton Shoals), Horn 1 bl. ev. 30 s., Ht. 84′, Rge. R. 10 mi., W. 14 mi., (41-28-51N/70-35-59W)

Nobska Point Lt., Woods Hole E. Entr. – Fl. W. ev. 6 s., R. Sect. 263°-289° (covers Hedge Fence and L'hommedieu Shoal), Horn 2 bl. ev. 30 s., Ht. 87′, Rge. R. 11 mi., W. 13 mi., ltd 24 hrs., (41-30-57N/70-39-18W)

Tarpaulin Cove Lt., SE side Naushon Is. – Fl. W. ev. 6 s., Ht. 78′, Rge. 9 mi., (41-28-08N/70-45-27W)

Menemsha Creek Entr. Jetty Lt. #3 – Fl. G. ev. 4 s., Ht. 25′, Rge. 5 mi., (41-21-16N/70-46-07W)

Gay Head Lt., W. point of Martha's Vineyard – Alt. W. and R. ev. 15 s., Ht. 170′, Rge. W. 24 mi., R. 20 mi., Obscured 342°-359° by Nomans Land, ltd. 24 hrs., (41-20-54N/70-50-06W)

Cuttyhunk East Entr. Ltd. Bell By. "CH" – Mo(A)W, Rge. 5 mi., RWSRST, (41-26-34N/70-53-22W)

BUZZARDS BAY

Canapitsit Ch. Entr. Bell By. "CC", – RWSRST, (41-25-01N/70-54-23W)

Narragansett - Buzzards Bay App. Ltd. Wh. By. "A" – Mo(A)W, Rge. 6 mi., Racon (N), RWSRST, (41-06-00N/71-23-22W)

Buzzards Bay Entr. Lt., W. Entr. – Fl. W. ev. 2.5 s., Horn 2 bl. ev. 30 s., Ht. 67′, Rge. 17 mi., Racon (B), (41-23-49N/71-02-05W)

Dumpling Rocks Lt. #7, off Round Hill Pt. – Fl. G. ev. 6 s., Ht. 52′, Rge. 8 mi., (41-32-18N/70-55-17W)

Buzzards Bay Midch. Ltd. Bell By. "BB" (east of Wilkes Ledge) – Mo(A)W, Rge. 4 mi., RWSRST, (41-30-33N/70-49-54W)

New Bedford West Barrier Lt. – Q.G., Horn 1 bl. ev. 10 s., Ht. 48′, Rge. 8 mi., (41-37-27N/70-54-22W)

New Bedford East Barrier Lt. – Q. R., Ht. 48′, Rge. 5 mi., (41-37-29N/70-54-19W)

Padanaram Breakwater Lt. #8 – Fl. R. ev. 4 s., Ht. 25′, Rge. 5 mi., (41-34-27N/70-56-21W)

Butler Flats Lt. – Fl. W. ev. 4 s., Ht. 25′, (41-36-12N/70-53-40W)

Cleveland East Ledge Lt., Cape Cod Canal App. E. side of S. Entr. – Fl. W. ev. 10 s., Horn 1 bl. ev. 15 s., Ht. 74′, Rge. 15 mi., Racon (C), (41-37-51N/70-41-39W)

Ned Point Lt. – Iso. W. ev. 6 s., Ht. 41′, Rge. 12 mi., (41-39-03N/70-47-44W)

Westport Harbor Entr. Lt. #7, W. side – Fl. G. ev. 6 s., Ht. 35′, Rge. 9 mi., (41-30-27N/71-05-17W)

Westport Harbor App. Ltd. Bell By. "WH", Mo(A)W, Rge. 4 mi., RWSRST, (41-29-15N/71-04-04W)

RHODE ISLAND

Sakonnet River Entr. Ltd. Wh. By. "SR" – Mo(A)W, Rge. 4 mi., RWSRST, (41-25-45N/71-13-23W)

Sakonnet Lt. – Fl. W. ev. 6 s., R. sect. 195°-350°, Ht. 58′, Rge. W. 7 mi., R. 5 mi., (41-27-11N/71-12-09W)

Sakonnet Breakwater Lt. #2, Entr. to hbr. – Fl. R. ev. 4 s., Ht. 34′, Rge. 8 mi., (41-28-00N/71-11-42W)

Narragansett Bay Entr. Ltd. Wh. By. "NB" – Mo(A)W, Rge. 6 mi., Racon (B), RWSRST, (41-23-00N/71-23-21W)

Beavertail Lt. – Narrag. Bay E. passage – Fl. W. ev. 10 s., Obscured 175°-215°, Horn 1 bl. ev. 30 s., Ht. 64′, Rge. 15 mi., ltd. 24 hrs., (41-26-58N/71-23-58W)

Castle Hill Lt. – Iso R. 6 s., Keyed (VHF 83A) Horn 1 bl. ev. 10 s., Ht. 40′, Rge. 12 mi., (41-27-44N/71-21-47W)

For abbreviations see footnote p. 167

Fort Adams Lt. #2, Narrag. Bay E. passage – Fl. R. ev. 6 s., Horn 1 bl. ev. 15 s., Ht. 32′, Rge. 7 mi., (41-28-54N/71-20-12W)

Newport Harbor Lt., N. end of breakwater – F.G., Ht. 33′, Rge. 11 mi., (41-29-36N/71-19-38W)

Rose Is. Lt., Fl W. ev. 6 s., Ht. 48′, (41-29-44N/71-20-34W)

Prudence Is. Lt. (Sandy Pt.), Narrag. Bay E. passage – Fl. G. ev. 6 s., Ht. 28′, Rge. 6 mi., (41-36-21N/71-18-13W)

Hog Island Shoal Lt., N. side Entr. to Mt. Hope Bay – Iso. W. ev. 6 s., Keyed (VHF 83A) Horn 2 bl. ev. 30 s., Ht. 54′, Rge. 12 mi., (41-37-56N/71-16-24W)

Musselbed Shoals Lt.#6A, Mt. Hope Bay Ch. – Fl. R. ev. 6 s., Ht. 26′, Rge. 6 mi., (41-38-11N/71-15-36W)

Castle Is. Lt. #2, N. of Hog Is. – Fl. R. ev. 6 s., Ht. 26′, Rge. 3 mi., (41-39-14N/71-17-10W)

Bristol Harbor Lt. #4 – F.R., Ht. 25′, Rge. 11 mi., (41-39-58N/71-16-42W)

Conimicut Lt., Providence R. App. – Fl. W. ev. 2.5 s., R. Sect. 322°-349°, Horn 2 bl. ev. 30 s., Ht. 58′, Rge. W. 8 mi., R. 5 mi., (41-43-01N/71-20-42W)

Bullock Point Lt. "BP", Prov. R. – Oc.W. ev. 4 s., Ht. 29′, Rge. 6 mi., (41-44-16N/71-21-51W)

Pomham Rocks Lt., Prov. R. – F.R., Ht. 67′, Rge. 6 mi., (41-46-39N/71-22-10W)

Providence River Ch. Lt. #42, off rock – Iso. R. ev. 6 s., Ht. 31′, Rge. 4 mi., (41-47-39N/71-22-47W)

Mt. Hope Bay Jct. Ltd. Gong By. "MH" – Fl(2+1) R. 6 s., Rge., 3 mi., R. & G. Bands, (41-39-32N/71-14-03W)

Borden Flats Lt., Mt. Hope Bay – Fl. W. ev. 2.5 s., Horn 1 bl. ev. 10 s., Ht. 47′, Rge. 11 mi., (41-42-16N/71-10-28W)

Wickford Harbor Lt. #1, Narrag. Bay W. passage – Fl. G. ev. 6 s., Ht. 40′, Rge. 6 mi., (41-34-21N/71-26-13W)

Warwick Lt., Greenwich Bay App. – Oc.G. ev. 4 s., Horn 1 bl. ev. 15 s., Ht. 66′, Rge. 12 mi., ltd. 24 hrs., (41-40-02N/71-22-42W)

Point Judith Lt., Block Is. Sd. Entr. – Oc(3)W. ev. 15 s., Horn 1 bl. ev. 15 s., Ht. 65′, Rge. 16 mi., (41-21-40N/71-28-53W)

Block Island North Lt., N. end of Is. – Fl. W. ev. 5 s., Ht. 58′, (41-13-39N/71-34-33W)

Block Island Southeast Lt., SE end of Is. – Fl. G. ev. 5 s., Horn 1 bl. ev. 30 s., Ht. 261′, Rge. 20 mi., ltd. 24 hrs., (41-09-10N/71-33-04W)

Block Island Breakwater Lt. #3, – F. G., Keyed (VHF 83A) Horn 2 bl. ev. 30 s., Ht. 27′, Rge. 11 mi., (41-10-38N/71-33-15W)

Pt. Judith Harbor of Refuge W. Entr. Lt. #3 – Fl. G. ev. 6 s., Horn 1 bl. ev. 30 s., Ht. 35′, Rge. 5 mi., (41-21-56N/71-30-53W)

Watch Hill Lt., Fishers Is. Sd. E. Entr. – Alt. W. and R. ev. 5 s., Horn 1 bl. ev. 30 s., Ht. 61′, Rge. 14 mi., ltd. 24 hrs., (41-18-14N/71-51-30W)

FISHERS ISLAND SOUND

Latimer Reef Lt., Fishers Is. Sd. main ch. – Fl. W. ev. 6 s., Bell 2 strokes ev. 15 s., Ht. 55′, Rge. 9 mi., (41-18-16N/71-56-00W)

N. Dumpling Lt., Fishers Is. Sd. main ch. – F.W., Horn 1 bl. ev. 30 s., R. Sect. 257°-023°, Ht. 94′, Rge. R. 7 mi., F.W. 9 mi., (41-17-17N/72-01-10W)

Stonington Outer Breakwater Lt. #4 – Fl. R. ev. 4 s., Horn 1 bl. ev. 10 s., Ht. 46′, Rge. 5 mi., (41-19-00N/71-54-28W)

LONG ISLAND SOUND, NORTH SIDE

Race Rock Lt., SW end of Fishers Is. – Fl. R. ev. 10 s., Horn 2 bl. ev. 30 s., Ht. 67′, Rge. 16 mi., (41-14-37N/72-02-50W)

Bartlett Reef Lt., S. end of reef – Fl. W. ev. 6 s., Keyed (VHF 79A may change to 83A) Horn 2 bl. ev. 60 s., Ht. 35′, Rge. 8 mi., (41-16-28N/72-08-14W)

New London Ledge Lt., W. side of Southwest ledge –Fl(3) W.R. ev. 30 s., Horn 2 bl. ev. 20 s., Ht. 58′, Rge. W. 17 mi., R. 14 mi., (41-18-21N/72-04-39W)

For abbreviations see footnote p. 167

New London Harbor Lt., W. side Entr. – Iso. W. ev. 6 s., R. Sect. 000°-041°, Ht. 89′, Rge. W. 17 mi., R. 14 mi., (41-19-00N/72-05-23W)

Saybrook Breakwater Lt., W. jetty – Fl. G. ev. 6 s., Horn 1 bl. ev. 30 s., Ht. 58′, Rge. 14 mi., (41-15-48N/72-20-34W)

Lynde Pt. Lt., Conn. R. mouth W. side – F.W., Ht. 71′, Rge. 14 mi., (41-16-17N/72-20-35W)

Twenty-Eight Foot Shoal Ltd. Wh. By. "TE" – Fl(2+1) R. ev. 6 s., , Rge. 4 mi., Racon (T), R&G Bands, (41-09-16N/72-30-25W)

Falkner Is. Lt., off Guilford Hbr. – Fl. W. ev. 10 s., Ht. 94′, Rge. 13 mi., (41-12-43N/72-39-13W)

Branford Reef Lt., SE Entr. New Haven – Fl. W. ev. 6 s., Ht. 22′, Rge. 7 mi., (41-13-17N/72-48-19W)

New Haven Hbr. Ltd. Wh. By. "NH" – Mo(A)W, Rge. 4 mi., RWSRST, (41-12-07N/72-53-47W)

Southwest Ledge Lt., E. side Entr. New Haven – Fl. R. ev. 5 s., Horn 1 bl. ev. 15 s., Ht. 57′, Rge. 14 mi., (41-14-04N/72-54-44W)

New Haven Lt. – Fl. W. ev. 4 s., Ht. 27′, Rge. 7 mi., (41-13-16N/72-56-32W)

Stratford Pt. Lt., W. side Entr. Housatonic R. –Fl(2)W. ev. 20 s., Ht. 52′, Rge. 16 mi., (41-09-07N/73-06-12W)

Stratford Shoal Lt., Middle Ground – Fl. W. ev. 5 s., Horn 1 bl. ev. 15 s., Ht. 60′, Rge. 13 mi., (41-03-35N/73-06-05W)

Tongue Pt. Lt., at Bridgeport Breakwater – Fl. G. ev. 4 s., Ht. 31′, Rge. 5 mi., (41-10-00N/73-10-39W)

Penfield Reef Lt., S. side Entr. to Black Rock – Fl. R. ev. 6 s., Horn 1 bl. ev. 15 s., Ht. 51′, Rge. 15 mi., (41-07-02N/73-13-20W)

Peck Ledge Lt., E. App. to Norwalk – Fl. G. ev. 2.5 s., Ht. 61′, Rge. 5 mi., (41-04-39N/73-22-11W)

Greens Ledge Lt., W. end of ledge – Alt. Fl. W. and R. ev. 24 s., Horn 2 bl. ev. 20 s., Ht. 62′, Rge. W. 18 mi., R. 15 mi., (41-02-30N/73-26-38W)

Stamford Harbor Ledge Obstruction Lt., on SW end of Harbor Ledge – Fl. W. ev. 4 s., (41-00-49N/73-32-34W)

Great Captain Is. Lt., SE Pt. of Is. – Alt. W. R. ev. 12 s., Horn 1 bl. ev. 15 s., Ht. 62′, Rge. W. 17 mi., R. 14 mi., (40-58-57N/73-37-23W)

Larchmont Harbor Lt. #2, East Entr. – Fl. R. ev. 4 s., Ht. 26′, Rge. 4 mi., (40-55-05N/73-43-52W)

LONG ISLAND SOUND, SOUTH SIDE

Little Gull Is. Lt., E. Entr. L.I. Sd. – Fl(2) W. ev. 15 s., Horn 1 bl. ev. 15 sec., Ht. 91′, Rge. 18 mi., (41-12-23N/72-06-25W)

Plum Gut Lt. – Fl. W. ev. 2.5 s., Ht. 21′ Rge. 5 mi., (41-10-26N/72-12-42W)

Plum Island Ltd. Wh. By. "PI" – Mo(A)W, Rge. 4 mi., RWSRST, (41-13-17N/72-10-48W)

Plum Is. Hbr. West Dolphin Lt., W. end of Is. – F.G., Horn 1 bl. ev. 10 s., (Maintained by U.S. Agr. Dept.), (41-10-17N/72-12-24W)

Orient Pt. Lt., outer end of Oyster Pond Reef – Fl. W. ev. 5 s., Horn 2 bl. ev. 30 s., Ht. 64′, Rge. 17 mi., (41-09-48N/72-13-25W)

Horton Pt. Lt., NW point of Horton Neck – Fl. G. ev. 10 s., Ht. 103′, Rge. 14 mi., (41-05-06N/72-26-44W)

Mattituck Breakwater Lt. "MI" – Fl. W. ev. 4 s., Ht. 25′, Rge. 6 mi., (41-00-55N/72-33-40W)

Old Field Pt. Lt. – Alt. Fl. R. and Fl. G. ev. 24 s., Ht. 74′, Rge. 14 mi., (40-58-37N/73-07-07W)

Eatons Neck Lt., E. side Entr. Huntington Bay – F. W., Horn 1 bl. ev. 30 s., Ht. 144′, Rge. 18 mi., (40-57-14N/73-23-43W)

Cold Spring Hbr. Lt., on Pt. of shoal – F.W., R. Sect. 039°-125°, Ht. 37′, Rge. W. Sect. 8 mi., R. Sect. 6 mi., (40-54-51N/73-29-35W)

For abbreviations see footnote p. 167

Glen Cove Breakwater Lt. #5, E. side Entr. to hbr. – Fl. G. ev. 4 s., Ht. 24′, Rge. 5 mi., (40-51-43N/73-39-37W)

Port Jefferson App. Ltd. Wh. By. "PJ" – Mo(A)W, Rge. 4 mi., RWSRST, (40-59-16N/73-06-27W)

Huntington Harbor Lt. – Iso. W. ev. 6 s., Horn 1 bl. ev. 15 s., Ht. 42′, Rge. 9 mi., (40-54-39N/73-25-52W)

LONG ISLAND, OUTSIDE

Montauk Pt. Lt., E. end of L.I. – Fl. W. ev. 5 s., Horn 1 bl. ev. 15 s., Ht. 168′, Rge. 18 mi., (41-04-15N/71-51-26W)

Montauk Hbr. Entr. Ltd. Bell By. "M" – Mo(A)W, Rge. 4 mi., (41-05-07N/71-56-23W)

Shinnecock Inlet App. Ltd. Wh. By. "SH" – Mo(A)W, Rge. 4 mi., RWSRST, (40-49-00N/72-28-35W)

Moriches Inlet App. Ltd. Wh. By. "M" – Mo(A)W, Rge. 6 mi., RWS, (40-44-08N/72-45-12W)

Shinnecock Lt., W. side of Inlet – Fl(2) W. ev. 15 s., Ht. 75′, Rge. 11 mi., (40-50-31N/72-28-42W)

Jones Inlet Lt., end of breakwater – Fl. W. ev. 2.5 s., Ht. 33′, Rge. 6 mi., (40-34-24N/73-34-32W)

Jones Inlet Ltd. Wh. By. "JI" – Mo(A)W, Rge. 4 mi., RWSRST, (40-33-37N/73-35-13W)

E. Rockaway Inlet Ltd. Bell By. "ER" – Mo(A)W, Rge. 5 mi., RWSRST, (40-34-17N/73-45-49W)

Fire Is. Lt., 5.5 mi. E. of inlet – Fl. W. ev. 7.5 s., Ht. 167′, ltd. 24 hrs., (40-37-57N/73-13-07W)

Rockaway Point Breakwater Lt. #4, end of breakwater – Fl. R. ev. 4 s., Ht. 34′, Rge. 5 mi., (40-32-25N/73-56-27W)

NEW YORK HARBOR & APPROACHES

Execution Rocks Lt. – Fl. W. ev. 10 s., Ht. 62′, Rge. 15 mi., Racon (X), (40-52-41N/73-44-16W)

Hart Is. Lt. #46, off S. end of Is. – Fl. R. ev. 4 s., Ht. 23′, Rge. 6 mi., (40-50-42N/73-46-00W)

Stepping Stones Lt., outer end of reef – Oc.G. ev. 4 s., Ht. 46′, Rge. 8 mi., (40-49-28N/73-46-29W)

Throgs Neck Lt., Fort Schuyler – F. R., Ht. 60′, Rge. 11 mi., (40-48-16N/73-47-26W)

Whitestone Pt. Lt. #1, East R. main ch. – Q.G., Ht. 56′, Rge. 3 mi., (40-48-06N/73-49-10W)

Kings Pt. Lt. – (Private Aid), Iso. W. ev. 2 s., (40-48-42N/73-45-48W)

Hell Gate Lt. #15, East R. Hallets Pt. – Fl. G. ev. 2.5 s., Ht. 33′, Rge. 4 mi., (40-46-41N/73-56-05W)

Mill Rock South Lt. #16, East R., main ch. – Fl. R. ev. 4 s., Ht. 37′, Rge. 4 mi., (40-46-46N/73-56-22W)

Governors Is. Extension Lt., SW Pt. of Is. – F. R., Horn 1 bl. ev. 15 s., Ht. 47′, Rge. 9 mi., (40-41-09N/74-01-35W)

Governors Is. Lt. #2 – NW pt of Is. – 2 F.R. arranged vertically, Lower Lt. Obscured from 240°-243°, 254°-256°, 264°-360°, Horn 2 bl. ev. 20 s., Ht. 75′, Rge. 7 mi., (40-41-35N/74-01-11W)

Verrazano-Narrows Bridge Sound Signal – (Private Aid), 2 Horns on bridge 1 bl. ev. 15 s., (none given)

Coney Is. Lt., N.Y. Hbr. main ch. – Fl. R. ev. 5 s., Ht. 75′, Rge. 16 mi., ltd. 24 hrs., (40-34-36N/74-00-42W)

Romer Shoal Lt., N.Y. Hbr. S. App. – Fl(2) W. ev. 15 s., Horn 2 bl. ev. 30 s., Ht. 54′, Rge. 15 mi., (40-30-47N/74-00-49W)

For abbreviations see footnote p. 167

West Bank (Range Front) Lt., Ambrose Ch. outer sect. – Iso. W. ev. 6 s., R. Sect. 004°-181° and W from 181° - 004°, Horn 2 bl. ev. 20 s., Ht. 69′, ltd. 24 hrs., (40-32-17N/74-02-34W)

Staten Island (Range Rear) Lt., Ambrose Ch. outer sect. – F. W. , Visible on range line only, Ht. 234′, ltd. 24 hrs., (40-34-34N/74-08-28W)

Old Orchard Shoal Lt., N.Y. Hbr. – Fl. W. ev. 6 s., R. Sect. 087°-203°, Ht. 51′, Rge. W. 7 mi., R. 5 mi., (40-30-44N/74-05-55W)

Sandy Hook Lt. – F. W., Ht. 88′, Rge. 19 mi., ltd. 24 hrs., (40-27-42N/74-00-07W)

Sandy Hook Ch. (Range Front) Lt. – Q. W., G., and R. sectors, Red from 063°-073° and Green from 300.5°-315.5°, Ht. 45′, Rge. 5 mi., Racon (C), (40-29-15N/73-59-35W)

Southwest Spit Jct. Ltd. Gong By. "SP" – Fl(2+1) R. ev. 6 s., Rge. 3 mi., R. & G. Bands, (40-28-46N/74-03-18W)

Sandy Hook Pt. Lt. – Iso W. ev. 6 s., Ht. 38′, Rge. 13 mi., Horn 1 bl. ev. 10 s., "NB" on Skeleton Tower, (40-28-15N/74-01-07W)

Scotland Ltd. Wh. By. "S", Sandy Hook Ch. App. – Mo(A)W, Rge. 7 mi., Racon (M), RWSRST, (40-26-33N/73-55-01W)

Ambrose Ch. Ltd. Wh. By. "A" – Mo(A)W, Rge. 7 mi., Racon (N), RWSRST, (40-27-28N/73-50-12W)

NEW JERSEY

Highlands Lt. – Oc. W. ev. 10 s., Obscured 334°-140°, (40-23-48N/73-59-09W)

Atlantic Highlands Breakwater Lt. – Fl. W. ev. 4 s., Ht. 33′, Rge. 7 mi., (40-25-07N/74-01-10W)

Kill Van Kull Ch. Jct. Ltd. Wh. By. "KV"– Fl (2+1) R. ev. 6 s., Rge. 3 mi., R. & G. Bands, Racon (K), (40-39-02N/74-03-51W)

Kill Van Kull Ch. Jct. Ltd. By. "A"– Fl (2+1) G. ev. 6 s., Rge. 3 mi., G. & R. Bands (40-38-45N/74-10-07W)

Kill Van Kull Ch. East Jct. Ltd. By. "E"– Fl (2+1) G. ev. 6 s., Rge. 3 mi. G. & R. Bands (40-38-31N/74-09-15W)

Manasquan Inlet Lt. #3 - Fl. G. ev. 6 s., Horn 1 bl. ev. 30 s., Ht. 35′ Rge. 8 mi., (40-06-01N/74-01-54W)

Shark River Inlet Ltd. Wh. By. "SI" – Mo(A)W, Rge. 6 mi., RWSRST, (40-11-09N/74-00-03W)

Barnegat Inlet S. Breakwater Lt. #7 – Q. G., Ht. 37′, Rge. 8 mi., Horn 1 bl. ev. 30 s., (39-45-26N/74-05-36W)

Barnegat Ltd. By. "B" – Fl. Y. ev. 6 s., Rge. 7 mi., Racon (B), Yellow, (39-45-48N/73-46-04W)

Barnegat Inlet Outer Ltd. Wh. By. "BI" – Mo(A)W, Rge. 6 mi., RWSRST, (39-44-28N/74-03-51W)

Little Egg Inlet Outer Ltd. Wh. By. "LE" – Mo(A)W, Rge. 6 mi., RWSRST, (39-27-56N/74-16-27W)

Brigantine Inlet Wreck Ltd. By. "WR2" (100 yards, 090° from wreck) – Q. R., Rge. 5 mi., Red, (39-24-48N/74-13-47W)

Great Egg Harbor Inlet Outer Ltd. Wh. By. "GE" – Mo(A)W, Rge. 5 mi., RWSRST, (39-16-14N/74-31-56W)

Hereford Inlet Lt., S. side – Fl. W. ev. 10 s., Ht. 57′, Rge. 24 mi., (39-00-24N/74-47-28W)

Five Fathom Bank Ltd. By. "F", Delaware Bay Entr. – Fl. Y. ev. 2.5 s., Rge. 7 mi., Racon (M), Yellow, (38-46-49N/74-34-32W)

Cape May Lt. – Fl. W. ev. 15 s., Ht. 165′, Rge. 24 mi., (38-55-59N/74-57-37W)

NEW JERSEY, DELAWARE AND MARYLAND

Delaware Ltd. By. "D", Del. Bay – Fl. Y. ev. 6 s., Rge. 7 mi., Racon (K), Yellow, (38-27-18N/74-41-47W)

For abbreviations see footnote p. 167

Delaware Traffic Lane Ltd. By. "DA" – Fl. Y. ev. 2.5 s., Rge. 6 mi., Yellow, (38-32-45N/74-46-56W)

Delaware Traffic Lane Ltd. By. "DB" – Fl. Y. ev. 4 s., Rge. 6 mi., Yellow, (38-38-12N/74-52-11W)

Delaware Traffic Lane Ltd. By. "DC" – Fl. Y. ev. 2.5 s., Rge. 7 mi., Yellow, (38-43-47N/74-57-33W)

Brown Shoal Lt. – Fl. W. ev. 2.5 s., Ht. 23′, Rge. 7 mi., Racon (B), (38-55-21N/75-06-01W)

Brandywine Shoal Lt., Del. Bay main ch. 9 mi. from S. end of shoal – Fl. W. ev. 10 s., R. Sect. 151°-338°, Horn 1 bl. ev. 15 s. (Mar. 15 - Dec. 15), Ht. 60′, Rge. W. 19 mi., R. 13 mi., (38-59-10N/75-06-47W)

Harbor of Refuge Lt., Del. Bay main ch., S. end of breakwater – Fl. W. ev. 10 s., 2 R. Sect. 325°-351° and 127°-175°, Horn 2 bl. ev. 30 s., Ht. 72′, Rge. W. 19 mi., R. 16 mi., (38-48-52N/75-05-33W)

Fourteen Foot Bank Lt., Del. Bay main ch. W. side – Fl. W. ev. 9 s., R. Sect. 332.5°-151°, Horn 1 bl. ev. 30 s., Ht. 59′, Rge. W. 13 mi., R. 10 mi., (39-02-54N/75-10-56W)

Miah Maull Shoal Lt., Del Bay main ch. – Oc. W. ev. 4 s., R. Sect. 137.5°-333°, Horn 1 bl. ev. 10 s., (Mar. 15-Dec. 15). Ht. 59′, Rge. W. 15 mi., R. 12 mi., Racon (M), (39-07-36N/75-12-31W)

Elbow of Cross Ledge Lt., Del. Bay main ch. – Iso. W. ev. 6 s., Horn 2 bl. ev. 20 s., (Mar. 15-Dec. 15), Ht. 61′, Rge. 15 mi., (39-10-56N/75-16-06W)

Ship John Shoal Lt., Del. Bay main ch. – Fl. W. ev. 5 s., R. Sect. 138°-321.5°, Horn 1 bl. ev. 15 s. (Mar. 15 - Dec. 15) , Ht. 50′, Rge. W. 16 mi., R. 12 mi., Racon (O), (39-18-19N/75-22-36W)

Egg Island Point Lt. – Fl. W. ev. 4 s., Ht. 27′, Rge. 7 mi., (39-10-21N/75-07-55W)

Ben Davis Pt. Lt. "BD" – Fl. W. ev. 6 s., Ht. 30′, Rge. 6 mi., (39-17-27N/75-17-18W)

Old Reedy Is. Lt. – Iso. W. ev. 6 s., R. Sect. 353°-014°, Ht. 20′, Rge. W. 8 mi., R. 6 mi., (39-30-03N/75-34-08W)

Fenwick Is. Lt. – Oc.W. ev. 13 s., Ht. 83′, (Rge. none given), (38-27-06N/75-03-18W)

Ocean City Inlet Jetty Lt., on end of Jetty – Iso. W. ev. 6 s., Ht. 38′, Rge. 6 mi., (38-19-27N/75-05-06)

VIRGINIA

Assateague Lt., S. side of Is. – Fl(2) W. ev. 5 s., Ht. 154′, Rge. 22 mi., (37-54-40N/75-21-22W)

Wachapreague Inlet Ltd. Wh. By. "W" – Mo(A)W, Rge. 6 mi., RWSRST, (37-34-54N/75-33-37W)

Quinby Inlet Ltd. Wh. By. "Q" – Mo(A)W, Rge. 5 mi., RWSRST, (37-28-06N/75-36-05W)

Great Machipongo Inlet Ltd. Wh. By. "GM" – Mo(A)W, Rge. 5 mi., RWSRST, (37-23-36N/75-39-06W)

Great Machipongo Inlet Lt. #5, S. side – Fl. G. ev. 4 s., Ht. 15′, Rge. 4 mi., (37-21-40N/75-44-06W)

Cape Charles Lt., N. side of Entr. to Ches. Bay – Fl. W. ev. 5 s., Ht. 180′, Rge. 18 mi., (37-07-23N/75-54-23W)

Chesapeake Lt., off Entr. to Ches. Bay – Fl(2) W. ev. 15 s., Horn 1 bl. ev. 10 s., Ht. 117′, Rge. 19 mi., Racon (N), (36-54-17N/75-42-46W)

Chesapeake Bay Entr. Ltd. Wh. By. "CH" – Mo(A)W, Rge. 7 mi., Racon (C), RWSRST, (36-56-08N/75-57-27W)

Cape Henry Lt., S. side of Entr. to Ches. Bay – Mo (U) W ev. 20 s., R. Sect. 154°-233°, Ht. 164′, Rge. W. 17 mi., R. 15 mi., (36-55-35N/76-00-26W)

For abbreviations see footnote p. 167

Distance Table in Nautical Miles

*Approximate

Bar Harbor to
- Halifax, N.S. 259
- Yarmouth, N.S. 101
- Saint John, N.B. 122
- Machiasport 52
- Rockland 62
- Boothbay Harbor 86
- Portland 115
- Marblehead 169

Rockland to
- Boothbay Harbor 42
- Belfast 22
- Bucksport 33

Boothbay Harbor to
- Kennebec River 11
- Monhegan 15
- Portland 36

Portland Ltd. Buoy "P" to
- Biddeford 17
- Portsmouth 54
- Cape Cod Light 99
- Cape Cod Canal (E. Entr.) 118
- Pollock Rip Slue 141

Portsmouth (Whaleback) to
- York River 7
- Biddeford Pool 30
- Newburyport Entr. 15
- Gloucester – via Annisquam 28

Gloucester to
- Boston 26
- Scituate 26
- Plymouth 43
- Cape Cod Canal (E. Entr.) 52
- Provincetown 45

Marblehead to
- Portsmouth 43
- Biddeford Pool 68
- Portland 87
- Boothbay Harbor 104
- Rockland 133
- Plymouth 38
- Cape Cod Canal (E. Entr.) 47

Boston (Commonwealth Pier)
- Marblehead 17
- Isles of Shoals 52
- Portsmouth 58
- Portland 95
- Kennebec River 107
- Boothbay Harbor 116
- Rockland 149
- North Haven 148
- Bangor 194
- St. John, N.B. 286
- Halifax, N.S. 380
- Cohasset 14
- Cape Cod Canal, E. Entr. 50
- Provincetown 50
- Vineyard Haven 77
- New Bedford 81
- Fall River 107
- Newport 122
- New London 140
- New York 234

****Western Entr., Cape Cod Canal to**
- East Entrance 8
- Woods Hole 15
- Quicks Hole 20
- New Bedford 24
- Newport 50
- New London 83

Woods Hole to
- Hyannis 19
- Chatham 32
- Cuttyhunk 14
- Marion 11

Vineyard Haven to
- Edgartown 9
- Marblehead – around Cape 114
- Canal – via Woods Hole 20
- Newport 45
- New London 77
- New Haven 114
- South Norwalk 140
- City Island 153

***Each distance is by the shortest route that safe navigation permits between the two ports concerned.**

*******Western entr.*, The beginning of the "land cut" at Bourne Neck, 7.3 nautical miles up the channel from Cleveland Ledge Lt.**

Continued p. 194

CHESAPEAKE BAY

Thimble Shoal Lt., Thimble Shoal Ch. – Fl. W. ev. 10 s., Ht. 55', Rge. 18 mi., (37-00-52N/76-14-23W)

Worton Pt. Lt., Fl. W. ev. 6 s., Ht. 93' Rge. 6 mi., (39-19-06N/76-11-11W)

Old Point Comfort Lt., N. side Entr. to Hampton Roads – Fl(2) R. ev. 12 s., W. Sect. 265°-038°, Ht. 54', Rge. W. 16 mi., R. 14 mi., (37-00-06N/76-18-23W)

York Spit Lt., N. side Entr. to York R. – On pile, Fl. W. ev. 6 s., Ht. 30', Rge. 8 mi., (37-12-35N/76-15-15W)

Wolf Trap Lt., Ches. Ch. – Fl. W. ev. 15 s., Ht. 52', Rge. 14 mi., (37-23-26N/76-11-22W)

Stingray Pt. Lt., Ches. Ch. – Fl. W. ev. 4 s., Ht. 34', Rge. 9 mi., (37-33-41N/76-16-12W)

Windmill Pt. Lt., Ches. Ch. – On pile. Fl. W. ev. 6 s., 2 R. Sectors 293°-082° and 091.5°-113°, Ht. 34', Rge. W. 9 mi., R. 7 mi., (37-35-49N/76-14-10W)

Tangier Sound Lt., Ches. Ch. – Fl. W. ev. 6 s., R. Sect. 115°-193°, Ht. 45', Rge. W. 12 mi., R. 9 mi., (37-47-17N/75-58-24W)

Smith Pt. Lt., Ches. Ch. – Fl. W. ev. 10 s., Ht. 52', Rge. 15 mi., (37-52-48N/76-11-01W)

Point Lookout Lt., Ches. Ch. – Fl(2) W. ev. 5 s., Ht. 39', Rge. 8 mi., (38-01-30N/76-19-25W)

Holland Is. Bar Lt., Ches. Ch. – Fl. W. ev. 2.5 s., Horn 1 bl. ev. 30 s. (operates continuously Sept. 15 - June 1), Ht. 37', Rge. 7 mi., (38-04-07N/76-05-45W)

Point No Point Lt., Ches. Ch. – Fl. W. ev. 6 s., Ht. 52', Rge. 9 mi., (38-07-41N/76-17-25W)

Hooper Is. Lt., Ches. Ch. – Fl. W. ev. 6 s., Horn 1 bl. ev. 30 s. (operates continuously Sept. 15 - June 1), Ht. 63', Rge. 9 mi., (38-15-23N/76-14-59W)

Drum Pt. Lt.#4, Ches. Ch. – Fl. R. ev. 2.5 s., Ht. 17', Rge. 5 mi., (38-19-08N/76-25-15W)

Cove Pt. Lt., Ches. Ch. – Fl. W. ev. 10 s., Ht. 45', Rge. 12 mi., Obscured from 040°-110°, ltd. 24 hrs, (38-23-11N/76-22-54W)

Bloody Point Bar Lt., Ches. Ch. – Fl. W. ev. 6 s., 2 R. Sectors 003°-022° and 183°-202°, Ht. 54', Rge. W. 9 mi., R. 7 mi., (38-50-02N/76-23-30W)

Thomas Pt. Shoal Lt., Ches. Ch. – Fl. W. ev. 5 s., 2 R. Sectors 011°-051.5° and 096.5°-202°, Horn 1 bl. ev. 15 s., Ht. 43', Rge. W. 16 mi., R. 11 mi., (38-53-56N/76-26-09W)

Sandy Pt. Shoal Lt., Ches. Ch. – Fl. W. ev. 6 s., Ht. 51', Rge. 9 mi., (39-00-57N/76-23-04W)

Baltimore Lt. – Fl. W. ev. 2.5 s., R. Sector 082°- 160°, Ht. 52', Rge. W. 7 mi., R. 5 mi., (39-03-33N/76-23-56W)

Wm. P. Lane, Jr. Bridge West Ch. Fog Signal, on main ch. span – Horn 1 bl. ev. 15 s., 5 s. bl., Horn Points 017° & 197°, (38-59-36N/76-22-53W)

Wm. P. Lane, Jr. Bridge East Ch. Fog Signal, on main ch. span – Horn 1 bl. ev. 20 s., 2 s. bl., (38-59-18N/76-21-30W)

NORTH CAROLINA

Currituck Beach Lt. – Fl. W. ev. 20 s., Ht. 158', Rge. 18 mi., (36-22-37N/75-49-47W)

Bodie Is. Lt. – Fl(2) W. ev. 30 s., Ht. 156', Rge. 18 mi., (35-49-07N/75-33-48W)

Oregon Inlet Jetty Lt. – Iso. W. ev. 6 s., Ht. 28', Rge. 7 mi., (35-46-26N/75-31-30W)

Cape Hatteras Lt., – Fl. W. ev. 7.5 s., Ht. 192', Rge. 24 mi., (35-15-02N/75-31-44W)

Hatteras Inlet Lt. – Iso. W. ev. 6 s., Ht. 48', Rge. 10 mi., (35-11-52N/75-43-56W)

Ocracoke Lt., on W. part of island – F.W., Ht. 75', Rge. 15 mi., (35-06-32N/75-59-10W)

Cape Lookout Lt., on N. pt. of cape – Fl. W. ev. 15 s., Ht. 156', Rge. 25 mi., (34-37-22N/76-31-28W)

Beaufort Inlet Ch. Ltd. Wh. By. "BM" – Mo(A)W, Rge. 6 mi., Racon (M), RWSRST, (34-34-49N/76-41-33W)

New River Inlet Ltd. Wh. By. "NR" – Mo(A)W, Rge. 6 mi., RWSRST, (34-31-02N/77-19-33W)

For abbreviations see footnote p. 167

Distance Table in Nautical Miles

*Approximate

Continued from p. 192

Nantucket Entr. Bell NB to
- Boston – around Cape 105
- Boston – via Canal 94
- Chatham 23
- Edgartown 23
- Hyannis 21
- Woods Hole 30
- Cape Cod Canal (W. Entr.) 45
- Newport 71

New Bedford (State Pier) to
- Woods Hole 14
- Newport 38
- New London 74
- New York (Gov. Is.) 166

Newport to
- Providence 21
- Stonington 34
- New London 48
- New Haven 84
- City Island 122

Block Island (FR Horn) to
- Nantucket 79
- Vineyard Haven 52
- Cleveland Ledge Lt 50
- New Bedford 44
- Newport 22
- Race Point Lt 21
- New London 29

New London to
- Greenport 25
- New Haven 49
- Bridgeport 60
- City Island 86

Port Jefferson to
- Larchmont 30
- So. Norwalk 15
- Milford 14
- Old Saybrook 43
- New London 53

City Island to
- Governors Island 17
- Execution Rocks 3

Execution Rocks to
- Port Chester 8
- Stamford 12
- Oyster Bay Harbor 14
- So. Norwalk 19
- Bridgeport 29
- Port Jefferson 30
- Milford 37
- New Haven 49
- Conn. River 69
- Mystic 84
- Montauk Point 87

New York (Battery) to
- Jones Inlet 34
- Fire Island Inlet 47
- Moriches Inlet 74
- Shinnecock Inlet 88
- Montauk Point 117
- Keyport 22
- Asbury Park 35
- Manasquan 40
- Little Egg Inlet 81
- Atlantic City 97
- Philadelphia 235
- Chesapeake Lt. Stn 247
- Cape Henry Lt. 262
- Norfolk 288
- Baltimore 418

Brielle-Manasquan to
- E. Rockaway Inlet 32
- Jones Inlet 35
- Fire Island Inlet 45
- Montauk Point 117
- Barnegat Inlet 21
- Atlantic City 51

Delaware Breakwater to
- Reedy Pt. Entr. (C&D Canal) .. 51
- Annapolis – via Canal 97
- Norfolk 167
- New York 150
- New London 242
- Providence 275
- New Bedford 278
- Boston (outside) 399
- Portland (outside) 443

Old Point Comfort to
- Baltimore 163
- Philadelphia 240
- New York 276
- New London 363
- Providence 392
- New Bedford 397
- Boston (outside) 512
- Portland 553

Oak Is. Lt., on SE pt. of island – Fl(4) W. ev. 10 s., Ht. 169′, Rge. 24 mi., (33-53-34N/78-02-06W)

Cape Fear River Entr. Ltd. Wh. By. "CF" – Mo(A)W, Rge. 6 mi., Racon (C), RWSRST, (33-46-17N/78-03-02W)

SOUTH CAROLINA

Little River Inlet Entr. Ltd. Wh. By. "LR" – Mo(A)W, Rge. 5 mi., RWSRST, (33-49-49N/78-32-27W)

Little River Inlet North Jetty Lt. #2 – Fl. R. ev. 4 s., Ht. 24′, Rge. 5 mi., (33-50-31N/78-32-39W)

Winyah Bay Ltd. Wh. By. "WB" – Mo(A)W, Rge. 6 mi., RWSRST, (33-11-37N/79-05-11W)

Georgetown Lt., E. side Entr. to Winyah Bay – Fl(2) W. ev. 15 s., Ht. 85′, Rge. 15 mi., (33-13-21N/79-11-06W)

Charleston Entr. Ltd. By. "C" – Mo(A)W, Rge. 6 mi., Racon (K), RWSRST, (32-37-05N/79-35-30W)

Charleston Lt., S. side of Sullivans Is. – Fl(2) W. ev. 30 s., Ht. 163′, Rge. 26 mi., (32-45-29N/79-50-36W)

GEORGIA

Tybee Lt., NE end of Is. – F. W., Ht. 144′, Rge. 19 mi., ltd. 24 hrs., (32-01-20N/80-50-44W)

Tybee Lighted Buoy "T" – Mo(A)W, Rge. 6 mi., Racon (G), RWSRST, (31-57-52N/80-43-10W)

St. Simons Ltd. By. "STS" – Mo(A)W, Rge. 7 mi., Racon (B), RWSRST, (31-02-49N/81-14-25W)

St. Simons Lt., N. side Entr. to St. Simons Sd. – F. Fl. W. ev. 60 s., Ht. 104′, Rge. F. W. 18 mi., Fl. W. 23 mi., (31-08-03N/81-23-37W)

FLORIDA

Amelia Is. Lt., 2 mi. from N. end of Is. – Fl. W. ev. 10 s., R. Sect. 344°-360°, Ht. 107′, Rge. W. 23 mi., R. 19 mi., (30-40-23N/81-26-33W)

St. Johns Lt., on shore – Fl(4) W. ev. 20 s., Obscured 179°-354°, Ht. 83′, Rge. 19 mi., (30-23-10N/81-23-53W)

St. Johns Ltd. By. "STJ" – Mo(A)W, Rge. 6 mi., Racon (M), RWSRST, (30-23-35N/81-19-08W)

St. Augustine Lt., N. end of Anastasia Is. – F. Fl. W. ev. 30 s., Ht. 161′, Rge. F. W. 19 mi., Fl. W. 24 mi., (29-53-08N/81-17-19W)

Ponce De Leon Inlet Lt., S. side on inlet – Fl(6) W. ev. 30 s., Ht. 159′, (29-04-50N/80-55-41W)

Cape Canaveral Lt., on Cape – Fl(2) W. ev. 20 s., Ht. 137′, Rge. 24 mi., (28-27-37N/80-32-36W)

Sebastian Inlet N. Jetty Lt. – Fl. W. ev. 4 s., R. Sect. 104°-154°, Ht. 27′, Rge. W. 5 mi., R. 4 mi., (27-51-41N/80-26-51W)

Jupiter Inlet Lt., N. side of inlet – Fl(2) W. ev. 30 s., Obscured 231°-234°, Ht. 146′, Rge. 25 mi., (26-56-55N/80-04-55W)

Hillsboro Inlet Entr. Lt., N. side of inlet – Fl. (2) W. ev. 20 s., Obscured 114°-119°, Ht. 136′, Rge. 28 mi., (26-15-33N/80-04-51W)

Port Everglades Ltd. By. "PE" - Mo(A)W, Rge. 7 mi., Racon (T), RWSRST, (26-05-30N/80-04-46W)

Miami Ltd. By. "M" – E. end of Miami Beach, Mo(A)W, Rge. 7 mi., Racon (M), RWSRST, (25-46-06N/80-05-00W)

Fowey Rocks Lt., Hawk Ch. – Fl. W. ev. 10 s., 2 R. Sect., W. Sectors 188°-359° and 022°-180°, R. in intervening sectors, Ht. 110′, Rge. W. 15 mi., R. 10 mi. Racon (O), (25-35-26N/80-05-48W)

For abbreviations see footnote p. 167

Carysfort Reef Lt., outer line of reefs – Fl(3) W. ev. 60 s., 3 W. Sectors 211°-018° and 049°-087° and 145°-184°, R. in intervening sectors, Ht. 100′, Rge. W. 15 mi., R. 13 mi., Racon (C), (25-13-19N/80-12-41W)

Alligator Reef Lt., outer line of reefs – Fl(4) W. ev. 60 s., 2 R. sectors 223°-249° and 047°-068°, Ht. 136′, Rge. W. 16 mi., R. 13 mi., Racon (G), (24-51-06N/80-37-08W)

Sombrero Key Lt., outer line of reefs – Fl(5) W. ev. 60 s., 3 W. Sectors 222°-238° and 264°-066° and 094°-163°, R. in intervening sectors, Ht. 142′, Rge. W. 15 mi., R. 12 mi., Racon (M), (24-37-40N/81-06-39W)

American Shoal Lt., outer line of reefs – Fl(3) W. ev. 15 s., W. sectors 270°-067° and 125°-242°, Obscured 90°-125°, R. in intervening sectors, Ht. 109′, Rge. R. 10 mi., Racon (Y), (24-31-30N/81-31-10W)

Sand Key Lt., Fl(2) W. ev. 15 s., 2 Red Sectors 072°-086° and 248°-270°, Ht. 109′, Rge. W. 14 mi., R. 11 mi., Racon (N), (24-27-14N/81-52-39W)

Dry Tortugas Lt., on Loggerhead Key – Fl. W. ev. 20 s., Ht. 151′, Rge. 20 mi., Racon (K), (24-38-00N/82-55-14W)

BERMUDA – APPROACH LIGHTS FROM SEAWARD

North Rock Beacon – Fl(4)W. ev. 20 s. yellow, Ht. 70′, Rge. 12 mi., RaRef, (32-28.5N/64-46.1W)

North East Breaker Beacon – Fl. W. ev. 2.5 s., Ht. 45′, Rge. 12 mi., Racon (N), RaRef, (Red tower on red tripod base reading "Northeast," (32-28.7N/64-41.0W)

Kitchen Shoal Beacon – Fl(3)W. ev. 15 s., Ht. 45′, Rge. 12 mi., RaRef, RWS, Red "Kitchen" on White background, (32-26.1N/64-37.6W)

Eastern Blue Cut Beacon – Fl. W. Mo(U) ev. 10 s., Ht. 60′, Rge. 12 mi., RaRef, B&W Tower "Eastern Blue Cut" on white band, (32-23.9N/64-52.6W)

Chub Heads – Q. Fl(9) W. ev. 15 s., Ht. 60′, Rge. 12 mi., RaRef, Yellow and Black Horizontal Stripe Tower with "Chub Heads" in White on Black Central band, Racon (C), (32-17.2N/64-58.9W)

Mills Breaker By. – Q. Fl(3)W. ev. 5 s., Black "Mills" on yellow background, (32-23.9N/64-36.9W)

Spit By. – Q. Fl(3) W. ev. 10 s., Black "Spit" on yellow, (32-22.7N/64-38.5W)

Sea By. –Mo(A)W ev. 6 s., RWS, Red "SB" in white on side, (32-22.9N/64-37.1W)

St. David's Is. Lighthouse – F. R. and G. Sectors below Fl(2) W. ev. 20 s., Ht. 212′, Rge. W. 15 mi., R. and G. 20 mi., (32-21.8N/64-39.1W) Your bearing from seaward of G. Sector is 221°-276° True; remaining Sector is R. and partially obscured by land 044°-135° True.

Kindley Field Aero Beacon – Alt. W and G.; 1 White, 1 Green (rotating Aero Beacon), Ht. 140′, Rge. 15 mi., (32-21.95N/64-40.55W)

Gibbs Hill Lighthouse – Fl. W. ev. 10 s., Ht. 354′, Rge. 26 mi., (32-15.2N/64-50.1W)

Foregoing information checked to date, September, 2013. See page 4 for free Supplement in May 2014.

Heaving the Lead

In tidal water where depths were doubtfully marked on the chart, or in thick weather off shore, soundings were made to determine the ship's position. The leadsman stood in the fore channels and swung the lead. [Aft] in the main and mizzen channels were other men who held the line as it led aft to the stern, where the mate stood by the line tub. The leadsman called, "All ready there?" to the next man, the mate shouted "Heave!" and the lead went spinning forward. Each man let go as the line tautened, and the mate grasped the line as it ran from the tub, and made the sounding. If the lead struck bottom before it reached him, one of the others took the sounding and called the marks. Markers on the line indicated the depth in fathoms, and an "arming" of tallow in the end of the lead showed the nature of the bottom.

Reprinted from Sail Ho! Windjammer Sketches Alow and Aloft, by Gordon Grant, 1931, William Farquhar Payson, Inc., NY

Traditional Markings for Leadlines

2 fathoms – a 2-ended scrap of leather
3 fathoms – a 3-ended scrap of leather
5 fathoms – a scrap of white calico
7 fathoms – a strip of red wool bunting
10 fathoms – leather with a round hole
13 fathoms – a piece of thick blue serge
15 fathoms – a piece of white calico
17 fathoms – a piece of red wool bunting
20 fathoms – a cord with 2 knots
30 fathoms – a cord with 3 knots

TIDAL HEIGHTS and DEPTHS

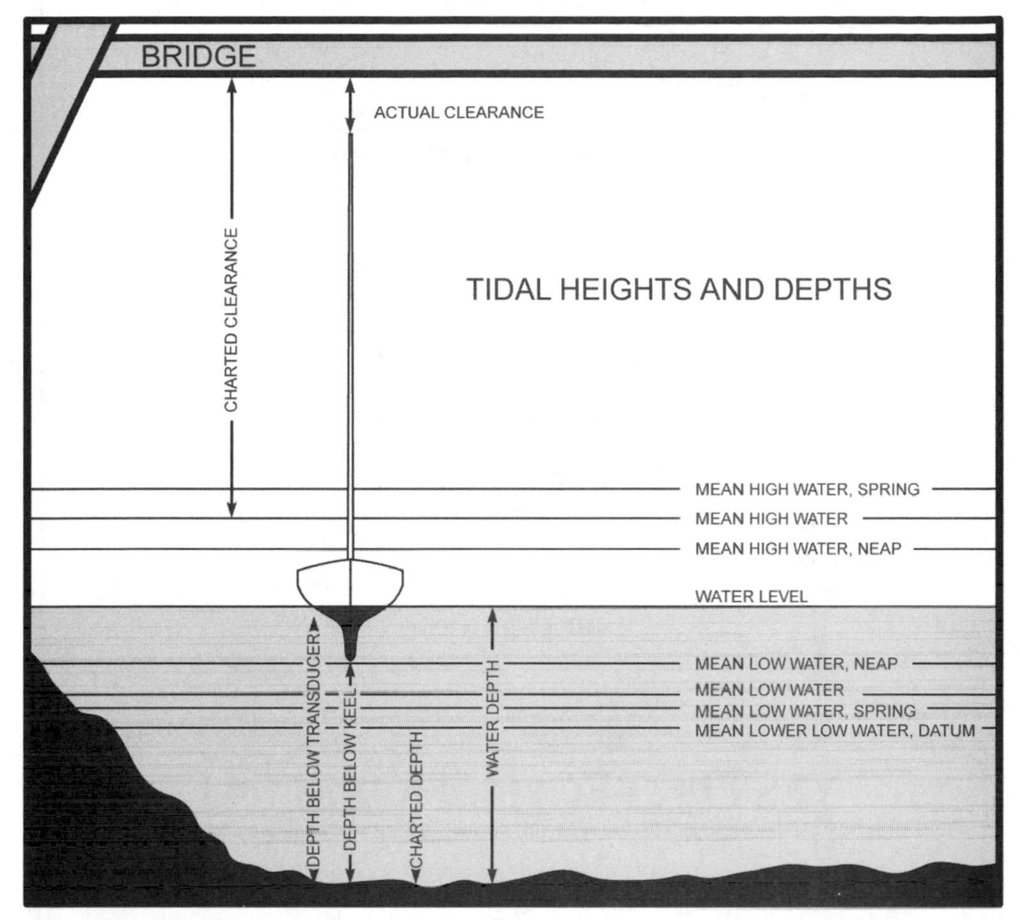

Mean High Water, Spring - the mean of high water heights of spring tides
Mean High Water - the mean of all high water heights; the charted clearance of bridges is measured from this height
Mean High Water, Neap - the mean of high water heights of neap tides
Mean Low Water, Neap - the mean of low water heights of neap tides
Mean Low Water - the mean of all low water heights
Mean Low Water, Spring - the mean of low water heights of spring tides
Mean Lower Low Water Datum - the mean of lower low water heights; charted depths originate from this reference height or datum

---

Spring Tides - tides of increased range, occurring twice a month, around the times of the new and full moons
Neap Tides - tides of decreased range, occurring twice a month, around the times of the half moons
Diurnal Inequality - the difference in height of the two daily low waters or the two daily high waters, a result of the moon's (and to a lesser extent the sun's) changing declination above and below the Equator

The Tide Cycle Simplified: The Rule of Twelfths

Since the average interval between high and low is just over six hours, we can divide the cycle into six segments of one hour each. On average the tide rises or falls approximately according to the fractions at right:

1st hour - 1/12
2nd hour - 2/12
3rd hour - 3/12
4th hour - 3/12
5th hour - 2/12
6th hour - 1/12

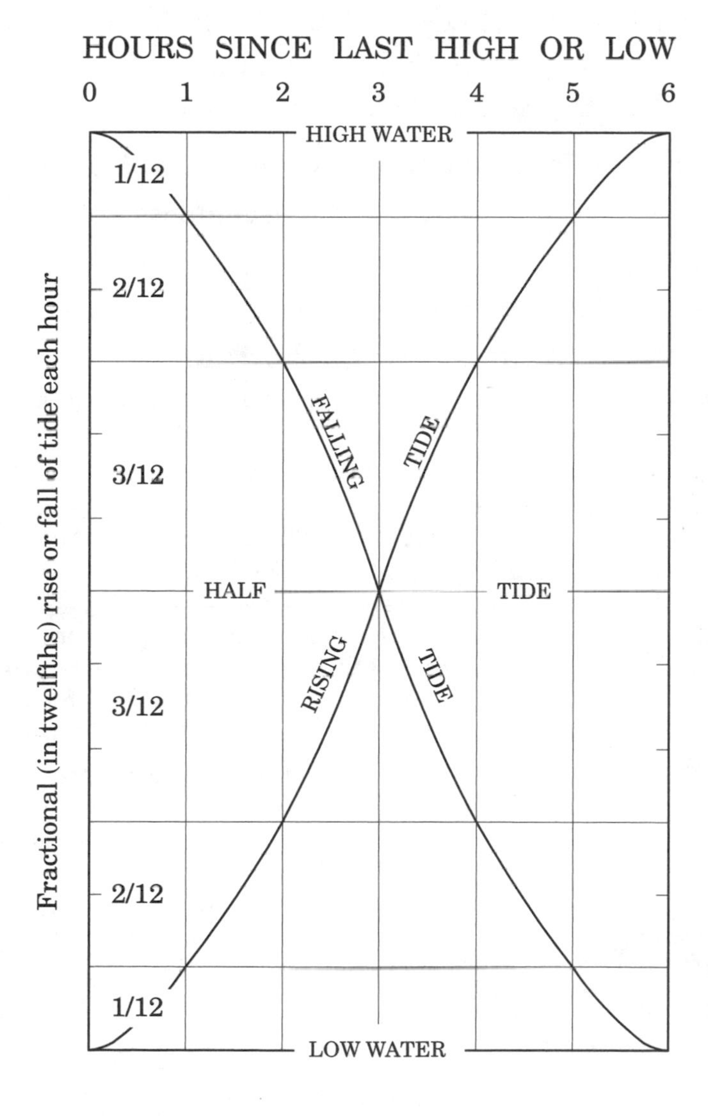

Mean tidal heights by the hour at five ports

Boston	Newport	New York	Charleston	Savannah
9.6	3.5	4.6	5.2	6.9
8.8	3.2	4.2	4.8	6.3
7.2	2.6	3.4	3.9	5.2
4.8	1.8	2.3	2.6	3.5
2.4	0.9	1.2	1.3	1.7
0.8	0.3	0.4	0.4	0.6
0.0	0.0	0.0	0.0	0.0

Using GPS to Create a Deviation Table

Most compasses are subject to onboard magnetic influences, called deviation. You can make your compass more trustworthy by using your GPS to create a deviation table.

Choose a day when the wind is light and sea as calm as possible. Find a large open area with little or no current and a minimum of boat traffic. Bring aboard an assistant. In a notebook create two columns: in pencil, label the left column GPS and the right column COMPASS. Down the right column, number each successive line using intervals of 15° [24 lines] up to 360°. You can concentrate on noting the four ***Cardinal*** and four ***Inter Cardinal*** headings (N, NE, E, SE, S, SW, W, NW) and safely interpolate and fill in the missing numbers for every 15°. Note that the Default setting on a GPS display is **TRUE**. For this exercise, make sure that your GPS is displaying **MAGNETIC-** **C**ourse **O**ver **G**round (COG) heading. This may require going into the GPS setup to insure that the COG is displaying a MAGNETIC heading.

Choose a speed which provides responsive steering and which will make any current or leeway a negligible factor. Proceed on any of the numbered courses for at least 30 seconds, giving the GPS time to report a consistent direction. Once you have held a steady course long enough to get a repeated reading, record it in the left column. Proceed to the next heading. Completing the circle results in a deviation table for your steering compass. Now, erase the penciled column headings and relabel the GPS column TO GO, and the COMPASS column STEER. *Example: TO GO 094°, STEER 090°.*

A deviation table admittedly falls far short of the ideal of a compensated compass; however, such a table will allow you to use your compass with a measure of confidence before an adjuster comes aboard. And that is much better than trying to steer by your GPS.

See p. 203 for how to adjust your compass.

Using GPS to Adjust Your Compass

Nothing can equal the expert services of a professional Compass Adjuster, but if these services are not available, the following information can help you adjust a compass yourself.

As you used the GPS to create a Deviation Table (see p. 202), you can also use it to correct your steering compass to eliminate deviation.

Built-in Correctors - Most modern compasses are fitted with a magnetic corrector system attached inside the bottom of the compass or the binnacle cylinder. Such "B.I.C.'s" (Built-in Correctors) are easy to use and are capable of removing virtually all the deviations of a well-located compass. B.I.C.'s consist of two horizontal shafts, slotted at each end, one running ***Athwartship*** (port and starboard) and one running ***Fore and Aft***. On each shaft are magnets. When these magnets are horizontal, they are in a neutral position. When a shaft is rotated to any angle, the magnets create correction. The usual B.I.C. can remove up to about 15° of deviation. The shaft which runs ***Athwartship*** corrects on **North and South** headings, and has zero effect on East and West. The ***Fore and Aft*** shaft corrects on **East and West** headings, and has zero effect on North and South.

Getting Started - Pick a quiet day and a swinging area with calm conditions. Have someone with a steady hand at the helm and the engine ticking over enough so there is good steering control. It is important to hold a steady heading for at least 30-45 seconds. You are looking at your GPS, and you are equipped with a non-magnetic screwdriver.

Adjusting your compass:

1. As steadily as possible steer within +/- 5° of a cardinal heading, let's say North. Slowly and with a non-magnetic screwdriver rotate the ***Athwartship*** B.I.C. until the compass reads the same as the GPS, creating zero error on North.
2. Turn 90° right to the next cardinal heading, let's say East. Turn the ***Fore/Aft*** B.I.C. to remove all of the existing error so the compass matches your GPS on East.
3. Turn 90° right to South. Compare your compass to the GPS. If the error is zero, move to step **#4**. If you have error, you will split the difference: if you have two (2) degrees of error, turn the ***Athwartship*** B.I.C. so you only have one (1) degree of error.
4. Turn 90° right to West. Compare your compass to the GPS. If the error is zero, move to step **#5**. If you have error, you will split the difference: if you have two (2) degrees of error, turn the ***Fore/Aft*** B.I.C. so you only have one (1) degree of error.
5. Return to North and confirm that you either have zero error, or if you had error on South and you split the difference to create an error of 1°, you have the same 1° of error on North.
6. Return to East and confirm that you either have zero error, or if you had error on West and you split the difference to create an error of 1°, you have the same 1° of error on East.

Check for deviation all around - Compare the steering compass to your GPS at least every 45° on all the cardinal (N, E, S, W) and intercardinal (NE, SE, SW, NW) points. If the steering compass reads too little (lower number of degrees), the deviation is Easterly on that heading, and the number of degrees must be subtracted when steering that magnetic course. If the steering compass reads too much (higher number of degrees), the deviation is Westerly on that heading, and the number of degrees must be added when steering that magnetic course.

Checking for Misalignment - After checking 8 headings, add up the total of Easterly deviations, subtract the total of Westerly, and divide by 8. The result tells you the amount by which your compass is misaligned. Let's say it is 1° Easterly. This means your lubbers line is off to port 1°. If you rotate the compass 1° clockwise, all your Easterly deviations will be reduced by 1°, your Westerly deviations will be increased by 1°, and your 0 deviations will become 1° Westerly. You will now find the total of your Easterly deviations is the very same as your Westerly total. You have eliminated misalignment error. Now you can trust your compass!

The Ship's Bell Code

Telling time by ship's bell has a romantic background that goes back hundreds of years. It is based in the workday routine of the ship's crew. A ship at sea requires a constant watch throughout the whole twenty-four hours of the day. To divide the duty, the day is broken up into six watches of four hours each and the crew into three divisions, or watches.

Each division of the crew stands two four-hour watches a day. In order to rotate the duty, so that a division does not have to stand the same watch day in and day out, the 4 to 8 watch in the afternoon is divided into two watches known as the dog watches.

The Mid-Watch - Midnight to 4 A.M.
The Morning Watch - 4 A.M. to 8 A.M.
The Forenoon Watch - 8 A.M. to 12 Noon
The Afternoon Watch - 12 Noon to 4 P.M.
The 1st Dog Watch - 4 P.M. to 6 P.M.
The 2nd Dog Watch - 6 P.M. to 8 P.M.
The First Watch - 8 P.M. to Midnight

To apprise the crew of the time, the ship's bell was struck by the watch officer at half hour intervals, the first half hour being one bell, the first hour two bells, hour and a half three bells, and so on up to eight bells, denoting time to relieve the watch. By this method of timekeeping eight bells marks 4, 8, or 12 o'clock.

8 Bells	4:00	8:00	12:00
1 Bell	4:30	8:30	12:30
2 Bells	5:00	9:00	1:00
3 Bells	5:30	9:30	1:30
4 Bells	6:00	10:00	2:00
5 Bells	6:30	10:30	2:30
6 Bells	7:00	11:00	3:00
7 Bells	7:30	11:30	3:30

Courtesy of Chelsea Clock Co., Chelsea, MA

What Your Marine GPS Can and Cannot Do

Marine GPS is a tremendous aid to today's boater. Most current GPS receivers include charting so a boater can not only determine position, but can also see chart information that pertains to the area around that position - buoys, depth, land masses, etc., all on one screen. In addition to waypoint and route navigation, GPS can do much more. But it also has some limitations. Here are a few of each.

What GPS can do

Marine GPS can pinpoint the spot where a person falls overboard. Simply press the MOB (man overboard) button. The GPS automatically saves the position and turns on the "Go To" function giving bearing and distance to that position. The one who is watching the person overboard should continue to do so. The GPS can help guide the boat back and aid the spotter if visual contact is lost.

GPS can find optimal sail trim by watching how the SOG (speed over ground) changes after adjusting sheets, travelers, etc. Don't forget to note the SOG before you start making changes.

GPS can find optimal trim tab setting. With the engine rpms set to a certain speed, watch the GPS readings of SOG as the trim tab angles are gradually changed.

GPS can determine effect of the current. Assuming the boat's knotmeter is correct, then the difference between the GPS reading of SOG and the knotmeter is the positive or negative effect of the current on your boat. Only if you are heading directly into or away from the current will this difference be the current's actual speed.

GPS can determine leeway. If your compass heading averages 150°, but your COG (course over ground) is 155°, then your leeway is 5° to starboard, and your heading should change to 145°. Do not confuse the GPS compass screen with a compass. GPS indicates the direction your boat is traveling; a compass indicates the direction your boat is pointing. Current and wind effects can make these very different values, especially at low boat speeds.

GPS can find compass deviation. (1) If you can hold a steady compass course for a period of time, and (2) if your GPS is set to read Magnetic, then the difference between what your compass has been reading and the GPS reading is the approximate deviation of your magnetic compass on that heading. Deviation will be different on other headings.

GPS can act as an anchor watch. Most have drift alarms, where you can set the drift distance above which the GPS alarm will go off. Make sure the drift distance will allow for anticipated anchor rode swing if current or wind shift.

What GPS cannot do

GPS cannot warn you of navigation dangers. Unless you have entered waypoints of hazards such as shoals, rocks, etc., your GPS will blindly point you to the next waypoint, possibly over impassable terrain.

GPS cannot tell you if you are on a collision course with another vessel. Keep watch at all times.

GPS will not normally warn you when you're off course. Some GPS have cross-track error alarms. They must be activated and set independently of your "Go To." The "highway" display on your screen may be most helpful for staying on a determined course.

GPS cannot tell you it's not working properly. If it loses adequate satellite contact, suffers battery failure, or experiences some other malfunction, you won't know unless you check it and compare the information displayed to what you should expect it to be.

GPS cannot recall your last position if it loses power. Plot its positions periodically, so when the power fails, you will know a recent position for dead-reckoning plotting.

Warning:

Your GPS cannot function optimally without your checking and possibly resetting the manufacturer's default settings. Recommended settings, if any, are underlined here. The Map Datum (probably <u>WGS-84</u>) must agree with your charts; <u>Nautical</u> or statute miles; True or <u>Magnetic</u> North; Degrees, minutes, and either <u>tenths of minutes</u> or seconds; GMT or <u>local time</u>; 24-hour or 12-hour clock; Grid preference should be <u>Latitude and Longitude</u>.

ATLANTIC COAST DGPS STATIONS

Revised as of September, 2013

	kHz	Rate	Signal Strength		Lat.N.	Lon.W.
CANADA						
Western Head, NS	312	200 bps			43-59	64-40
Hartlen Point, NS	298	200 bps			44-36	63-27
Fox Island, NS	307	200 bps			45-20	61-05
Pt. Escuminiac, NB	319	200 bps			47-04	64-48
Partridge Is., NB	295	200 bps			45-14	66-03
UNITED STATES		**microvolts/meter (uV)**				
Penobscot, ME	290	200 bps	100uV	at 435 km	44-27.10	68-46.33
Brunswick, ME	316	100 bps	75uV	at 322 km	43-53.4	69-56.8
Acushnet, MA	306	200 bps	100uV	at 370 km	41-44.57	70-53.19
Moriches Pt, NY	293	100 bps	75uV	at 241 km	40-47.40	72-44.83
Sandy Hook, NJ	286	200 bps	100uV	at 185 km	40-28.29	74-00.71
Reedy Point, DE	309	200 bps	100uV	at 113 km	39-33.69	75-34.19
Hagerstown, MD	307	100 bps	75uV	at 250 km	39-33.19	77-42.79
Annapolis, MD	301	200 bps	100uV	at 290 km	39-00.65	76-36.36
Driver, VA	289	100 bps	75uV	at 241 km	36-57.48	76-33.44
New Bern, NC	294	100 bps	75uV	at 259 km	35-10.50	77-02.91
Kensington, SC	292	100 bps	75uV	at 200 km	33-28.86	79-20.58
Savannah, GA	319	100 bps	75uV	at 298 km	32-08.40	81-42.00
Cape Canaveral, FL	289	100 bps	75uV	at 371 km	28-27.62	80-32.74
Card Sound, FL	314	200 bps	100uV	at 261 km	25-25.90	80-27.98
Key West, FL	286	100 bps	75uV	at 204 km	24-34.94	81-39.18
BERMUDA						
St. David's Head	323	BSD		150	32-22.0	64-39.0

These stations are land-based receivers and transmitters for Differential GPS (DGPS). The carrier of these Radiobeacons is modulated with a GPS correction (differential) signal, which may be used to greatly improve the accuracy of GPS. Mariners should see no degradation in the usablllty of the radiobeacon signal for direction finding although a warbling of the identification signal may be noticed. Correction broadcasts are changing from Type 1 format to Type 9-3. Your equipment may need upgrading. For an automated status report of GPS broadcasts: call NAVCEN (703) 313-5900 x5907.

Please see our May 2014 Free Supplement for further updates. To order see p. 4.
For further information visit U.S.C.G. Navigation Centers website at ***http://www.navcen.uscg.gov*** *or Canadian CG at* ***http://www.ccg-gcc.gc.ca/eng/CCG/DGPS_Beacon_Information***

RACONS

RACONS are Radar Beacons operating in the marine radar frequency bands, 2900-3100 MHz (s-band) and 9300-9500 MHz (x-band). When triggered by a vessel's radar signal they provide a bearing by sending a coded reply (e.g. "T": –). This signal received takes the form of a single line or narrow sector extending radially towards the circumference of the radarscope from a point slightly beyond the spot formed by the echo from the lighthouse, buoy, etc. at the Racon site. Thus may be measured to the point at which the Racon coded flash begins. (The figure obtained will be a few hundred feet greater than the actual distance of the ship from the Racon due to the slight response delay in the Racon apparatus.)

Hours of transmission are continuous and coverage is all around the horizon unless otherwise stated. Their ranges depend on the effective range of the ship's radar and on the power and elevation of the Racon apparatus. Under conditions of abnormal radio activity, reliance should only be put on a Racon flash that is consistent and when the ship is believed to be within the area of the Racon's quoted range. Mariners are advised to turn off the interference controls of their radar when wishing to receive a Racon signal or else the signal may not come through to the ship.

***See p. 209* for list of Racons.**

ATLANTIC COAST RACONS

Location	RACON SITE	SIGNAL	LAT. N	LONG. W
NS	Cranberry Islands	– • • • (B)	45-19-29.6	60-55-38.2
	Bear Cove Lt. & Bell By "H6"	– • (N)	44-32-36.3	63-31-19.6
	Chebucto Head	– – • • (Z)	44-30-26.6	63-31-21.8
	Cape Sable	– • – • (C)	43-23-24	65-37-16.9
	Cape Forchu	– • • • (B)	43-47-38.8	66-09-19.3
	Lurcher Shoal Bifurcation Lt. By. "NM"	– • – (K)	43-48-57.2	66-29-58
NB	St. John Harbour. Lt. & Wh. By. "J"	– • (N)	45-12-55.3	66-02-36.9
	Gannet Rock	– – • (G)	44-30-37.1	66-46-52.9
ME	Frenchman Bay Ltd. By. "FB"	– • • • (B)	44-19-21	68-07-24
	Portland Ltd. Wh. By. "P"	– – (M)	43-31-36	70-05-28
MA	Boston Ltd. Wh. By. "B"	– • • • (B)	42-22-42	70-46-58
	Boston North Ch.Entr. Ltd. Wh. By. "NC"	– • (N)	42-22-32	70-54-18
	Cleveland East Ledge Lt.	– • – • (C)	41-37-51	70-41-39
	Buzzards Bay Entr. Lt., Horn	– • • • (B)	41-23-49	71-02-05
RI	Narrag. Bay Entr. Ltd. Wh. By. "NB"	– • • • (B)	41-23-00	71-23-21
	Narrag.-Buzz. Bay Appr. Ltd. Wh. By. "A"	– • (N)	41-06-00	71-23-22
	Newport - Pell Bridge Fog Signal	– • (N)	41-30-18	71-20-55
	Mount Hope Bay Bridge Racon. "MH"	– – • • • • (MH)	41-38-24	71-15-28
CT	Twenty-Eight Ft. Sh. Ltd. Wh. By. "TE"	– (T)	41-09-16	72-30-25
NY	Ambrose Ch. Ltd. Wh. By. "A"	– • (N)	40-27-28	73-50-12
	Tappen Zee Bridge	– – • • (Z)	41-04-12	73-52-47
	Kill van Kull Ch. Ltd. Jct. By. "KV"	– • – (K)	40-39-02	74-03-51
	Southwest Ledge Ltd. Wh. By. #2	– • • • (B)	41-06-23	71-40-14
	Execution Rocks Lt.	– • • – (X)	40-52-41	73-44-16
NJ	Scotland Ltd. Wh. By. "S"	– – (M)	40-26-33	73-55-01
	Sandy Hook Ch. Rge. Front Lt.	– • – • (C)	40-29-15	73-59-35
	Barnegat Ltd. By. "B"	– • • • (B)	39-45-48	73-46-04
DE	Del. Bay Appr. Ltd. Wh. By. "CH"	– • – (K)	38-46-14	75-01-20
	Del. Ltd. By. "D"	– • – (K)	38-27-18	74-41-47
	Del. River, Pea Patch Is.	– • • – (X)	39-36-42	75-34-54
	Five Fathom Bank Ltd. By. "F"	– – (M)	38-46-49	74-34-32
	Brown Shoal Lt.	– • • • (B)	38-55-21	75-06-01
	Miah Maull Shoal Lt.	– – (M)	39-07-36	75-12-31
	Ship John Shoal Lt.	– – – (O)	39-18-19	75-22-36
VA	Chesapeake Light Tower	– • (N)	36-54-17	75-42-46
	Chesapeake Bay Ent. Ltd. Wh. By. "CH"	– • – • (C)	36-56-08	75-57-27
	Ches. Ch. Ltd. By. #78	– – • – (Q)	38-33-19	76-25-39
	Ches. Ch. Ltd. Bell By. #68	– • • – (X)	37-59-53	76-11-49
	Ches. Ch. Ltd. By. #42	– – – (O)	37-25-37	76-05-07
NC	Beaufort Inlet Ch. Ltd. Wh. By. "BM"	– – (M)	34-34-49	76-41-33
	Cape Fear River Ent. Ltd. Wh. By. "CF"	– • – • (C)	33-46-17	78-03-02
SC	Charleston Entr. Ltd. By. "C"	– • – (K)	32-37-05	79-35-30
GA	Tybee Ltd. By. "T"	– – • (G)	31-57-52	80-43-10
	St. Simons Ltd. By. "STS"	– • • • (B)	31-02-49	81-14-25
FL	St. Johns Ltd. By. "STJ"	– – (M)	30-23-35	81-19-08
	Port Everglades Ltd. By. "PE"	– (T)	26-05-30	80-04-46
	Miami Ltd. By. "M"	– – (M)	25-46-06	80-05-00
	Fowey Rocks Lt.	– – – (O)	25-35-26	80-05-48
	Carysfort Rf. Lt.	– • – • (C)	25-13-19	80-12-41
	Alligator Rf. Lt.	– – • (G)	24-51-06	80-37-08
	Sombrero Key Lt.	– – (M)	24-37-40	81-06-39
	American Shoal Lt.	– • – – (Y)	24-31-30	81-31-10
	Sand Key Lt.	– • (N)	24-27-14	81-52-39
	Dry Tortugas Lt.	– • – (K)	24-38-00	82-55-14
Bermuda	North East Breaker Beacon	– • (N)	32-28.7	64-41.0
	Chub Heads Beacon	– • – • (C)	32-17.2	64-58.9

Range: Canada under 10 mi., US under 16 mi. ***See p. 208** for more on RACONS.*

DIAL-A-BUOY SERVICE

SEA-STATE & WEATHER CONDITIONS BY TELEPHONE

If you are planning a coastwise voyage, you can rely on a number of sources for weather. A possible source is **Dial-A-Buoy**, offering reports of conditions at numerous coastal and offshore locations along the Atlantic Coast, as well as the coasts of the Gulf of Mexico, the Pacific, and the Great Lakes. In all there over 100 buoy and 60 Coastal-Marine Automated Network (C-MAN) stations. The system is operated by the National Data Buoy Center (NDBC), with headquarters at the Stennis Space Center in Mississippi. The NDBC is part of the National Weather Service (NWS).

The reports from offshore buoys include wind speed, gusts, and direction, wave heights and periods, water temperature, and barometric pressure as recorded within the last hour or so. Reports from land stations cover wind speed and direction, temperature and pressure; some land stations also add water temperature, visibility, and dew point.

The value of this information is apparent. Say someone in your boating party is susceptible to seasickness, and the Dial-A-Buoy report says wave heights are six feet with a period (interval) of eight seconds. Maybe that person would rather stay ashore and experience the gentler motion of a rocking chair. (Wave heights of six feet with a period of twenty seconds, on the other hand, might be tolerable.) Surfers, too, can benefit greatly from wave height reports. Likewise, since actual conditions frequently differ dramatically from forecasts, someone sailing offshore might be interested to know that a Data Buoy ahead is reporting squalls, giving time to shorten sail. And bathers and fishermen might gain from hearing the water temperature reports.

On the next page, we give the station or buoy identifier, location name, and lat/long in degrees and hundredths, as provided by the NWS. To find the station or buoy locations and identifiers using the Internet, you can see maps with station identifiers at **www.ndbc.noaa.gov/**. To find locations by telephone, you can enter a latitude and longitude to receive the locations and identifiers of the closest stations.

<u>To access Dial-A-Buoy</u> using any touch-tone or cell phone, here are the steps:

1. Call 888-701-8992.
2. If you know the identifier of the station or buoy, press 1. Press 2 to get station locations by entering the approximate lat/long of the area you want.
3. Enter the five-digit (or character) station identifier. To enter a Character press the key containing the character.
4. Press 1 to confirm that your entry was correct.
5. If, after hearing the latest report, you wish to hear a forecast for that same location, press 2 then 1. You can jump to the forecast before the end of the station report by pressing 2 then 1 during the reading of the station conditions.
6. If you want to hear the report for another station, press 2 then 2.
7. You do not have to wait for the prompts. For example, you can press "1440271" as soon as you begin to hear the welcome message to hear the report from station 44027.

NOTE: In some cases a buoy may become temporarily unavailable. You should try again later to see if it has come back online. Please be aware that stations that may be adrift and not at the stated location are not reported via the telephone feature. This information is only available on the website by typing in the station identifier at:
www.ndbc.noaa.gov/dial.shtml

DIAL-A-BUOY and C-MAN STATION LOCATIONS

Station ID	Location Name	Latitude	Longitude
44027	JONESPORT, ME	44.28N	67.31W
MDRM1	MT DESERT ROCK, ME	43.97N	68.13W
MISM1	MATINICUS ROCK, ME	43.78N	68.86W
44007	PORTLAND, ME	43.53N	70.14W
44005	GULF OF MAINE	43.20N	69.13W
IOSN3	ISLE OF SHOALS, NH	42.97N	70.62W
44013	BOSTON, MA	42.35N	70.65W
BUZM3	BUZZARDS BAY, MA	41.40N	71.03W
44018	E. CAPE COD, MA	42.13N	69.63W
44011	GEORGES BANK, MA	41.11N	66.60W
44017	MONTAUK POINT, NY	40.69N	72.05W
44008	NANTUCKET, MA	40.50N	69.25W
ALSN6	AMBROSE LIGHT, NY	40.45N	73.80W
44025	LONG ISLAND, NY	40.25N	73.17W
TPLM2	THOMAS POINT, MD	38.90N	76.44W
44066	TEXAS TOWER #4, NJ	39.58N	72.60W
44009	DELAWARE BAY, NJ	38.46N	74.70W
CHLV2	CHESAPEAKE LIGHT, VA	36.91N	75.71W
44014	VIRGINIA BEACH, VA	36.61N	74.84W
DUKN7	DUCK PIER, NC	36.18N	75.75W
41025	DIAMOND SHOALS	35.01N	75.40W
41001	E. HATTERAS, NC	34.56N	72.63W
CLKN7	CAPE LOOKOUT, NC	34.62N	76.53W
41013	FRYING PAN SHOAL, NC	33.44N	77.74W
41004	EDISTO, SC	32.50N	79.10W
41002	S. HATTERAS, SC	31.86N	74.84W
41008	GRAYS REEF, GA	31.40N	80.87W
41012	ST. AUGUSTINE, FL	30.04N	80.53W
SAUF1	ST AUGUSTINE, FL	29.86N	81.27W
41010	CANAVERAL EAST, FL	28.91N	78.47W
41009	CANAVERAL, FL	28.52N	80.17W
LKWF1	LAKE WORTH, FL	26.61N	80.03W
FWYF1	FOWEY ROCKS, FL	25.59N	80.10W
MLRF1	MOLASSES REEF, FL	25.01N	80.38W
LONF1	LONG KEY, FL	24.84N	80.86W
SMKF1	SOMBRERO KEY, FL	24.63N	81.11W
SANF1	SAND KEY, FL	24.45N	81.88W

Most stations have added the ability to access information via RSS feed using your Internet browser. For information regarding how to use this feature please go to: www.ndbc.noaa.gov/rss_access.shtml

IALA BUOYAGE SYSTEM

Lateral Aids marking the sides of channels seen when entering from Seaward

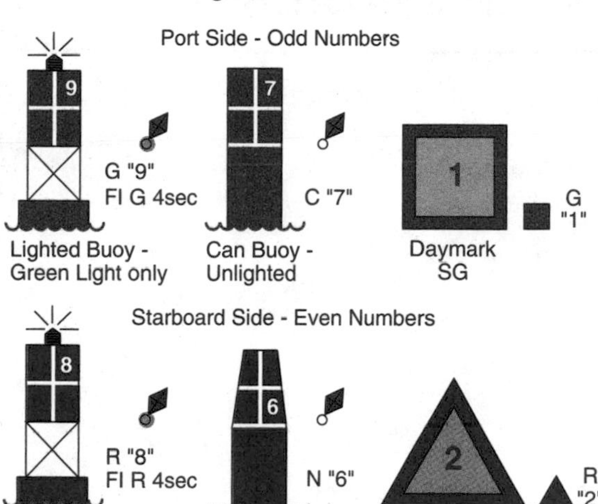

Port- hand aids are Green, some with Flashing Green Lights.
Daymarks:
1st letter "S" = Square
2nd letter "G" = color Green

Starboard-hand aids remain Red, some with Flashing Red Lights.
Daymarks:
1st letter "T" = Triangle
2nd letter "R"= color Red

Safe Water Aids Marking Mid-Channels & Fairways - No Numbers - May Be Lettered:

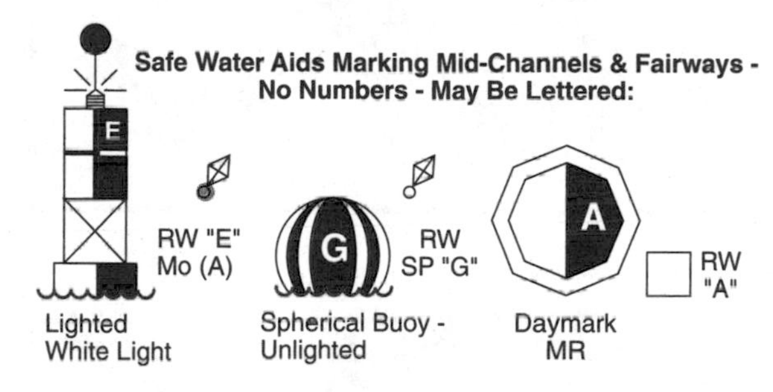

Red and White replaces Black and White. Buoys are spherical; or have a Red spherical topmark. Flashing White Light only: Mo (A).
Daymarks:
1st letter "M" = Octagon
2nd letter "R" = color Red

Preferred Channel Aids - Mark Bifurcations - No Numbers - Preferred Ch. to Starboard (Aid to Port):

M
GR "M"
CGpFl G
Lighted Buoy -
Green Light only

F
GR
C "F"
Can Buoy -
Unlighted

A
Daymark
JG

GR
"A"

Green replaces Black. Flashing Light (Red or Green) is Composite Gp. Fl. (2 + 1).
Daymarks:
1st letter "J" = Square or Triangle
2nd letter "R" or "G" is color of top band

Preferred CH. to Port (Aid to Starboard):

D
RG "D"
CGpFl R
Lighted Buoy -
Red Light only

L
RG
N "L"
Nun Buoy -
Unlighted

B
Daymark
JR

RG
"B"

Note: ISOLATED DANGER BUOYS, Black and Red with two Black spherical topmarks - no numbers, may be lettered (if lighted, white light only, Fl (2) 5s). Stay Clear. SPECIAL AIDS BUOYS will be all YELLOW (if lighted, with yellow light only, Fixed Flashing): Anchorage Areas, Fish Net Areas, Spoil Grounds, Military Exercise Zones, Dredging Buoys (where conventional markers would be confusing), Ocean Data Systems, some Traffic Separations Zone Mid-Channel Buoys.

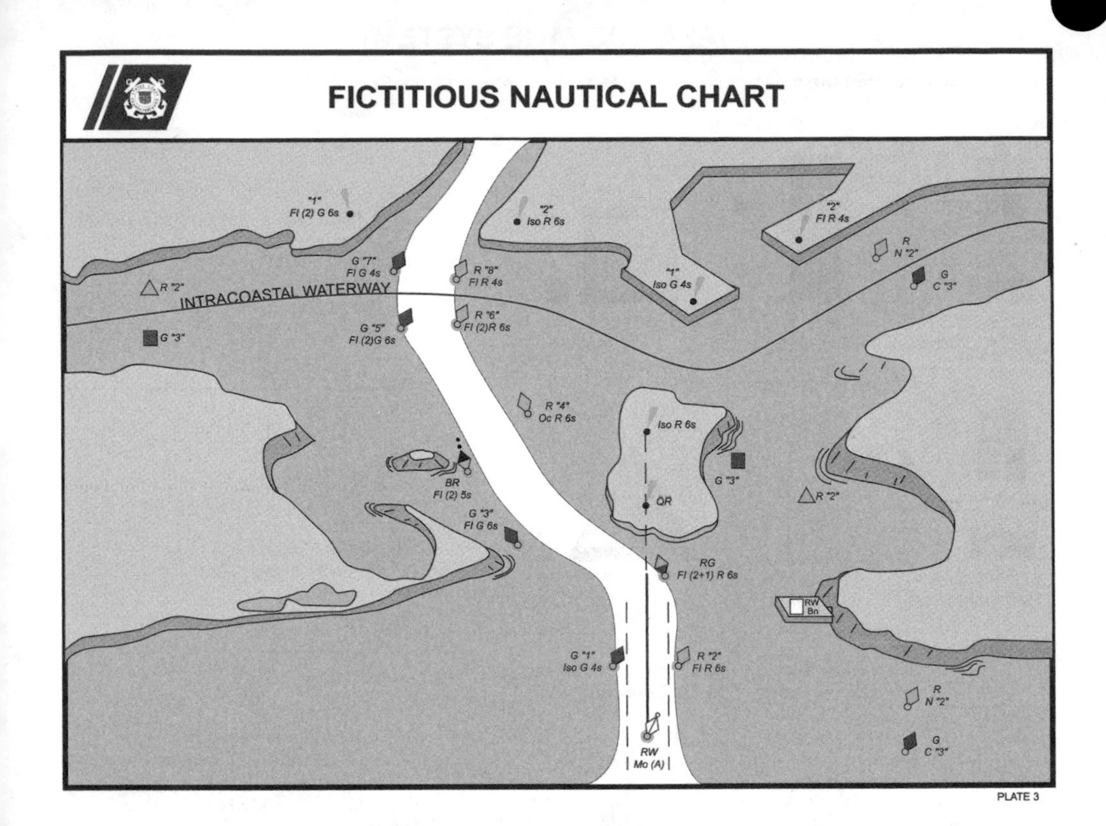

FICTITIOUS NAUTICAL CHART
"1"
Fl (2) G 6s
"2"
Iso R 6s
"2"
Fl R 4s
R
N "2"
G
C "3"
G "7"
Fl G 4s
R "8"
Fl R 4s
"1"
Iso G 4s
R "2"
INTRACOASTAL WATERWAY
G "5"
Fl (2)G 6s
R "6"
Fl (2)R 6s
G "3"
R "4"
Oc R 6s
Iso R 6s
G "3"
BR
Fl (2) 5s
QR
R "2"
G "3"
Fl G 6s
RG
Fl (2+1) R 6s
RW
Bn
G "1"
Iso G 4s
R "2"
Fl R 6s
R
N "2"
G
C "3"
RW
Mo (A)
PLATE 3

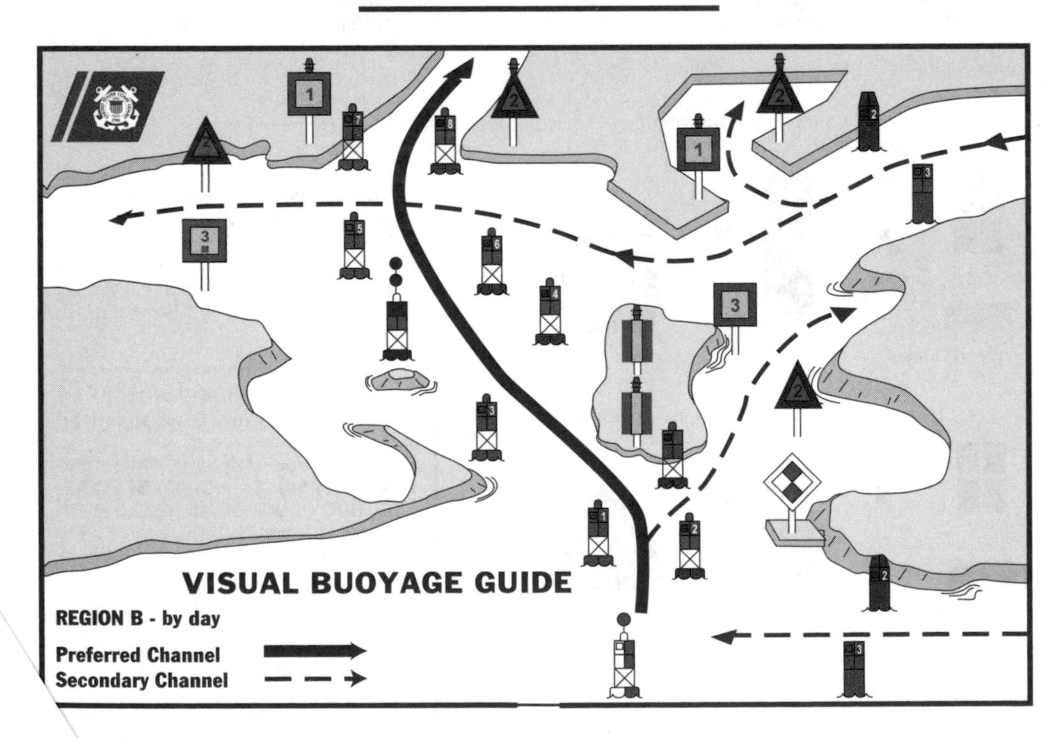

VISUAL BUOYAGE GUIDE
REGION B - by day
Preferred Channel
Secondary Channel

Photo by Jeremy D'Entremont, www.lighthouse.cc

Portland Head Light

Perched on a headland at Cape Elizabeth, just south of the main entrance to Portland Harbor, this historic lighthouse was completed in 1791, the oldest in Maine. It was constructed of rubblestone lined with brick, and the original sixteen lamps were fueled by whale oil. Originally planned for a height of 58', the tower was raised 20' to be visible over headlands to the south. A cast iron staircase was installed in 1864, along with a 4th order Fresnel lens.

A long list of lightkeepers kept the light burning for almost 150 years. In 1939 the United States Coast Guard was given the responsibility for aids to navigation, including lighthouses. For the next 50 years lightkeepers came from the Coast Guard. In 1989, the 200th anniversary of the Lighthouse Service, the light and horn were finally automated.

Today Portland Head Light stands 80' above ground and 101' above water. A DCB 224 airport aerobeacon with a 200,000 candlepower metal halide lamp gives the light 24-mile visibility, flashing white every 4 seconds. The foghorn blasts once every 15 seconds. Its position is 43° 37' 23"N, 070° 12' 28"W.

The adjacent ninety-acre site is Fort Williams Park, where visitors can enjoy facilities for sports, recreation, hiking, and picnics, while taking in views of the nearby rugged coast and Casco Bay.

A museum and gift shop, open June through October, 10 a.m. to 4 p.m., occupy the former Keepers' Quarters. www.portlandheadlight.com

Yacht Flags and How To Fly Them

U.S. Ensign: 8 a.m. to sundown only. Not flown while racing.
At the stern staff of all vessels at anchor, or under way by power or sail.
At the leech of the aftermost sail, approximately 2/3 of the leech above the clew.
When the aftermost sail is gaff-rigged, the Ensign is flown immediately below the peak of the gaff.

U.S. Power Squadron Ensign: 8 a.m. to sundown when flown at the stern staff in place of the U.S. Ensign; otherwise, day and night from the starboard spreader. In either case it is flown only when a Squadron member is in command.

Club Burgee: Day and night. Not flown while racing.
At the bow staff of power vessels with one mast.
At the main peak of yawls, ketches, sloops, cutters, and catboats.
At the fore peak of schooners and power vessels with two masts.

Private Signal: Day and night.
At the bow staff of power vessels without a mast.
At the masthead of power and sailing vessels with one mast.
At the mizzen peak of yawls and ketches.
At the main peak of schooners and power vessels with two masts.

Flag Officers' Flags: Day and night. Flown in place of the private signal on all rigs except single-masted sailboats, when it is flown in place of the club burgee at the masthead.

Union Jack: 8 a.m. to sundown, only at anchor, and only on Sundays, holidays, or occasions for dressing ship, at the bow staff. Sailboats without a bow staff may fly it from the forestay a few feet above the stem head.

INTERNATIONAL SIGNAL FLAGS AND MORSE CODE

CODE FLAG AND ANSWERING PENNANT

A	Alpha	•–
B	Bravo	–•••
C	Charlie	–•–•
D	Delta	–••
E	Echo	•
F	Foxtrot	••–•
G	Golf	––•
H	Hotel	••••
I	India	••
J	Juliet	•–––
K	Kilo	–•–
L	Lima	•–••
M	Mike	––
N	November	–•
O	Oscar	–––
P	Papa	•––•
Q	Quebec	––•–
R	Romeo	•–•
S	Sierra	•••
T	Tango	–
U	Uniform	••–
V	Victor	•••–
W	Whiskey	•––
X	XRay	–••–
Y	Yankee	–•––
Z	Zulu	––••

NUMERAL PENNANTS

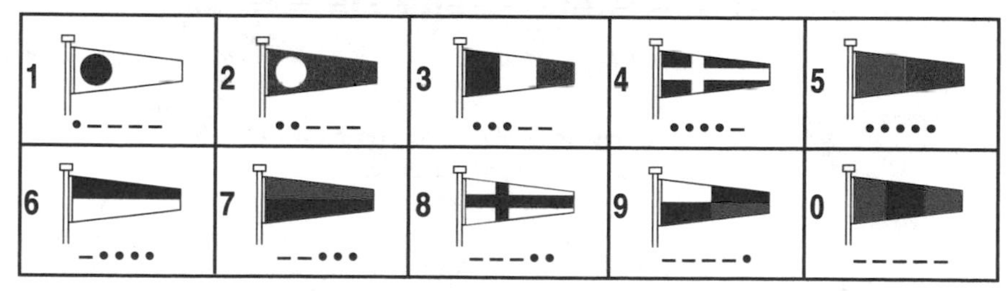

REPEATERS

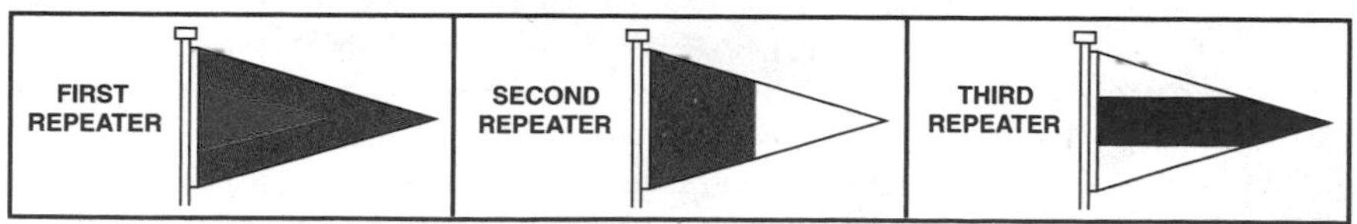

THE INTERNATIONAL CODE OF SIGNALS

The Code comprises 40 flags: 1 Code Flag; 26 letters; 10 numerals; 3 repeaters. With this Code it is possible to converse freely at sea with ships of different countries.

Single Flag Signals

A :: I have a diver down; keep well clear at slow speed.
B :: I am taking in, or discharging, or carrying dangerous goods.
C :: Yes
D :: Keep clear of me; I am maneuvering with difficulty.
E :: I am altering my course to starboard.
F :: I am disabled; communicate with me.
G :: I require a pilot. (When made by fishing vessels when operating in close proximity on the fishing grounds it means; "I am hauling nets.")
H :: I have a pilot on board.
I :: I am altering my course to port.
J :: I am on fire and have dangerous cargo on board; keep well clear of me.
K :: I wish to communicate with you.
L :: You should stop your vessel instantly.
M :: My vessel is stopped and making no way through water.
N :: No
O :: Man overboard.
P :: *In harbor*; All persons should report on board as the vessel is about to proceed to sea.
At sea; It may be used by fishing vessels to mean "My nets have come fast upon an obstruction."
Q :: My vessel is healthy and I request free pratique.
R :: *nothing currently assigned*
S :: My engines are going astern.
T :: Keep clear of me; I am engaged in pair trawling.
U :: You are running into danger.
V :: I require assistance.
W :: I require medical assistance.
X :: Stop carrying out your intentions and watch for my signals.
Y :: I am dragging my anchor.
Z :: I require a tug. (When made by fishing vessels operating in close proximity on the fishing grounds it means : "I am shooting nets.")

FLAGS SHOWING "DIVER DOWN"

There are two flags that may be flown to indicate diving operations, and each has a distinct meaning.

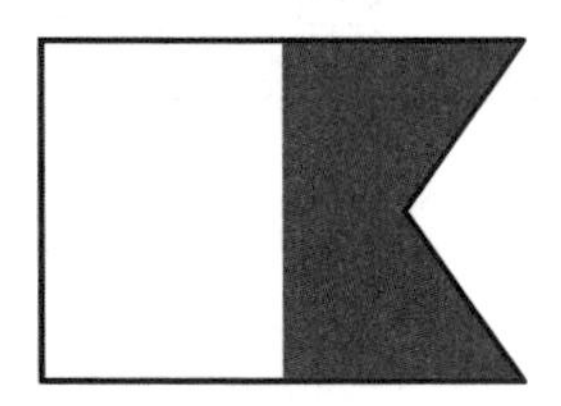

The **Alpha or "A" flag**, according to the U.S. Coast Guard, is to be flown on small vessels engaged in diving operations (1) whenever these vessels are restricted in their ability to maneuver (2) if divers are attached to the vessel. Generally, only vessels to which the divers are physically connected by communication lines, air hoses, or the like are affected by this requirement. The Alpha flag is a signal intended to *protect the vessel from collision*.

In sports diving, where divers are usually free-swimming, the Alpha flag does not have to be shown. The Coast Guard encourages the use of the traditional sports diver flag. The **sports diver flag** is an unofficial signal that, through custom, has come to be used to *protect the diver in the water*. To be most effective, the sports diver flag should be exhibited on a float in the water to mark the approximate location of the diver. Restrictions for nearby vessels vary from state to state, but typically they include a zone of 100' radius around the flag where no other boats are allowed, and a second larger zone in which speed is limited.

U.S. STORM SIGNALS

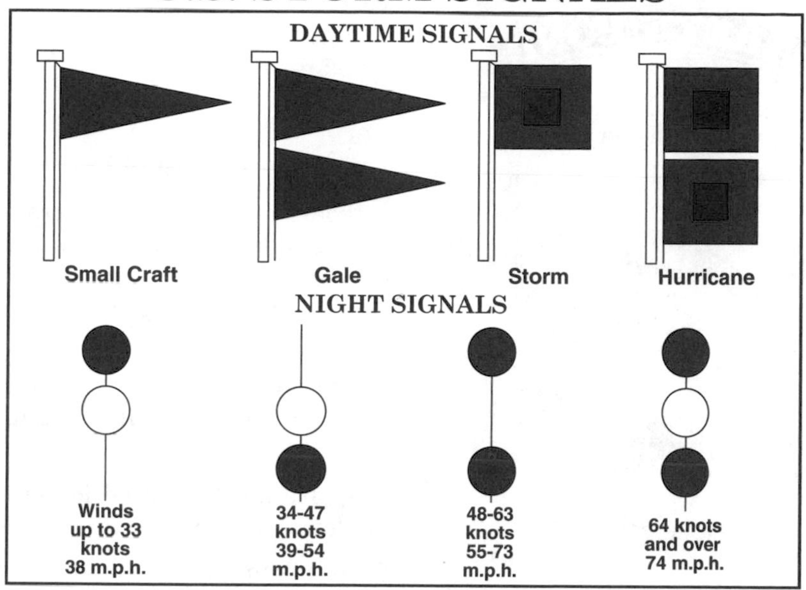

The above signals are displayed regularly on Light Vessels, at Coast Guard shore stations, and at many principal lighthouses. Each Coast and Geodetic Survey Chart lists those locations which appear within the area covered by that chart.

When the Rain before the Wind,
Topsail Sheets and Halliards Mind,
But when the Wind before the Rain,
Then you may set sail again.

Rain before 7
Clear before 11

Distance of Visibility

Given the curvature of the earth, can you see a 200' high headland from 20 miles away? (Answer below.) How far you can see depends on visibility, which we will assume here to be ideal, and the heights above water of your eye and the object.

To find the theoretical maximum distance of visibility, use the Table below. First, using your height of eye above water (say, 8'), the Table shows that at that height, your horizon is 3.2 n.m. away. Then, from our Lights, Fog Signals and Offshore Buoys (pp. 167-197), your chart, or the Light List, find the height of the object (say, 200'). The Table shows that object can be seen 16.2 n.m. from sea level. Add the two distances: 3.2 + 16.2 = 19.4 n.m. *Answer: not quite!* (Heights below in feet, distance in nautical miles)

Ht.	Dist.	Ht.	Dist.	Ht.	Dist.	Ht.	Dist.	Ht.	Dist.
4	2.3	30	6.3	80	10.3	340	21.1	860	33.6
6	2.8	32	6.5	90	10.9	380	22.3	900	34.4
8	3.2	34	6.7	100	11.5	420	23.5	1000	36.2
10	3.6	36	6.9	120	12.6	460	24.6	1400	42.9
12	4.0	38	7.1	140	13.6	500	25.7	1800	48.6
14	4.3	40	7.3	160	14.5	540	26.7	2200	53.8
16	4.6	42	7.4	180	15.4	580	27.6	2600	58.5
18	4.9	44	7.6	200	16.2	620	28.6	3000	62.8
20	5.1	46	7.8	220	17.0	660	29.4	3400	66.9
22	5.4	48	8.0	240	17.8	700	30.4	3800	70.7
24	5.6	50	8.1	260	18.5	740	31.1	4200	74.3
26	5.9	60	8.9	280	19.2	780	32.0	4600	77.7
28	6.1	70	9.6	300	19.9	820	32.8	5000	81.0

CATBOATS

Catboats are distinctive craft with their single mast far forward, large and usually gaff-rigged sail, great beam, and gently curved upright stem.

The great beam brings stability and creates a large cockpit for family outings.

They are ideally suited for sailing in shallow, relatively sheltered waters.

There's something about a catboat that beckons the eye and makes one want to smile.

ELDRIDGE STORY CONTEST for 2015

We are seeking true stories about **an unforgettable boating experience.** The winning entry will be published in our 2015 edition, and the author will receive $200. Submitted material becomes the property of the Publishers and will not be returned. See 2014 winner on facing page.

Submit your entry:

EMAIL: ebb2flood@gmail.com
Subject: Story Contest 2015

Mail:

Eldridge Tide and Pilot Book
Story Contest 2015
P.O. Box 775
Medfield, MA 02052

Deadline: August 25, 2014
Winner Notification: October 1, 2014
Limit: 600 words

Authors provide: name, address, daytime phone number and email address
Finalists: may be asked to send an electronic version

A Dog and a Dinghy Story

It was lovely, as lovely as it gets on a July morning: blue sky and a gentle breeze at 6 a.m. in beautiful, protected Coecles Harbor in Shelter Island, New York – one of our favorite anchorages. There are about a dozen other widely scattered boats. I am sitting in the cockpit of our Cape Dory 36, enjoying my cup of tea while Captain Billy is, well, fully engaged with his morning routine in the head.

I am on harbor patrol, making sure all is well. I see a dinghy moving about – not unusual for this time of day. Then, it circles around. "Oh, I thought, "must have forgotten something on the boat." But this dinghy circled back again. Okay, time for the binoculars.

In the dinghy was a beautiful, frightened Wheaton terrier... alone. Really! I called to Billy, who heard the urgency in my voice. By now I could see there was another dinghy pursuing the runaway dinghy. This was like an Indiana Jones movie right here in Shelter Island! As the man approached, the poor, scared dog panicked and jumped into the water.

This man was able to capture the runaway dinghy. I thought Captain Billy should go after the dog, but he smartly went to find the owner first. "That dog barked at a stranger and jumped into the water when he came near, so I doubt he will happily climb into my dinghy if I try to rescue him." Hmmm, what would Indy do? We noticed a boat where there was a lot of commotion, and Billy hopped into our dinghy and zipped over. "Are you the owner of the dog?" he asked. A quick nod was the response by the worried man. Billy said, "Jump in!" and off they sped.

A relieved and happy dog jumped into his master's arms and climbed into our dinghy. Owner and dog were safely returned to their Nordic Tug 38'. Job well done! But how did this dog come to be alone in a moving dinghy in the first place?

We later learned it was the habit of the woman on board the trawler to let her husband sleep in while she took their dog ashore first thing every morning. And every time, from Maine to here, her trip had been uneventful, until this morning.

When she was returning to her boat with the dog, her husband sleeping way up in the bow, she pushed the kill button on the motor and started to climb onto the swim platform. This motor, though, had other plans. The outboard didn't stop, just paused, and as it kicked back on she was thrown into the water and the dinghy took off ... with the dog! Oh, but, you already know this story.

So, there she was, in the water, screaming for her husband. She had no way to get onto the boat and could only hang on. But remember the man I mentioned earlier? The one chasing after the dog? Well, he was on a nearby boat and, hearing her cries, quickly came over. He helped her onto her boat before going off in pursuit of the dog. Isn't this what boaters do? We all look out for each other.

So in the end, all was well, and with a good story to tell. And, you know, I think I'll remember to wear a tether to the kill switch on the dinghy motor next time.

Susan O'Leary

See p. 220 for 2015 Story Contest rules.

Coping with Currents

See also p. 21, Piloting in a Cross Current

When going directly with or against a current, our piloting problems are simple. (See Smarter Boating, p. 36.) There is no change in course, and our speed over the bottom is easily figured. However, we tend to guess a bit when the current is at some other angle. Where these currents are strong, as between New York and Nantucket, it will be vital to figure the factors carefully, especially in haze or fog.

The Table below tells 1) how many degrees to change your course; 2) by what percent your speed is decreased, with the current off the Bow; 3) or by what percent it is increased, with the current off the Stern.

First: estimate your boat's speed through the water. Then refer to the appropriate TIDAL CURRENT CHART (see pp. 66-77 or pp. 92-97) and estimate the current's speed. Put these two in the form of a ratio, for example: boat speed is 8 kts, current 2 kts; ratio is 4 to 1.

Second: using the same CURRENT CHART, estimate the relative direction of the current to the nearest 15°. Example: your desired course is 60°, the current is from the East, or a relative angle of 30 on your starboard bow.

Third: Enter the Tables under Ratio of 4.0; drop down to the 30° block of numbers (indicated in the left margin). The top figure in the block shows you must change your course 7°, always toward the current, and in this example, to 67°. The middle figure, 22%, is the amount by which your speed over the bottom will be decreased if the current is off your bow, i.e. from 8 kts down to 6.25 kts. Had the figure been 30% off your stern, instead of your bow, you would apply the third figure, 21%, adding it to your 8 kts, making your true speed about 9.7 kts.

RATIOS OF BOAT SPEED TO CURRENT SPEED

Relative Angle of Current		2	2½	3	3½	4	5	6	7	8	10	12	15	20
0° from	°	0	0	0	0	0	0	0	0	0	0	0	0	0
Bow	−%	50	40	33	29	25	20	17	14	12	10	8.3	6.7	5.0
Stern	+%	50	40	33	29	25	20	17	14	12	10	8.3	6.7	5.0
15° from	°	7.0	6.0	5.0	4.0	3.5	3.0	2.5	2.0	1.5	1.5	1.0	1.0	0.5
Bow	−%	49	39	33	28	24	20	16	14	12	10	8.0	6.4	4.8
Stern	+%	48	38	32	27	24	19	16	14	12	10	8.0	6.4	4.8
30° from	°	14	11	9.5	8.0	7.0	5.5	4.5	4.0	3.0	2.5	2.0	2.0	1.0
Bow	−%	46	36	30	26	22	18	15	13	11	8.8	7.3	5.9	4.3
Stern	+%	40	33	28	24	21	17	14	12	11	8.6	7.1	5.7	4.3
45° from	°	20	16	13	11	10	8.0	7.0	5.5	5.0	4.0	3.0	2.5	1.5
Bow	−%	42	32	26	22	19	15	12	11	9.2	7.4	6.1	4.9	3.6
Stern	+%	29	24	21	18	16	13	11	10	8.4	6.8	5.7	4.5	3.4
60° from	°	25	20	16	14	12	9.5	8.0	7.0	6.0	4.5	3.5	3.0	2.0
Bow	−%	34	26	21	18	15	11	9.3	7.8	6.7	5.4	4.4	3.5	2.6
Stern	+%	16	14	13	11	10	8.6	7.3	6.4	5.7	4.6	3.8	3.1	2.4
75° from	°	29	23	18	16	14	10	9.0	7.5	6.5	5.0	4.0	3.5	2.5
Bow	−%	25	18	14	11	9.1	6.8	5.5	4.5	3.8	3.0	2.5	1.9	1.4
Stern	+%	0.8	2.5	3.7	3.8	3.6	3.6	3.1	2.9	2.6	2.2	1.9	1.5	1.2
90°	°	30	24	19	17	14	11	9.5	8.0	7.0	5.5	4.5	3.5	2.5
Abeam	−%	13	8.6	5.4	4.1	3.0	1.8	1.4	1.0	0.7	0.4	0.3	0.2	0.1

Note: In general, while rounding a headland where head current is strong, hug the shore as far as safety will permit or go well out. (Current is usually apt to be strongest between these two points.)

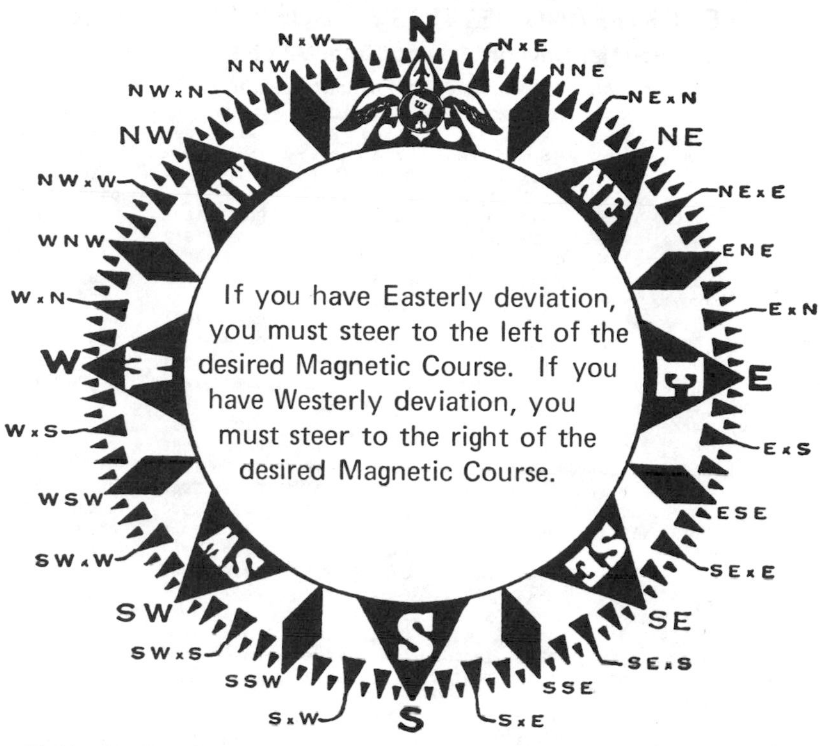

Table for Turning Compass Points into Degrees, and the Contrary

MERCHANT MARINE PRACTICE

Point	Degrees	Point	Degrees	Point	Degrees	Point	Degrees
NORTH	**0**	**EAST**	**90**	**SOUTH**	**180**	**WEST**	**270**
N. 1/4E.	2 3/4	E. 1/4S.	92 3/4	S. 1/4W.	182 3/4	W. 1/4N.	272 3/4
N. 1/2E.	5 3/4	E. 1/2S.	95 3/4	S. 1/2W.	185 3/4	W. 1/2N.	275 3/4
N. 3/4E.	8 1/2	E. 3/4S.	98 1/2	S. 3/4W.	188 1/2	W. 3/4N.	278 1/2
N. by E.	11 1/4	E. by S.	101 1/4	S. by W.	191 1/4	W. by N.	281 1/4
N. by E. 1/4E.	14	E. by S. 1/4S.	104	S. by W. 1/4W.	194	W. by N. 1/4N.	284
N. by E. 1/2E.	17	E. by S. 1/2S.	107	S. by W. 1/2W.	197	W. by N. 1/2N.	287
N. by E. 3/4E.	19 3/4	E. by S. 3/4S.	109 3/4	S. by W. 3/4W.	199 3/4	W. by N. 3/4N.	289 3/4
N.N.E.	**22 1/2**	**E.S.E.**	**112 1/2**	**S.S.W.**	**202 1/2**	**W.N.W.**	**292 1/2**
N.E. by N. 3/4N.	25 1/4	S.E. by E. 3/4E.	115 1/4	S.W. by S. 3/4S.	205 1/4	N.W. by W. 3/4W.	295 1/4
N.E. by N. 1/2N.	28 1/4	S.E. by E. 1/2E.	118 1/4	S.W. by S. 1/2S.	208 1/4	N.W. by W. 1/2W.	298 1/4
N.E. by N. 1/4N.	31	S.E. by E. 1/4E.	121	S.W. by S. 1/4S.	211	N.W. by W. 1/4W.	301
N.E. by N.	33 3/4	S.E. by E.	123 3/4	S.W. by S.	213 3/4	N.W. by W.	303 3/4
N.E. 3/4N.	36 1/2	S.E. 3/4E.	126 1/2	S.W. 3/4S.	216 1/2	N.W. 3/4W.	306 1/2
N.E. 1/2N.	39 1/2	S.E. 1/2E.	129 1/2	S.W. 1/2S.	219 1/2	N.W. 1/2W.	309 1/2
N.E. 1/4N.	42 1/4	S.E. 1/4E.	132 1/4	S.W. 1/4S.	222 1/4	N.W. 1/4W.	312 1/4
N.E.	**45**	**S.E.**	**135**	**S.W.**	**225**	**N.W.**	**315**
N.E. 1/4E.	47 3/4	S.E. 1/4S.	137 3/4	S.W. 1/4W.	227 3/4	N.W. 1/4N.	317 3/4
N.E. 1/2E.	50 3/4	S.E. 1/2S.	140 3/4	S.W. 1/2W.	230 3/4	N.W. 1/2N.	320 3/4
N.E. 3/4E.	53 1/2	S.E. 3/4S.	143 1/2	S.W. 3/4W.	233 1/2	N.W. 3/4N.	323 1/2
N.E. by E.	56 1/4	S.E. by S.	146 1/4	S.W. by W.	236 1/4	N.W. by N.	326 1/4
N.E. by E. 1/4E.	59	S.E. by S. 1/4S.	149	S.W. by W. 1/4W.	239	N.W. by N. 1/4N.	329
N.E. by E. 1/2E.	62	S.E. by S. 1/2S.	152	S.W. by W. 1/2W.	242	N.W. by N. 1/2N.	332
N.E. by E. 3/4E.	64 3/4	S.E. by S. 3/4S.	154 3/4	S.W. by W. 3/4W.	244 3/4	N.W. by N. 3/4N.	334 3/4
E.N.E.	**67 1/2**	**S.S.E.**	**157 1/2**	**W.S.W.**	**247 1/2**	**N.N.W.**	**337 1/2**
E. by N. 3/4N.	70 1/4	S. by E. 3/4E.	160 1/4	W. by S. 3/4S.	250 1/4	N. by W. 3/4W.	340 1/4
E. by N. 1/2N.	73 1/4	S. by E. 1/2E.	163 1/4	W. by S. 1/2S.	253 1/4	N. by W. 1/2W.	343 1/4
E. by N. 1/4N.	76	S. by E. 1/4E.	166	W. by S. 1/4S.	256	N. by W. 1/4W.	346
E. by N.	78 3/4	S. by E.	168 3/4	W. by S.	258 3/4	N. by W.	348 3/4
E. 3/4N.	81 1/2	S. 3/4E.	171 1/2	W. 3/4S.	261 1/2	N. 3/4W.	351 1/2
E. 1/2N.	84 1/2	S. 1/2E.	174 1/2	W. 1/2S.	264 1/2	N. 1/2W.	354 1/2
E. 1/4N.	87 1/4	S. 1/4E.	177 1/4	W. 1/4S.	267 1/4	N. 1/4W.	357 1/4
EAST	**90**	**SOUTH**	**180**	**WEST**	**270**	**NORTH**	**0**

2014 SUN'S RISING AND SETTING AT BOSTON - 42° 20'N 71°W

Add one hour for Daylight Saving Time, March 9 - November 2

Times shown in table are first tip of Sun at Sunrise and last tip at Sunset.

	JANUARY		FEBRUARY		MARCH		APRIL		MAY		JUNE		
	Rise	Set	Rise	Set	Rise	Set	Rise	Set	Rise	Set	Rise	Set	
Day	h m	h m	h m	h m	h m	h m	h m	h m	h m	h m	h m	h m	Day
1	0713	1622	0658	1658	0619	1734	0527	1810	0439	1844	0410	1914	1
2	0713	1623	0656	1659	0618	1735	0525	1811	0438	1845	0409	1915	2
3	0713	1624	0655	1701	0616	1736	0523	1812	0437	1846	0409	1916	3
4	0713	1625	0654	1702	0615	1738	0521	1813	0435	1847	0409	1916	4
5	0713	1626	0653	1703	0613	1739	0520	1814	0434	1848	0408	1917	5
6	0713	1627	0652	1705	0611	1740	0518	1815	0433	1849	0408	1918	6
7	0713	1628	0651	1706	0610	1741	0516	1817	0432	1850	0408	1918	7
8	0713	1629	0650	1707	0608	1742	0515	1818	0430	1851	0407	1919	8
9	0713	1630	0648	1709	0606	1743	0513	1819	0429	1852	0407	1920	9
10	0712	1631	0647	1710	0605	1745	0511	1820	0428	1853	0407	1920	10
11	0712	1632	0646	1711	0603	1746	0510	1821	0427	1854	0407	1921	11
12	0712	1633	0644	1712	0601	1747	0508	1822	0426	1855	0407	1921	12
13	0711	1634	0643	1714	0559	1748	0506	1823	0425	1857	0407	1922	13
14	0711	1636	0642	1715	0558	1749	0505	1824	0424	1858	0407	1922	14
15	0710	1637	0640	1716	0556	1750	0503	1826	0423	1859	0407	1922	15
16	0710	1638	0639	1718	0554	1752	0502	1827	0422	1900	0407	1923	16
17	0709	1639	0638	1719	0553	1753	0500	1828	0421	1901	0407	1923	17
18	0709	1640	0636	1720	0551	1754	0458	1829	0420	1902	0407	1923	18
19	0708	1642	0635	1721	0549	1755	0457	1830	0419	1903	0407	1924	19
20	0708	1643	0633	1723	0547	1756	0455	1831	0418	1904	0407	1924	20
21	0707	1644	0632	1724	0546	1757	0454	1832	0417	1905	0407	1924	21
22	0706	1645	0630	1725	0544	1758	0452	1833	0416	1906	0408	1924	22
23	0706	1647	0629	1726	0542	1800	0451	1835	0416	1906	0408	1925	23
24	0705	1648	0627	1728	0540	1801	0449	1836	0415	1907	0408	1925	24
25	0704	1649	0626	1729	0539	1802	0448	1837	0414	1908	0408	1925	25
26	0703	1650	0624	1730	0537	1803	0446	1838	0413	1909	0409	1925	26
27	0702	1652	0623	1731	0535	1804	0445	1839	0413	1910	0409	1925	27
28	0701	1653	0621	1733	0533	1805	0444	1840	0412	1911	0410	1925	28
29	0700	1654	....	....	0532	1806	0442	1841	0411	1912	0410	1925	29
30	0700	1656	....	....	0530	1808	0441	1842	0411	1913	0411	1925	30
31	0659	1657	....	....	0528	1809	....	....	0410	1913	....	....	31

	JULY		AUGUST		SEPTEMBER		OCTOBER		NOVEMBER		DECEMBER		
	Rise	Set	Rise	Set	Rise	Set	Rise	Set	Rise	Set	Rise	Set	
Day	h m	h m	h m	h m	h m	h m	h m	h m	h m	h m	h m	h m	Day
1	0411	1924	0437	1903	0509	1818	0541	1726	0617	1637	0653	1612	1
2	0412	1924	0438	1902	0510	1816	0542	1724	0618	1636	0654	1612	2
3	0412	1924	0439	1901	0511	1815	0543	1722	0620	1635	0655	1612	3
4	0413	1924	0440	1900	0512	1813	0544	1720	0621	1634	0656	1612	4
5	0413	1924	0441	1859	0513	1811	0545	1719	0622	1632	0657	1612	5
6	0414	1923	0442	1857	0514	1809	0547	1717	0623	1631	0658	1611	6
7	0415	1923	0443	1856	0515	1808	0548	1715	0625	1630	0659	1611	7
8	0415	1923	0444	1855	0517	1806	0549	1714	0626	1629	0700	1611	8
9	0416	1922	0445	1853	0518	1804	0550	1712	0627	1628	0701	1611	9
10	0417	1922	0446	1852	0519	1802	0551	1710	0628	1627	0702	1611	10
11	0417	1921	0447	1851	0520	1801	0552	1709	0630	1626	0703	1611	11
12	0418	1921	0448	1849	0521	1759	0553	1707	0631	1625	0704	1612	12
13	0419	1920	0449	1848	0522	1757	0554	1705	0632	1624	0705	1612	13
14	0420	1919	0450	1846	0523	1755	0556	1704	0633	1623	0705	1612	14
15	0421	1919	0451	1845	0524	1754	0557	1702	0635	1622	0706	1612	15
16	0421	1918	0452	1844	0525	1752	0558	1700	0636	1621	0707	1613	16
17	0422	1917	0453	1842	0526	1750	0559	1659	0637	1620	0707	1613	17
18	0423	1917	0454	1841	0527	1748	0600	1657	0638	1620	0708	1613	18
19	0424	1916	0455	1839	0528	1747	0601	1656	0640	1619	0709	1614	19
20	0425	1915	0457	1837	0529	1745	0603	1654	0641	1618	0709	1614	20
21	0426	1914	0458	1836	0530	1743	0604	1653	0642	1617	0710	1615	21
22	0427	1914	0459	1834	0531	1741	0605	1651	0643	1617	0710	1615	22
23	0428	1913	0500	1833	0532	1740	0606	1650	0644	1616	0711	1616	23
24	0429	1912	0501	1831	0533	1738	0607	1648	0646	1616	0711	1616	24
25	0430	1911	0502	1830	0535	1736	0609	1647	0647	1615	0711	1617	25
26	0431	1910	0503	1828	0536	1734	0610	1646	0648	1614	0712	1617	26
27	0432	1909	0504	1826	0537	1733	0611	1644	0649	1614	0712	1618	27
28	0433	1908	0505	1825	0538	1731	0612	1643	0650	1614	0712	1619	28
29	0434	1907	0506	1823	0539	1729	0614	1641	0651	1613	0713	1620	29
30	0435	1906	0507	1821	0540	1727	0615	1640	0652	1613	0713	1620	30
31	0436	1905	0508	1820	....	....	0616	1639	....	....	0713	1621	31

2014 SUN'S RISING AND SETTING AT NEW YORK - 40° 42'N 74°W

Add one hour for Daylight Saving Time, March 9 - November 2

Times shown in table are first tip of Sun at Sunrise and last tip at Sunset.

	JANUARY		FEBRUARY		MARCH		APRIL		MAY		JUNE		
	Rise	Set	Rise	Set	Rise	Set	Rise	Set	Rise	Set	Rise	Set	
Day	h m	h m	h m	h m	h m	h m	h m	h m	h m	h m	h m	h m	Day
1	0720	1639	0706	1714	0630	1747	0540	1821	0455	1852	0427	1921	1
2	0720	1640	0705	1715	0628	1748	0538	1822	0453	1853	0427	1922	2
3	0720	1641	0704	1716	0627	1750	0536	1823	0452	1854	0426	1922	3
4	0720	1642	0703	1717	0625	1751	0535	1824	0451	1855	0426	1923	4
5	0720	1643	0702	1719	0624	1752	0533	1825	0450	1856	0426	1924	5
6	0720	1644	0701	1720	0622	1753	0532	1826	0449	1857	0425	1924	6
7	0720	1645	0700	1721	0621	1754	0530	1827	0447	1858	0425	1925	7
8	0720	1646	0659	1722	0619	1755	0528	1828	0446	1859	0425	1925	8
9	0720	1647	0657	1724	0617	1756	0527	1829	0445	1900	0425	1926	9
10	0719	1648	0656	1725	0616	1757	0525	1830	0444	1901	0425	1927	10
11	0719	1649	0655	1726	0614	1758	0524	1831	0443	1902	0424	1927	11
12	0719	1650	0654	1727	0613	1759	0522	1832	0442	1903	0424	1928	12
13	0719	1651	0652	1728	0611	1801	0520	1833	0441	1904	0424	1928	13
14	0718	1652	0651	1730	0609	1802	0519	1834	0440	1905	0424	1928	14
15	0718	1653	0650	1731	0608	1803	0517	1835	0439	1906	0424	1929	15
16	0717	1655	0649	1732	0606	1804	0516	1836	0438	1907	0424	1929	16
17	0717	1656	0647	1733	0604	1805	0514	1837	0437	1908	0424	1930	17
18	0716	1657	0646	1734	0603	1806	0513	1839	0436	1909	0425	1930	18
19	0716	1658	0645	1736	0601	1807	0511	1840	0435	1910	0425	1930	19
20	0715	1659	0643	1737	0559	1808	0510	1841	0435	1911	0425	1930	20
21	0715	1700	0642	1738	0558	1809	0508	1842	0434	1912	0425	1931	21
22	0714	1702	0640	1739	0556	1810	0507	1843	0433	1913	0425	1931	22
23	0713	1703	0639	1740	0554	1811	0506	1844	0432	1914	0426	1931	23
24	0713	1704	0637	1742	0553	1812	0504	1845	0432	1914	0426	1931	24
25	0712	1705	0636	1743	0551	1813	0503	1846	0431	1915	0426	1931	25
26	0711	1706	0634	1744	0550	1814	0501	1847	0430	1916	0426	1931	26
27	0710	1708	0633	1745	0548	1815	0500	1848	0430	1917	0427	1931	27
28	0710	1709	0631	1746	0546	1817	0459	1849	0429	1918	0427	1931	28
29	0709	1710	....	....	0545	1818	0457	1850	0429	1919	0428	1931	29
30	0708	1711	....	....	0543	1819	0456	1851	0428	1919	0428	1931	30
31	0707	1712	....	....	0541	1820	....	....	0428	1920	....	....	31

	JULY		AUGUST		SEPTEMBER		OCTOBER		NOVEMBER		DECEMBER		
	Rise	Set	Rise	Set	Rise	Set	Rise	Set	Rise	Set	Rise	Set	
Day	h m	h m	h m	h m	h m	h m	h m	h m	h m	h m	h m	h m	Day
1	0429	1931	0453	1911	0523	1828	0553	1738	0626	1652	0701	1629	1
2	0429	1931	0454	1910	0524	1827	0554	1736	0627	1651	0702	1629	2
3	0430	1931	0455	1909	0525	1825	0555	1735	0629	1650	0703	1629	3
4	0430	1930	0456	1908	0526	1823	0556	1733	0630	1649	0704	1629	4
5	0431	1930	0457	1907	0527	1822	0557	1732	0631	1648	0704	1629	5
6	0431	1930	0458	1906	0528	1820	0558	1730	0632	1647	0705	1629	6
7	0432	1930	0459	1904	0529	1818	0559	1728	0633	1646	0706	1628	7
8	0433	1929	0459	1903	0530	1817	0600	1727	0635	1645	0707	1628	8
9	0433	1929	0500	1902	0531	1815	0601	1725	0636	1644	0708	1628	9
10	0434	1928	0501	1901	0532	1813	0602	1724	0637	1643	0709	1629	10
11	0435	1928	0502	1859	0533	1812	0603	1722	0638	1642	0710	1629	11
12	0435	1927	0503	1858	0534	1810	0604	1720	0639	1641	0711	1629	12
13	0436	1927	0504	1857	0535	1808	0605	1719	0640	1640	0711	1629	13
14	0437	1926	0505	1855	0536	1807	0606	1717	0642	1639	0712	1629	14
15	0438	1926	0506	1854	0537	1805	0607	1716	0643	1638	0713	1629	15
16	0439	1925	0507	1853	0538	1803	0608	1714	0644	1637	0713	1630	16
17	0439	1925	0508	1851	0539	1802	0609	1713	0645	1636	0714	1630	17
18	0440	1924	0509	1850	0540	1800	0610	1711	0646	1636	0715	1630	18
19	0441	1923	0510	1848	0541	1758	0611	1710	0647	1635	0715	1631	19
20	0442	1922	0511	1847	0542	1757	0613	1708	0649	1634	0716	1631	20
21	0443	1922	0512	1845	0543	1755	0614	1707	0650	1634	0716	1632	21
22	0444	1921	0513	1844	0544	1753	0615	1705	0651	1633	0717	1632	22
23	0444	1920	0514	1842	0545	1751	0616	1704	0652	1633	0717	1633	23
24	0445	1919	0515	1841	0546	1750	0617	1703	0653	1632	0718	1633	24
25	0446	1918	0516	1839	0546	1748	0618	1701	0654	1632	0718	1634	25
26	0447	1917	0517	1838	0547	1746	0619	1700	0655	1631	0719	1635	26
27	0448	1916	0518	1836	0548	1745	0621	1659	0656	1631	0719	1635	27
28	0449	1915	0519	1835	0549	1743	0622	1657	0657	1630	0719	1636	28
29	0450	1914	0520	1833	0551	1741	0623	1656	0659	1630	0719	1637	29
30	0451	1913	0521	1831	0552	1740	0624	1655	0700	1630	0720	1638	30
31	0452	1912	0522	1830	....	....	0625	1654	....	....	0720	1638	31

2014 SUN'S RISING AND SETTING AT JACKSONVILLE - 30° 20'N 81° 37'W

Add one hour for Daylight Saving Time, March 9 - November 2

Times shown in table are first tip of Sun at Sunrise and last tip at Sunset.

	JANUARY		FEBRUARY		MARCH		APRIL		MAY		JUNE		
	Rise	Set	Rise	Set	Rise	Set	Rise	Set	Rise	Set	Rise	Set	
Day	h m	h m	h m	h m	h m	h m	h m	h m	h m	h m	h m	h m	Day
1	0723	1737	0717	1803	0653	1825	0616	1845	0543	1904	0525	1924	1
2	0723	1738	0717	1804	0652	1826	0614	1846	0542	1905	0525	1924	2
3	0724	1739	0716	1805	0650	1827	0613	1847	0541	1906	0525	1925	3
4	0724	1739	0715	1806	0649	1827	0612	1847	0541	1906	0525	1925	4
5	0724	1740	0715	1807	0648	1828	0611	1848	0540	1907	0524	1926	5
6	0724	1741	0714	1808	0647	1829	0610	1848	0539	1908	0524	1926	6
7	0724	1742	0713	1808	0646	1830	0609	1849	0538	1908	0524	1927	7
8	0724	1743	0712	1809	0645	1830	0607	1850	0537	1909	0524	1927	8
9	0724	1743	0712	1810	0644	1831	0606	1850	0537	1910	0524	1928	9
10	0724	1744	0711	1811	0642	1832	0605	1851	0536	1910	0524	1928	10
11	0724	1745	0710	1812	0641	1832	0604	1852	0535	1911	0524	1928	11
12	0724	1746	0709	1812	0640	1833	0603	1852	0534	1912	0524	1929	12
13	0724	1747	0708	1813	0639	1833	0602	1853	0534	1912	0524	1929	13
14	0724	1748	0708	1814	0638	1834	0600	1853	0533	1913	0524	1930	14
15	0724	1748	0707	1815	0636	1835	0559	1854	0532	1914	0524	1930	15
16	0724	1749	0706	1816	0635	1835	0558	1855	0532	1914	0524	1930	16
17	0723	1750	0705	1816	0634	1836	0557	1855	0531	1915	0524	1930	17
18	0723	1751	0704	1817	0633	1837	0556	1856	0531	1915	0525	1931	18
19	0723	1752	0703	1818	0632	1837	0555	1857	0530	1916	0525	1931	19
20	0723	1753	0702	1819	0630	1838	0554	1857	0530	1917	0525	1931	20
21	0722	1754	0701	1820	0629	1839	0553	1858	0529	1917	0525	1931	21
22	0722	1754	0700	1820	0628	1839	0552	1859	0529	1918	0525	1932	22
23	0722	1755	0659	1821	0627	1840	0551	1859	0528	1919	0526	1932	23
24	0721	1756	0658	1822	0625	1840	0550	1900	0528	1919	0526	1932	24
25	0721	1757	0657	1822	0624	1841	0549	1900	0527	1920	0526	1932	25
26	0720	1758	0656	1823	0623	1842	0548	1901	0527	1920	0526	1932	26
27	0720	1759	0655	1824	0622	1842	0547	1902	0527	1921	0527	1932	27
28	0719	1800	0654	1825	0621	1843	0546	1902	0526	1922	0527	1932	28
29	0719	1801	....	....	0619	1844	0545	1903	0526	1922	0527	1932	29
30	0718	1801	....	....	0618	1844	0544	1904	0526	1923	0528	1932	30
31	0718	1802	....	....	0617	1845	....	....	0525	1923	....	....	31

	JULY		AUGUST		SEPTEMBER		OCTOBER		NOVEMBER		DECEMBER		
	Rise	Set	Rise	Set	Rise	Set	Rise	Set	Rise	Set	Rise	Set	
Day	h m	h m	h m	h m	h m	h m	h m	h m	h m	h m	h m	h m	Day
1	0528	1932	0545	1920	0603	1849	0620	1812	0641	1739	0705	1726	1
2	0529	1932	0546	1920	0604	1848	0621	1811	0642	1738	0706	1726	2
3	0529	1932	0546	1919	0604	1847	0621	1809	0642	1737	0707	1726	3
4	0529	1932	0547	1918	0605	1846	0622	1808	0643	1737	0708	1726	4
5	0530	1932	0547	1917	0605	1844	0622	1807	0644	1736	0709	1726	5
6	0530	1932	0548	1916	0606	1843	0623	1806	0645	1735	0709	1726	6
7	0531	1932	0549	1915	0607	1842	0624	1805	0646	1734	0710	1726	7
8	0531	1932	0549	1915	0607	1841	0624	1803	0646	1734	0711	1726	8
9	0532	1932	0550	1914	0608	1839	0625	1802	0647	1733	0711	1726	9
10	0532	1931	0550	1913	0608	1838	0625	1801	0648	1733	0712	1726	10
11	0533	1931	0551	1912	0609	1837	0626	1800	0649	1732	0713	1727	11
12	0533	1931	0552	1911	0609	1836	0627	1759	0650	1731	0714	1727	12
13	0534	1930	0552	1910	0610	1834	0627	1758	0650	1731	0714	1727	13
14	0534	1930	0553	1909	0610	1833	0628	1757	0651	1730	0715	1727	14
15	0535	1930	0553	1908	0611	1832	0629	1755	0652	1730	0716	1728	15
16	0535	1929	0554	1907	0611	1831	0629	1754	0653	1729	0716	1728	16
17	0536	1929	0555	1906	0612	1829	0630	1753	0654	1729	0717	1729	17
18	0537	1929	0555	1905	0613	1828	0631	1752	0655	1728	0717	1729	18
19	0537	1928	0556	1904	0613	1827	0631	1751	0656	1728	0718	1729	19
20	0538	1928	0556	1903	0614	1826	0632	1750	0656	1728	0718	1730	20
21	0538	1927	0557	1902	0614	1824	0633	1749	0657	1727	0719	1730	21
22	0539	1927	0558	1901	0615	1823	0633	1748	0658	1727	0719	1731	22
23	0540	1926	0558	1900	0615	1822	0634	1747	0659	1727	0720	1731	23
24	0540	1926	0559	1858	0616	1821	0635	1746	0700	1726	0720	1732	24
25	0541	1925	0559	1857	0616	1819	0636	1745	0701	1726	0721	1732	25
26	0541	1924	0600	1856	0617	1818	0636	1744	0701	1726	0721	1733	26
27	0542	1924	0600	1855	0618	1817	0637	1743	0702	1726	0722	1734	27
28	0543	1923	0601	1854	0618	1816	0638	1742	0703	1726	0722	1734	28
29	0543	1922	0602	1853	0619	1814	0639	1742	0704	1726	0722	1735	29
30	0544	1922	0602	1852	0619	1813	0639	1741	0705	1726	0723	1736	30
31	0544	1921	0603	1850	....	....	0640	1740	....	....	0723	1736	31

2014 SUN'S SETTING AT OTHER LOCATIONS FOR FLAG USE

Add one hour for Daylight Saving Time, March 9 - November 2

Times shown in tables pp. 224-226 are first tip of Sun at Sunrise and last tip at Sunset.

Vernal Equinox: March 20th, 11:57 a.m. E.S.T. Summer Solstice: June 21st, 5:51 a.m. E.S.T.
Autumnal Equinox: Sept. 22nd, 9:29 p.m. E.S.T. Winter Solstice: Dec. 21st, 6:03 p.m. E.S.T.

Add to or subtract from the referenced table

	1/15	2/15	3/15	4/15	5/15	6/15	7/15	8/15	9/15	10/15	11/15	12/15
BOSTON p. 224												
New London, CT	+7	+6	+4	+2	0	-1	0	+1	+2	+5	+6	+7
Newport, RI	+4	+3	+1	-1	-2	-3	-2	-1	0	+2	+4	+5
New Bedford, MA	+3	+2	0	-1	-2	-3	-2	-1	0	+1	+2	+3
Vineyard Haven, MA	+1	-1	-2	-4	-5	-6	-5	-4	-3	-2	0	+1
Nantucket, MA	-1	-2	-4	-6	-7	-8	-7	-6	-5	-3	-2	-1
Portland, ME	-8	-6	-3	-1	+1	+2	+1	0	-2	-4	-6	-7
Rockland, ME	-14	-12	-8	-6	-4	-2	-4	-5	-7	-10	-12	-14
Bar Harbor, ME	-18	-15	-11	-8	-5	-3	-5	-7	-9	-13	-17	-18
NEW YORK p. 225												
Hampton Roads, VA	+18	+15	+9	+2	0	-1	-1	+2	+7	+13	+18	+20
Oxford, MD	+14	+12	+9	+5	+4	+3	+3	+5	+8	+11	+14	+15
Annapolis, MD	+14	+13	+10	+7	+6	+5	+5	+7	+9	+12	+14	+15
Cape May, NJ	+8	+7	+4	+1	0	-1	-1	+1	+3	+6	+8	+9
Atlantic City, NJ	+5	+4	+2	0	-1	-2	-2	0	+2	+4	+6	+6
Mannasquan, NJ	+2	+1	0	-1	-2	-2	-2	-1	0	+1	+2	+2
Port Jefferson, NY	-5	-4	-4	-3	-3	-3	-3	-3	-4	-4	-5	-5
Bridgeport, CT	-4	-4	-3	-2	-2	-1	-1	-2	-3	-4	-4	-5
New Haven, CT	-7	-6	-4	-3	-3	-3	-3	-3	-4	-5	-6	-7
JACKSONVILLE p. 226												
Morehead City, NC	-28	-24	-20	-14	-10	-7	-9	-13	-19	-24	-28	-29
Wilmington, NC	-22	-18	-14	-10	-5	-3	-5	-8	-14	-18	-22	-23
Myrtle Beach, SC	-16	-14	-10	-6	-3	-1	-3	-5	-7	-12	-17	-17
Charleston, SC	-11	-9	-6	-3	+1	0	-1	-3	-7	-9	-11	-12
Savannah. GA	-1	0	+2	+4	+6	+6	+5	+4	+3	0	-2	-2
Brunswick, GA	+1	-1	0	+1	+2	+2	+1	+1	-1	+1	+2	+2
Ponce Inlet, FL	+1	-1	-1	-2	-4	-5	-5	-4	-2	0	+1	+1
Melbourne, FL	+1	-2	-4	-5	-7	-9	-9	-8	-5	-1	0	+1
North Palm Beach, FL	+2	-2	-6	-8	-11	-14	-14	-11	-7	-4	0	+2
Miami, FL	+6	0	-3	-8	-13	-15	-15	-11	-7	-2	+3	+5
Key West, FL	+13	+7	+2	-3	-8	-12	-12	-6	-1	+6	+11	+14

Celestial Navigation - Not Yet a Lost Art

Andy Sumberg

For the past few years I've been teaching celestial navigation classes. I begin every course by saying, "If you give me a sextant, Nautical Almanac, a watch, paper, pencil, and a GPS and I can tell your exact location." Generally I get a small chuckle out of this joke. Of course the ringer in the list was GPS, which alone is enough to find one's position. GPS has become so inexpensive that no one should ever be lost again.

So why do we still teach and yes, even fill celestial navigation classes? When asked why they take the course, my students usually offer several answers. The fewest say they want to become better offshore navigators. In fact only about 15 to 20 percent even plan to boat offshore. Before GPS, this would have been the bulk of the students in a celestial class. Most are genuinely curious as to how our mariner predecessors could navigate, and are willing to work a bit to find out. Yes, while not rocket science, celestial navigation is a bit of work, but for the effort it is very satisfying when learned.

One of the two primary tools a celestial navigator needs is a watch, which at one time was society's "killer app." As we have recently learned to take GPS for granted, we have for a few decades expected to have affordable, accurate timepieces. Why does this matter? One of the first things we learn in celestial navigation is that a celestial sight with a one second time error could result in a position error of a quarter mile. A minute off in time could be a 15-mile position error. Imagine missing an island by fifteen miles. It's just over the horizon, and yet you don't see it. You get the point. Until very recently the celestial navigator had to put more care into maintaining the timepiece, a key-wind chronometer, than he did the sextant. Studying celestial navigation hints at how important portable accurate clocks are. What do you give up if you don't have accurate time? Longitude. Fortunately, without a clock you can still use a sextant to find latitude, and that is far better than nothing.

The second tool needed is a sextant, a somewhat arcane instrument by today's electronic standards, which precisely measures the angle between two objects. A remarkable instrument, the sextant has a special feel like all other great precision hand tools. People naturally want to hold it in their hands and try it. Its operation is quite simple. By using mirrors and an adjusting screw the sextant allows the celestial navigator to see two objects at the same time through a single eyepiece. One of the objects sighted is the horizon, and the other is one of the navigation bodies for which position tables are printed in the Nautical Almanac: the Sun, Moon, Venus, Jupiter, Mars, Saturn, or one of fifty-seven stars. When the objects line up, the angle between them can be read off the scale. The measured angle, tables in the nautical almanac, and a few calculations yield the sought-after latitude and longitude.

Some of us find that sextants are also pretty to look at. While they all come in a box designed for carrying, when mine is not in use, it sits in a Lucite display case doubling as a work of art.

Celestial navigation is a necessary skill to have when boating offshore and your electronic systems fail. But it is also an intellectual and emotional connection with our seafaring past, our industrial history, and our stargazing ancestors.

Andy Sumberg is a member of the Unites States Power Squadrons where he has obtained the rank of Senior Navigator. He teaches Celestial Navigation and is currently the District Educational Officer for USPS District 12. He is actively involved in developing USPS e-learning boating courses and seminars, and will play hooky to go sailing on a moment's notice.

TIME OF LOCAL APPARENT NOON (L.A.N.) 2014
FOR THE CENTRAL MERIDIAN OF ANY TIME ZONE

	JAN.	FEB.	MAR.	APR.	MAY	JUN.	JUL.	AUG.	SEP.	OCT.	NOV.	DEC.
	h:m:s	h:m:s	h:m:s	h:m:s	h:m:s	h:m:s	h:m:s	h:m:s	h:m:s	h:m:s	h:m:s	h:m:s
1	12:03:39	12:13:35	12:12:17	12:03:49	11:57:04	11:57:51	12:03:54	12:06:20	11:59:58	11:49:37	11:43:35	11:49:05
2	12:04:07	12:13:42	12:12:05	12:03:31	11:56:57	11:58:01	12:04:05	12:06:16	11:59:38	11:49:18	11:43:34	11:49:28
3	12:04:34	12:13:49	12:11:53	12:03:14	11:56:51	11:58:11	12:04:16	12:06:11	11:59:19	11:48:59	11:43:34	11:49:51
4	12:05:02	12:13:55	12:11:40	12:02:56	11:56:46	11:58:21	12:04:27	12:06:06	11:58:59	11:48:41	11:43:35	11:50:15
5	12:05:29	12:14:00	12:11:27	12:02:39	11:56:41	11:58:31	12:04:38	12:06:00	11:58:39	11:48:23	11:43:36	11:50:40
6	12:05:55	12:14:04	12:11:13	12:02:22	11:56:36	11:58:42	12:04:48	12:05:53	11:58:19	11:48:05	11:43:38	11:51:05
7	12:06:21	12:14:07	12:10:59	12:02:05	11:56:32	11:58:53	12:04:58	12:05:46	11:57:58	11:47:47	11:43:41	11:51:30
8	12:06:46	12:14:10	12:10:44	12:01:49	11:56:29	11:59:05	12:05:07	12:05:38	11:57:37	11:47:30	11:43:45	11:51:56
9	12:07:11	12:14:11	12:10:29	12:01:32	11:56:26	11:59:17	12:05:16	12:05:30	11:57:17	11:47:14	11:43:50	11:52:23
10	12:07:36	12:14:12	12:10:13	12:01:16	11:56:23	11:59:29	12:05:24	12:05:21	11:56:55	11:46:57	11:43:56	11:52:50
11	12:07:59	12:14:13	12:09:58	12:01:00	11:56:22	11:59:41	12:05:32	12:05:11	11:56:34	11:46:42	11:44:02	11:53:17
12	12:08:22	12:14:12	12:09:42	12:00:44	11:56:20	11:59:53	12:05:40	12:05:01	11:56:13	11:46:27	11:44:09	11:53:45
13	12:08:45	12:14:10	12:09:25	12:00:29	11:56:20	12:00:05	12:05:47	12:04:50	11:55:52	11:46:12	11:44:17	11:54:13
14	12:09:07	12:14:08	12:09:09	12:00:14	11:56:20	12:00:18	12:05:54	12:04:39	11:55:30	11:45:58	11:44:26	11:54:42
15	12:09:28	12:14:05	12:08:52	11:59:59	11:56:20	12:00:31	12:06:00	12:04:27	11:55:09	11:45:44	11:44:36	11:55:11
16	12:09:48	12:14:02	12:08:35	11:59:45	11:56:21	12:00:43	12:06:05	12:04:15	11:54:48	11:45:31	11:44:47	11:55:40
17	12:10:08	12:13:58	12:08:17	11:59:31	11:56:23	12:00:56	12:06:10	12:04:02	11:54:26	11:45:19	11:44:58	11:56:09
18	12:10:27	12:13:53	12:08:00	11:59:17	11:56:25	12:01:09	12:06:15	12:03:49	11:54:05	11:45:07	11:45:11	11:56:39
19	12:10:46	12:13:47	12:07:42	11:59:04	11:56:28	12:01:22	12:06:19	12:03:35	11:53:43	11:44:56	11:45:25	11:57:08
20	12:11:03	12:13:41	12:07:25	11:58:52	11:56:31	12:01:35	12:06:23	12:03:21	11:53:22	11:44:46	11:45:39	11:57:38
21	12:11:20	12:13:34	12:07:07	11:58:40	11:56:35	12:01:49	12:06:26	12:03:06	11:53:01	11:44:36	11:45:54	11:58:08
22	12:11:36	12:13:26	12:06:49	11:58:28	11:56:39	12:02:02	12:06:28	12:02:51	11:52:40	11:44:27	11:46:09	11:58:38
23	12:11:52	12:13:18	12:06:31	11:58:16	11:56:44	12:02:15	12:06:30	12:02:36	11:52:19	11:44:19	11:46:26	11:59:08
24	12:12:06	12:13:09	12:06:13	11:58:06	11:56:50	12:02:28	12:06:31	12:02:20	11:51:58	11:44:11	11:46:43	11:59:38
25	12:12:20	12:13:00	12:05:55	11:57:55	11:56:56	12:02:41	12:06:32	12:02:03	11:51:37	11:44:04	11:47:01	12:00:08
26	12:12:33	12:12:50	12:05:37	11:57:45	11:57:02	12:02:53	12:06:32	12:01:46	11:51:17	11:43:58	11:47:20	12:00:37
27	12:12:46	12:12:40	12:05:18	11:57:36	11:57:09	12:03:06	12:06:32	12:01:29	11:50:56	11:43:52	11:47:40	12:01:07
28	12:12:57	12:12:29	12:05:00	11:57:27	11:57:17	12:03:18	12:06:31	12:01:12	11:50:36	11:43:47	11:48:00	12:01:36
29	12:13:08		12:04:42	11:57:19	11:57:25	12:03:30	12:06:29	12:00:54	11:50:16	11:43:43	11:48:21	12:02:05
30	12:13:18		12:04:24	11:57:11	11:57:33	12:03:42	12:06:27	12:00:35	11:49:57	11:43:40	11:48:43	12:02:34
31	12:13:27		12:04:07		11:57:42		12:06:24	12:00:17		11:43:37		12:03:03

Explanatory Notes: The noon sight and the Sun's Declination (p. 231) result in the vessel's parallel of latitude. It is taken at the time of the sun's meridian passage, when the sun is at maximum altitude.

The moment of meridian passage is called Local Apparent Noon (L.A.N.), and only rarely is it the same time as noon Standard Time or Local Mean Time. Instead, as this Table shows, the sun is either ahead of or behind its theoretical schedule.

Two corrections are involved. 1) To correct for your difference in longitude from the central meridian of your time zone (i.e. 75° for U.S. Atlantic Coast), either a) add 4 minutes of time for each degree West or b) subtract 4 minutes of time for each degree East. 2) If necessary, convert from Daylight Savings Time to Standard Time by subtracting 1 hour from your watch.

Thus for Boston, at 71° West longitude (or 4° East of 75°), L.A.N. occurs 16 minutes before the times listed in the Table.

For New York, at 74° West (1° East of 75°), L.A.N. occurs 4 minutes earlier than times shown.

Converting arc to time:

360° = 24 hours
15° = 1 hour
1° = 4 minutes
15' = 1 minute
1' = 4 seconds

SUN'S TRUE BEARING AT RISING AND SETTING

To find compass deviation using the Sun.
Figures are correct for all Longitudes

Sun's Decl.	38° N Rise	38° N Set	40° N Rise	40° N Set	42° N Rise	42° N Set	44° N Rise	44° N Set	Sun's Decl.
N 23°	60.3°	299.7°	59.3°	300.7°	58.3°	301.7°	57.1°	302.9°	N 23°
22	61.6	298.4	60.7	299.3	59.7	300.3	58.6	301.4	22
21	63.0	297.0	62.1	297.9	61.2	298.8	60.1	299.9	21
20	64.3	295.7	63.5	296.5	62.6	297.4	61.6	298.4	20
19	65.6	294.4	64.9	295.1	64.0	296.0	63.1	296.9	19
18	66.9	293.1	66.2	293.8	65.4	294.6	64.6	295.4	18
17	68.2	291.8	67.6	292.4	66.8	293.2	66.0	294.0	17
16	69.5	290.5	68.9	291.1	68.2	291.8	67.5	292.5	16
15	70.8	289.2	70.3	289.7	69.6	290.4	68.9	291.1	15
14	72.1	287.9	71.6	288.4	71.0	289.0	70.4	289.6	14
13	73.4	286.6	72.9	287.1	72.4	287.6	71.8	288.2	13
12	74.7	285.3	74.3	285.7	73.8	286.2	73.2	286.8	12
11	76.0	284.0	75.6	284.4	75.1	284.9	74.6	285.4	11
10	77.3	282.7	76.9	283.1	76.5	283.5	76.0	284.0	10
9	78.6	281.4	78.2	281.8	77.9	282.1	77.4	282.6	9
8	79.8	280.2	79.5	280.5	79.2	280.8	78.9	281.1	8
7	81.1	278.9	80.9	279.1	80.6	279.4	80.3	279.7	7
6	82.4	277.6	82.2	277.7	81.9	278.1	81.7	278.3	6
5	83.7	276.3	83.5	276.5	83.3	276.7	83.0	277.0	5
4	84.9	275.1	84.8	275.2	84.6	275.4	84.4	275.6	4
3	86.2	273.8	86.1	273.9	86.0	274.0	85.8	274.2	3
2	87.5	272.5	87.4	272.6	87.3	272.7	87.2	272.8	2
N 1°	88.7	271.3	88.7	271.3	88.7	271.3	88.6	271.4	N 1°
0	90.0	270.0	90.0	270.0	90.0	270.0	90.0	270.0	0
S 1°	91.3	268.7	91.3	268.7	91.3	268.7	91.4	268.6	S 1°
2	92.5	267.5	92.6	267.4	92.7	267.3	92.8	267.2	2
3	93.8	266.2	93.9	266.1	94.0	266.0	94.2	265.8	3
4	95.1	264.9	95.2	264.8	95.4	264.6	95.6	264.4	4
5	96.3	263.7	96.5	263.5	96.7	263.3	97.0	263.0	5
6	97.6	262.4	97.8	262.2	98.1	261.9	98.3	261.7	6
7	98.9	261.1	99.1	260.9	99.4	260.6	99.7	260.3	7
8	100.2	259.8	100.5	259.5	100.8	259.2	101.1	258.9	8
9	101.4	258.6	101.8	258.2	102.1	257.9	102.6	257.4	9
10	102.7	257.3	103.1	256.9	103.5	256.5	104.0	256.0	10
11	104.0	256.0	104.4	255.6	104.9	255.1	105.4	254.6	11
12	105.3	254.7	105.7	254.3	106.2	253.8	106.8	253.2	12
13	106.6	253.4	107.1	252.9	107.6	252.4	108.2	251.8	13
14	107.9	252.1	108.4	251.6	109.0	251.0	109.6	250.4	14
15	109.2	250.8	109.7	250.3	110.4	249.6	111.1	248.9	15
16	110.5	249.5	111.1	248.9	111.8	248.2	112.5	247.5	16
17	111.8	248.2	112.4	247.6	113.2	246.8	114.0	246.0	17
18	113.1	246.9	113.8	246.2	114.6	245.4	115.4	244.6	18
19	114.4	245.6	115.1	244.9	116.0	244.0	116.9	243.1	19
20	115.7	244.3	116.5	243.5	117.4	242.6	118.4	241.6	20
21	117.0	243.0	117.9	242.1	118.8	241.2	119.9	240.1	21
22	118.4	241.6	119.3	240.7	120.3	239.7	121.4	238.6	22
S 23°	119.7	240.3	120.7	239.3	121.7	238.3	122.9	237.1	S 23°

Instructions: (1) Knowing the date, find the Sun's Declination from the facing page. Find that Declination down the left column on this page. (2) Find the column with your Latitude, and choose either Rise or Set to determine the True Bearing. (3) Add the local Westerly Variation to the figure. (4) If you are a couple of minutes after sunrise or before sunset, the Sun's bearing changes about 1° each 6 minutes during the first hour after sunrise and before sunset. (5) The deviation found will be correct only for the heading you are on at that time.

THE SUN'S DECLINATION 2014

For celestial navigators, the "noon sight" reading of the Sun's height above the horizon, together with the Sun's Declination from this table, determines latitude.

THE SUN'S DECLINATION 2014

MEAN NOON – 75° MERIDIAN **(1700 G.M.T.)**

	JAN.	FEB.	MAR.	APR.	MAY	JUN.	JUL.	AUG.	SEPT.	OCT.	NOV.	DEC.	
	South	South	South	North	North	North	North	North	North	South	South	South	
	° '	° '	° '	° '	° '	° '	° '	° '	° '	° '	° '	° '	
1	-22 58	-16 59	- 7 26	+ 4 42	+15 12	+22 06	+23 04	+17 54	+ 8 07	- 3 21	-14 33	-21 52	1
2	-22 52	-16 41	- 7 03	+ 5 05	+15 30	+22 14	+23 00	+17 39	+ 7 46	- 3 44	-14 52	-22 00	2
3	-22 47	-16 24	- 6 40	+ 5 28	+15 48	+22 22	+22 55	+17 23	+ 7 24	- 4 07	-15 11	-22 09	3
4	-22 40	-16 06	- 6 17	+ 5 51	+16 05	+22 29	+22 50	+17 07	+ 7 02	- 4 30	-15 30	-22 17	4
5	-22 34	-15 48	- 5 54	+ 6 14	+16 22	+22 35	+22 44	+16 51	+ 6 39	- 4 53	-15 48	-22 25	5
6	-22 26	-15 29	- 5 30	+ 6 36	+16 39	+22 41	+22 38	+16 35	+ 6 17	- 5 16	-16 06	-22 32	6
7	-22 19	-15 10	- 5 07	+ 6 59	+16 56	+22 47	+22 32	+16 18	+ 5 55	- 5 39	-16 24	-22 39	7
8	-22 11	-14 51	- 4 44	+ 7 21	+17 12	+22 53	+22 25	+16 01	+ 5 32	- 6 02	-16 41	-22 45	8
9	-22 02	-14 32	- 4 20	+ 7 44	+17 28	+22 58	+22 18	+15 44	+ 5 09	- 6 25	-16 58	-22 51	9
10	-21 53	-14 13	- 3 57	+ 8 06	+17 44	+23 02	+22 11	+15 26	+ 4 47	- 6 48	-17 15	-22 57	10
11	-21 44	-13 53	- 3 33	+ 8 28	+17 59	+23 07	+22 03	+15 08	+ 4 24	- 7 10	-17 32	-23 01	11
12	-21 34	-13 33	- 3 09	+ 8 50	+18 14	+23 10	+21 54	+14 50	+ 4 01	- 7 33	-17 48	-23 06	12
13	-21 24	-13 13	- 2 46	+ 9 12	+18 29	+23 14	+21 46	+14 32	+ 3 38	- 7 55	-18 04	-23 10	13
14	-21 14	-12 53	- 2 22	+ 9 33	+18 43	+23 17	+21 37	+14 14	+ 3 15	- 8 18	-18 20	-23 14	14
15	-21 03	-12 32	- 1 59	+ 9 55	+18 58	+23 19	+21 27	+13 55	+ 2 52	- 8 40	-18 35	-23 17	15
16	-20 51	-12 11	- 1 35	+10 16	+19 12	+23 21	+21 17	+13 36	+ 2 29	- 9 02	-18 50	-23 20	16
17	-20 39	-11 50	- 1 11	+10 37	+19 25	+23 23	+21 07	+13 17	+ 2 06	- 9 24	-19 05	-23 22	17
18	-20 27	-11 29	- 0 47	+10 58	+19 38	+23 25	+20 57	+12 58	+ 1 42	- 9 46	-19 19	-23 24	18
19	-20 15	-11 08	- 0 24	+11 19	+19 51	+23 25	+20 46	+12 38	+ 1 19	-10 07	-19 33	-23 25	19
20	-20 02	-10 46	+ 0 00	+11 39	+20 04	+23 26	+20 35	+12 18	+ 0 56	-10 29	-19 47	-23 26	20
21	-19 48	-10 25	+ 0 24	+12 00	+20 16	+23 26	+20 23	+11 58	+ 0 33	-10 50	-20 00	-23 26	21
22	-19 35	-10 03	+ 0 47	+12 20	+20 28	+23 26	+20 11	+11 38	+ 0 09	-11 12	-20 13	-23 26	22
23	-19 21	- 9 41	+ 1 11	+12 40	+20 39	+23 25	+19 59	+11 18	- 0 14	-11 33	-20 25	-23 25	23
24	-19 06	- 9 18	+ 1 35	+13 00	+20 50	+23 24	+19 46	+10 57	- 0 38	-11 53	-20 37	-23 24	24
25	-18 51	- 8 56	+ 1 58	+13 19	+21 01	+23 22	+19 33	+10 37	- 1 01	-12 14	-20 49	-23 23	25
26	-18 36	- 8 34	+ 2 22	+13 39	+21 12	+23 20	+19 20	+10 16	- 1 24	-12 35	-21 01	-23 21	26
27	-18 21	- 8 11	+ 2 45	+13 58	+21 22	+23 18	+19 07	+ 9 55	- 1 48	-12 55	-21 12	-23 18	27
28	-18 05	- 7 49	+ 3 09	+14 17	+21 31	+23 15	+18 53	+ 9 34	- 2 11	-13 15	-21 22	-23 15	28
29	-17 49		+ 3 32	+14 35	+21 41	+23 12	+18 39	+ 9 12	- 2 34	-13 35	-21 32	-23 12	29
30	-17 33		+ 3 56	+14 54	+21 50	+23 08	+18 24	+ 8 51	- 2 58	-13 55	-21 42	-23 08	30
31	-17 16		+ 4 19		+21 58		+18 09	+ 8 29		-14 14		-23 04	31

Vernal Equinox: March 20th, 11:57 a.m. E.S.T.
Summer Solstice: June 21st, 5:51 a.m. E.S.T.

Autumnal Equinox: September 22nd, 9:29 p.m. E.S.T.
Winter Solstice: December 21st, 6:03 p.m. E.S.T.

To find Sun's Declination in the Atlantic Time Zone (1 hour earlier than E.S.T.), take 1/24 of the difference between Day 1 and Day 2. Add or subtract this figure from Day 2 to find the Declination for Day 2.

If Declination is increasing (N. or S.), *subtract*. If Declination is decreasing (N. or S.), *add*.

2014 MOONRISE AND MOONSET
BOSTON, MA
Add one hour for Daylight Saving Time, March 9 - November 2

	JANUARY		FEBRUARY		MARCH		APRIL		MAY		JUNE		
	Rise	Set	Rise	Set	Rise	Set	Rise	Set	Rise	Set	Rise	Set	
Day	h m	h m	h m	h m	h m	h m	h m	h m	h m	h m	h m	h m	Day
1	0617	1746	0730	1941	0603	1822	0653	2001	0705	2032	0817	2132	1
2	0714	1852	0816	2042	0649	1923	0739	2057	0755	2122	0908	2212	2
3	0807	1957	0859	2141	0733	2022	0826	2151	0845	2210	0957	2250	3
4	0855	2059	0943	2238	0818	2120	0915	2242	0935	2254	1046	2327	4
5	0940	2159	1026	2333	0903	2216	1004	2331	1025	2335	1135	....	5
6	1023	2256	1110	....	0949	2310	1054	....	1115	....	1225	0004	6
7	1105	2352	1155	0027	1036	....	1144	0016	1205	0015	1316	0041	7
8	1146	....	1241	0119	1123	0002	1234	0059	1254	0052	1410	0119	8
9	1228	0047	1329	0209	1212	0051	1323	0139	1344	0129	1505	0201	9
10	1312	0140	1417	0256	1302	0138	1413	0218	1435	0207	1603	0246	10
11	1357	0232	1507	0342	1351	0222	1503	0256	1528	0245	1704	0335	11
12	1444	0323	1557	0425	1441	0303	1554	0333	1624	0326	1805	0430	12
13	1532	0412	1647	0506	1531	0343	1647	0411	1721	0409	1905	0529	13
14	1621	0500	1736	0545	1621	0421	1740	0451	1820	0457	2002	0631	14
15	1711	0544	1826	0623	1711	0459	1836	0533	1920	0549	2056	0735	15
16	1801	0626	1917	0700	1803	0536	1934	0617	2020	0645	2145	0839	16
17	1851	0706	2008	0737	1855	0615	2032	0706	2117	0744	2232	0941	17
18	1940	0745	2100	0815	1949	0654	2130	0758	2211	0845	2316	1041	18
19	2030	0822	2153	0854	2044	0736	2228	0854	2301	0947	2359	1140	19
20	2120	0858	2249	0937	2141	0821	2322	0952	2348	1048	....	1237	20
21	2211	0935	2346	1022	2238	0910	....	1053	....	1148	0041	1334	21
22	2304	1013	....	1112	2336	1002	0014	1153	0033	1247	0124	1429	22
23	2358	1054	0044	1207	....	1058	0103	1254	0116	1345	0208	1524	23
24	....	1138	0142	1305	0032	1158	0149	1354	0158	1442	0254	1617	24
25	0056	1226	0239	1407	0126	1259	0233	1453	0241	1538	0342	1709	25
26	0155	1320	0334	1511	0217	1401	0317	1551	0325	1634	0431	1758	26
27	0256	1418	0427	1616	0306	1502	0400	1649	0410	1729	0521	1845	27
28	0356	1521	0516	1719	0353	1604	0444	1747	0457	1823	0611	1929	28
29	0454	1627	....	....	0438	1704	0530	1844	0546	1914	0702	2010	29
30	0550	1733	....	....	0523	1804	0617	1939	0636	2003	0751	2049	30
31	0642	1838	....	....	0607	1903	....	....	0727	2049	....	....	31

	JULY		AUGUST		SEPTEMBER		OCTOBER		NOVEMBER		DECEMBER		
	Rise	Set	Rise	Set	Rise	Set	Rise	Set	Rise	Set	Rise	Set	
Day	h m	h m	h m	h m	h m	h m	h m	h m	h m	h m	h m	h m	Day
1	0840	2126	0953	2154	1121	2249	1203	2329	1319	0023	1331	0116	1
2	0929	2203	1045	2234	1217	2342	1256	....	1405	0124	1415	0214	2
3	1018	2239	1138	2317	1314	....	1347	0030	1450	0224	1501	0312	3
4	1108	2316	1233	....	1410	0039	1437	0131	1535	0324	1548	0410	4
5	1159	2355	1330	0004	1505	0140	1525	0233	1621	0424	1638	0508	5
6	1252	....	1429	0056	1558	0243	1612	0336	1708	0524	1729	0603	6
7	1348	0037	1527	0153	1648	0347	1658	0438	1758	0623	1821	0657	7
8	1446	0123	1625	0254	1736	0452	1744	0540	1848	0720	1913	0747	8
9	1546	0214	1720	0358	1824	0556	1832	0641	1940	0815	2005	0833	9
10	1646	0310	1812	0504	1910	0658	1920	0741	2032	0907	2055	0916	10
11	1746	0411	1902	0610	1956	0800	2010	0839	2124	0955	2145	0957	11
12	1842	0515	1949	0714	2044	0900	2101	0935	2214	1040	2234	1035	12
13	1936	0621	2035	0816	2132	0957	2152	1027	2304	1121	2323	1112	13
14	2025	0726	2120	0917	2220	1053	2243	1116	2353	1200	....	1148	14
15	2112	0829	2205	1016	2310	1146	2333	1202	....	1237	0012	1224	15
16	2157	0931	2251	1113	....	1236	....	1245	0042	1314	0102	1302	16
17	2240	1030	2338	1208	0000	1323	0023	1325	0131	1351	0153	1343	17
18	2324	1128	....	1301	0050	1407	0112	1403	0221	1428	0247	1426	18
19	....	1225	0026	1352	0140	1448	0201	1440	0313	1508	0342	1514	19
20	0008	1320	0115	1440	0229	1527	0250	1517	0406	1550	0440	1606	20
21	0053	1414	0204	1526	0318	1605	0340	1554	0501	1636	0539	1703	21
22	0140	1506	0254	1609	0407	1642	0431	1633	0558	1726	0637	1804	22
23	0228	1555	0344	1649	0457	1718	0523	1713	0655	1820	0733	1906	23
24	0317	1643	0433	1728	0546	1756	0616	1757	0753	1917	0826	2009	24
25	0407	1727	0522	1805	0637	1835	0711	1843	0848	2017	0916	2111	25
26	0457	1809	0611	1841	0729	1915	0807	1934	0941	2117	1003	2211	26
27	0547	1849	0700	1918	0822	1959	0904	2028	1031	2218	1048	2311	27
28	0637	1927	0750	1955	0917	2046	0959	2124	1119	2318	1132	....	28
29	0726	2004	0841	2034	1012	2137	1052	2223	1204	....	1215	0009	29
30	0814	2040	0933	2116	1108	2232	1144	2323	1248	0017	1259	0107	30
31	0903	2116	1026	2200	....	....	1233	....	....	....	1345	0204	31

Time meridian 75˘ W. 0000 is midnight. 1200 is noon. Standard Time.

2014 MOONRISE AND MOONSET
NEW YORK, NY

Add one hour for Daylight Saving Time, March 9 - November 2

	JANUARY		FEBRUARY		MARCH		APRIL		MAY		JUNE		
	Rise	Set	Rise	Set	Rise	Set	Rise	Set	Rise	Set	Rise	Set	
Day	h m	h m	h m	h m	h m	h m	h m	h m	h m	h m	h m	h m	Day
1	0707	1722	0753	1947	0622	1832	0641	2041	0640	2123	0757	2214	1
2	0759	1835	0829	2058	0658	1943	0720	2143	0728	2213	0852	2248	2
3	0843	1949	0904	2206	0734	2051	0803	2241	0819	2259	0948	2319	3
4	0923	2100	0938	2311	0810	2157	0849	2334	0912	2339	1044	2349	4
5	0958	2210	1014	....	0847	2300	0938	....	1007	....	1141	....	5
6	1032	2317	1051	0014	0927	2358	1029	0021	1103	0015	1239	0018	6
7	1105	....	1131	0113	1010	....	1123	0104	1159	0048	1339	0047	7
8	1138	0021	1214	0209	1057	0053	1218	0142	1256	0119	1440	0118	8
9	1213	0124	1301	0301	1146	0142	1314	0216	1354	0148	1543	0152	9
10	1251	0224	1351	0348	1238	0227	1410	0248	1453	0218	1648	0230	10
11	1331	0321	1444	0431	1332	0308	1508	0319	1554	0248	1753	0313	11
12	1416	0415	1539	0509	1427	0344	1607	0349	1658	0321	1857	0403	12
13	1504	0505	1635	0545	1524	0418	1708	0419	1802	0357	1956	0501	13
14	1555	0551	1732	0617	1621	0449	1810	0450	1908	0437	2049	0606	14
15	1649	0632	1829	0648	1720	0519	1914	0524	2011	0524	2136	0715	15
16	1745	0709	1928	0717	1819	0549	2018	0602	2112	0617	2218	0826	16
17	1841	0743	2027	0746	1920	0619	2121	0644	2206	0717	2255	0937	17
18	1938	0814	2127	0816	2021	0650	2222	0732	2255	0822	2331	1047	18
19	2036	0844	2229	0848	2124	0725	2319	0827	2339	0930	....	1155	19
20	2134	0913	2331	0923	2227	0803	....	0927	....	1039	0004	1301	20
21	2233	0942	....	1002	2329	0846	0010	1031	0018	1148	0038	1405	21
22	2334	1012	0034	1047	....	0936	0056	1139	0053	1255	0113	1508	22
23	....	1045	0135	1139	0028	1031	0138	1247	0127	1402	0151	1609	23
24	0037	1121	0234	1238	0122	1133	0215	1356	0201	1508	0232	1707	24
25	0141	1204	0329	1344	0212	1239	0251	1504	0235	1613	0316	1800	25
26	0245	1253	0419	1454	0258	1348	0325	1612	0312	1716	0404	1849	26
27	0348	1350	0504	1607	0339	1459	0400	1719	0351	1816	0455	1934	27
28	0448	1455	0545	1720	0416	1609	0436	1824	0433	1913	0549	2013	28
29	0543	1606	....	....	0452	1720	0514	1927	0520	2006	0644	2049	29
30	0631	1720	....	....	0528	1829	0555	2027	0610	2053	0740	2121	30
31	0714	1834	....	....	0603	1936	....	....	0702	2136	....	....	31

	JULY		AUGUST		SEPTEMBER		OCTOBER		NOVEMBER		DECEMBER		
	Rise	Set	Rise	Set	Rise	Set	Rise	Set	Rise	Set	Rise	Set	
Day	h m	h m	h m	h m	h m	h m	h m	h m	h m	h m	h m	h m	Day
1	0836	2152	1016	2152	1208	2225	1252	2305	1346	0016	1335	0133	1
2	0932	2220	1115	2225	1308	2315	1342	....	1423	0125	1410	0240	2
3	1029	2249	1216	2301	1405	....	1427	0010	1458	0235	1448	0346	3
4	1126	2319	1317	2342	1459	0013	1509	0119	1535	0344	1529	0452	4
5	1225	2350	1419	....	1549	0117	1548	0230	1612	0453	1614	0554	5
6	1326	....	1520	0030	1634	0226	1626	0341	1653	0601	1703	0653	6
7	1429	0025	1618	0126	1716	0339	1703	0453	1736	0707	1755	0747	7
8	1533	0104	1712	0229	1755	0452	1740	0605	1824	0809	1850	0836	8
9	1637	0150	1801	0338	1833	0606	1820	0714	1914	0906	1946	0919	9
10	1738	0243	1845	0451	1910	0718	1902	0822	2007	0957	2042	0957	10
11	1835	0344	1925	0606	1948	0829	1947	0925	2102	1042	2138	1031	11
12	1927	0452	2003	0720	2028	0937	2035	1024	2158	1123	2234	1102	12
13	2012	0604	2039	0832	2110	1041	2126	1118	2254	1158	2331	1132	13
14	2053	0718	2116	0942	2156	1141	2219	1205	2350	1231	....	1201	14
15	2131	0830	2153	1049	2244	1236	2314	1248	....	1301	0027	1230	15
16	2206	0942	2232	1153	2335	1326	....	1325	0046	1331	0125	1300	16
17	2241	1050	2315	1254	....	1411	0009	1359	0143	1400	0224	1333	17
18	2316	1157	....	1351	0028	1450	0105	1431	0241	1430	0325	1410	18
19	2353	1301	0000	1443	0122	1526	0201	1501	0340	1502	0427	1452	19
20	....	1403	0049	1530	0217	1559	0257	1530	0440	1537	0529	1541	20
21	0033	1502	0140	1613	0313	1630	0355	1600	0542	1617	0630	1637	21
22	0115	1556	0233	1651	0410	1700	0453	1631	0644	1702	0727	1739	22
23	0202	1647	0328	1726	0507	1729	0553	1704	0745	1754	0819	1846	23
24	0251	1732	0424	1758	0604	1759	0654	1741	0844	1851	0906	1956	24
25	0344	1813	0520	1828	0703	1830	0755	1822	0937	1954	0949	2106	25
26	0438	1850	0616	1857	0802	1904	0856	1909	1025	2100	1027	2216	26
27	0533	1924	0713	1926	0902	1942	0954	2001	1109	2208	1103	2325	27
28	0629	1955	0810	1956	1002	2024	1049	2059	1148	2317	1138	....	28
29	0725	2024	0909	2027	1101	2112	1140	2202	1225	....	1213	0032	29
30	0822	2053	1008	2102	1158	2206	1226	2308	1300	0025	1250	0138	30
31	0919	2122	1108	2141	....	....	1308	....	....	....	1329	0243	31

Time meridian 75` W. 0000 is midnight. 1200 is noon. Standard Time.

PHASES OF THE MOON 2014 E.S.T.

● New Moon, ◐ 1st Quarter, ○ Full Moon, ◑ Last Quarter, A in Apogee

P in Perigee, N, S Moon farthest North or South of Equator, E on Equator

January	February	March	April	May	June
● 1 6am	E 2 noon	● 1 3am	N 5 3am	N 2 noon	A 2 11pm
P 1 4pm	◐ 6 2pm	E 1 11pm	◐ 7 4am	A 6 5am	◐ 5 4pm
E 6 1am	N 9 11am	◐ 8 8am	A 8 9am	◐ 6 10pm	E 6 8am
◐ 7 11pm	A 12 midn	N 8 6pm	E 12 2pm	E 9 11pm	○ 12 11pm
N 13 4am	○ 14 7pm	A 11 2pm	○ 15 3am	○ 14 2pm	S 13 2am
A 15 8pm	E 17 midn	E 16 7am	S 19 8am	S 16 4pm	P 14 10pm
○ 16 midn	◑ 22 noon	○ 16 noon	◑ 22 3am	P 18 6am	E 19 6am
E 20 5pm	S 23 9pm	S 23 3am	P 22 7pm	◑ 21 8am	◑ 19 2pm
◑ 24 midn	P 27 2pm	◑ 23 9pm	E 25 5pm	E 22 11pm	N 26 4am
S 27 noon		P 27 1pm	● 29 1am	● 28 2pm	● 27 3am
P 30 4am		E 29 noon		N 29 8pm	A 30 2pm
● 30 5pm		● 30 2pm			

July	August	September	October	November	December
E 3 5pm	◐ 3 8pm	◐ 2 6am	◐ 1 3pm	P 2 7pm	○ 6 7am
◐ 5 7am	S 7 midn	S 3 9am	P 6 4am	E 3 6am	N 7 5am
S 10 1pm	P 10 noon	P 7 10pm	E 6 9pm	○ 6 5pm	A 12 6pm
○ 12 6am	○ 10 1pm	○ 8 9pm	○ 8 6am	N 8 7pm	◑ 14 8am
P 13 3am	E 13 midn	E 9 11am	N 13 9am	◑ 14 10am	E 14 3pm
E 16 2pm	◑ 17 7am	◑ 15 9pm	◑ 15 2pm	A 14 8pm	S 21 2pm
◑ 18 9pm	N 19 6pm	N 16 1am	A 18 1am	E 17 5am	● 21 9pm
N 23 11am	A 24 1am	A 20 9am	E 20 9pm	● 22 8am	P 24 11am
● 26 6pm	● 25 9am	E 23 2pm	● 23 5pm	S 24 4am	E 26 7pm
A 27 10pm	E 27 7am	● 24 1am	S 27 8pm	P 27 6pm	◐ 28 2pm
E 31 1am		S 30 3pm	◐ 30 10pm	◐ 29 5am	
				E 30 1pm	

Midnight is the *beginning* of the day.

see p. 235 for daily moon phases throughout the year

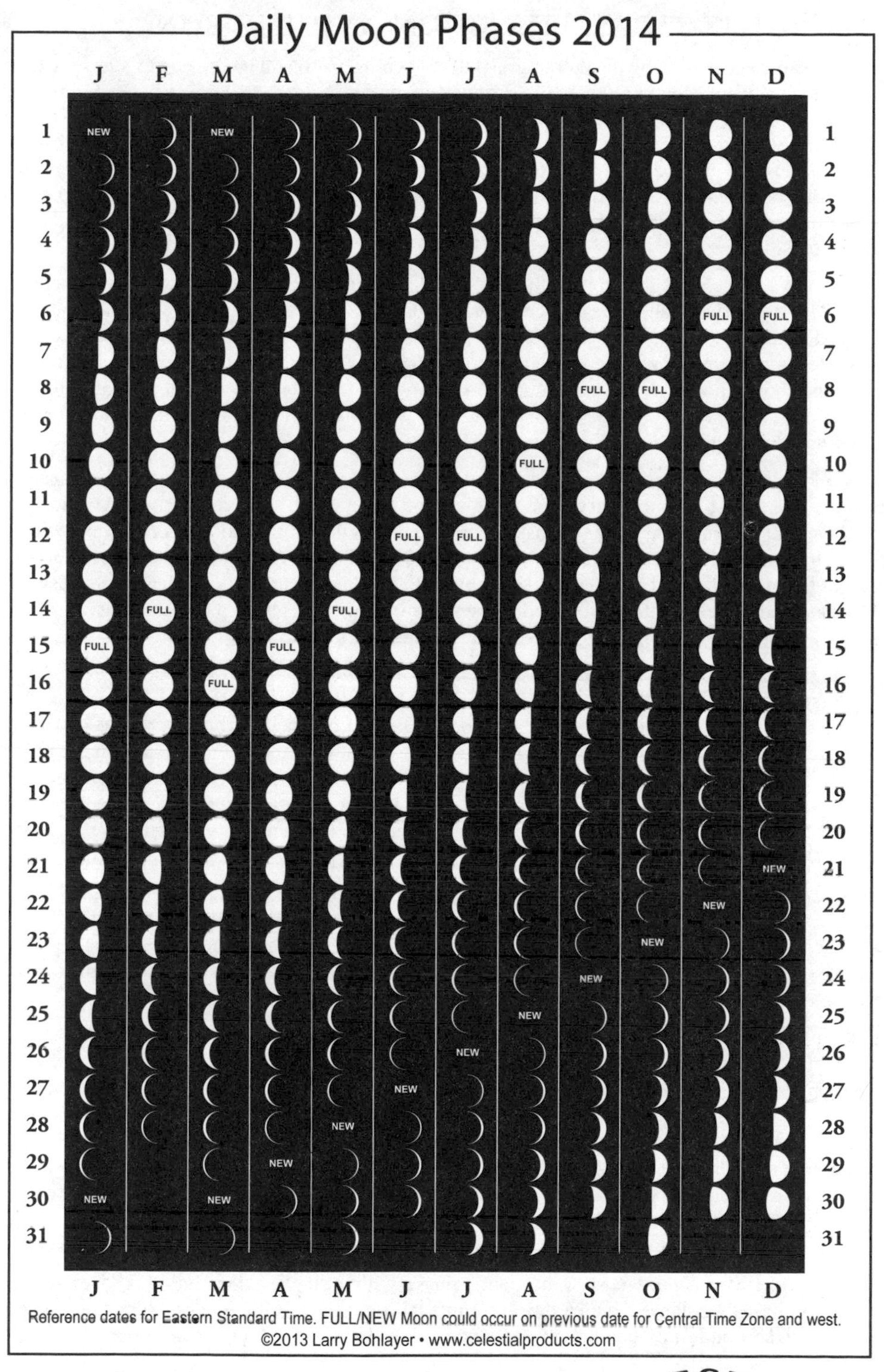
Daily Moon Phases 2014
J F M A M J J A S O N D
1 2 3 4 5 6 7 8 9 10 11 12 13 14 15 16 17 18 19 20 21 22 23 24 25 26 27 28 29 30 31
Reference dates for Eastern Standard Time. FULL/NEW Moon could occur on previous date for Central Time Zone and west.
©2013 Larry Bohlayer • www.celestialproducts.com

THE TIDES, THE MOON AND THE SUN

Tides are created on the earth by the pull of gravity between the earth and moon, and to a lesser extent the sun. Since the moon's pull weakens with distance, its pull is stronger on water located on the near side of the earth than it is on the earth's center. This creates a bulge of water on the side facing the moon. Similarly, the moon's pull on the earth's center is stronger than it is on the water on the earth's far side. This tends to pull the earth away from the water, creating another bulge of water of equal size on the far side of the earth. High tides are where the bulges are. The two bulges can also be explained as the moon's gravity being dominant on the earth's near side, and centrifugal force being dominant on the earth's far side.

The earth rotates in the same direction as the moon orbits, but much more rapidly, with a period of 24 hours vs. 27.3 days. The tidal bulges, or ocean tides, follow the slowly orbiting moon, which takes 24 hours and 50 minutes to reappear above the same part of the earth, so that each day the tides occur 50 minutes later than the previous day. As there are usually two highs and two lows per day, highs and lows average about 6 hours 12 1/2 minutes apart. A handy fact for coastwise planners: in the course of 7 days, the tides are about the *reverse* of the previous week: if on a Sunday it is *low* tide at about noon, the following Sunday it will be *high* at about noon.

The time of high tide usually does not usually coincide exactly with the time the moon is overhead or underneath. The largest astronomical reason for this is the effect of the sun, which has its own tidal effect on the earth. Although the sun has a mass 27 million times that of the moon, it is the moon which dominates by being on average 390 times closer to earth. Since the sun's effect on the tides is about one-half that of the moon, the sun can shift tidal times by up to one hour or more, depending on its position. Tidal times are also greatly affected by land masses that impede the current flows necessary to create the tides, the speeds of traveling ocean waves, and underwater topography.

How much the ocean tides rise and fall depends basically on three conditions. (See Phases of the Moon, p. 234.) First, when the sun and moon are in a line with the earth, their gravitational forces work together to produce a greater range of tide than usual. This occurs both at full moon, when the moon is opposite the earth from the sun, and at new moon, when the moon is between the earth and sun. These higher tides are called "spring tides." But when the moon and sun are at right angles to the earth (first and last quarter, or half moon), their forces are working against each other, and the result is a lower range of tide than usual. These are called "neap tides." As each year has about 13 "lunar" months, we have 26 spring tides and 26 neap tides in the year.

Second, the moon's orbit around the earth is elliptical, ranging from 252,000 miles at apogee (A) down to 221,000 miles at perigee (P), so the moon's effect on the earth is greater at "P" than at "A." Note again in the High and Low Water Tables how much higher the tide is when the Full Moon is at "P" than when the Full Moon is at "A." The position of the moon along its elliptical path is very important to the height of the tides.

Third, the plane of the moon's orbit about the earth is inclined to the plane of the earth's equator. The moon therefore travels above and below the earth's equator, and sits directly above the equator only twice a month. When it is over the equator, the day's two high tides will be about the same height. The rest of the time the moon is either above the northern hemisphere (northern declination) or the southern hemisphere (southern declination). When the moon is north of the equator (N) or south (S), the two high water marks on the same day will differ in height. This effect is known as "semidiurnal inequality." When the moon has northern declination and is over the U.S., this high tide will be the greater of our two daily highs. Our second high tide of the day, when the moon is over Asia, will be less high, because the far side tidal bulge will be greatest over South America, and less over us in the northern hemisphere.

The height of tides is influenced most by the moon's phase, with the highest tides at Full and New Moon; second by the moon's distance from earth in its elliptical orbit, tides being highest when the moon is closest, at perigee; and last by the moon's declination, north or south, which creates tides of different heights on the same day.

For a more complete discussion, see NOAA's website: http://tidesandcurrents.noaa.gov/restles1.html

The Publishers thank Nelson Caldwell, of the Smithsonian Astrophysical Observatory, Cambridge, MA, and Hale Bradt, Department of Physics, M.I.T., for their valuable contributions to this article.

Visibility of the Planets, 2014

MERCURY can only be seen low in the east before sunrise, or low in the west after sunset. It is visible in the mornings from February 22 to April 18, June 29 to August 1, and October 23 to November 22. It is brighter toward the end of each period. It is visible in the evenings from January 13 to February 9, May 4 to June 10, August 18 to October 11, and December 25 to December 31. It is brighter at the beginning of each period.

VENUS is a brilliant object in the evening sky until the end of the first week of January, when it becomes too close to the Sun for observation. It reappears in the third week of January as a morning star and can be seen in the morning sky until mid-September, when it again becomes too close to the Sun for observation. It is visible in the evening sky from early December until the end of the year. Venus is in conjunction with Jupiter on August 18.

MARS rises around midnight at the beginning of the year in Virgo. It is at opposition on April 8, when it is visible through the night as a bright reddish object. From mid-July until the end of the year it is visible only in the evening sky. It remains in Virgo until mid-August, and moves through Libra, Scorpius, Ophiuchus, Sagittarius, and into Capricornus in early December. Mars is in conjunction with Saturn on August 27.

JUPITER can be seen in Gemini from the beginning of the year for most of the night. On January 5 it is at opposition where it can be seen through the night. From early April it can only be seen in the evening sky. It moves into Cancer in early July. In the second week of July it becomes too close to the Sun for observation until early August, when it reappears in the morning sky. It passes into Leo in mid-October and from mid-November can be seen for more than half the night. Jupiter is in conjunction with Venus on August 18.

SATURN rises well after midnight at the beginning of the year in Libra, where it remains throughout the year. It is at opposition on May 10, where it can be seen through the night. From early August to the start of November it is visible only in the evening sky. It then becomes too close to the Sun for observation. It reappears in early December, where it can be seen in the morning sky for the rest of the year. Saturn is in conjunction with Mars on August 27.

Conjunction occurs when a body has the same horizontal bearing from Earth as another. When Venus is in conjunction with Jupiter on August 18, they appear one over the other, in the same sector of the sky.

Opposition occurs when a body, farther than Earth from the Sun, appears opposite the Sun. On a line drawn from the Sun through the Earth and beyond, the body lies on that extension. It is brightest at that time.

Elongation is apparent motion eastward or westward (relative to the Sun) across the sky. When a planet has 0° elongation, it lies on a line from Earth to the Sun, is in conjunction and not visible; when it has 90° elongation, it is in eastern quadrature; when it has 180° elongation, it is in opposition and has the best visibility; when it has 270° elongation, it is in western quadrature.

Visibility of Planets in Morning and Evening Twilight

	Morning	Evening
VENUS	---	January 1 – January 5
	January 17 – September 17	December 5 – December 31
MARS	January 1 – April 8	April 8 – December 31
JUPITER	January 1 – January 5	January 5 – July 11
	August 8 – December 31	---
SATURN	January 1 – May 10	May 10 – November 1
	December 6 – December 31	---

RADIO TELEPHONE INFORMATION – VHF SYSTEM

Calling Guidelines: Avoid excessive calling. Make calls as brief as possible. Give name of called vessel first, then "This is (name of your vessel)," your call sign (if you have a Station License), and the word "Over." If station does not answer, delay your repeat call for 2 minutes. At the end of your message, sign off with "This is (your vessel's name)," your call sign, and "Out."

Range and Power: Operation is essentially line-of-sight. Since the elevation of antennas at both communications points extends the "horizon," range may be 20 to 50 miles on a 24-hour basis between a boat and a land station. Effective range between boats will be less because of lower antenna heights. 25 watts is the maximum power permitted.

Interference factor: Most VHF-FM equipment has 6 or more channels, so it is possible to shift to a clear channel. Like the FM in your home radio, the system is practically immune to interference from ignition noise, static, etc., except under unusual conditions.

Channelization: A minimum of 3 channels is required by the FCC. Two are mandatory: Channel 16 (156.800 MHz), the International Distress frequency; and Channel 06 (156.300 MHz), the Intership Safety Frequency. The Coast Guard *strongly recommends* that you have Channel 22A as your third channel.

Channel	Purpose and Comments
16 156.800 MHz	**Distress and Safety**: Ship to Shore and Intership. Guarded 24 hours by the Coast Guard. No routine messages allowed other than to establish the use of a working channel. *See page 239* for Distress calling procedure. **Calling**: Ship to Shore and Intership. Use Channel 16 to establish contact, then switch to a working channel (see below). Calling Channel: New England waters. Commercial and pleasure.
09 156.450 MHz	**Boater Calling:** Commercial and Non-Commercial
06 156.300 MHz	**Intership Safety:** No routine messages allowed. 06 is limited to talking with the Coast Guard and others at the scene of an emergency, and to information on the movement of vessels.
22A 157.100 MHz (**21** in Canada) 161.65 MHz	**Maritime Safety Information** channel. Not guarded by the CG, but after a vessel makes contact with the CG for non-distrress calls on Channel 16, they will tell you to switch to and use *only* 22A for communicating. Channel 22A is also used for CG weather advisories and Notices to Mariners. *Times* of these broadcasts given on Channel 16.
12, 14, 20A, 65A, 66A, 73, 74, 77	**Ship to Shore and Intership:** Port operations, harbormasters, etc. (Your electronics dealer should have local frequencies.)
08, 67, 88A	**Commercial (intership only):** For ocean vessels, dredges, tugs, etc.
07A, 10, 11, 18A, 19A, 79A, 80A	**Commercial only**
13 156.650 MHz	**Intership Navigation Safety:** (bridge to bridge). Ships > 20 m length maintain a listening watch on this channel in US waters.
68, 69, 71, 72, 78A	**Ship to Shore and Intership, Pleasure craft only:** Shore stations, marinas, etc. The best channels for general communication.
70 156.525 MHz	**Digital Selective Calling (DSC)**. Special equipment required. See p. 241 Marine Communications.
81A 157.075 MHz **83A** 157.175 MHz ***for Keyed Fog Signals,*** Check Notice to Mariners	**If** fog signal is radio activated. During times of reduced visibility, within ½ mile of the fog signal, turn VHF marine radio to channel 81A or 83A. Key microphone 5 – 10 times consecutively to activate fog signal for 45 minutes. The CG is currently using **83A**, **81A** if interference.
AIS 1 161.975 MHz	Automatic Identification System (AIS)
AIS 2 162.025 MHz	Automatic Identification System (AIS)

MARINE EMERGENCY AND DISTRESS CALLS

See p. 241, Marine Communications, for information on why you should update your VHF Radio for Digital Selective Calling (DSC).

If you have DSC, have an MMSI number and your unit is properly installed with a GPS connection, follow your manufacturer's instructions.

If you do not have a DSC-equipped radio, use the following:

Speak slowly and clearly. Use: VHF Ch. 16 (156.800 MHz) or MF/HF 2182 kHz

1. DISTRESS SIGNAL (top priority)

If you are in distress (i.e. when threatened by grave and imminent danger) transmit the International distress call on either 2182 kHz or 156.800 MHz (Channel 16) — "MAYDAY MAYDAY MAYDAY, THIS IS (Your vessel's name and call sign repeated three times)"

IF CALLING FROM A VESSEL IN TROUBLE — give:

1. WHO you are (Your vessel's name, registration number or call sign).
2. WHERE you are (Your vessel's position in latitude/longitude or true bearing and distance in nautical miles from a known geographical point. Local names known only in the immediate vicinity are confusing).
3. WHAT is wrong (Nature of distress or difficulty, if not in distress).
4. Kind of assistance desired.
5. Number of persons aboard and condition of any injured.
6. Present seaworthiness of your vessel.
7. Description of your vessel (length, type cabin, masts, power, color of hull, superstructure and trim).
8. Your listening frequency and schedule.

IF CALLING WHILE OBSERVING ANOTHER VESSEL IN DIFFICULTY — give:

1. Your position and the bearing and distance of the vessel in difficulty.
2. Nature of distress or difficulty.
3. Description of the vessel in distress or difficulty, (see item 7 above).
4. Your intentions, course and speed, etc.
5. Your radio call sign, name of your vessel, listening frequency and schedule.

If there is no immediate response, repeat appropriate messages above; if still no response, you may send on any other available frequency until you make contact.

IF YOU HEAR A MAYDAY CALL — Immediately discontinue any transmission. Note details in your radio log right away. Do not make any transmission on this distress channel until MAYDAY condition is lifted by the Coast Guard, unless you are in a position to be of assistance.

2. URGENCY SIGNAL (second in priority)

If you have an urgent message to send (threat to a vessel's safety or to someone on board, overboard or within sight), use the same procedure as above but say the word "PAN" three times. "PAN" (pronounced "PAWN") is also used as a warning signal that a Distress Signal may be sent out at a later stage. Morse Code signal is – (T) – (T) – (T)

3. SAFETY SIGNAL (third priority)

If you wish to report navigation or weather warnings (ice, derelicts, tropical storms, etc.) use the same procedure as above but say the word "SECURITY" (pronounced SAY-CUR-I-TAY) three times. Morse Code signal is – • • – (X) – • • – (X) – • • – (X)

See pp. 8-9 for a list of visual and audible distress signals

How to Contact the U.S. Coast Guard

U.S. Coast Guard Rescue Coordination Centers (RCCs)

24-hour Regional Contacts for Emergencies

RCC Boston, MA – 617-223-8555 New England south to northern New Jersey
RCC Norfolk, VA – 757-398-6231 New Jersey to border of N. Carolina and S. Carolina
RCC Miami, FL – 305-415-6800 S. Carolina to Key West including much of Caribbean

USCG Navigation Information Service (NIS). Watchstander, (24/7): 703-313-5900

USCG Suspected Terrorist Incidents Hotline, National Response Center (NRC): 800-424-8802

USCG Maritime Safety Line: 800-682-1796

INTERNET: USCG - www.navcen.uscg.gov/ Canada - www.notmar.gc.ca/

U.S. Coast Guard Stations – (monitoring VHF Ch. 16)

1st District – Boston – (800) 848-3942

Eastport, ME (207) 853-2845
Jonesport, ME (207) 497-2134
Southwest Harbor, ME (207) 244-4250
Rockland, ME (207) 596-6667
Boothbay Hbr., ME (207) 633-2661
S. Portland, ME (207) 767-0364
Sector Northern New England (207) 767-0302
Portsmouth, NH (603) 436-4415
Merrimac-Newburyport, MA (978) 465-0731
Gloucester, MA (978) 283-0705
Boston, MA (617) 223-3224
Point Allerton-Hull, MA (781) 925-0166
Scituate, MA (781) 545-3801
Cape Cod Canal-E. Entr. (508) 888-0020
Provincetown, MA (508) 487-0077
Chatham, MA (508) 945-3830
Brant Pt.-Nantucket, MA (508) 228-0388
Woods Hole, MA (508) 457-3219
Menemsha, MA (508) 645-2662
Castle Hill, Newport, RI (401) 846-3676
Point Judith, RI (401) 789-0444
New London, CT (860) 442-4471
New Haven, CT (203) 468-4498
Sector NY, NY (718) 354-4353
Fire Island, NY (631) 661-9101
New York Station, NY (718) 354-4099
Eatons Neck, NY (631) 261-6959
Kings Point, NY (516) 466-7135
Jones Beach, NY (516) 785-2995
Moriches, NY (631) 395-4400
Shinnecock, NY (631) 728-0078
Montauk, NY (631) 668-2773
Sandy Hook, NJ (732) 872-3429

5th District – Portsmouth, VA (757) 398-6486

Manasquan Inlet, NJ (732) 899-0887
Shark River, NJ (732) 776-6730
Barnegat, NJ (609) 494-2661
Atlantic City, NJ (609) 344-6594
Cape May, NJ (609) 898-6995
Great Egg, NJ (609) 399-0144
Indian River Inlet, DE (302) 227-2440
Ocean City, MD (410) 289-1905

5th District, cont.

St. Inigoes, MD (301) 872-4344
Crisfield, MD (410) 968-0323
Annapolis, MD (410) 267-8108
Oxford, MD (410) 226-0581
Curtis Bay-Baltimore, MD (410) 576-2625
Stillpond, MD (410) 778-2201
Chincoteague, VA (757) 336-2874
Little Creek-Norfolk, VA (757) 464-9371
Wachapreague, VA (757) 787-9526
Portsmouth, VA (757) 483-8527
Cape Charles, VA (757) 331-2000
Milford Haven, VA (804) 725-3732
Oregon Inlet, NC (252) 441-6260
Hatteras Inlet, NC (252) 986-2176
Hobucken, NC (252) 745-3131
Ocracoke, NC (252) 928-4731
Fort Macon, NC (252) 247-4581
Elizabeth City, NC (252) 335-6086
Wrightsville Beach, NC (910) 256-4224
Emerald Isle, NC (252) 354-2719
Oak Island, NC (910) 278-1133

7th District – Miami, FL (305) 415-6800

Georgetown, SC (843) 546-2052
Charleston, SC (843) 720-7727
Tybee, GA (912) 786-5440
Brunswick, GA (912) 267-7999
Mayport, FL (904) 564-7592
Ponce de Leon Inlet, FL (386) 428-9085
Cape Canaveral, FL (321) 853-7601
Fort Pierce Inlet, FL (772) 464-6100
Lake Worth Inlet, FL (561) 840-8503
Ft. Lauderdale, FL (954) 927-1611
Miami Beach, FL (305) 535-4368
Islamorada, FL (305) 664-4404
Marathon, FL (305) 743-1945
Key West, FL (305) 292-8856

Canada-Nova Scotia

Canadian Coast Guard
Joint Rescue Coordination Center
Halifax, NS (902) 427-2110

MARINE COMMUNICATIONS

Emergencies: The Coast Guard is required to monitor Channel 16; they are not required to answer the telephone. In an emergency, use your VHF radio to call the Coast Guard on Channel 16 (156.80 MHz). Digital Selective Calling (DSC) is on Channel 70 on your VHF. The Coast Guard urges, in the strongest terms possible, that you take the time to interconnect your GPS and DSC-equipped radio. Doing so may save your life in a distress situation!

DSC: As part of the Global Maritime Distress and Safety System (GMDSS), Rescue 21 is the Coast Guard system that provides the emergency response made possible by DSC-equipped VHF radios. It has been active for a while, and if you don't yet have a DSC-VHF radio, you need to know the significant advantages it offers.

Rescues initiated by DSC-equipped radios are far quicker and more successful. Why? With the push of one button, an automated digital distress alert is sent to other DSC-equipped vessels and rescue facilities. This transmission includes your vessel's unique, 9-digit MMSI (Marine Mobile Service Identity) number, which contains your vessel's description for easier identification by response teams. If connected to a compatible GPS, the signal will give your vessel's latitude and longitude for faster and more efficient assistance or rescue. For more information go to: www.navcen.uscg.gov/?pageName=mtDsc. Domestic users (non-commercial) who do not travel outside of the US can be issued an MMSI number without applying for an FCC Station License. You can register for an MMSI online at www.boatus.com/mmsi/instruct.asp, or www.seatow.com/boating-safety

Non-emergency: Near shore (range will vary) a cell phone can be used successfully for non-emergency calls. The usable distance assumes line-of-sight, so an antenna which is higher may help communicate farther. That distance may be less where there are fewer cell towers.

MARINE WEATHER FORECASTS

VHF-FM, NOAA All-Hazards Weather Radio - Continuous broadcasts 24 hours a day are provided by the National Weather Service with taped messages repeated every 4-6 minutes. These are updated every 3-6 hours and include weather and radar summaries, wind observations, visibility, sea conditions and detailed local forecasts. NOAA VHF-FM broadcasts can be received 20-40 miles from transmitting site.

	MHz		MHz
WX-1	162.550	WX-5	162.450
WX-2	162.400	WX-6	162.500
WX-3	162.475	WX-7	162.525
WX-4	162.425		

Jonesboro, ME (5)	Riverhead, NY (3)	Cape Hatteras, NC (3)	Melbourne, FL (1)
Ellsworth, ME (2)	Philadelphia, PA (3)	New Bern, NC (2)	Fort Pierce, FL (4)
Dresden, ME (3)	Atlantic City, NJ (2)	Georgetown, SC (6)	W. Palm Bch., FL (3)
Gloucester, MA (4)	Lewes, DE (1)	Charleston, SC (1)	Miami, FL (1)
Boston, MA (3)	Baltimore, MD (2)	Beaufort, SC (5)	Key West, FL (2)
Hyannis, MA (1)	Hagerstown, MD (3)	Brunswick, GA (4)	
Providence, RI (2)	Norfolk, VA (1)	Jacksonville, FL (1)	
New York, NY (1)	Mamie, NC (4)	Daytona Bch., FL (2)	

TIME SIGNALS

Bureau of Standards Time Signals: WWV, Ft. Collins, Col., every min. on 2500, 5000, 10000, 15000, 20000, 25000 kHz. **Canadian Time Signals:** CHU, (frequently easier to get than WWV) 45° 17' 47" N, 75° 45' 22" W. Continuous transmission on 3330 kHz, 7850 kHz and 14670 kHz. For more information on time visit the following websites. http://tf.nist.gov/timefreq/, http://nist.time.gov/

Omission of a tone indicates the 29th second of each minute. The new minute is marked by the full tone *immediately* following the voice announcement. Five sets of two short tones mark the first five seconds of the next minute. The hour is identified by a pulse of one full second followed by 12 seconds of silence.

HYPOTHERMIA
and Cold Water Immersion
What You Need To Know

It is not uncommon for a boater to fall off a boat or dock. Most are rescued immediately. However, when rescue is delayed and conditions are present which threaten survival, all who go boating should know what to do.

Hypothermia is a state of low body core temperature - specifically below 95° F. This loss of body heat may be caused by exposure to cold air or cold water. Since water conducts heat away 25 times more quickly than air, time is critical for rescue. There are many variables beyond water temperature that combine to determine survival time: whether a life jacket is on, body size and composition, type of clothing, movement in the water, etc. Wearing a Personal Flotation Device (PFD) greatly extends survival time by keeping your head above water and by allowing you to float without expending energy.

What a person in the water should do:

1. If at all possible, get out of the water, or at least grab hold of anything floating. If the boat is swamped, stay with it and crawl as far out of the water as possible.
2. Do not try to swim, unless a boat or floating object is very nearby and you are certain you can get to it.
3. Control heat loss by keeping clothing on as partial insulation. In particular, keep the head out of water. To protect the groin, sides, and chest from heat loss, use the H.E.L.P. (heat escape lessening position), a fetal position with hands clasped around the legs, which extends survival time.
4. Conserve energy by remaining as still as possible. Physical effort promotes heat loss. Swimming, or even treading water, reduces survival time.

The states of hypothermia:

1. Mild: victim feels cold, exhibits violent shivering, lethargy, slurred speech
2. Medium: loss of some muscle control, incoherence or combativeness, stupor, and exhaustion
3. Severe: unconsciousness, respiratory distress, possible cardiac arrest

What a rescuer should do:

1. Move the victim to a warm place, position on his/her back, and check breathing and heartbeat.
2. Start CPR (p. 243) if necessary.
3. Carefully remove wet clothing, cutting it away if necessary.
4. Take steps to raise the body temperature gradually: cover the victim with blankets or a sleeping bag, and apply warm moist towels to the neck, chest, and groin.
5. Provide warm oral fluids and sugar sources after uncontrolled shivering stops and the patient shows evidence of ability to swallow and of rewarming.

What NOT to do:

1. Do not give alcohol, coffee, tea, or nicotine. If the victim is not fully conscious, do not attempt to provide food or water.
2. Do not massage arms or legs or handle the patient roughly, as this could cause cold blood from the periphery to circulate to the body's core, which needs to be warm first.

See **Emergency First Aid, pp. 243-245.**

EMERGENCY FIRST AID

These are guidelines to be used only when professional help is not readily available.

Good Samaritan laws were enacted to encourage people to help others in emergency situations. Laws vary from state to state, but all require that the caregiver use common sense and a reasonable level of skill.

Before giving care to a conscious victim you must first get consent. If the victim does not give consent call 911. Consent may be implied if a victim is unconscious, confused, or seriously ill.

Prevent disease transmission by avoiding contact with bodily fluids, using protective equipment such as disposable gloves and thoroughly washing hands after giving care.

PRIMARY ASSESSMENT

Check for: 1. Unresponsiveness 2. Breathing - Look, listen and feel. 3. Pulse (any movement or sign of life) - If pulse and breathing are present, check for and control any severe bleeding.

* If no sign of life or breathing, call for help and then begin CPR. For children, do 2 minutes of CPR, then call for help while continuing CPR.
* If pulse is present but no breathing, begin Rescue Breathing.
* If airway is obstructed, do Heimlich to clear airway. Do not use Heimlich if drowning is suspected; go to Rescue Breathing.

CPR - Cardiopulmonary Resuscitation

CPR* - Use only when there is no sign of breathing and no sign of movement or life. First, call or get someone to call for help. CPR has two components: compressions and giving breaths. Pushing hard and fast on the chest is the most important part of CPR because you are pumping blood to the brain and heart.

Roll victim onto back as a unit, being careful to keep spine in alignment. Move clothing out of the way. Put the heel of 1 hand on the lower half of the breastbone. Put the heel of your other hand on top of the first hand. Push straight down at least 2 inches at a rate of at least 100 compressions a minute. After each compression, let the chest come back up to its normal position. Giving compressions is tiring. If someone else is available, take turns being careful to make pauses in giving compressions as short as possible.

Compressions are the most important part of CPR. If you are also able to give breaths, you will help even more. To give breaths, tilt the head back while lifting the chin up. Pinch nose shut, seal your lips tight around victim's mouth, GIVE 2 FULL BREATHS for 1 to 1 ½ seconds each, checking for chest rise. Do not take more than 10 seconds away from compressions to give breaths. It is better to have two rescuers to do both compressions and breaths.

** To perform CPR you should be trained. Courses are available through the American Red Cross and the American Heart Association.* ***If you are unable or uncomfortable doing Rescue Breathing, the American Heart Association states that performing Chest Compressions alone can be effective in helping to circulate oxygenated blood through the body. Follow these two important steps: 1) first call 911 and 2) using both hands pump on center of chest between the nipples hard, fast and continuously. For internet help for the traditional instruction of CPR, use www.heart.org and click on CPR. For the new hands only go to www.handsonlycpr.org.***

Continued p. 244

EMERGENCY FIRST AID
continued

RESCUE BREATHING - no obstruction. Call or get someone to call for help.

Pulse present, unresponsive, no breathing.

Roll victim onto back and open airway. Tilt head back and lift chin. Look, listen and feel for breath for 3-5 seconds.

If no breath, keep head tilted back, pinch nose shut, seal your lips tight around victim's mouth, GIVE 2 NORMAL FULL BREATHS for 1 to 1 ½ seconds each until chest rises. Feel for pulse at side of neck for 5-10 seconds.

If pulse present, begin Rescue Breathing for 1 minute. Keep head tilted back, pinch nose. Give 1 breath every 5 seconds. Look, listen and feel for breath between breaths.

RECHECK PULSE EVERY MINUTE. If victim has pulse but is not breathing, continue rescue breathing.

If victim has no sign of life or breath, go to CPR. pg. 243.

OBSTRUCTED AIRWAY - If victim cannot cough, breathe, or speak, use HEIMLICH.

If drowning suspected, use Rescue Breathing.

Do not try to clear water from lungs. Roll to side if vomiting occurs so victim won't choke.

HEIMLICH - If victim is conscious, stand behind him. Wrap your arms around victim's waist. Place your fist (thumbside) against the victim's stomach in the midline, just above the navel and well below the rib margin. Grasp your fist with other hand. Press into stomach with a quick upward thrust.

If victim is unconscious, lay victim on back, do finger sweep on adult (on child only if you can see object). Attempt rescue breathing. If airway remains blocked, give 6-10 abdominal thrusts and repeat as necessary.

BLEEDING - Apply pressure directly over wound with a dressing, until bleeding stops or until EMS rescuers arrive. If possible, press edges of a large wound together before using dressing and bandage.

If bleeding continues, apply additional bandages and continue to maintain pressure.

If possible, elevate wounded area, apply ice wrapped in cloth to wound and keep the patient warm.

BURNS, SCALDS - No open blisters: Use cool water, then cover with a moist sterile dressing.

Open blisters - Heat: Cover with dry sterile dressing. Do not put water on burn or remove clothing sticking to burn. Treat for shock.

Open blisters - Chemical: Flush all chemical burns with water for 15 to 30 minutes. Remove all clothing on which chemical has spilled. Cover with dry sterile dressing and treat for shock. Eyes: Flush with cool water only for 15 minutes.

SHOCK - Confused behavior, rapid pulse and breathing, cool moist skin, blue tinge to lips and nailbeds, weakness, nausea and vomiting, etc.

Keep patient lying down with legs elevated. Remove wet clothing. Maintain normal body temperature. Do not give victim food or drink.

FRACTURES - Do not move victim or try to correct any deformity. Immobilize the area. If bone penetrates the skin use a sterile dressing and control bleeding before splinting.

Splint a broken arm to the trunk or a broken leg to the other leg. A padded board or pole can be used along the side, front or back of a broken limb. A pillow or a rolled blanket can be used around the arm or leg.

For an injured shoulder put a pillow between the arm and chest and bind arm to body. For an injured hip, place pillow between legs and bind legs together.

HEAD, NECK and **SPINE INJURIES** - Do not move victim or try to correct any deformity. Stabilize head and neck as you found them.

POISONING - Call for help immediately: Poison Control Center 800-222-1222.

Have poison container available. Antidotes listed on the label may be wrong. Keep syrup of ipecac and activated charcoal available, but do not administer unless advised to do so.

HEAT PROSTRATION - Strip victim. Move to shaded area. Wrap in cool, wet sheet. Treat for shock.

EXPOSURE TO COLD - Provide a warm dry bunk and warm drink, not coffee, tea or alcohol.

Frostbite: Rewarm slowly, beginning with the body core rather than the extremities. Elevate and protect affected area. Do not rub frozen area, break blisters or use dry heat to thaw.

Treat for shock. **See Hypothermia and Cold Water Immersion article p. 242.**

SUNBURN – Treat heat prostration if present. Take the heat out of the skin by using a cool damp cloth laid over the area. Do not apply ice as this may damage the skin further.

Painkillers like acetaminophen (Tylenol) or ibuprofen may be used for pain. Use topical lotions to keep the skin moist and reduce dehydration. Those containing aloe work well.

If the skin is blistering, prevent secondary infection by keeping the area clean and by applying an antibacterial cream.

Victim should rest and keep hydrated. Seek medical help if area does not improve.

SEASICKNESS – This form of motion sickness is characterized by headache, drowsiness, nausea and vomiting often brought on by sailing in rough or inconsistent seas. Seasickness can be difficult to control.

Preventive: Medications taken before you get on the boat: Dramamine® and Bonine® are the two most common over the counter seasickness remedies. Transderm Scop® Scopolamine patch is a common prescription medicine. *Talk with your doctor or pharmacist about which approach might be right for you.* Avoid strong odors, greasy, spicy, high-fat foods as well as alcohol and excessive sugars as they can make you queasy or light-headed. Avoid reading books and computer screens.

Coping with seasickness: Take ginger, chew gum, look at the horizon, stay on deck, get fresh air, try to sleep, stay as close to the center of the boat as possible. Anti-seasickness wristbands are also known to relieve symptoms. Keep hydrated.

KEYS TO PREDICTING THE WEATHER

Trend and rate of change – The most important point to remember about barometric pressure is that the trend (up, down, or steady) and the rate of change are far more predictive of coming weather than the position of the pointer at any one time. Tap your barometer periodically; the pointer's direction should indicate the pressure trend. Pay little attention to the words Stormy, Rain, Change, Fair, Very Dry; they are traditional, decorative, and often inaccurate.

State of the air – In addition to changes in barometric pressure, the state of the air (cool, dry, warm, moist) and the appearance of the sky foretell coming weather. See Weather Signs on the next page.

Long foretold – long last; short notice – soon past. This handy saying has much truth in it. Slow changes last longer; sudden changes are quickly over. A steady barometer with dry air indicates continuing fine weather. A rapid rise or fall of barometric pressure indicates unsettled or stormy weather for a short period of time.

FORECASTING

with Wind Direction and Barometric Pressure

Wind Dir.	Pressure	Trend	Likely Forecast
SW to **NW**	30.1-30.2	Steady	Fair, little temp. change
	30.1-30.2	Rising rapidly	Fair, perhaps warmer with rain
	30.2+	Steady	Fair, no temp. change
	30.2+	Falling	Fair, gradual rise in temp.
S to **SW**	30.0	Rising slowly	Clearing, then fair
S to **SE**	30.2	Falling rapidly	Increasing wind, rain to follow
S to **E**	29.8	Falling rapidly	Severe NE gale, heavy rain/snow
SE to **NE**	30.1-30.2	Falling slowly	Rain
	30.1-30.2	Falling rapidly	Increasing wind and rain
	30.0	Falling slowly	Rain continuing
	30.0	Falling rapidly	Rain, high wind, then clearing and cooler
E to **NE**	30.0+	Falling slowly	Rain with light winds
	30.1	Falling rapidly	Rain or snow, increasing wind
Shifting W	29.8	Rising rapidly	Clearing and cooler

WEATHER SIGNS IN THE SKY

Signs of Good Weather

- A gray sky in the morning or a "low dawn" – when the day breaks near the horizon, with the first streaks of light low in the sky – brings fair weather.
- Light, delicate tints with soft, undefined clouds accompany fine weather.
- Seabirds flying out early and far to sea suggest moderate wind, fair weather.
- A rosy sky at sunset, clear or cloudy: "Red sky at night, sailor's delight."
- High, wispy cirrus clouds, or even high cumulus, indicate immediate fair weather, with a possible change from a front within 24 hours.
- High contrails disappearing quickly show dry air aloft.
- Steady mild-to-moderate winds from the same direction indicate continuing fair weather.
- A low dew point relative to temperature means dry air. (see p. 256)

Signs of Bad Weather

- "Red sky at morning, sailor take warning." Poor weather, wind, maybe rain.
- A "high dawn" – when the first streaks of daylight appear above a bank of clouds – often precedes a turn for worse weather.
- Light scud clouds driving across higher, heavy clouds show wind and rain.
- Hard-edged, inky clouds foretell rain and strong wind.
- Seabirds hanging over the land or headed inland suggest wind and rain.
- Remarkable clearness of atmosphere near the horizon, when distant hills or vessels are raised by refraction, are signs of an Easterly wind and indicate coming wet weather.
- Long-lasting contrails indicate humid air aloft.
- Low-level clouds, and clouds at several heights
- Rising humidity, dewpoint close to temperature
- Strong early-morning winds

Signs of Wind

- Soft-looking, delicate clouds indicate light to moderate wind.
- Stronger wind is suggested by hard-edged, oily-looking, ragged clouds, or a bright yellow sky at sunset.
- A change in wind is indicated by high clouds crossing the sky in a different direction from that of lower clouds.
- Increasing wind and possibly rain are preceded by greater than usual twinkling of stars, indistinctness of the moon's horns, "wind dogs" (fragments of rainbows) seen on detached clouds, and the rainbow.
- "First rise after very low, indicates a stronger blow."

How To Become Wind-Wise

Wind is the most dynamic of weather factors, and both power and sail boaters ignore it to their peril. A calm day can postpone or cancel a sailing outing, while powerboat skippers love it. A windy day can thrill a sailor but keep the power-boater at home.Our suggestion: watch the wind before you get to the shore.

The **MAESTRO** gives you:

- current wind speed from 0-100 mph
- memory of the highest gust
- wind direction from 16 points
- choice of silver or black dial
- brass case (nickel or chrome extra)
- 60' wires, mounting hardware
- speed & direction sensors
- full 5-year warranty
- made in the USA

The **MAESTRO** is handsome and practical – a great gift!

ROBERT E. WHITE INSTRUMENTS, INC.
www.robertwhite.com
617-482-8460

Wind Chill Chart

Wind (mph)	Temperature (°F)																	
Calm	40	35	30	25	20	15	10	5	0	-5	-10	-15	-20	-25	-30	-35	-40	-45
5	36	31	25	19	13	7	1	-5	-11	-16	-22	-28	-34	-40	-46	-52	-57	-63
10	34	27	21	15	9	3	-4	-10	-16	-22	-28	-35	-41	-47	-53	-59	-66	-72
15	32	25	19	13	6	0	-7	-13	-19	-26	-32	-39	-45	-51	-58	-64	-71	-77
20	30	24	17	11	4	-2	-9	-15	-22	-29	-35	-42	-48	-55	-61	-68	-74	-81
25	29	23	16	9	3	-4	-11	-17	-24	-31	-37	-44	-51	-58	-64	-71	-78	-84
30	28	22	15	8	1	-5	-12	-19	-26	-33	-39	-46	-53	-60	-67	-73	-80	-87
35	28	21	14	7	0	-7	-14	-21	-27	-34	-41	-48	-55	-62	-69	-76	-82	-89
40	27	20	13	6	-1	-8	-15	-22	-29	-36	-43	-50	-57	-64	-71	-78	-84	-91
45	26	19	12	5	-2	-9	-16	-23	-30	-37	-44	-51	-58	-65	-72	-79	-86	-93
50	26	19	12	4	-3	-10	-17	-24	-31	-38	-45	-52	-60	-67	-74	-81	-88	-95
55	25	18	11	4	-3	-11	-18	-25	-32	-39	-46	-54	-61	-68	-75	-82	-89	-97
60	25	17	10	3	-4	-11	-19	-26	-33	-40	-48	-55	-62	-69	-76	-84	-91	-98

Frostbite Times: 30 minutes, 10 minutes, 5 minutes

BEAUFORT SCALE

Beaufort Force	Knots	Wind Condition	Conditions at Sea	Conditions Ashore
0	0-1	**Calm**	Smooth, mirror-like sea	Calm, smoke rises vertically
1	1-3	**Light Air**	Scaly ripples, no foam crests	Smoke drifts at an angle, leaves move
2	4-6	**Light Breeze**	Small wavelets, crests glassy, not breaking	Leaves rustle, flags begin to move
3	7-10	**Gentle Breeze**	Large wavelets, some crests break, scattered whitecaps	Small branches move, light flags extended
4	11-16	**Moderate Breeze**	Small waves 1-4 ft. getting longer, numerous whitecaps	Leaves, loose paper lifted, larger flags flapping
5	17-21	**Fresh Breeze**	Moderate waves 4-8 ft., many whitecaps	Small trees in leaf begin to sway, flags extended
6	22-27	**Strong Breeze**	Larger waves 8-13 ft., more whitecaps, spray	Larger tree branches and small trees in motion
7	28-33	**Near Gale**	Sea heaps up, waves 13-20 ft., white foam streaks	Whole trees moving, resistance in walking
8	34-40	**Gale**	Waves 13-20 ft. of greater length, crests break, spindrift	Large trees in motion, small branches break
9	41-47	**Strong Gale**	High waves, 20+ ft., dense streaks of foam, spray reduces visibility	Slight structural damage, roof shingles may blow off, signs in motion
10	48-55	**Storm**	Very high waves, 20-30 ft., overhanging crests, lowered visibility, sea white with densely blown foam	Trees broken or uprooted, considerable structural damage, very high tides
11	56-63	**Violent Storm**	Exceptionally high waves, 30-45 ft., foam patches cover sea, visibility limited	Widespread damage, light structures in peril, coastal flooding
12	64+	**Hurricane**	Air filled with foam, waves 45+ ft., wind shrieks, sea white with spray, visibility poor	Storm surge at coast, serious beach erosion, extensive flooding, trees and wires down

NOTES:

- When the wind speed doubles, the pressure of the wind on an object *quadruples*. Example: the wind pressure at 40 kts.is *four times* what it is at 20 kts.
- In many tidal waters wave heights are apt to increase considerably in a very short time, and conditions can be more dangerous near land than in the open sea.

Hurricane SANDY — October 2012

Every now and then a storm comes along that defies prediction. In late October 2012, Sandy was in that rare category, for it had a trick up its sleeve.

When just slightly south of Jamaica on October 24, Tropical Storm Sandy became a hurricane with maximum sustained winds of 74 mph. Warm Caribbean waters further fueled its strength and it slammed into Cuba as a Category 3 hurricane. Weakening, it became a Category 1 and then a tropical storm on the 27th. Marching north-northeast parallel to the Atlantic coast, it regained strength and size, now with a record diameter of 1100 miles.

On October 29 Sandy, now a Category 1 storm, made its surprising and fatal move to the northwest, a very rare left turn, coming ashore just northeast of Atlantic City, NJ with winds of 80 mph. Its central pressure was about 27.8 inHg (945 mb), so low it is well off the scale of most barometers. It was not so much the wind strength that caused devastating damage as the combination of extra-high full moon tides and the T-bone approach that pushed a phalanx of ocean water into the vulnerable coast of New Jersey, New York City, Long Island and Connecticut. True to form, the right side of the storm path sustained the greatest damage.

The catastrophic storm surge was dramatic and deadly. Houses along the shore were demolished. On Long Island alone over 100,000 homes were severely damaged or destroyed. Yachts at marinas were thrown inland. Flooding was unprecedented. New Jersey recorded tide levels up to 9 feet above normal and severe coastal devastation. Staten Island and parts of Manhattan had flooding 4-9 feet above ground level. Connecticut reported similar tide surges. The storm quickly dissipated over land, with winds down to 30-35 mph on Oct. 31.

Only a Category 1 storm when it came ashore, Hurricane Sandy was responsible for about 150 deaths, 650,000 houses damaged or destroyed, and $50-$60 billion in damage. After Katrina, it was the second most costly storm in U.S. history.

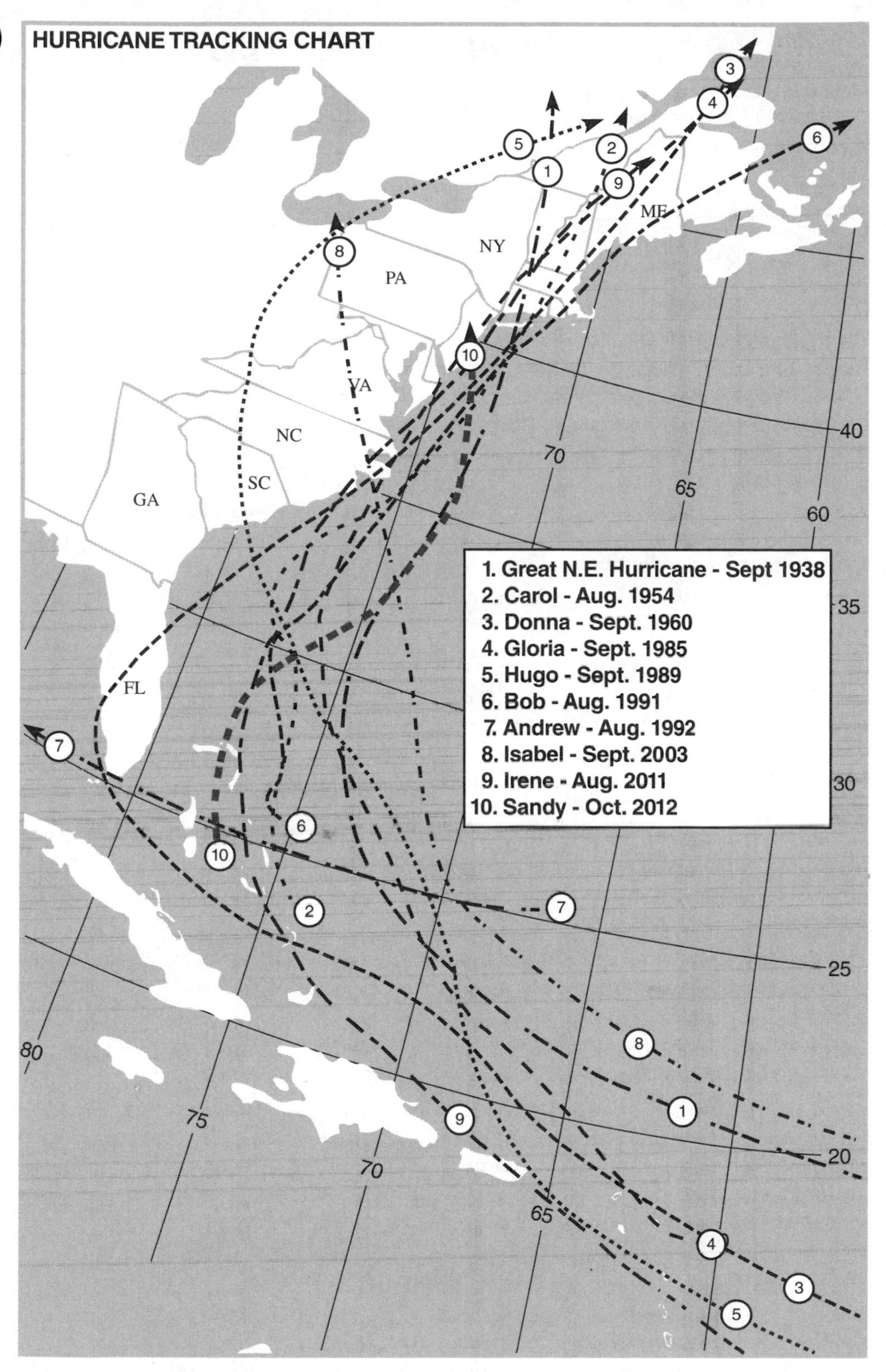

http://www.nhc.noaa.gov/

The Saffir-Simpson Hurricane Wind Scale

CATEGORY ONE: Winds 74-95 mph: Very dangerous winds will produce some damage. Falling or flying debris. Damage primarily to power lines, mobile homes, shopping center roofs, shrubbery, and trees. Also, some coastal road flooding and minor pier damage.

CATEGORY TWO: Winds 96-110 mph: Extremely dangerous winds will cause extensive damage. Some roofing material, door, and window damage to buildings. Considerable damage to vegetation, mobile homes, and piers. Small craft in unprotected areas break moorings. Near-total power outages. Some water systems fail.

CATEGORY THREE: Winds 111-130 mph: Devastating damage will occur. Some structural damage to small residences. Mobile homes destroyed. Many trees snapped or uprooted. Coastal flooding may extend inland, destroying smaller structures, damaging larger structures. Electricity and water may be unavailable for days or weeks.

CATEGORY FOUR: Winds 131-155 mph: Catastrophic damage will occur. More extensive failures including roofs on small residences. Major erosion of beach areas. Major damage to lower floors of structures near the shore. Power poles down. Terrain may be flooded well inland. Long-term water shortages.

CATEGORY FIVE: Winds greater than 155 mph: Catastrophic damage will occur. Complete roof failure on many residences and industrial buildings. Some complete building failures with small utility buildings destroyed. Major damage to most structures located near the shoreline. Massive evacuation of residential areas may be required. Most of the area will be uninhabitable for weeks or months.

The Effects of Storm Surge in Hurricanes

While high winds can be very destructive, it is often water that causes the greatest damage in a hurricane. Storm surge is the rise of water in a storm above predicted tide levels. When combined with high tide, storm surge can cause extreme flooding.

It is the swirling winds around a hurricane's eye, as well as the much smaller effect of low pressure, that produce storm surge. How great the surge is depends on a host of factors: storm intensity, forward speed, approach angle, and underwater topography. A steeply rising slope in the continental shelf will tend to minimize water levels, and a gradual slope will maximize them.

The weight of water, about 1700 pounds per cubic yard, when added to violent wave action, can cause extensive damage very quickly. When the battering takes place over a period of several or many hours, there can be devastation on a great scale. Beaches erode, seawalls crumble, structures are demolished, boats are smashed or hurled inland, and living creatures are in great peril. During Hurricane Katrina, the damage estimate is about $75 billion along the Gulf coast, with storm surge of 25-28 feet above normal tide levels.

With population density increasing along U.S. coasts and over half the nation's economic productivity located within coastal zones, it is no wonder that hurricanes in general, and storm surge in particular, receive such attention.

HURRICANES

For their awesome power to wreak havoc by wind and water, hurricanes have always been fascinating. Early warnings have all but eliminated surprise, yet these storms often defy attempts to prepare. Always vulnerable, we must know what to expect.

Hurricanes affecting the East Coast are born as tropical depressions in the Atlantic west of Africa, move westward through the eastern Caribbean, and eventually veer northwest and then north and northeast up our coast (see p. 251). Counter-clockwise winds spiral inward and accelerate toward the eye, the center of lowest pressure. The sharper the drop in pressure, the more violent the winds. Hurricanes lose power as they move north out of the tropics because warm ocean water, the energy source which helped create them, turns cooler.

A hurricane's forward motion, which can vary from 5 to 50+ knots, means that the winds are stronger on the right side. Winds of 100 knots spiraling around the eye, when you add a forward speed of 25 knots, create a speed of 125 knots on the right side, but only 75 knots on the left side, a dramatic difference. Note: a doubling of wind speed means the force on an object is increased four times, so that a wind of 100 knots has four times the power it does at 50 knots.

If the eye is moving directly toward you, the wind direction will remain fairly constant and the velocity will increase until the eye arrives. When the eye passes, the velocity will suddenly increase, rather stronger than before, from the opposite direction. These factors make the vicinity of the eye most dangerous.

In our diagram the hurricane is approaching, and vessels A, B, and C are at positions A1, B1, and C1 relative to the storm. When the storm passes, these vessels will be at positions A2, B2, and C2. Each will have experienced very different wind speeds and directions:

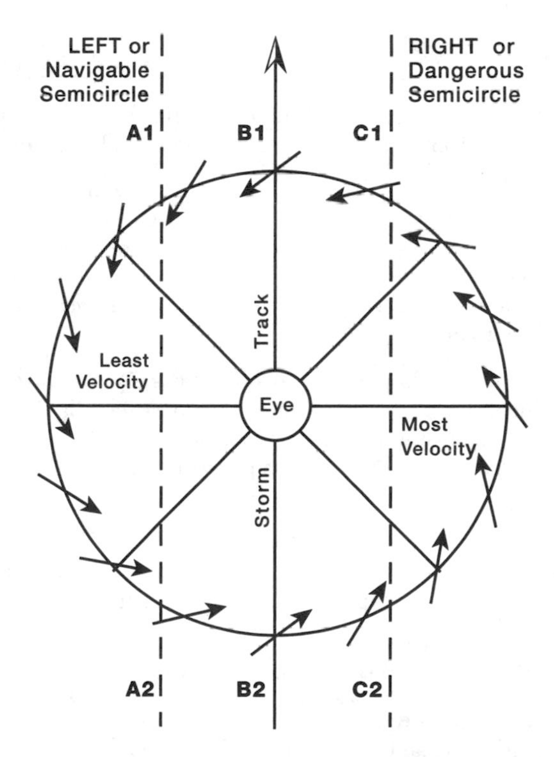

Vessel A, in the least dangerous semi-circle, will experience winds from the NE (at A1), backing to N (least velocity), NW, and finally W (at A2) as the storm passes.

Vessel B (at B1) will have ENE winds, increasing until the eye arrives. After the deceptive calm of the eye passes, the wind will rise, stronger than before, from the WSW (at B2), gradually decreasing.

Vessel C, in the most dangerous semi-circle, has the strongest winds, beginning (at C1) from the E, veering to SE, S (greatest velocity), and finally to SW (at C2).

If space and time permit, try to reduce your vulnerability by proceeding at right angles to the storm track. Which way to go depends on a number of factors, including how far away the storm is, its speed, the speed of the vessel, and sea conditions or sea room on either side of the expected storm path.

Hurricane Precautions Alongshore

Extremely high tides accompany hurricanes. If the storm arrives anywhere near the usual time of high water, low-lying areas will be flooded. Especially high tides will occur in all bays or V-mouth harbors if they are facing the wind direction. High water in all storm areas will remain a much longer time.

At or near the coastline, pull small craft well above the high water mark, dismast sailboats, remove outboard motors, and remove or lash down loose objects.

Seek the most protected anchorage possible, considering possible wind direction reversals, extreme tides, and other vessels. If on a mooring or at anchor, use maximum scope, allowing for room to swing clear of other boats. In a real blow it is easy to slack off, but not to shorten scope. Use as much chain in your anchor rode as possible. Another piece of chain or a weight, attached halfway along your mooring or anchor line, will help absorb sudden strains. Use chafing gear liberally at bits and bow chocks to minimize fraying of lines. Rig fenders to minimize damage from/to other boats.

Shut off gas, stove tanks, etc. Douse any fires in heating stoves. Secure all portholes, skylights, ventilators, hatch covers, companionways, etc. Pump the boat dry.

At a wharf or pier, use fenders liberally. If possible, rig one or more anchors abreast of the boat in the event the tide rises above pilings.

Boats are replaceable: don't wait for the last moment to get ashore!

Hurricane Precautions Offshore

Monitor storm reports on your radio. The U.S. Coast Guard warns all vessels offshore to seek shelter at least 72 hours ahead of a hurricane.

But, if caught offshore with no chance to reach shelter, watch the wind most carefully. First, note that if you face the wind, the eye is about 10 points or 112° to your right. If the wind "backs" (moves counter-clockwise), you are already on the less dangerous side of the storm track. If the wind "veers" (moves clockwise), you are on the right or more dangerous side. If the wind direction is constant, the hurricane's eye is headed directly at you, so make haste to get to the left side of the track.

Use your radiotelephone to advise the Coast Guard and other vessels of your position. Have a liferaft and safety equipment (flares, flashlights, EPIRB, etc.) ready. Put on life jackets. If it is impossible to hold your intended course, head your powerboat directly into the wind and sea, using only enough power to maintain steerageway. If power fails, rig a sea anchor or drogue to keep the bow to the wind.

Sailing vessels heaving to should consider doing so on the starboard tack (boom to port) in the more dangerous semicircle, or on the port tack (boom to starboard) in the less dangerous semicircle, to keep the wind drawing aft.

NOTE: This information is necessarily very general, the diagram (p. 253) is over-simplified, and the suggestions assume a straight storm track. If the storm track curves to the right, vessels A and B will have an easier time of it, but C may wind up on the track. The best advice: monitor weather reports continuously and seek shelter well ahead of time.

National Weather Service – www.weather.gov/
National Hurricane Center – www.nhc.noaa.gov
NOAA Hurricane Research – www.aoml.noaa.gov/hrd

WEATHER NOTES
From Maine to the Chesapeake

Sea Fog
There is always invisible moisture in the air, and the warmer the air, the more moisture it can contain invisibly. But when such a mass of moist air is cooled off, as it does when passing over a body of cooler water, the moisture often condenses into visible vapor, or fog. The fog clears when the air temperature rises, from the sun or a warm land mass, or by a warm, dry wind.

To predict fog accurately, you can use a "sling psychrometer." This instrument uses two thermometers side by side, one of which has a wick fastened to the bulb end. After wetting the wick on the "wet bulb" thermometer, the user swings the instrument in a circle for 60-90 seconds. This causes water to evaporate from the wet bulb thermometer, lowering its reading. The dry bulb thermometer simply tells air temperature. The difference in readings between the dry bulb and wet bulb thermometers determines the relative humidity of the air, and - especially valuable for determining the likelihood of fog - the dew point. The dew point is the (lower) temperature to which air must be cooled for condensation, or fog, to occur. (see Dew Point Table, p. 256)

Eastport, Maine to Cape Cod
Cold water (48°-55°) off the northern New England coast often causes heavy fog conditions in the spring and summer, when a warm moist southwesterly flow of air passes over it. East of Portland to the Bay of Fundy, fog is not apt to occur when the dew point is under 55°, unless there is a very warm moist wind. The effect of the cold water on the warm air is reduced if the winds become brisk, as they are apt to do in the afternoon. Visibility should then improve.

Long Island Sound and the New Jersey Coast
Summertime warm water (in the 70s) in this area rarely cools down any warm air mass enough to produce fog. This is not the case farther offshore, where cooler water temperatures (in the 50s) can produce fog.

On the south coast of Long Island, when the southwest wind blows toward the shore at the same time as the ebb tide, inlets can become dangerous with short, steep seas. Also, offshore swells can become very high near the mouths of inlets.

On the New Jersey shore, prevailing winds in summer are southerly, increasing in mid-morning to rarely more than 20 knots and usually dying down at dusk. Occasional summer thunderstorms can be expected. Any brisk winds from the east, northeast, or southeast can produce dangerous conditions along this lee shore and at the mouths of inlets. When the wind is from offshore like this, inlets should be entered on a flood tide.

At the mouth of Delaware Bay, seas can build up to a hazardous degree when there is a southeast wind at the same time as an ebb current at the mouth of the Bay.

Chesapeake Bay
There is little chance of fog in this region because of the warmth of the water. The Bay has quirks of its own in weather and sea conditions. It is a narrow and fairly shallow body of water, and winds tend to blow up or down it. Sharp seas can result, depending on the direction of the current and the wind. Opposing forces make for rough water.

Prevailing winds in spring and summer are southerly, freshening in the afternoon after a morning of calm. Summer thunderstorms occur frequently in afternoon and early evening, usually from the west. In the fall, after a cold front passes through, the winds will shift into the north or northeast, usually for three days, and increase in velocity, causing seas to build up. Calm follows for a day or so until the wind shifts to the southwest.

Dew Point and Humidity Afloat

Relative humidity (RH) is the measure of the air's capacity to hold water vapor at a certain temperature. At higher temperatures the air can hold more moisture: a 50%RH at 60°F is pleasant, but 50%RH at 80°F is unpleasant because the air is holding far more moisture. High humidity makes fog more likely and tends to make everyone uncomfortable.

Dew point is slightly different. It is the temperature to which air must be cooled for suspended (invisible) water vapor to condense into (visible) water. Fog and rain are examples. Dew on the deck in the morning means the night temperatures were low enough that the air could no longer hold all its daytime moisture. If the dew point and air temperature are quite close, then fog is more likely, and you might wait before heading to your next destination. Dew points under 55°F are comfortable, but those above 64°F are sticky to oppressive.

Marine weather forecasts often give dew point readings. If we want to measure dew point, there are some handheld digital instruments for under $100 which measure temperature, humidity, and dew point. If the outside temperature is 75°F and the dew point is in the 50s, go boating. If there's little spread between air temperature and dew point, you might want to postpone your voyage.

The table below is for those who have a sling psychrometer to measure dry-bulb and wet-bulb temperatures. A dry bulb reading of 70°F and a wet-bulb reading that is 4°F lower yields a dew point of 64°F, which means uncomfortably damp with possible fog. Periodic measurements which show an increase in the difference between dry- and wet-bulb readings mean fog should dissipate and visibility increase.

Sling Psychrometers and Hygrometers are available at www.robertwhite.com.

DEW POINT

Dry-bulb temp. F	Difference between dry-bulb and wet-bulb temperatures														Dry-bulb temp. F
	1°	2°	3°	4°	5°	6°	7°	8°	9°	10°	11°	12°	13°	14°	
+50	+48	+46	+44	+42	+40	+37	+35	+32	+29	+25	+21	+17	+12	+5	+50
52	50	48	46	44	42	40	37	35	32	29	25	21	17	11	52
54	52	50	49	47	44	42	40	37	35	32	28	25	21	16	54
56	54	53	51	49	47	45	42	40	37	35	32	28	25	21	56
58	56	55	53	51	49	47	45	43	40	38	35	32	28	25	58
+60	+58	+57	+55	+53	+51	+49	+47	+45	+43	+40	+38	+35	+32	+28	+60
62	60	59	57	55	54	52	50	48	45	43	41	38	35	32	62
64	62	61	59	57	56	54	52	50	48	46	43	41	38	35	64
66	64	63	61	60	58	56	54	52	50	48	46	44	41	39	66
68	67	65	63	62	60	58	57	55	53	51	49	46	44	42	68
+70	+69	+67	+66	+64	+62	+61	+59	+57	+55	+53	+51	+49	+47	+45	+70
72	71	69	68	66	64	63	61	59	58	56	54	52	50	47	72
74	73	71	70	68	67	65	63	62	60	58	56	54	52	50	74
76	75	73	72	70	69	67	66	64	62	61	59	57	55	53	76
78	77	75	74	72	71	69	68	66	65	63	61	59	57	55	78
+80	+79	+77	+76	+74	+73	+72	+70	+68	+67	+65	+64	+62	+60	+58	+80
82	81	79	78	77	75	74	72	71	69	67	66	64	62	61	82
84	83	81	80	79	77	76	74	73	71	70	68	67	65	63	84
86	85	83	82	81	79	78	76	75	74	72	70	69	67	66	86
88	87	85	84	83	81	80	79	77	76	74	73	71	70	68	88
+90	+89	+87	+86	+85	+84	+82	+81	+79	+78	+76	+75	+73	+72	+70	+90
92	91	89	88	87	86	84	83	82	80	79	77	76	74	73	92
94	93	92	90	89	88	86	85	84	82	81	79	78	76	75	94
96	95	94	92	91	90	88	87	86	84	83	82	80	79	77	96
98	97	96	94	93	92	91	89	88	87	85	84	82	81	80	98
+100	+99	+98	+96	+95	+94	+93	+91	+90	+89	+87	+86	+85	+83	+82	+100

Beachcombing

by Peter H. Spectre

First you have to learn the beach.

Okay, I confess. I'm a compulsive beachcomber. Colored glass is my principal target. I've spent hours searching a beach down the road from my home for what is generally known as sea glass. It's actually a mixed beach -- some sand, mostly small stones -- that years ago had been used by commercial fishermen to launch their small boats. Lots of bottles were broken on that beach, and over the years the action of the waves has broken the pieces smaller and smaller, and made the edges smoother and smoother.

I won't say where that beach is (monomaniacal beachcombers keep their secrets secret), but I will say that the best place to look isn't necessarily a beach; it could be the rugged shore around a lighthouse or a decommissioned lifesaving station. With no nearby sanitary disposal facility -- a.k.a., a town dump --the keepers and boat crews in the old days would toss their empty bottles overboard and let wave action go to work.

Some beaches are good for one color, others for another. Generally, I look for blue glass, green, white, and a rare translucent mauve, and pieces the size of my pinky fingernail and smaller. I don't look for brown, or big, honking pieces of any color. Why? I don't know. I just don't.

For a long time I didn't even know why I looked. There's a limit to what you can do with sea glass. Put it in bottles on the mantelpiece. Construct little crafty things. Make a set of earrings or a pendant for the true love. But then what?

I store my glass in coffee cans out in the workshop, and when the collection gets too overwhelming I give it away and start all over again.

One day I was searching my best beach, the one I think of as the Mother Lode. There was another fellow, a lone figure, way down where the sand ended and the ledge began. He was engaged in what I think of as the Sea Glass Shuffle -- head down, shoulders hunched, eyes glued to the ground, a slow, erratic walking motion with random starting and stopping and lots of stooping and studying. Eventually I caught up with him, because he was on my hot spot and that was my destination. (All beaches have a "good" side and a "bad" side, and at least one really hot spot, mostly the result of longshore currents.)

It didn't take me long after the usual "Howdy-do's" and "Nice days" and "How's it goings" to figure out that this fellow was a high priest of beachcombing. In no time we were talking about beaches we had trod without revealing where they were, and our particular color obsessions, and our theories about why this spot was hot and that one over there was not, and what we did with the glass we found (he had his own coffee cans), and then I asked him the fundamental question, the one with an elusive answer: Why do you do what you do?

"Therapy," he said. "It frees my mind.

Peter H. Spectre of Spruce Head, Maine, is a freelance writer and editor. This article originally appeared in Maine Boats, Homes & Harbors magazine and appears here with the author's permission.

What Time Is It?

by Jan Adkins

THIS VENERABLE BOOK has a heritage of practical value to mariners. In its yearly incarnations it has hurried commerce, kept sailors safe, and has offered assurance to those in little boats on the large ocean.

It seems like a book of technical certainties, scientific data, and mathematical inevitability. In great part this is true. But anyone exploring the science of time and tides must acknowledge that time is a fuzzy concept, and that tides move in strange ways.

All the tables in this book deal with time that was originally measured by three oscillators – separate clocks: the earth's daily spin, the earth's journey around the sun, and the phases of the moon. None of these oscillators pays a bit of attention to the other two.

We say our day is 24 hours. We call this a *solar day*. But earth revolves once every 23 hours 56 minutes and 4.091 seconds in relation to distant stars (a *sidereal day*). Our planetary orbit changes our angle to the sun about 1° every day, and this shift accounts for our noon-to-noon measure of 24 hours. Tell your paymaster that you're putting in an extra 3 minutes and 55 seconds.

But even this is not completely accurate, because the earth's orbit around the sun isn't a circle, but an ellipse with a difference of 3,107,000 miles (5,000,000 km) between *perhelion* (our closest approach to the sun) and *aphelion* (farthest). Newtonian mechanics insists that a planet describing an ellipse goes faster on the tighter curves, slower on the broad curves, and there goes accuracy. More bad news: the shape of the ellipse itself rotates around the sun. Add – or subtract – *precession* from the accuracy: the earth wobbles just a bit like a top. The length of our apparent days is the result of a jolt given to the earth long ago, knocking its *spin axis* 23.4° out of the *ecliptic* (the plane of our ellipse around the sun). In the summer the north pole points toward the sun, in the winter it points away, and our time in the sun is subject to the season and to our latitude. Disgusting.

What we blithely assume is a 24-hour day is actually an approximation, an average.

In the last century science determined that the big oscillators were simply too ragged for close work. In the 1930's careful astronomical observatories shifted from pendulum clocks to quartz crystal clocks with a frequency of 100,000 Hz. This improved accuracy to about .005 second a day. Most contemporary quartz wristwatches are set to 32,768 Hz and are accurate to about half a second a day.

Still not good enough. The measure of time has been determined by the frequency of Cesium 133 at 9,192,631,770 pulses a second. At this rate the Cesium atomic clock might lose or gain a second in 1,400,000 years.

This is a new age of time. High frequency oscillators and the electronic ability to count the pulses gives us the accuracy to measure the passage of radio waves between our GPS satellites and our position, to calculate the solid geometry, and to place ourselves within a meter or two on the heaving sea.

Although the *Eldridge Tide and Pilot Book* is conveniently notated yearly in months and even in Daylight Savings Time (a New Deal innovation of the 30's), science long ago gave up on months and even years. Scientific time is measured in (average, theoretical) days beginning at a relatively arbitrary point. The first day of 2015 will be TJD 17,023 (*Truncated Julian Day*, the updated, shorter version used by NASA).

But even with precise time, tide is a challenge of fantastic complexity. It's affected by so many elements! The main factors are the sun and the moon. The moon is close, about 270,000 miles away. The sun is 93,000,000 miles away (give or take, see above) but enormously larger. The tidal influence of the moon is slightly more than twice that of the sun. These forces have their own positional geometry, however: they can multiply effects for a *spring tide* or nearly cancel effects for a *neap tide*. These gravitational forces also cause a "land tide" – the earth's exterior shell buckles up and down about 12 inches twice a day. But the ocean is a fluid, at the whim of hydrography, shore geography, ocean currents, the Coriolis force due to Earth's spin, the constrictions of shores and rivers, and even wind. A reliable tide chart can't be generated by a simple formula. The tables in this small book are miracles of long empirical observation, intelligent estimation, and mathematical calculation. Over many years we've achieved a remarkable accuracy.

Professor Einstein established the curious fact that time for objects traveling near the speed of light is elastic. But we're simple mariners moving at antique speeds. We keep our time as well as we can, we reckon currents and wind and a bit of intuition in our courses. We're closer to Captain Eldridge than to Captain Kirk.

A version of this article, which the author adapted for Eldridge, first appeared in Maine Boats, Homes & Harbors.

Jan Adkins, author of many articles and books, is a frequent contributer to Eldridge, with hand-drawn graphics and entertaining articles.

FEDERAL POLLUTION REGULATIONS

Revisions to MARPOL V (Marine Pollution) regulations have set stricter standards, prohibiting the disposal of almost all kinds of garbage at sea.

Federal Regulations for Waste Disposal

Prohibited in all waters: The discharge of plastic or garbage mixed with plastic, including synthetic ropes, fishing nets, plastic bags, bottles, paper, cardboard, animal fat, cooking oil, cleaning agents, dunnage, packing materials.

Prohibited within 3 n.m. of land: The discharge of any garbage or ground food waste.

Permitted beyond 3 n.m. of land: Food waste must be ground up to less than 1 inch.

Prohibited within 12 n.m. of land: The discharge of unground garbage larger than 1 inch.

Permitted beyond 12 n.m. of land: Unground food waste.

The Damage Caused by Pollution

Sewage is not just a repulsive visual pollutant. The microorganisms in sewage, including pathogens and bacteria, degrade water quality by introducing diseases like hepatitis, cholera, typhoid fever and gastroenteritis, which can contaminate shellfish beds. Shellfish are filter feeders that eat tiny food particles filtered through their gills into their stomachs, along with bacteria from sewage. Nearly all waterborne pathogens can be conveyed by shellfish to humans.

Marine Sanitation Devices (MSD)

Vessels under 65' may install type I, II or III MSD. Vessels over 65' must install a type II or III MSD. All installed MSD's must be U. S. Coast Guard certified.

- **Type I** MSDs are allowed only on vessels under 65'. They treat sewage with disinfectant chemicals before discharge. The discharge must not show any visible floating solids, and must have a fecal coliform bacterial count not greater than 1000 per 100 milliliters of water.
- **Type II** MSDs are allowed on vessels of any length. They provide a higher level of treatment than Type I, using greater levels of chemicals to create effuent having less than 200 per 100 milliliters and suspended solids not greater than 150 milligrams per liter.
- **Type III** MSDs are allowed on vessels of any length. They do not allow discharge of sewage, except through a Y-valve to discharge at a pumpout facility, or overboard when outside the 3 nautical miles. They include holding tanks, recirculating and incinerating units.
- **Portable toilets** or "porta-potties" are not considered installed toilets and are not subject to MSD regulations. They are, however, subject to the disposal regulations which prohibit the disposal of raw sewage within the three-mile limit or territorial waters of the U.S.

No Discharge Areas (NDAs)

NDAs are water bodies where the Environmental Protection Agency (EPA) and local communities prohibit the discharge of all vessel sewage. Many States are adding NDAs. **It is the boater's responsibility to be aware of where those NDAs are.** For NDAs by state: http://water.epa.gov/polwaste/vwd/vsdnozone.cfm. See p. 261 for Pumpout Information.

When operating vessel in NDAs, the operator must secure each Type I or Type II MSD in a manner which prevents discharge of treated or untreated sewage.

Type III MSDs, or holding tanks, must also be secured in a manner that prevents discharge of sewage. Acceptable methods of securing the device include: closing appropriate valves, removing the handle, padlocking each valve, or using a non-reusable wire-tie to hold each valve in a closed position. Sewage held in Type III MSDs can be removed by making arrangements with pumpout stations or pumpout boats. Call Harbormaster for details.

Pumpout Information - State Sources

Please be sure to call or radio in advance for rates and availability. While we have taken all possible care in compiling this list, changes may have occurred and we cannot guarantee accuracy. For more current information check the state website or call the agency listed. See p. 260 for Federal Pollution Regulations and No Discharge Areas (NDAs)

Look for [CVA symbol] Clean Vessel Act (CVA):

http://wsfrprograms.fws.gov/Subpages/GrantPrograms/CVA/CVA.htm

Most major harbors now have a pumpout boat.
Contact the local Harbormaster. Many monitor VHF channel 09.

MAINE: ME Dept. of Environ. Protection, 207-485-3038
www.maine.gov/dep/water/wd/vessel/pumpout/index.html

NEW HAMPSHIRE: NH Environ. Serv., 603-670-5130
des.nh.gov/organization/divisions/water/wmb/cva/dir_map.htm

MASSACHUSETTS: MA Coastal Zone Mgmt., 617-626-1200
www.mass.gov/eea/agencies/czm/program-areas/coastal-water-quality/clean-boating/pumpout-list.html

RHODE ISLAND: RI Environmental Mgmt., 401-222-6800
www.dem.ri.gov/programs/benviron/water/shellfsh/pump/index.htm

CONNECTICUT: CT Environ. Protection, 860-424-3034, 860-424-3652
www.ct.gov/deep/cwp/view.asp?A=2705&Q=323708

NEW YORK: NY State Environmental Facilities Corp., 518-402-7461
www.efc.ny.gov/cvap

NEW JERSEY: NJ Fish & Wildlife, 732-872-1300 ext. 29
NJBoating.org
Due to Superstorm Sandy, boaters should contact marinas in advance of their first visit to confirm the pumpout facility or ramp is operational for the season.

DELAWARE: DE Fish & Wildlife, 302-739-9915
www.dnrec.delaware.gov/p2/Pages/PumpoutStations.aspx

MARYLAND: MD Natural Resources, 410-260-8772
dnr.maryland.gov/boating/pumpout/

VIRGINIA: VA Dept. of Health, 804-864-7468
www.vdh.virginia.gov/EnvironmentalHealth/Onsite/MARINA/

NORTH CAROLINA: NC Div. of Coastal Management, 888-472-6278
www.nccoastalmanagement.net/marinas/pumplist.htm

SOUTH CAROLINA: SC Dept. of Health & Env. Control, 843-953-9062
www.dnr.sc.gov/marine/vessel/index.html

GEORGIA: Georga Dept. of Natural Resources, 912-280-6926, contact local marinas

FLORIDA: FL Dept. of Environmental Protection, 850-245-2100
www.dep.state.fl.us/cleanmarina/CVA/default.htm

Got a Minute?
Angular and Linear Equivalents

Whether you are navigating purely by GPS or using a paper chart, it can be helpful to know how degrees, minutes, and seconds – or tenths or hundredths of a minute – translate into linear distance on the water. Knowing both is important because your GPS can display part of a coordinate as 41° 23' 25", or as 41° 23.42', where each is correct, but one is more accurate.

First, the basics. Latitude is the angular distance north or south of the Equator, and the parallels are equidistant. The latitude scale appears on the vertical edges of your chart. (Longitude, measured east and west of Greenwich and appearing along the top and bottom edges of your chart, is never used for distance measurement.) For practical purposes, the distance between parallels of latitude which are one degree (1°) apart is 60 nautical miles (n.m.).

- 1° (degree) = 60 nautical miles (Ex: from 42° North to 43° North is 60 n.m.)
- 1' (minute, or 1/60th of a degree) = 1 n.m., or 6076 feet)
- 1" (second, or 1/60th of a minute) = 101.3 feet (acceptable for general purposes)

The U.S. Coast Guard gives positions of buoys, lights, and lighthouses in degrees, minutes, and seconds, or within roughly 100 feet. (See pp. 166-197).

Sometimes minutes are divided into tenths or hundredths instead of seconds.

- 1' (minute) = 1 n.m., or 6076 feet
- 0.1' (1/10th of a minute) = 608 feet (acceptable tolerance at sea; not so near shore)
- 0.01' (1/100th of a minute) = 61 feet (acceptable for almost any purpose)

Use the Table below to convert seconds to tenths or hundredths of a minute.

Table for Converting Seconds to Decimals of a Minute

From many sources, including charts, Light Lists, and Notices to Mariners, positions are in degrees, minutes, and seconds. These are written either 34° 54' 24" or 34-54-24

However, for navigating with GPS, Loran, chart plotters, and celestial calculators, it can be useful to convert the last increment – seconds – to either tenths or hundredths of a minute. The numbers above become 34° 54.40' or 34-54.4'

Secs.	Tenths	Hundredths	Secs.	Tenths	Hundredths	Secs.	Tenths	Hundredths
1	.0	.02	**21**	.4	.35	**41**	.7	.68
2	.0	.03	**22**	.4	.37	**42**	.7	.70
3	.1	.05	**23**	.4	.38	**43**	.7	.72
4	.1	.07	**24**	.4	.40	**44**	.7	.73
5	.1	.08	**25**	.4	.42	**45**	.8	.75
6	.1	.10	**26**	.4	.43	**46**	.8	.77
7	.1	.12	**27**	.5	.45	**47**	.8	.78
8	.1	.13	**28**	.5	.47	**48**	.8	.80
9	.2	.15	**29**	.5	.48	**49**	.8	.82
10	.2	.17	**30**	.5	.50	**50**	.8	.83
11	.2	.18	**31**	.5	.52	**51**	.9	.85
12	.2	.20	**32**	.5	.53	**52**	.9	.87
13	.2	.22	**33**	.6	.55	**53**	.9	.88
14	.2	.23	**34**	.6	.57	**54**	.9	.90
15	.3	.25	**35**	.6	.58	**55**	.9	.92
16	.3	.27	**36**	.6	.60	**56**	.9	.93
17	.3	.28	**37**	.6	.62	**57**	1.0	.95
18	.3	.30	**38**	.6	.63	**58**	1.0	.97
19	.3	.32	**39**	.7	.65	**59**	1.0	.98
20	.3	.33	**40**	.7	.67	**60**	1.0	1.00

TABLE OF EQUIVALENTS
and other useful information

Length

English	Metric
1 inch	2.54 centimeters
1 foot	.30 meters
1 fathom	1.61 meters
1 statute mile	1.61 kilometers
1 nautical mile	1.85 kilometers

Metric	English
1 meter	39.37 inches
"	3.28 feet
"	.55 fathoms
1 kilometer	.62 statute miles
"	.54 nautical miles

Nautical	Terrestrial
1 fathom	6 feet
1 cable	608 feet
1 nautical mile	6076 feet
"	1.15 statute miles
1 knot	1.15 mph
7 knots	8 mph approx.

Capacity

English	Metric
1 quart	.95 liters
1 gallon	3.78 liters

Metric	English
1 liter	1.06 quarts
"	.26 US gallons

Weight

English	Metric
1 ounce	28.35 grams
1 pound	.45 kilograms
1 US ton	.907 metric tons
"	.893 long tons

Metric	English
1 gram	.035 ounces
1 kilogram	2.20 pounds
1 metric ton	2204.6 pounds

Weight of 1 US Gallon

Gasoline	6 pounds
Diesel fuel	7 pounds
Fresh water	8.3 pounds
Salt water	8.5 pounds

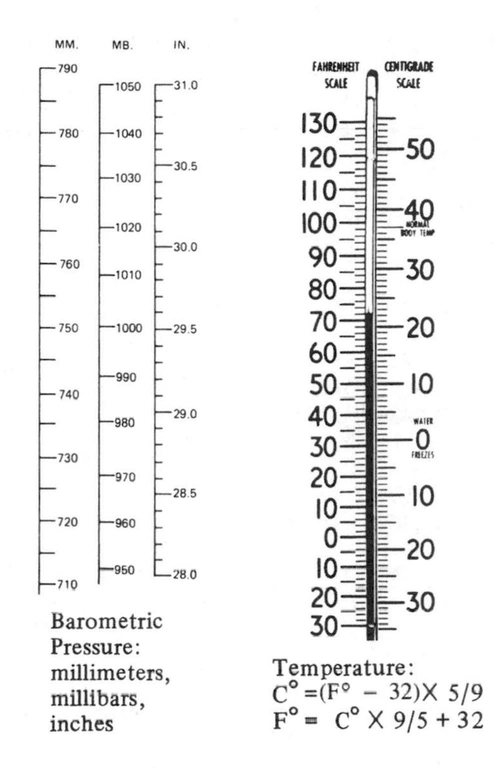

Barometric Pressure: millimeters, millibars, inches

Temperature:
C° =(F° − 32)X 5/9
F° = C° X 9/5 + 32

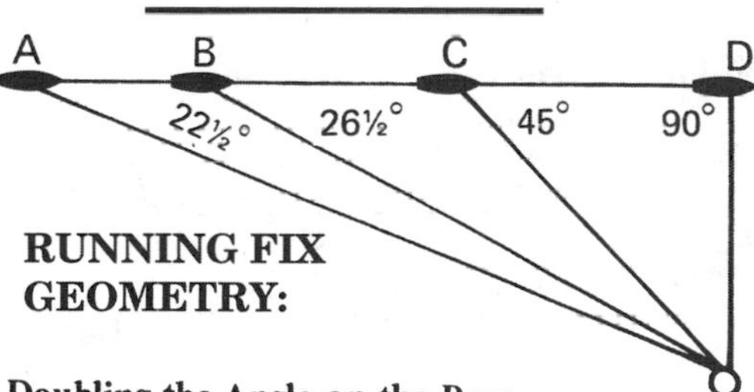

RUNNING FIX GEOMETRY:

Doubling the Angle on the Bow

1. Angle DCO = 45°; Angle CDO = 90°; True distance run (CD) = distance DO.
2. Angle DAO = 22½°; Angle DCO = 45°; True distance run (AC) = Distance CO.

Other Useful Bow Bearings

3. Angle DBO = 26½°; Angle DCO = 45°; True distance run (BC) = distance DO.
4. Example 3 also works with angles of 25° and 41°; 32° and 59°; 35° and 67°; 37° and 72° when distance run will be distance DO.

CRUISING CHARTS

DESIRABLE CRUISING CHARTS FROM CAPE BRETON I. TO KEY WEST, FL.

Numbers listed to the left are general coastal charts. Indented numbers refer to harbor charts. For USCG Local Notices to Mariners for Critical Chart updates: www.nauticalcharts.noaa.gov/mcd/updates/LNM_NM.html; for Canadian Notices to Mariners: www.notmar.gc.ca/

1:80(000), 1:40(000), etc. indicates scale

Canada

4013 Halifax to Sydney 1:350
4279 Bras d'Or Lake 1:60
4447 Pomquet and Tracadie Harbours 1:25
4385 Chebucto Hd. to Betty Is. 1:39
4335 Strait of Canso and Approaches 1:75
4321 Cape Canso to Liscomb Is. 1:108.8
4227 Country Hbr. to Ship Hbr. 1:50
4320 Egg Is. to W. Ironbound Is. 1:145
4012 Yarmouth to Halifax 1:300
4386 St. Margaret's Bay 1:39.4
4381 Mahone Bay 1:38.9
4384 Pearl Is. to Cape LaHave 1:39
4211 Cape LaHave to Liverpool Bay 1:37.5
4230 Little Hope Is. to Cape St. Mary's 1:50
4240 Liverpool Hbr. to Lockeport Hbr. 1:60
4241 Lockeport to Cape Sable 1:60
4242 Cape Sable to Tusket Is. 1:60
4243 Tusket Is. to Cape St. Marys 1:60
4010 Bay of Fundy (inner portion) 1:200
4011 Approaches to Bay of Fundy 1:300
4118 St. Marys Bay 1:60
4396 Annapolis Basin 1:24
4116 Approaches to St. John 1:60
4340 Grand Manan 1:60

U.S. East Coast

13325 Quoddy Narrows to Petit Manan Is. 1:80
13312 Frenchman & Blue Hill Bays & apprs. 1:80
  13315 Deer Is. Thoro. and Casco Pass. 1:20
13302 Penobscot Bay and apprs. 1:80
  13308 Fox Islands Thorofare 1:15
13288 Monhegan Is. to Cape Elizabeth 1:80
  13290 Casco Bay 1:40
13286 Cape Elizabeth to Portsmouth 1:80
  13283 Cape Neddick Hbr. to Isles of Shoals 1:20, Portsmouth Hbr. 1:10
13278 Portsmouth to Cape Ann 1:80, Hampton Harbor 1:30
  13281 Gloucester Hbr. and Annisquam R. 1:10
13267 Massachusetts Bay 1:80
  13275 Salem and Lynn Harbors 1:25, Manchester Harbor 1:10
  13276 Salem, Marblehead & Beverly Hbrs. 1:10
  13270 Boston Harbor 1:25
13246 Cape Cod Bay 1:80
  13253 Plymouth, Kingston and Duxbury Hbrs. 1:20, Green Hbr. 1:10
  13236 Cape Cod Canal and approaches 1:20
13237 Nantucket Sound and approaches 1:80
  13241 Nantucket Island 1:40
  13242 Nantucket Harbor 1:10
13218 Martha's Vineyard to Block Island 1:80
  13230 Buzzards Bay 1:40, Quicks Hole 1:20
  13233 Martha's Vineyard 1:40, Menemsha Pond 1:20
  13221 Narragansett Bay 1:40
  13219 Point Judith Harbor 1:15
13205 Block Island Sound and apprs. 1:80
  13217 Block Island 1:15
  13209 Block Is. Sd. & Gardiners Bay, Long Is., 1:40
  13214 Fishers Island Sound 1:20
  13212 Approaches to New London Hbr. 1:20
  13213 New London Harbor and Vicinity 1:10, Bailey Point to Smith Cove 1:5
  13211 North Shore of Long Is. Sd.-Niantic Bay & Vicinity 1:20
12354 Long Island Sound - eastern part 1:80
  12375 Connecticut R. -Long Is. Sd. to Deep R. 1:20
  12374 Duck Island to Madison Reef 1:20
  12373 Guilford Hbr to Farm R. 1:20
  12371 New Haven Harbor 1:20
  12370 Housatonic R. and Milford Hbr. 1:20
  12362 Port Jefferson & Mt. Sinai Hbrs. 1:10
12363 Long Island Sound - western part 1:80
  12369 Stratford to Sherwood Pt. 1:20
  12368 Sherwood Pt. to Stamford Hbr. 1:20
  12367 Greenwich Pt. to New Rochelle 1:20
  12366 L.I. Sd. and East R., Hempstead Hbr. to Tallman Is. 1:20
  12365 L.I. Sd. S. Shore, Oyster and Huntington Bays 1:20
12353 Shinnecock Light to Fire Island Light 1:80
  12352 Shinnecock B. to E. Rockaway In. 1:20; 1:40
  12339 East R. - Tallman I. to Queensboro Br. 1:10
  12331 Raritan Bay and Southern Part of Arthur Kill 1:15

12327 New York Harbor 1:40
12335 Hudson & E. Rs. - Governors I. to 67 St. 1:10
12326 Appr. to N.Y., Fire I. to Sea Girt 1:80
12350 Jamaica Bay and Rockaway In. 1:20
12323 Sea Girt to Little Egg In. 1:80
12324 Sandy Hook to Little Egg Harbor 1:40
12318 Little Egg In. to Hereford In. 1:80, Absecon In. 1:20
12316 Little Egg Harbor to Cape May 1:40
12304 Delaware Bay 1:80
12311 Delaware R.- Smyrna R. to Wilmington 1:40
12312 Wilmington to Philadelphia 1:40
12277 Chesapeake and Delaware Canal 1:20
12214 Cape May to Fenwick I. 1:80
12211 Fenwick I. to Chincoteague In.1:80, Ocean City In. 1:20
12210 Chincoteague In. to Great Machipongo In. 1:80, Chincoteague In. 1:20
12221 Chesapeake Bay Entrance 1:80
12222 Cape Charles to Norfolk Hbr. 1:40
12224 Cape Charles to Wolf Trap 1:40
12256 Chesapeake Bay-Thimble Shoal Channel 1:20
12225 Wolf Trap to Smith Point 1:80
12228 Pocomoke and Tangier Sds. 1:40
12230 Smith Point to Cove Point 1:80
12231 Tangier Sd.-northern part 1:40
12233 Chesapeake Bay to Piney Pt. 1:40
12285 Potomac River, DC 1:80, DC 1:20
12286 Piney Pt. to Lower Cedar Pt. 1:40
12288 Lower Cedar Pt. to Mattawoman Cr. 1:40
12289 Mattawoman Cr. to Georgetown 1:40; Washington Hbr. 1:20
12263 Cove Point to Sandy Point 1:80
12282 Severn and Magothy Rs. 1:25
12273 Sandy Point to Susquehanna River 1:80
12274 Head of Chesapeake Bay 1:40
12278 Appr. to Baltimore Harbor 1:40
12207 Cape Henry to Currituck Bch. Lt. 1:80
12253 Norfolk Hbr. and Elizabeth R. 1:20
12254 Cape Henry to Thimble Shoal Lt. 1:20
12245 Hampton Roads 1:20
12205 Cape Henry to Pamlico Sd. incl. Albemarle Sd. 1:40; 1:80
12204 Currituck Beach Lt. to Wimble Shoals 1:80
11555 Cape Hatteras-Wimble Shoals to Ocracoke In. 1:80
11548 Pamlico Sd.-western part 1:80
11550 Ocracoke In. and N. Core Sd. 1:40
11544 Portsmouth I. to Beaufort incl. Cape Lookout Shoals 1:80
11545 Beaufort In. and S. Core Sd. 1:40, Lookout Bight 1:20
11543 Cape Lookout to New R. 1:80
11539 New R. In. to Cape Fear 1:80
11536 Appr. to Cape Fear R. 1:80
11535 Little R. In. to Winyah Bay Entr.1:80
11531 Winyah Bay to Bulls Bay 1:80
11532 Winyah Bay 1:40
11521 Charleston Hbr. & Appr. 1:80
11513 St. Helena Sd. to Savanna R. 1:80
11509 Tybee I. to Doboy Sd. 1:80
11502 Doboy Sd. to Fernandina 1:80
11488 Amelia I. to St. Augustine 1:80
11486 St. Augustine Lt. to Ponce de Leon In. 1:80
11484 Ponce de Leon In. to Cape Canaveral 1:80
11476 Cape Canaveral to Bethel Shoal 1:80
11474 Bethel Shoal to Jupiter In. 1:80
11466 Jupiter In. to Fowey Rocks 1:80, Lake Worth In. 1:10
11469 Straits of FL.Fowey Rks., Hillsboro Inlet to Bimini Is. Bahamas 1:100
11462 Fowey Rocks to Alligator Reef 1:80
11452 Alligator Reef to Sombrero Key 1:80
11442 Florida Keys - Sombrero Key to Sand Key 1:80
11439 Sand Key to Rebecca Shoal 1:80
11438 Dry Tortugas 1:30

To find your nearest Canadian Chart Agent and and chart corrections: www.charts.gc.ca

To find your nearest NOAA Chart Agent: http://aeronav.faa.gov/agents.asp

Print-on-Demand Nautical Charts for up to date NOAA charts: www.OceanGrafix.com

NOAA has posted all 1000+ of its US Nautical charts on the internet. The charts can be viewed using any internet browser. Each chart is up-to-date with the most recent Notices to Mariners. Use these online charts as a ready reference or planning tool, not for actual navigation. Online charts can be viewed at: www.nauticalcharts.noaa.gov/mcd/OnLineViewer.html

Ferry Service Information

🚗 Vehicle reservations may be required

CANADA

Bay ferries 🚗, between St. John, NB and Digby, NS, (877) 762-7245. www.nfl-bay.com/

MAINE

-- *For all Maine Ferry Service information:* www.exploremaine.org/ferry/

Maine State Ferry Service, Rockland ME, (207) 596-5400.

General Schedule Information www.maine.gov/mdot/msfs/

Frenchboro Ferry 🚗 between Frenchboro and Bass Harbor, Bass Harbor (207) 244-3254

Islesboro Ferry 🚗 between Islesboro (207) 734-6935 and Lincolnville (207) 789-5611

Matinicus Island Ferry 🚗 between Matinicus Island and Rockland (207) 596-5400

North Haven Ferry 🚗 between North Haven (207) 596-5400 and Rockland (207) 596-5400

Vinalhaven Ferry 🚗 between Vinalhaven (207) 596-5400 and Rockland (207) 596-5400

Monhegan Island Ferry between Monhegan Island and Port Clyde (207) 372-8848. www.monheganboat.com

MASSACHUSETTS

Steamship Authority 🚗 www.steamshipauthority.com/ssa/

Between Woods Hole, MA (508) 548-3788 (information), (508) 477-8600 (car reservation) and Martha's Vineyard (508) 693-9130

Between Hyannis and Nantucket. Hyannis (508) 771-4000

Fast Ferry (508) 495-3278 (passenger reservation), Nantucket (508) 228-0262

Island Queen Ferry between Falmouth and Oak Bluffs, M.V. (508) 548-4800. www.islandqueen.com

SeaStreak between New Bedford and Martha's Vineyard (800) 262-8743. www.mvexpressferry.com/

Hy-Line Cruises, Nantucket, Martha's Vineyard, Hyannis loop (800) 492-8082. www.hy-linecruises.com/

-- *For more Martha's Vineyard ferry information: www.mvol.com/directory/transportation/Ferries/*

Freedom Cruise Line between Harwichport and Nantucket (508) 432-8999. www.nantucketislandferry.com

Cuttyhunk Is. Ferry between Cuttyhunk and New Bedford (508) 992-0200. www.cuttyhunkferryco.com/

RHODE ISLAND

Vineyard Fast Ferry between Quonsett, RI and M.V. (401) 295-4040. www.vineyardfastferry.com

Block Island Ferry 🚗 and High Speed Ferry between Block Island and Jerusalem (Pt. Judith) (866) 783-7996. www.blockislandferry.com/

Jamestown-Newport Ferry (401) 423-9900. www.jamestownnewportferry.com/

CONNECTICUT

Cross Sound Ferry 🚗 between New London, CT (860) 443-5281, Orient Pt., L.I., NY (631) 323-2525. www.longislandferry.com

Block Is. Express Ferry between New London (860) 444-4624 and Block Is. (401) 466-2212. www.goblockisland.com

Fishers Island Ferry 🚗 between Fishers Island, NY (631) 788-7744 and New London, CT (860) 442-0165. Car reservations must be made online only: www.fiferry.com

Bridgeport-Port Jefferson Ferry 🚗 between Port Jefferson, NY (631) 473-0286 and Bridgeport, CT (888) 443-3779. www.88844ferry.com

NEW YORK

Long Island *is served by two year-round ferry lines that cross Long Island Sound connecting Port Jefferson to Bridgeport, CT and Orient Point to New London, CT. For more information for ferries in Long Island Sound (Fire Island, Shelter Island, etc.)*: www.webscope.com/li/ferries.html

Seastreak serving NY, NJ, Martha's Vineyard, MA. (800) 262-8743. http://seastreak.com/

NEW JERSEY

Cape May-Lewes Ferry 🚗 between Cape May, NJ (800) 643-3779 and Lewes, DE., www.capemaylewesferry.com

NORTH CAROLINA

-- *To request all NC ferry routes and schedules (800) 293-3779 or download at* www.ncferry.org, including **Hatteras Inlet Ferry** 🚗 between Hatteras and Ocracoke;

Cedar Island and Swan Quarter Ferry between Cedar Island and Swan Quarter

Where To Buy The Eldridge Tide and Pilot Book

CANADA

NOVA SCOTIA

Halifax
Binnacle Yachting Equip.

ONTARIO

Toronto
Nautical Mind Bookstore

UNITED STATES

MAINE

Bar Harbor
Sherman's Book Store
Bath
Bath Bookshop
Blue Hill
Blue Hill Books
Boothbay
Sherman's Book Store
Brooksville
Buck's Harbor Marine
Camden
Owl and Turtle Bookshop
Sherman's of Camden
Damariscotta
Maine Coast Book Shop
Freeport
Sherman's Book Store
Kittery
Jackson Hardware
Northeast Harbor
F.T. Brown Co.
Portland
Chase Leavitt
Hamilton Marine
West Marine
Rockland
Reading Corner
Searsport
Hamilton Marine
South Freeport
Brewer's Yacht Yard
Stonington
Billings Diesel & Marine
Yarmouth
Landing Boat Supply

NEW HAMPSHIRE

Keene
Toadstool Bookshop
Portsmouth
West Marine
Seabrook
West Marine

MASSACHUSETTS

Beverly Farms
The Bookshop
Boston
Boston Hbr. Sailing Club
Boston Sailing Center
Boxell's Chandlery
Braintree
West Marine
Buzzards Bay
Red Top Sporting Goods
Cataumet
Kingman Yachting Center **
Parker's Boat Yard **
Chatham
Mayflower Shop
Stage Harbor Marine
Yellow Umbrella Books
Cohasset
Buttonwood Books
Concord
Concord Book Shop
Cotuit
Peck's Boats
Cuttyhunk
Island Market
Danvers
West Marine
Dedham
West Marine
Duxbury
Bayside Marine Corp.
East Sandwich
Titcomb's Bookshop
Edgartown
Edgartown Books
Edgartown Marine
Fairhaven
West Marine
Falmouth
Booksmith-Falmouth Plaza
Eastman's Sport & Tackle
MacDougalls **
West Marine
Gloucester
The Bookstore
Harbor Loop Gifts
Hanover
Sylvester & Co.
Harwichport
Allen Harbor Marine Serv.
Hingham
RNR Marine
Hyannis
Sports Port
West Marine
Marblehead
The Forepeak
F. L. Woods
Marblehead Outfitters
Spirit of '76 Bookstore
West Marine
Marion
Book Stall
Burr Bros. Boats **
Mattapoisett
Town Wharf General Store
Nantucket
Mitchell's Book Corner
Nantucket Book Works
Nantucket Ship Chandlery
New Bedford
Bay Fuels, Inc.
C.E. Beckman Co.
CMS Enterprises Inc.
Hercules SLR
Lighthouse Marine Supply
Luzo Fishing Gear
New Bedford Ship Supply
West Marine
Newton
Charles River Canoe & Kayak
North Dartmouth
Baker Books
North Falmouth
N. Fal. Hardware & Marine
Oak Bluffs
Dick's Bait and Tackle
Orleans
Goose Hummock Shop
Nauset Marine
Peabody
West Marine
Plymouth
West Marine
Provincetown
Land's End Mar. Supply
Raynham
Slip's Capeway Marine
Rockport
Toad Hall Bookstore
Sandwich
Sandwich Ship Supply
Scituate
Front St. Bookshop
Front St. Marine
Seekonk
West Marine
South Dartmouth
Cape Yachts
Concordia Co. **
South Yarmouth
Riverview Bait & Tackle
Swansea
Newsbreak, Inc.
Vineyard Haven
Bunch of Grapes Book Store
Gannon & Benjamin Marine
Martha's Vineyard Fuel & Ice
Martha's Vineyard Shipyard **
West Marine
Wakefield
Boats and Motors
West Dennis
Sportsman's Landing
Westport
Partners Village Store
Weymouth
Monahan's Marine
Winthrop
Woodside Hardware
Woburn
West Marine

RHODE ISLAND

Barrington
Barrington Books
Brewer Cove Haven Marina **

Bristol
Herreshoff Museum
Jamestown Distributors
East Greenwich
West Marine
Jamestown
Conanicut Marine
Jamestown Boat Yard
Narragansett
R.I. Engine Co.
West Marine
Newport
Newport Nautical Supply
NV.Charts
West Marine
Portsmouth
Ship's Store & Rigging
Providence
New England Marine Supply
Warren
West Marine
Wakefield
Ram Point Marina
Snug Harbor Marina

CONNECTICUT

Branford
West Marine
Clinton
Riverside Basin Marina
West Marine
Deep River
Brewer Deep River Marina
Essex
Boatique
Fairfield
West Marine
Guilford
Breakwater Books
Madison
R. J. Julia Booksellers
Milford
Ship's Store@MilfordBoat-Works
Mystic
Bank Square Books
Brewer Yacht Yard
Mystic Seaport Stores
West Marine
New London
West Marine
Noank
Spicer's Marinas
Norwalk
West Marine
Old Lyme
Emerson & Cook Book Co.
Kellog Marine Supply
Old Saybrook
River's End Tackle
West Marine
Portland
William J. Petzold Inc.
Portland Boat Works
South Norwalk
Rex Marine Center
Stamford
Brewer Yacht Haven
Hathaway Reiser & Raymond
Landfall**
West Marine
Stonington
Wilcox Marine Supply
Waterford
Defender Industries
Hillyer's Tackle Shop
Westbrook
Pilots Point Marine

NEW YORK

Babylon
West Marine
Brooklyn
Bernie's Fishing Tackle
Connelly
Rondout Yacht Basin
East Hampton
Seacoast Enterprises
Three Mile Harbor Boat Yard
Fisher's Island
Pirate's Cove Marine
Freeport
Freeport Marine Supply
Garden City
West Marine
Glen Cove
Brewer Glen Cove Marina
Greenport
Brewer Stirling Hbr. Marina
S.T. Preston **
White's Hardware
Huntington
Book Revue
Coney's Marine
Island Park
West Marine
Latham
West Marine
Montauk
Montauk Marine Basin
New Rochelle
Post Marine Supply
West Harbor Yacht Service
New York City
New York Kayak Co.
New York Nautical **
West Marine
Northport
Tidewater Marine
Oyster Bay
Nobman's Marine Hdwre.
Oyster Bay Marine Supply
Seawanhaka Boat Yard
Patchoque
B. Sack
West Marine
Port Jefferson
West Marine
Port Washington
Brewer Capri Marina
West Marine
Riverhead
West Marine
Sag Harbor
Emporium Hardware
Henry Persan & Sons
Sag Harbor Yacht Yard
Saugerties
Atlantic Kayak Tours
Shelter Island
Coecles Harbor Marina
Southold
Wego Bait and Tackle
Staten Island
Nautical Chart Supply
Westhampton Beach
Chesterfield Assoc.
West Haverstraw
West Marine
West Islip
West Marine

NEW JERSEY

Atlantic Highlands
West Marine
Bayonne
Ken's Marine Services
Belford
Mariner's Mart
Brick
West Marine
Cape May
South Jersey Marina
West Marine
Cherry Hill
West Marine
Eatontown
West Marine
Lodi
West Marine
Mt. Laurel
West Marine
Paramus
Ramsey Outdoor
Perth Amboy
West Marine
Somers Point
West Marine
South Amboy
Lockwood Boat Works
West Marine
Toms River
West Marine

PENNSYLVANIA

Bensalem
West Marine
Philadelphia
Pilot House Nautical Books
Pittsburgh
West Marine

DELAWARE

Bear
West Marine
New Castle
West Marine
Rehoboth Beach
West Marine

MARYLAND

Annapolis
Fawcett Boat Supplies
West Marine
Baltimore
Maryland Nautical Sales
West Marine
Chester
West Marine
Easton
West Marine
Edgewater
West Marine
Georgetown
Georgetown Yacht Basin
Glen Burnie
West Marine
Havre de Grace
West Marine
Middle River
West Marine
Ocean City
West Marine
Pasadena
West Marine
Rock Hall
West Marine
Solomons
West Marine
Tracey's Landing
West Marine

VIRGINIA

Alexandria
West Marine
Deltaville
West Marine
Glen Allen
West Marine
Gloucester Point
West Marine
Hampton
West Marine
Norfolk
W.T. Brownley
West Marine
Portsmouth
Tidewater Yacht Agency
Virginia Beach
West Marine
Woodbridge
West Marine

NORTH CAROLINA

Beaufort
Scuttlebutt
Charlotte
West Marine
Cornelius
West Marine
Morehead City
West Marine
Nags Head
West Marine
New Bern
West Marine
Oriental
West Marine
Raleigh
US Power Squadron
West Marine
Washington
West Marine
Wilmington
West Marine

SOUTH CAROLINA

Charleston
West Marine
Cherry Grove
West Marine
Hilton Head
West Marine

GEORGIA

Brunswick
West Marine
Savannah
West Marine

FLORIDA

Daytona
West Marine
Delray
West Marine
Ft. Lauderdale
Bluewater Books & Charts
West Marine
Fort Pierce
West Marine
Hollywood
West Marine
Jacksonville
West Marine
Key Largo
West Marine
Lake Park
West Marine
Marathon
West Marine
Miami
West Marine
North Palm Beach
West Marine
Port Charlotte
West Marine
St. Augustine
West Marine
Stuart
West Marine
Tequesta
West Marine
Titusville
West Marine
Vero Beach
West Marine
West Palm Beach
West Marine

ELDRIDGE 2014

can be purchased from the following on line:

Amazon.com

US Power Squadron members can purchase on line at: www.shopusps.org/books.html

Available at most East Coast West Marine. For store locations in each state see website: **WestMarine.com**

www.robertwhite.com
www.eldridgetide.com

For a current list of **ELDRIDGE** dealers please visit:

Robert E. White Instruments **
www.eldridgetide.com
www.robertwhite.com

***** Advertisers in book. Refer to page 271.***

What Would You Do?

Two Scenarios to Provoke Discussion

Your pride and joy, *Dad's Dream*, a 40' sloop with a 6' draft, is headed under power, sails furled, into an unfamiliar harbor for the night. As a result of distracted steering, you go hard aground on a sandy shoal at the harbor entrance. Your only path to safety seems to be astern, but backing the engine does no good. There is little boat traffic. You consult *Eldridge* and find that the tidal range is 5'. It is now 6 p.m. and low tide will come at around 11 p.m. The sun is getting low in the sky. The angle of heel at this time is about 10° to starboard. There is a moderate breeze of 12-15 knots from abeam over the port side.

On board with you are a 40 year-old college classmate, who played tackle on the football team, and your 11 year-old daughter. You have a dinghy with a 5-hp outboard and two anchors, each with plenty of rode. Shore is not far away, but the local boatyard has closed for the day. Not wanting to admit you might need help, you let some time slip by as you and your friend consider your options. Meanwhile your daughter is texting her friends that "Daddy has hit something and the boat is slanting!" She wants to go home.

Discuss the following. Are there other actions you might take?

1. *How could you make the boat heel farther to lift the keel off the bottom?*
2. *What methods could help to move the boat astern?*

After a lovely day's outing in your old 38' diesel powerboat, *Whistling Dixie*, you are returning to familiar waters, and in half an hour you will be back at your mooring. Calm seas and blue skies have made the day delightful. All is well until your 20 year-old son calls up from below, "Hey Dad, there's water over the floorboards!" You immediately throttle down to an idle, put her in neutral, and race below to confirm his report. You pull up a floorboard and realize that water is gushing in at an alarming rate. Reaching down, you discover that the saltwater intake hose has become disconnected from the through-hull fitting. You can tell by feeling that the hose clamps are no longer on the hose. They must have rusted and fallen off and are now out of sight, underwater in the bilge. You wish you'd thought to carry spares. The automatic bilge pump is on, but you see it is no match for the inflow. So far the water level has yet to rise to the battery terminals, but it is just over the engine's oil pan. You tell your son to put on a lifejacket, and you do the same. You hand him a bucket and tell him to empty as much as he can into the galley sink, hoping that is the best place.

The nearest shore is two miles away. At twelve knots, your top speed, you could reach it in ten minutes, but is that soon enough? You don't have a dinghy. A Mayday call through your DSC-equipped radio would bring help, but probably not right away. Your vessel is well equipped with all the recommended safety gear.

What would you do? Possible options are below. Are there better choices?

1. *Stop the engine and find some way to reconnect the hose to the fitting. Send out a Mayday from your radio.*
2. *Keep the engine running. Shut off the through-hull fitting. Tell your son to make sure the intake hose stays immersed in the bilge water, keeping the engine cooled and helping to empty the bilge. What could you use as a strainer? Head for the nearest dock, while wondering if the nearest beach might be a better choice.*

INDEX TO ADVERTISERS

For more information about **ELDRIDGE** advertisers and links to their websites visit:
www.eldridgetide.com

Almeida & Carlson Insurance
Sandwich, MA 250

Andrews Compass Service
Mattapoisett, MA 202

Arey's Pond Boatyard
S. Orleans, MA........................ 220

Burr Bros. Boats
Marion, MA............................ 200

Cape Cod Shipbuilding
Wareham, MA......................... 212

Concordia Co.
South Dartmouth, MA 216

Crosby Yacht Yard
Osterville, MA 182

Edgartown Marine
Edgartown, MA 212

Fairhaven Shipyard
Fairhaven, MA........................ 196

Kingman Yacht Center
Cataumet, MA........................ 272

Landfall
Stamford, CT 172

MacDougalls' Boatyard
Falmouth, MA... inside front cover

Mack Boring
www.mackboring.com 190

Maine Yacht Center
Portland, ME 174

Maptech
www.maptech.com.................. 188

Marblehead Trading Co.
Marblehead, MA.......... back cover

Marina Bay
Quincy, MA 206

Marshall Marine Corp.
S. Darthmouth, MA................ 176

Martha's Vineyard Shipyard
Vineyard Haven, MA 180

Massachusetts Maritime Academy
Center for Maritime Training
Bourne, MA............................ 206

Nantucket Boat Basin
Nantucket, MA 178

New England Boatworks
Portsmouth, RI 184

New York Nautical
Instruments and Service
NYC, NY 186

Parker's Boat Yard
Cataumet, MA........................ 208

Preston's
Nautical Supplies
Greenport, NY 184

Squeteague Sailmakers
Cataumet, MA........................ 176

Thames Shipyard
New London, CT..................... 186

Tisbury Wharf Co.
Vineyard Haven, MA 180

Virginia Dept. of Public Health
VA.. 190

White, Robert E. Instruments, Inc.
.............................inside back cover
Marine and Weather
Instruments and Service
Clocks.................................... 204
Sextants 234
Weather Instruments............. 248
Yestertech Hospital................ 200

A destination all on its own!
KINGMAN
YACHT KYC CENTER
Nestled in scenic, sheltered Red Brook Harbor on the eastern shore of Buzzards Bay, Kingman Yacht Center welcomes seasonal and transient guests in a resort community atmosphere.
Our one-stop destination offers:
• Chart Room Restaurant, Shops & Galleries
• Events Every Weekend All Summer Long
• Pristine Mile-long Beach & Hiking Trails
• Fuel, Ice, Provisioning & Ships Chandlery
• Complimentary Launch Service, WiFi & Pumpout
• Factory Authorized, Full Service Marine Repair
KINGMAN
YACHT KYC CENTER
Since 1932... Brand New Every Day!
Shipyard Ln | Red Brook Harbor | Cataumet (Bourne) | MA 02534 | (508) 563-7136
www.KingmanYachtCenter.com